Communications
in Computer and Information Science

Ana Fred   Jan L.G. Dietz   Kecheng Liu
Joaquim Filipe (Eds.)

Knowledge Discovery, Knowledge Engineering and Knowledge Management

4th International Joint Conference, IC3K 2012
Barcelona, Spain, October 4-7, 2012
Revised Selected Papers

 Springer

Volume Editors

Ana Fred
IST - Technical University of Lisbon, Portugal
E-mail: afred@lx.it.pt

Jan L.G. Dietz
Delft University of Technology, The Netherlands
E-mail: j.l.g.dietz@tudelft.nl

Kecheng Liu
Henley Business School, University of Reading, UK
E-mail: k.liu@henley.reading.ac.uk

Joaquim Filipe
INSTICC and IPS, Estefanilha, Setúbal, Portugal,
E-mail: joaquim.filipe@estsetubal.ips.pt

ISSN 1865-0929   e-ISSN 1865-0937
ISBN 978-3-642-54104-9   e-ISBN 978-3-642-54105-6
DOI 10.1007/978-3-642-54105-6
Springer Heidelberg New York Dordrecht London

Library of Congress Control Number: 2013957791

CR Subject Classification (1998): H.3, H.2.8, H.2, I.2, J.1, K.4

Typesetting: Camera-ready by author, data conversion by Scientific Publishing Services, Chennai, India

Printed on acid-free paper

Springer is part of Springer Science+Business Media (www.springer.com)

Preface

The present book includes extended and revised versions of a set of selected papers from the 4th International Joint Conference on Knowledge Discovery, Knowledge Engineering and Knowledge Management (IC3K 2012), held in Barcelona, Spain, during October 4-7, 2012. IC3K was sponsored by the Institute for Systems and Technologies of Information Control and Communication (INSTICC), it was organized in cooperation with the Association for the Advancement of Artificial Intelligence (AAAI) and in collaboration with the Informatics Research Centre (IRC), University of Reading, UK.

The purpose of IC3K is to bring together researchers, engineers, and practitioners in the areas of Knowledge Discovery, Knowledge Engineering and Knowledge Management to foster scientific and technical advances in these areas.

IC3K is composed of three concurrent and co-located conferences, each specialized in at least one of the aforementioned main knowledge areas, namely:

- KDIR (International Conference on Knowledge Discovery and Information Retrieval). Knowledge discovery is an interdisciplinary area focusing on methodologies for identifying valid, novel, potentially useful, and meaningful patterns from data, often based on underlying large data sets. A major aspect of knowledge discovery is data mining, i.e., applying data analysis and discovery algorithms that produce a particular enumeration of patterns (or models) over the data. Knowledge discovery also includes the evaluation of patterns and identification of which ones add to knowledge. This has proven to be a promising approach for enhancing the intelligence of software systems and services. The ongoing rapid growth of online data due to the Internet and the widespread use of large databases has created an important need for knowledge discovery methodologies. The challenge of extracting knowledge from data draws on research in a large number of disciplines including statistics, databases, pattern recognition, machine learning, data visualization, optimization, and high-performance computing, to deliver advanced business intelligence and Web discovery solutions. Information retrieval (IR) is concerned with gathering relevant information from unstructured and semantically fuzzy data in texts and other media, searching for information within documents and for metadata about documents, as well as searching relational databases and the Web. Automation of information retrieval enables the reduction of what has been called "information overload."

Information retrieval can be combined with knowledge discovery to create software tools that empower users of decision support systems to better understand and use the knowledge underlying large data sets.

- KEOD (International Conference on Knowledge Engineering and Ontology Development). Knowledge Engineering (KE) refers to all technical, scientific, and social aspects involved in building, maintaining, and using knowledge-based systems. KE is a multidisciplinary field, bringing in concepts and methods from several computer science domains such as artificial intelligence, databases, expert systems, decision support systems and geographic information systems. From the software development point of view, KE uses principles that are strongly related to software engineering. KE is also related to mathematical logic, as well as strongly involved in cognitive science and socio-cognitive engineering where the knowledge is produced by humans and is structured according to our understanding of how human reasoning and logic works. Currently, KE is strongly related to the construction of shared knowledge bases or conceptual frameworks, often designated as ontologies.

Ontology development aims at building reusable semantic structures that can be informal vocabularies, catalogs, glossaries as well as more complex finite formal structures representing the entities within a domain and the relationships between those entities. Ontologies, have been gaining interest and acceptance in computational audiences: Formal ontologies are a form of software, thus software development methodologies can be adapted to serve ontology development. A wide range of applications are emerging, especially given the current Web emphasis, including library science, ontology-enhanced search, e-commerce and configuration.

- KMIS (International Conference on Knowledge Management and Information Sharing). Knowledge management (KM) is a discipline concerned with the analysis and technical support of practices used in an organization to identify, create, represent, distribute, and enable the adoption and leveraging of good practices embedded in collaborative settings and, in particular, in organizational processes. Effective knowledge management is an increasingly important source of competitive advantage and a key to the success of contemporary organizations, bolstering the collective expertise of its employees and partners.

There are several perspectives on KM, but all share the same core components, namely: people, processes, and technology. Some take a techno centric focus, in order to enhance knowledge integration and creation; some take an organizational focus, in order to optimize organizational design and workflows; some take an ecological focus, where the important aspects are related to people interaction, knowledge, and environmental factors as a complex adaptive system similar to a natural ecosystem.

Information sharing (IS) is a term used for a long time in the information technology (IT) lexicon, related to data exchange, communication protocols, and technological infrastructures. Although standardization is indeed an essential element for sharing information, IS effectiveness requires going beyond the syntactic nature of IT and delving into the human functions involved in the semantic, pragmatic, and social levels of organizational semiotics. The two areas are intertwined as information sharing is the foundation for knowledge management.

IC3K received 347 paper submissions from 59 countries in all continents. To evaluate each submission, a double-blind paper review was performed by the Program Committee. After a stringent selection process, 44 papers were accepted to be published and presented as full papers. From these, we further selected 29 papers, based not only on the reviewers' classifications and comments but also on the session chairs, assessment of oral presentations, whose extended and revised versions are included in this publication.

On behalf of the conference Organizing Committee, we would like to thank all participants. First of all the authors, whose quality work is the essence of the conference and the members of the Program Committee, who helped us with their expertise and diligence in reviewing the papers. As we all know, producing a conference requires the effort of many individuals. We wish to also thank all the members of our Organizing Committee, whose work and commitment were invaluable.

December 2012

Ana Fred
Jan L.G. Dietz
Kecheng Liu
Joaquim Filipe

Organization

Conference Chair

Joaquim Filipe — Polytechnic Institute of Setúbal/INSTICC, Portugal

Program Co-chairs

KDIR

Ana Fred — Technical University of Lisbon/IT, Portugal

KEOD

Jan L.G. Dietz — Delft University of Technology, The Netherlands

KMIS

Kecheng Liu — University of Reading, UK

Organizing Committee

Helder Coelhas — INSTICC, Portugal
Vera Coelho — INSTICC, Portugal
Andreia Costa — INSTICC, Portugal
Patrícia Duarte — INSTICC, Portugal
Bruno Encarnação — INSTICC, Portugal
Liliana Medina — INSTICC, Portugal
Raquel Pedrosa — INSTICC, Portugal
Vitor Pedrosa — INSTICC, Portugal
Cláudia Pinto — INSTICC, Portugal
Susana Ribeiro — INSTICC, Portugal
José Varela — INSTICC, Portugal
Pedro Varela — INSTICC, Portugal

KDIR Program Committee

Yasser Abdelhamid Abdelfattah,
  Egypt
Muhammad Abulaish, Saudi Arabia
Andrea Addis, Italy
Samad Ahmadi, UK
Shawkat Ali, Australia
Francisco Martínez Álvarez, Spain
Eva Armengol, Spain
Daniel Barbará, USA
Pierpaolo Basile, Italy
Florian Boudin, France
Marc Boullé, France
Maria Jose Aramburu Cabo, Spain
Rui Camacho, Portugal
Luis M. de Campos, Spain
Annalina Caputo, Italy
Keith C. C. Chan, Hong Kong
Meng Chang Chen, USA
Camelia Chira, Romania
Juan Manuel Corchado, Spain
Antonio Corradi, Italy
Jerome Darmont, France
Spiros Denaxas, UK
Dejing Dou, USA
Antoine Doucet, France
Floriana Esposito, Italy
Iaakov Exman, Israel
Katti Facelli, Brazil
Philippe Fournier-Viger, Canada
Ana Fred, Portugal
Susan Gauch, USA
Marco de Gemmis, Italy
Antonella Guzzo, Italy
Yaakov Hacohen-Kerner, Israel
Greg Hamerly, USA
Jianchao Han, USA
Nima Hatami, USA
Fumio Hattori, Japan
José Hernández-Orallo, Spain
Kaizhu Huang, China
Yo-Ping Huang, Taiwan
Leo Iaquinta, Italy

Beatriz de la Iglesia, UK
Szymon Jaroszewicz, Poland
Liu Jing, China
Estevam Hruschka Jr., Brazil
Mouna Kamel, France
Rajkumar Kannan, India
Mehmed Kantardzic, USA
Ron Kenett, Israel
Steven Kraines, Japan
Cristian Lai, Italy
Hagen Langer, Germany
Anne Laurent, France
Carson K. Leung, Canada
Chun Hung Li, Hong Kong
Xia Lin, USA
Berenike Litz, Germany
Jun Liu, UK
Rafael Berlanga Llavori, Spain
Pasquale Lops, Italy
Alicia Troncoso Lora, Spain
Devignes Marie-Dominique, France
Edson T. Matsubara, Brazil
Paul McNamee, USA
Misael Mongiovi, USA
Maurizio Montagnuolo, Italy
Stefania Montani, Italy
Pierre Morizet-Mahoudeaux, France
Claude Moulin, France
Henning Müller, Switzerland
Giorgio Maria Di Nunzio, Italy
Mitsunori Ogihara, USA
Byung-Won On, Korea, Republic of
Nuno Pina, Portugal
Agostino Poggi, Italy
Luigi Pontieri, Italy
Francois Poulet, France
Ronaldo Prati, Brazil
Marcos Gonçalves Quiles, Brazil
Zbigniew W. Ras, USA
Luís Paulo Reis, Portugal
Eduarda Mendes Rodrigues, Portugal
Sebastian Rodriguez, Argentina

Arun Ross, USA
Henryk Rybinski, Poland
Hesham Salman, Saudi Arabia
Ovidio Salvetti, Italy
Paul Schmidt, Germany
Filippo Sciarrone, Italy
Fabricio Silva, Brazil
Fabrizio Silvestri, Italy
Dominik Slezak, Poland
Manas Somaiya, USA
Alessandro Soro, Italy
Marcin Sydow, Poland
Kosuke Takano, Japan
Ying Tan, China

Andrew Beng Jin Teoh, Korea,
    Republic of
Ulrich Thiel, Germany
Kar Ann Toh, Korea, Republic of
Yannick Toussaint, France
Panagiotis Tsaparas, Greece
Evelyne Tzoukermann, USA
Joaquin Vanschoren, Belgium
Jeen-Shing Wang, Taiwan
Yang Xiang, USA
Yiyu Yao, Canada
Daoqiang Zhang, China
Yonggang Zhang, USA
Haibin Zhu, Canada

KDIR Auxiliary Reviewers

Maisa Duarte, Brazil
Irina Illina, France
Surya Kallumadi, USA
Ilona Nawrot, Poland
Siwipa Pruitikanee, France

Coralie Reutenauer, France
Tiemi Sakata, Brazil
Shing Chiang Tan, Malaysia
Claudio Taranto, Italy

KEOD Program Committee

Alia Abdelmoty, UK
Alessandro Agostini, Italy
Antonia Albani, Switzerland
Carlo Allocca, UK
Frederic Andres, Japan
Francisco Antunes, Portugal
Sören Auer, Germany
Marie-aude Aufaure, France
David Aveiro, Portugal
Eduard Babkin, Russian Federation
Costin Badica, Romania
Janaka Balasooriya, USA
Claudio de Souza Baptista, Brazil
Jean-Paul Barthes, France
Teresa M. A. Basile, Italy
Sonia Bergamaschi, Italy
Patrick Brezillon, France

Giacomo Bucci, Italy
Vladimír Bureš, Czech Republic
Doina Caragea, USA
Núria Casellas, USA
Helder Castro, Portugal
Jin Chen, USA
Yixin Chen, USA
Ruth Cobos, Spain
James Crawford, USA
Fabiano Dalpiaz, Italy
Jan Dietz, The Netherlands
John Edwards, UK
Anna Fensel, Austria
Dieter A. Fensel, Austria
Salvatore Gaglio, Italy
Raul Garcia-Castro, Spain
Faiez Gargouri, Tunisia

Serge Garlatti, France
Rosario Girardi, Brazil
Matteo Golfarelli, Italy
Sven Groppe, Germany
Ronghuai Huang, China
Philip Huysmans, Belgium
Junichi Iijima, Japan
John Josephson, USA
Asanee Kawtrakul, Thailand
Katia Lida Kermanidis, Greece
Pieter De Leenheer, The Netherlands
Ming Li, China
Xiao-Lin Li, China
Antoni Ligeza, Poland
Elena Lloret, Spain
Xudong Luo, China
Rocio Abascal Mena, Mexico
Munir Merdan, Austria
Riichiro Mizoguchi, Japan
Andres Montoyo, Spain
Claude Moulin, France
Ana Maria Moura, Brazil
João Moura-Pires, Portugal
Hans Mulder, The Netherlands
Kazumi Nakamatsu, Japan
William Nelson, USA
Erich Neuhold, Austria
Catalina Nicolin, Romania
Jørgen Fischer Nilsson, Denmark
Nan Niu, USA

Jivka Ovtcharova, Germany
Manuel Palomar, Spain
Enric Plaza, Spain
Mihail Popescu, USA
Violaine Prince, France
Juha Puustjärvi, Finland
Amar Ramdane-Cherif, France
Domenico Redavid, Italy
Ramana Reddy, USA
Thomas Reineking, Germany
M. Teresa Romá-Ferri, Spain
Martín Serrano, Ireland
Nuno Silva, Portugal
Deborah Stacey, Canada
Anna Stavrianou, France
Heiner Stuckenschmidt, Germany
Mari Carmen Suárez-Figueroa, Spain
Christos Tatsiopoulos, Greece
Orazio Tomarchio, Italy
Shengru Tu, USA
Rafael Valencia-Garcia, Spain
Iraklis Varlamis, Greece
Cristina Vicente-Chicote, Spain
Bruno Volckaert, Belgium
Sebastian Wandelt, Germany
Martin Wolpers, Germany
Yue Xu, Australia
Gian Piero Zarri, France
Jinglan Zhang, Australia
Catherine Faron Zucker, France

KEOD Auxiliary Reviewers

Domenico Beneventano, Italy
Nesrine Ben Mustapha, Tunisia
Grzegorz J. Nalepa, Poland
Daniele Peri, Italy
Valeria Seidita, Italy

Serena Sorrentino, Italy
Yves Vanrompay, France
Johanna Völker, Germany
Luyi Wang, USA

KMIS Program Committee

Marie-Helene Abel, France
Adriano Albuquerque, Brazil
Miriam C. Bergue Alves, Brazil
Ioannis Anagnostopoulos, Greece
Rangachari Anand, USA
Alessio Bechini, Italy
Malgorzata Bugajska, Switzerland
Elsa Cardoso, Portugal
Marcello Castellano, Italy
Xiaoyu Chen, China
Ying Chen, USA
Reynold Cheng, Hong Kong
David Cheung, China
Dickson K. W. Chiu, Hong Kong
Giulio Concas, Italy
Dominique Decouchant, Mexico
Mariagrazia Dotoli, Italy
Zamira Dzhusupova, Macau
Alan Eardley, UK
Elsa Estevez, Macau
Susana Falcão, Portugal
Joan-Francesc
    Fondevila-Gascón, Spain
Anna Goy, Italy
Le Gruenwald, USA
Felix Hamza-Lup, USA

Jan Hidders, The Netherlands
Anca Daniela Ionita, Romania
Dan Kirsch, USA
Mieczyslaw Klopotek, Poland
Elise Lavoué, France
Juhnyoung Lee, USA
Kecheng Liu, UK
Anthony Masys, Canada
Nada Matta, France
Sonia Mendoza, Mexico
Christine Michel, France
Taneli Mielikainen, USA
Owen Molloy, Ireland
Patricia Ordóñez de Pablos,
    Spain
Augusta Maria Paci, Italy
Hye-young Paik, Australia
Nuno Pina, Portugal
Marina Ribaudo, Italy
John Rohrbaugh, USA
Ricardo da S. Torres, Brazil
Maggie Minhong Wang, Hong Kong
Robert Warren, Canada
Leandro Krug Wives, Brazil
Andreas Wombacher, The Netherlands
Yonggang Zhang, USA

KMIS Auxiliary Reviewers

Kimberly García, Mexico
Yifan Jin, Hong Kong
Avila Mora Ivonne Maricela, Mexico

Luyi Mo, Hong Kong
Genaro Saucedo-Tejada, Mexico
John Shepherd, Australia

Invited Speakers

Daniel O'Leary
Sophia Ananiadou
Alfred Inselberg
Peter F. Patel-Schneider
Florian Michahelles

University of Southern California, USA
University of Manchester, UK
Tel Aviv University, Israel
USA
ETH Zürich, Switzerland

Table of Contents

Part I: Knowledge Discovery and Information Retrieval

Statistical Approach for Term Weighting in Very Short Documents
for Text Categorization 3
  Mika Timonen and Melissa Kasari

An Approach to Model Selection in Spectral Clustering
with Application to the Writing Style Determination Problem 19
  Renata Avros, Avi Soffer, Zeev Volkovich, and Orly Yahalom

A Seed-Based Inter-Domain Supervised Framework to Cluster Mixed
Data Types 37
  Artur Abdullin and Olfa Nasraoui

Strategies for Guided Exploratory Search on the Mobile Web 53
  Günter Neumann and Sven Schmeier

Towards a Unified Thematic Model for Recommending Context-
Sensitive Content 68
  Mihaela Dinsoreanu and Rodica Potolea

ILP Characterization of 3D Protein-Binding Sites and FCA-Based
Interpretation 84
  Emmanuel Bresso, Renaud Grisoni, Marie-Dominique Devignes,
  Amedeo Napoli, and Malika Smail-Tabbone

Combination of Lexical and Structure-Based Similarity Measures
to Match Ontologies Automatically 101
  Thi Thuy Anh Nguyen and Stefan Conrad

Toponym Extraction and Disambiguation Enhancement Using Loops
of Feedback 113
  Mena B. Habib and Maurice van Keulen

Keyword Extraction from Short Documents Using Three Levels
of Word Evaluation 130
  Mika Timonen, Timo Toivanen, Melissa Kasari, Yue Teng,
  Chao Cheng, and Liang He

Part II: Knowledge Engineering and Ontology Development

Applying Simple Ontology Relations for Receiving Better
Recommendations.. 149
  Lamiaa Abdelazziz and Khaled Nagi

A Knowledge-Based Approach to the User-Centered Design Process 165
  Stefan Negru and Sabin Buraga

Towards Automatic Ontology Alignment for Enriching Sensor Data
Analysis 179
  Marjan Alirezaie and Amy Loutfi

Performing Ontology Alignment via a Fuzzy-Logic Multi-layer
Architecture... 194
  Susel Fernández, Ivan Marsa-Maestre, and Juan R. Velasco

An Answer Set Programming Solution for Supply Chain Traceability ... 211
  Monica L. Nogueira and Noel P. Greis

Modelling Services with DEMO 228
  *Carlos Mendes, Mário Almeida, Nuno Salvador, and
  Miguel Mira da Silva*

A Method for Reengineering Healthcare Using Enterprise Ontology
and Lean 243
  David Galego Dias, Carlos Mendes, and Miguel Mira da Silva

Interactive Exploration of Structural Concepts in Code 260
  Paul Heckmann and Daniel Speicher

ROM: An Approach to Self-consistency Verification of a Runnable
Ontology Model 271
  Iaakov Exman and Reuven Yagel

Topology Labeling: An Indexing Structure to Find Complex
Relationships within Ontologies.................................... 284
  Karina Robles, Alejandro Ruiz, Anabel Fraga, and Juan Llorens

Part III: Knowledge Management and Information Sharing

Reference Operation Generation Method on Project Manager Skill-up
Simulator 297
  Keiichi Hamada, Masanori Akiyoshi, and Masaki Samejima

Exploiting and Reusing Collaborative Traces to Facilitate Sharing
Experiences in Groups 308
  Qiang Li, Marie-Hélène Abel, and Jean-Paul A. Barthès

Towards Value-Oriented Use of Social Media for Knowledge
Management in SME 323
  Ulrike Borchardt

Semantically Enriched Obligation Management: An Approach
for Improving the Handling of Obligations Represented in Contracts.... 337
  Barbara Thönssen and Jonas Lutz

Knowledge Management for Innovation and Product Development
in Supply Chains 350
  Lixin Wang and Athanassios Kourouklis

Soft Systems Methodology: A Conceptual Model of Knowledge
Management System Initiatives 377
  *Nor Hasliza Md Saad, Hasmiah Kasimin, Rose Alinda Alias, and
  Azizah Abdul Rahman*

Towards a Procedure for Assessing Supply Chain Risks Using Semantic
Technologies 393
  *Sandro Emmenegger, Knut Hinkelmann, Emanuele Laurenzi, and
  Barbara Thönssen*

Pervasive Ensemble Data Mining Models to Predict Organ Failure
and Patient Outcome in Intensive Medicine 410
  *Filipe Portela, Manuel Filipe Santos, Álvaro Silva,
  António Abelha, and José Machado*

Watch Out and Improve IT: Adapting COBIT 5.0 Framework Based
on External Context Discovery 426
  Eduardo Costa Ramos, Flávia Maria Santoro, and Fernanda Baião

Integration of Event Data from Heterogeneous Systems to Support
Business Process Analysis....................................... 440
  Alejandro Vera Baquero and Owen Molloy

Author Index 455

Part I

Knowledge Discovery
and Information Retrieval

Statistical Approach for Term Weighting in Very Short Documents for Text Categorization

Mika Timonen[1] and Melissa Kasari[2]

[1] VTT Technical Research Centre of Finland, P.O. 1000, FI-02044 VTT, Finland
[2] Department of Computer Science, P.O. 68, FI-00014 University of Helsinki, Finland

Abstract. In this paper, we propose a novel approach for term weighting in very short documents that is used with a Support Vector Machine classifier. We focus on market research and social media documents. In both of these data sources, the average length of a document is below twenty words. As the documents are short, each word occurs usually only once within a document. This is known as *hapax legomenon* and in our previous work as *Term Frequency=1 challenge*. For this reason, the traditional term weighting approaches become less effective with short documents. In this paper we propose a novel approach for term weighting that does not use term frequency within a document but substitutes it with other word statistics. In the experimental evaluation and comparison against several other term weighting approaches the proposed method produced promising results by out-performing the competition.

Keywords: Feature Weighting, Hapax Legomenon, Short Document Categorization, Support Vector Machine, Text Categorization.

1 Introduction

In this paper we propose a novel term weighting approach for short documents. This approach is used with a *Support Vector Machine* classifier to categorize text. Text categorization is a text mining task that aims to label the documents into predefined classes. This task is usually supervised, i.e., the classes are known beforehand and the dataset contains examples of which document belongs to which category.

The first step of the categorization process is often term weighting where each word of a document is given a weight that indicates its importance for the classification. Words that give a strong indication that the document belongs or does not belong to a specific category are given a strong weight. Words that contain only little information are given a small weight and they may be excluded from the classification process.

Many traditional term weighting approaches, such as *Term Frequency - Inverse Document Frequency (TF-IDF)* [1], are based on the word occurrence count within the document. That is, often occurring words may indicate that the word is important for the document. However, when the documents are very short, for example, under 20 words, each word occurs usually only once within a single document. This is called *hapax legomenon* and we call it *TF=1 challenge* [2].

With TF=1 challenge, TF-IDF reduces to IDF as term frequency is the same for each word within the document. As Inverse Document Frequency emphasizes words

A. Fred et al. (Eds.): IC3K 2012, CCIS 415, pp. 3–18, 2013.

that occur only in a few documents, TF-IDF gives the most emphasis on the rarest words when used in short documents. We have found that this approach is not good for categorization of short documents [2].

In this paper we propose a novel term weighting approach designed to be used with very short documents. We tackle the TF=1 challenge by replacing term frequency with other word statistics including word's distribution within the category and distribution among all the categories. This approach is designed to be used with labeled data. The approach proposed in this paper was originally introduced in our previous article [3]. In this paper, we extend our previous work by evaluating the use of the proposed approach in a real world setting of market research data categorization.

The starting point of this project came from a market research company that wanted to find a method for an automatic categorization of their market research data; especially the open ended questions. As they have several questionnaires that they use repeatedly, they have a lot of training data that is relevant for their future research also. The aim was to implement a classifier that can categorize the answers of the repeating questionnaires effectively by utilizing the data from the previous surveys.

As the traditional term weighting approaches have trouble identifying important words from short documents, we implemented an approach we call *Fragment Length Weighted Category Distribution* ($FLWCD$) that is based on our previous work [2]. We have modified the previous approach by substituting two components with a method called *Bi-Normal Separation* (*BNS*) [4]. BNS is an approach that compares the distribution of a feature among positive and negative samples. We combine BNS with two components of our previous work: the average fragment length where the word appears in, and the distribution of the word among all categories.

We experimented the approach using several different datasets of market research data. This data is very short as it holds less than 15 words per document on average. In addition to market research data, we also evaluated our approach using Twitter data. This data is also very short as it can contain at most 140 characters. In our testset the average length of a tweet was 15 words. We compared the results against a wide variety of traditional term weighting methods such as TF-IDF, Mutual Information, Chi-Squared, Residual IDF, and Information Gain. The results are promising as $FLWCD$ out-performed the competition.

The main finding of this work is that term frequency based methods have trouble identifying the relevant words within short documents. However, when the document is short enough, around five words, the selected weighting method does not have as much impact as it does when the documents contain over ten words.

This paper makes the following contributions: (1) a description of a novel approach on feature weighting for very short documents, (2) evaluation of traditional term weighting approaches for text categorization using two relevant types of very short documents (Twitter and market research data), and (3) evaluation of the use of the proposed approach in a real world setting in a market research company.

This paper is organized as follows. In Section 2 we give a brief survey of the related approaches on feature weighting and text categorization. In Section 3 we present our approach on categorizing short documents. In Section 4 we evaluate our approach and compare it against other relevant methods. We conclude the paper in Section 5.

2   Text Categorization

The process of text categorization can be divided into three steps: preprocessing which includes term weighting, classifier training, and classification. In most cases, the documents are transformed into feature vectors that are used for training the classifier. Each word within a document corresponds to a feature. However, not all features have the same impact within a document. Therefore each feature is weighted using a feature weighting approach. After the features have been weighted, a classifier is built using the feature vectors. There are numerous approaches for classification; the most notable ones include Support Vector Machine, Naive Bayes and k-Nearest Neighbors classifiers. In this section we describe each step of the text categorization process and present the related work.

2.1   Term Weighting

When transforming documents to feature vectors each term (word or set of words such as noun phrases) of the document is used as a feature. Weighting these terms is an important part of classification as without weighting each word would have the same impact for the classification process. The aim of the process is to find features that are important and remove the unimportant ones. In most cases the process takes a set of documents as its input and outputs a set of weighted terms for each document. The weights indicate the importance of each term. In this section we provide a quick survey of the most notable approaches for feature weighting.

Term Frequency - Inverse Document Frequency (TF-IDF) [1] is the most traditional term weighting method and it is used, for example, in information retrieval. The idea is to find the most important terms for the document within a corpus by assessing how often the word occurs within a document (TF) and how often in other documents (IDF):

$$\text{TF-IDF}(t, d) = -\log \frac{df(t)}{N} \times \frac{tf(t, d)}{|d|} \, , \tag{1}$$

where $tf(t, d)$ is the term frequency of word t within the document d, $|d|$ is the number of words in d, $df(t)$ is the document frequency within the corpus, and N is the number of documents in the corpus.

There are also other approaches that are based on either TF or IDF. Rennie and Jaakkola [5] have surveyed several of them and their use for named entity recognition. In their experiments, *Residual IDF* produced the best results. Residual IDF is based on the idea of comparing the word's observed IDF against predicted IDF ($\widehat{IDF}$) [6]. Predicted IDF is calculated using the term frequency and assuming a random distribution of the term in the documents (Poisson model). If the difference between IDF and $\widehat{IDF}$ is large, the word is informative. Equation 2 presents how the residual IDF ($RIDF$) is calculated using observed IDF and predicted IDF:

$$RIDF(t) = IDF(t) - \widehat{IDF}(t) = -\log \frac{df(t)}{N} + \log\left(1 - e^{-\frac{ctf(t)}{N}}\right), \tag{2}$$

where $ctf(t)$ is the collection term frequency; $ctf(t) = \sum_d tf(t, d)$.

Other traditional approaches include *Odds Ratio (OR), (Pointwise) Mutual Information (MI), Information Gain (IG)*, and *Chi-squared* (χ^2). Odds Ratio is an approach used for relevance ranking in information retrieval [7]. It is calculated by taking the ratio of positive samples and negative samples; i.e., the odds of having a positive instance of the word when compared to the negative [4]:

$$OR(t) = \log \frac{N_{t,c} \times N_{\neg t, \neg c}}{N_{t,\neg c} \times N_{\neg t, c}}, \tag{3}$$

where $N_{t,c}$ denotes the number of times term t occurs in category c, $N_{t,\neg c}$ is the number of times t occurs in other categories than c, $N_{\neg t, c}$ is the number of times c occurs without term t, $N_{\neg t, \neg c}$ is the number of times neither c nor t occurs.

Information Gain, shown in Equation 5, is often used with decision trees such as C4.5. It measures the change in entropy when the feature is given as opposed of being absent [4]. This is estimated as the difference in observed entropy $H(C)$ and the expected entropy $E_T(H(C|T))$:

$$
\begin{aligned}
IG(t) &= H(C) - E_T(H(C|T)) \\
&= H(C) - (P(t) \times H(C|t) + P(\neg t) \times H(C|\neg t)) \\
&= -\sum_{i=1}^{m} P(c_i) \log P(c_i) + P(t) \sum_{i=1}^{m} P(c_i|t) \log P(c_i|t) + P(\neg t) \sum_{i=1}^{m} P(c_i|\neg t) \log P(c_i|\neg t),
\end{aligned}
\tag{4}
$$

where $\neg t$ indicates the absence of t, m is the number of all categories, c_i is the ith category.

Chi-squared (χ^2) is a traditional statistical test method. It is used in text categorization to assess the dependence of the feature - category pairs. The idea is to do a χ^2 test by assuming that the feature and the category are independent. If the score is large, they are not independent which indicates that the feature is important for the category:

$$\chi^2(t,c) = \frac{N \times ((N_{t,c} \times N_{\neg t, \neg c}) - (N_{\neg t, c} \times N_{t, \neg c}))^2}{(N_{t,c} + N_{\neg t, c}) \times (N_{t,c} + N_{t, \neg c}) \times (N_{t, \neg c} + N_{\neg t, \neg c}) \times (N_{\neg t, c} + N_{\neg t, \neg c})}, \tag{5}$$

where the notation is the same as in Odds Ratio.

Pointwise Mutual Information is similar with the Chi-squared weighting. The idea is to score each feature - category pair and see how much a feature contributes to the category:

$$MI(t,c) = \log \frac{N_{t,c} \times N}{(N_{t,c} + N_{\neg t, c}) \times (N_{t,c} + N_{t, \neg c})}. \tag{6}$$

Forman [4] has proposed a feature weighting approach called Bi-Normal Separation (BNS). The approach scores the top n features by comparing standard normal distribution's inverse cumulative probability functions of positive and negative examples:

$$BNS(t,c) = |F^{-1}(\frac{N_{t,c}}{N_{t,c} + N_{t, \neg c}}) - F^{-1}(\frac{N_{\neg t, c}}{N_{\neg t, c} + N_{\neg t, \neg c}})|, \tag{7}$$

where F^{-1} is the inverse Normal cumulative distribution function. As the inverse Normal would go to infinity at 0 and 1, to avoid this problem Forman limited both distributions to the range [0.0005,0.9995].

The idea of BNS is to compare the two distributions; the larger the difference between them, more important the feature. Forman later compared its performance against several other feature weighting approaches including OR, IG, and χ^2 [8]. In his experiments BNS produced the best results with IG performing the second best.

Yang and Pedersen [9] have compared the performance of χ^2, MI, and IG. They reported that χ^2 and Information Gain are the most effective for text categorization of Reuters news articles. Mutual Information on the other hand performed poorly.

Finally, we have previously studied short document categorization and proposed a term weighting approach called *Term-Corpus Relevance* × *Term-Category Relevance* (*TT*) [2]. The idea is to assess the word's importance on two levels: corpus level and category level. The weighting was used in a classifier that resembled a Naive Bayes classifier. This approach is based on the word statistics that measure the word's relevance within the categories and within the corpus:

$$
\begin{aligned}
TT(t) &= (P(c|t) + P(t|c)) \times (ifl(t) + |c_t|^{-1}) \\
&= (\frac{N_{t,c}}{N_{t,c} + N_{t,\neg c}} + \frac{N_{t,c}}{N_{t,c} + N_{\neg t,c}}) \times (ifl(t) + |c_t|^{-1}),
\end{aligned}
\tag{8}
$$

where $P(c|t)$ is the probability for the category given the word, $P(t|c)$ probability for the word appearing in the given category, $ifl(t)$ is inverted average fragment length, and $|c_t|$ is the number of categories in which the term t appears in. In this paper we propose a modified version of this approach.

2.2 Classification

Classification is a task that aims to build a model for categorizing unlabeled documents under predefined labels. The training process takes the set of feature vectors with their labels as its input and outputs the model. The classification process takes the model and the unlabeled vectors as its input and outputs the classes for each of the vectors.

Document classification, which is often called text categorization, is a well researched area that has several traditional methods available. Yang compared several different approaches using Reuters news article data [10,11]. In these experiments k-Nearest Neighbors (kNN) and Linear Least Squares Fit (LLSF) produced the best results. In several other studies Support Vector Machine (SVM) has been reported to produce the best results [12,13].

Naive Bayes classification has also been able to produce good results. Rennie et al. [14] describe an approach called Transformed Weight-normalized Complement Naive Bayes (TWCNB) that can produce similar results as SVM. They base the term weighting mostly on term frequency but they also use an idea to assess term's importance by comparing its distribution among categories. Kibriya et al. [15] extended this idea by using TF-IDF instead of TF in their work.

In text categorization, most research has been done with news articles that contain over 150 words on average. However, there are a few instances that focus on short documents, such as tweets, where the most researched domain is possibly spam detection and opinion mining from tweets. Pak and Paroubek [16], for example, use linguistic analysis on the corpus and conclude that when using part of speech tagging it is possible to find strong indicator for emotion in text.

Even though text categorization methods have been successfully used, for example, to detect spam e-mails, Irani et al. [17] believe that the traditional methods produce poor results with Twitter message categorization. Classifiers, such as Support Vector Machines, perform better in this case as presented by Benevenuto et al. [18].

3   Short Document Categorization

Categorization of short documents differs from traditional text categorization mainly in the term weighting step. This is due to the TF=1 Challenge; the fact that each word of a document occurs usually only once within the document. We have taken our previous work [2] as the starting point and use similar components in this approach. We also include Bi-Normal Separation introduced by Forman [4] to our approach. In this section we describe the proposed approach for term weighting that is designed to tackle the challenges of short documents.

3.1   Term Weighting

The challenge with term weighting in short document categorization is the fact that each word occurs usually only once within a single document. This is problematic for the approaches that are based on term frequency. We have tackled this problem by using relevance values called Term-Corpus Relevance and Term-Category Relevance (TT) that were assessed using the following statistics [2]: *inverse average fragment length* where the word appears in, *category probability of the word*, *document probability within the category*, and *inverse category count*.

Bi-Normal Separation, presented by Forman [4], is based on the idea of comparing the distribution of a feature in the positive and negative examples; i.e., documents within a category and outside of the category. By combining term frequency with his approach, Forman was able to get better results from BNS [8].

Instead of combining BNS with term frequency, we combine BNS with two components from TT; inverse average fragment length and category probability of a word (i.e., distribution among categories). We chose this approach as the idea behind BNS is sound but alone inefficient when used with short documents.

Inverse average fragment length indicates the importance of a word by using the length of a fragment where the word occurs. A fragment is a part of the text that is broken from the document using predefined breaks. We break the text into fragments using stop words and break characters. For English, we include the following stop words: *and, or, both*. We use the following characters to break the text: comma (,), exclamation mark (!), question mark (?), and full stop (.). For example, sentence "The car is new, shiny and pretty" is broken into fragments "The car is new", "shiny", "pretty".

The idea behind average fragment length is based on an assumption that important words require fewer surrounding words than unimportant ones. Consider the previous example. Words *shiny* and *pretty* are alone where as words *the*, *car*, *is*, and *new* have several surrounding words. As the words *new*, *shiny*, and *pretty* form a list, they can appear in any order (i.e., in any fragment) where as the words *the*, *car*, and *is* cannot. By taking the average fragment length, the three words (*new*, *shiny*, *pretty*) will stand out from the less important ones (*the*, *car*, *is*).

The inverse average fragment length $ifl(t)$ for the word t is calculated as follows:

$$ifl(t) = \frac{1}{\frac{1}{|f_t|} \sum l_f(t)} \, , \tag{9}$$

where f_t is the collection of fragments where the word t occurs in, and l_f is the length of the fth fragment where the word t occurs in. In other words, we take the average length of the fragments the word t occurs in. If the word occurs always alone, $ifl(t) = 1$.

For example, if the given sentence occurs two additional times as "The car is shiny, pretty and new", the word *shiny* would have occurred alone once, word *new* two times, and the word *pretty* three times. The unimportant words occur with three other words in every instance making their inverse average fragment length smaller. In this example the inverse average fragment length for each of the words are: $ifl(car) = 0.25$, $ifl(is) = 0.25$, $ifl(new) = 0.5$, $ifl(shiny) = 0.33$, and $ifl(pretty) = 1.0$. As can be seen, the emphasis is on the words that require fewer surrounding words.

This component was selected mainly because it represents the characteristics of market research data well. We believe that this component is good with other types of data also but we leave the evaluation of its effects out of scope of this paper.

In addition to fragment length, we use the distribution of the feature among all categories. A feature is important if it occurs often in a single category and seldom in others. This is assessed by estimating the probability $P(c|t)$, i.e., the probability for the category c given the word t. The probability is estimated simply by taking the number of documents in the category's document set that contain the word t ($|d : t \in d, d \in c|$) and dividing it by the total number of documents where the word t appears in ($|d : t \in d|$):

$$P(c|t) = \frac{|\{d : t \in d, d \in c\}|}{|\{d : t \in d\}|} = \frac{N_{t,c}}{N_{t,c} + N_{t,\neg c}} \, . \tag{10}$$

Here we also include the notation from Section 2, where $N_{t,c}$ is the number of times word occurs within the category c, and $N_{t,\neg c}$ is the number of times word occurs in other categories.

We substituted the other two statistics used in TT with Bi-Normal Separation. The idea with BNS is to compare the distribution of a word within a category and the word outside of the category. This weight is estimated as described in Equation 7.

BNS has similarities with the probability within a category ($P(t|c)$), i.e., the ratio of documents with t ($|d : t \in d, d \in c|$) against all documents $|d : d \in c|$ within the category c, used in our previous work [2]. However, by using BNS we get more information and give more weight in the cases when the word occurs often within the category and when it occurs seldom in other categories.

We call the resulting weight *Fragment Length Weighted Category Distribution* ($FLWCD$). For a word t in the category c it is calculated as follows:

$$FLWCD(t,c) = w(t,c) = BNS(t,c) \times P(c|d) \times \mathit{ifl}(t)$$

$$= |F^{-1}(\frac{N_{t,c}}{N_{t,c} + N_{\neg t,c}}) - F^{-1}(\frac{N_{t,\neg c}}{N_{t,\neg c} + N_{\neg t,\neg c}})| \times \frac{N_{d,c}}{N_{d,c} + N_{d,\neg c}} \times \frac{1}{\frac{1}{|f_t|}\sum l_f(t)},$$

$$(11)$$

where $N_{t,c}$ is the number of times word t occurs in category c, $N_{d,c}$ is the number of times document with term t occurs in category c, and $N_{\neg d,\neg c}$ is the number of documents that are neither in the category c nor contain the word t. For convenience, we shorten $FLWCD(t,c)$ to $w(t,c)$ to denote the weight in the equations in the following sections.

3.2 Classification

The first step of training is breaking the text into fragments. The fragmentation step was described in the previous section. After the text has been fragmented, each fragment is preprocessed and tokenized. Preprocessing includes stemming and stop word removal. The stop words are removed only after the fragmentation as some of the fragment breaks include stop words. For English, we use a standard stop word list.

Next, the statistics of each token is calculated. We get the following statistics for each word (w): for each category c the number of documents where the word w occurs in ($N_{w,c}$), number of documents with the word w not within c ($N_{w,\neg c}$), number of documents within the category c where w does not occur in ($N_{\neg w,c}$), and number of documents without w and that are not within c ($N_{\neg w,\neg c}$). In addition, the average length of the fragment the word appears in ($\mathit{ifl}(t)$) is calculated. All of these are simple statistics that can be estimated from the training data with one pass ($\mathcal{O}(n)$, where n is the number of documents).

The feature weight for each feature is calculated using Equation 11. After each of the features for each of the documents is weighted the feature vectors are then created for each of the categories. The feature vectors are normalized using the l^2-normalization (vector length norm):

$$w_{l^2}(t,v) = \frac{w(t,v)}{\sqrt{\sum_{w \in v} w(w,v)^2}}. \tag{12}$$

The normalized weight $w_{l^2}(t,v)$ for the feature t in the feature vector v is calculated by dividing the old weight $w(t,v)$ of the feature t with the length of the feature vector v.

When a training document has several categories, we use the document for all the categories. That is, if the document d has categories c_1 and c_2, there will be two feature vectors v_1 and v_2 created from d, where v_1 is used in category c_1 and v_2 in c_2. However, the weights of the features will be different as the weighting process may weight each feature differently for each category.

The feature vectors can be used with several types of classifiers. In this paper we focus on a Support Vector Machine classifier called SVM^{light} [19]. We do not focus on finding the optimal SVM kernel in this paper as we use the default linear kernel from SVM^{light}.

The model for each category is created using SVM which takes the training vectors as input and outputs the model. As the version of SVM^{light} we use does not support multi-label classification, the process is done by using a binary classifier where each document is compared to each category separately. When training the classifier the training data is divided into two sets for each category: set of positive examples and set of negative examples. The former set contains all the documents with the given category and the latter contains the rest. When a vector has several categories it is not included as a negative example to any of its categories.

The categories for each unlabeled document are predicted separately: for each category predict if the document belongs to this category. At the beginning of the categorization process, the unlabeled document is turned into a feature vector where each feature is weighted. The weight of a feature is the maximum weight of the word among all the categories it appears in. If the word is new, i.e., it does not occur in the training set, it is not included into the feature vector.

The feature vector is then used to predict if it is a positive or negative example of the category. This is done for each of the categories. When the prediction is positive, the document is classified into the given category. Finally, the process returns the set of positive categories for the document which are then used as its labels. That is, the document may receive several labels if the classifier gives positive predictions to several categories. More detailed descriptions of this approach can be found from our previous paper [3].

4   Evaluation

We compare the categorization and term weighting methods using two different datasets that are common in the field of short document categorization. The first dataset consists of market research data containing 12 different sets of answers from four different polls. This data was received from a market research company and it is actual real life data from their archives. The second datasets consists of tweets that were downloaded from Twitter. We built three testsets from the downloaded tweets.

We ran the test for each testset 10 times and report the average $F_{0.5}$-scores. The tests are done by randomly dividing the set of documents into training and testset with roughly 70 % - 30 % distribution, respectively. When dividing the data to training and testsets we check if there are enough instances of the class in the training set before including it in the testset. If the training set does not have at least two training documents for the category the document will not be included into the testset. We use the same training and testsets for the tests of each approach.

4.1   Data

An overview of the datasets is given in Table 1. The number of documents and the number of categories are the corpus level numbers, and average words and average categories are the averages per document in the dataset. When compared to the Reuters-21578 dataset, which has 160 words per document on average (before stop word removal) and where the average document term frequency is 1.57, we can see that the data we use in our experiments differs greatly from this traditional testset.

Table 1. Characteristics of the datasets used in our experiments

Dataset	Number of documents	Number of categories	Average words	Average categories
Yogurt Q1	1,030	40	3.62	1.16
Yogurt Q2	1,030	40	3.93	1.2
Commercial Q1	235	18	7.29	1.09
Commercial Q2	235	11	5.21	1.32
Commercial2 Q1	477	19	5.16	1.22
Commercial2 Q2	437	11	3.46	1.17
Commercial2 Q3	394	13	5.79	1.28
Vitamin Q1	742	28	5.96	2.07
Vitamin Q2	742	26	11.78	2.12
Vitamin Q3	742	14	6.26	1.61
Vitamin Q4	419	14	4.19	1.00
Vitamin Q5	742	17	5.56	1.32
Twitter Manual	1,810	5	14.45	1.00
Twitter HT	427	52	14.95	1.07
Twitter RMHT	427	52	14.95	1.07

We use real world datasets received from a market research company that have been collected from multiple surveys. We use twelve datasets from four different market research polls. The polls were: 1) feelings toward a yogurt product (Yogurt), 2) messages of a commercial (Commercial), 3) impression of another commercial (Commercial2), and 4) usage of dietary supplements such as vitamins (Vitamin). Yogurt poll contained two different questions, Commercial contained two and Commercial2 three different questions, and Vitamin contained five questions. The questions were about a specific product or a commercial shown to the respondents. All of the data was in Finnish. The average term frequency within the document with this data is 1.01.

Twitter data was collected using the Twitter4J[1] Java-library. We collected a set of tweets and used them to create three different datasets. We used only tweets that were written in English[2]. In addition, we only included data that contained hashtags, which is a Twitter keyword in the form of *#word*, to make the manual categorization easier. The first dataset was created by manually labeling approximately 1,000 tweets into 5 different categories. The categories we used were technology, sports, movies, music, and world. Each document (tweet) was given a single label. The tweets that do not fall under these hashtags were removed from this dataset. The dataset is referred to as *Twitter Manual* or *T M* later in this paper. The other two datasets were created by selecting tweets with a particular hashtag. We used a set of 30 predefined hashtags that we considered as an interesting or a current topic at the time.

We built the testset by using the predefined hashtags as the labels for the document. If the tweet held hashtags other than the ones in the predefined list, each of the new hashtags were included as a label for the document if the hashtag occurs at least 5 times in the whole dataset. Using this approach we built two different datasets: 1) Tweets

---

[1] http://twitter4j.org/en/index.html

[2] We used a language detector http://code.google.com/p/language-detection/

that contained the hashtag labels in the body text (*Twitter HT*), and 2) Tweets where the labels were removed from the body text (*Twitter RMHT*). That is, in the second case, we remove the hashtags that are used as the labels from the tweets in the Twitter RMHT dataset. For example, a tweet "What a great #golf round!" is used as "What a great round!". We did this to evaluate the performance of the approaches when there is no explicit information of the category.

Even though the labels are among the features in Twitter HT we decided to include this testset as it is similar with the case where the labels are the words occurring within the documents (which is often the case with market research data). In addition, as there are tweets with several labels the classification of this data is not as trivial as one might think. The average document term frequency in the Twitter data is 1.05.

Even though we do not have the permission to distribute the datasets in their original text form, the data can be made publicly available upon request in the form of feature vectors.

4.2 Evaluation Setup

For evaluation of text categorization approaches, we implemented kNN and the Naive Bayes approaches as described in [2]. We used SVM^{light} [19] for the SVM classification. Instead of using the normalization provided by SVM^{light} we normalize the feature vectors using the l^2-normalization described previously. We use Snowball stemmer[3] for stemming both English and Finnish words. To calculate Inverse Normal Cumulative Distribution Function used by BNS we use StatUtil for Java[4].

When using Naive Bayes with TT the threshold t_c was found for each test case by using a ten-fold cross validation process. This was done also for kNN to find the optimal k and the minimum similarity between the documents. The threshold and the minimum similarity between the documents were described in [2] and they are used since they produce better results for kNN and TT. The feature weighting approaches are implemented as described in Section 2.

We report the $F_{0.5}$-scores as the main requirement from the market research company is to emphasize precision over recall. We discuss this requirement in more detail in Section 4.5

4.3 Comparison of Classifiers

Table 2 shows the results of classifier comparison. We use Fragment Length Weighted Category Distribution for feature weighting in SVM. NB (TT) is our old approach [2]. TWCNB is the Naive Bayes approach described by Rennie et al. [14] and kNN is the k-Nearest Neighbors approach.

These results differ slightly from the earlier results we have reported [2] where the difference between kNN and TT was smaller. In our experiments kNN does not produce as good results as before. SVM clearly out-performs the competition in both test cases, as was expected. In our opinion, the poor performance of TWCNB is due to its

---

[3] http://snowball.tartarus.org/
[4] http://home.online.no/~pjacklam/notes/invnorm/

Table 2. Comparison of the $F_{0.5}$-scores between the different classification approaches. NB is the Naive Bayes like method described in [2]. Total average is the average of the two test cases (Average Market and Average Twitter) and not the average of all the testsets.

Dataset	SVM	NB (TT)	TWCNB	kNN
Yogurt Q1	0.76	0.71	0.30	0.66
Yogurt Q2	0.76	0.71	0.21	0.65
Commercial Q1	0.50	0.53	0.21	0.38
Commercial Q2	0.70	0.65	0.28	0.56
Commercial2 Q1	0.72	0.66	0.15	0.54
Commercial2 Q2	0.73	0.67	0.30	0.50
Commercial2 Q3	0.71	0.63	0.28	0.56
Vitamin Q1	0.79	0.64	0.26	0.56
Vitamin Q2	0.68	0.54	0.26	0.45
Vitamin Q3	0.79	0.71	0.31	0.64
Vitamin Q4	0.75	0.71	0.33	0.69
Vitamin Q5	0.70	0.63	0.26	0.56
Average Market	**0.72**	**0.65**	**0.26**	**0.56**
Twitter Manual	0.84	0.74	0.30	0.74
Twitter HT	0.81	0.70	0.13	0.61
Twitter RMHT	0.55	0.42	0.14	0.35
Average Twitter	**0.73**	**0.62**	**0.19**	**0.57**
Total Average	*0.73*	*0.64*	*0.23*	*0.57*

strong relation to term frequency. In our opinion, as SVM produces considerably better results, we recommend using SVM over the other approaches when categorizing short documents. However, we believe that a comparison against a wider range of classifiers should be done before making any final conclusions.

4.4   Comparison of Feature Weighting Methods

In this section we compare the feature weighting approaches presented in Section 2. We use the same datasets as in previous section. The results of the tests can be found from Table 3. As can be seen from the results, *FLWCD* performs the best with TT coming in second. Odds Ratio, BNS and Chi-Squared also produce comparable results. All of these approaches perform well in both test cases. The difference between the feature weighting approaches is not great in the test with market research data but when using Twitter data several approaches perform considerably worse.

The results seem to support our hypothesis that approaches that rely on term frequency tend to perform poorly; especially when compared to approaches that use term distribution among positive and negative samples. Residual IDF, TF-IDF and TF all produce weak results, as expected. This is most evident with the Twitter testset. This may be due to the fact that tweets contain more features: 15.0 words on average versus 5.7 words on average found in market research data. In those cases, these approaches cannot distinguish the difference between the words well but instead distribute the weights some what equally among all words. That is, with TF-IDF the weight is IDF which emphasizes the rarest words. As there are several similar IDF scores within the document,

Table 3. Comparison of the feature weighting methods. Compared approaches are Fragment Length Weighted Category Distribution ($FLWCD$), TT, Bi-Normal Separation (BNS), Chi-squared (χ^2), Pointwise Mutual Information (MI), Information Gain (IG), Odds Ratio (OR), Residual IDF (RIDF), Term Frequency (TF), and Term Frequency - Inverse Document Frequency (TFIDF).

Dataset	$FLWCD$	TT	BNS	χ^2	MI	IG	OR	RIDF	TFIDF	TF
Y Q1	0.76	0.76	0.72	0.72	0.76	0.75	0.75	0.75	0.75	0.69
Y Q2	0.76	0.74	0.76	0.77	0.79	0.76	0.75	0.76	0.77	0.72
C Q1	0.50	0.50	0.46	0.44	0.46	0.43	0.46	0.37	0.40	0.43
C Q2	0.70	0.69	0.69	0.60	0.67	0.67	0.69	0.66	0.64	0.55
C2 Q1	0.72	0.68	0.67	0.63	0.66	0.58	0.69	0.62	0.62	0.58
C2 Q2	0.73	0.73	0.74	0.68	0.76	0.75	0.73	0.72	0.69	0.57
C2 Q3	0.71	0.71	0.71	0.68	0.71	0.67	0.69	0.62	0.64	0.58
V Q1	0.79	0.78	0.75	0.70	0.75	0.69	0.80	0.76	0.76	0.65
V Q2	0.68	0.67	0.65	0.63	0.60	0.42	0.67	0.58	0.58	0.60
V Q3	0.79	0.79	0.77	0.70	0.74	0.64	0.76	0.77	0.76	0.70
V Q4	0.75	0.76	0.74	0.72	0.74	0.74	0.74	0.73	0.73	0.71
V Q5	0.70	0.69	0.72	0.68	0.71	0.63	0.71	0.66	0.66	0.63
Avg Mrk	**0.72**	**0.70**	**0.70**	**0.66**	**0.70**	**0.65**	**0.70**	**0.67**	**0.67**	**0.62**
T M	0.84	0.84	0.76	0.73	0.71	0.80	0.75	0.81	0.80	0.80
T HT	0.81	0.77	0.76	0.73	0.68	0.61	0.77	0.37	0.42	0.48
T RM	0.55	0.46	0.43	0.58	0.36	0.36	0.50	0.21	0.19	0.24
Avg Tw	**0.73**	**0.69**	**0.64**	**0.70**	**0.58**	**0.59**	**0.67**	**0.46**	**0.47**	**0.51**
Ttl Avg	*0.73*	*0.70*	*0.67*	*0.68*	*0.64*	*0.62*	*0.69*	*0.57*	*0.57*	*0.57*

the weight becomes same for all words. This can also be seen with the Vitamin Q2 testset, which is the largest dataset among market research data. We suspect that when the documents are extremely short (around 5 words) the weighting has smaller impact than when the documents are longer.

4.5   Categorization of Market Research Data

In this section we discuss briefly the characteristics and challenges of the market research case and the utilization of the proposed method in a real world application. The term weighting approach and the classifier was developed in collaboration with a market research company. The starting point for this work was the desire to find out if the existing market research data can be utilized for analysis of their future surveys. A market research company has dozens of different surveys out at the same time and each survey holds several open ended questions that each receive hundreds, sometimes thousands of answers. This makes the manual categorization an arduous process.

As there are several repeating surveys, i.e., cases where the same questions are asked after a specific time period, the same questionnaires are used repeatedly. Our aim was to build a classifier for the repeating questionnaires using the existing data from previous surveys for training. As the aim was to use the classifier by the market research company in their work, there was one specific requirement for the classifier: it needs to be as precise as possible. Recall was less important as not all of the answers need to be, or even can be, categorized. The market research experts can go through the uncategorized answers as long as there are not too many of them.

The market research data has several characteristics that make it challenging. First, the documents are very short, as can be seen from Table 1. This is the main challenge we focus on in our work. Second, the text is informal and often misspelled. Therefore, some of the answers may be too difficult to categorize without using some form of text correction. Third, even though an answer is short, it is possible that it has several (three or more) labels. This kind of data is difficult to use for classifier training and it is almost very challenging to predict all the labels. Finally, some of the answers are difficult to categorize even by domain experts. Therefore, some of the training data may have contradictions or even errors. Addressing these issues may offer interesting research topics in the future.

We evaluated the use of the proposed approach in a real world setting by gathering feedback from the market research professionals regarding the results of the experimental evaluation presented in our previous paper [3]. They agreed that the results were quite good when considering precision. In almost all of the cases SVM was able to produce 80 % precision. However, in some of the cases, especially with the Commercial Q1 (C Q1) the recall was very low, under 30 %. In other testsets the recall was between 40 % and 60 % which is not good but still acceptable according to the market research professionals.

In the second evaluation a new dataset was independently by the market research experts and they used the classifier in their actual work. The training set contained 995 answers and the unlabeled set contained 3,409 answers. There were 18 labels to which to categorize the data. Most answers in the training set contained a single class.

From the 3,409 documents the classifier was able to categorize 2,112 documents (62 %). The rest did not receive any label. In this case the recall was quite good, especially when compared to some other datasets. The precision was 89.3 % which can be considered very high. This precision was regarded as very good even by the domain experts. However, the processing time, which was around six hours, was problematic. As we use a binary classifier of SVM^{light}, testing each of the over 3,000 documents with 18 categories is very time consuming. This problem was emphasized by the fact that the computers used in the market research company were old with limited processing power. This problem can be alleviated through optimization and with a more effective classifier that supports multi-class categorization.

The overall impression of the evaluation was that the use of the proposed approach with a SVM classifier is feasible. The precision is usually good enough but the low recall and the long processing times are the major drawbacks that should be addressed in the future.

5   Conclusions

In this paper we have proposed a novel approach for term weighting in very short documents and performed a comprehensive evaluation and the comparison of the results against a wide variety of traditional approaches. Our approach is designed to tackle the biggest challenge faced with short documents: the TF=1 challenge. The proposed approach uses Bi-Normal Separation with inverse average fragment length and word's distribution among categories.

The experimental evaluation supported our hypothesis that term frequency based methods struggle when weighting terms in short documents. In addition, we found that the best performance among the categorization methods was received using a Support Vector Machine classifier. This finding was not surprising as it has produced strong results in several other studies as well. When comparing the term weighting approaches, $FLWCD$ produced the best results. The classifier was also evaluated in a real world setting by market research professionals. In this evaluation the approach produced promising results as the precision was good. However, the real challenge is the processing time that should be reduced considerably.

In the future we focus on two improvements: optimization of the classifier to address the processing time issue, and improvement of recall which is in some cases too poor. We have shown that the classifier can be used effectively in a real world setting and we hope that our work can be used as stepping stones in the future work of the emerging field of text mining from very short documents.

Acknowledgements. The authors wish to thank Taloustutkimus Oy for supporting the work, and Prof. Hannu Toivonen and the anonymous reviewers for their valuable comments.

References

1. Salton, G., Buckley, C.: Term-weighting approaches in automatic text retrieval. Information Processing and Management 24, 513–523 (1988)
2. Timonen, M., Silvonen, P., Kasari, M.: Classification of short documents to categorize consumer opinions. In: Online Proceedings of 7th International Conference on Advanced Data Mining and Applications (ADMA 2011), China (2011),
   http://aminer.org/PDF/adma2011/session3D/adma11_conf_32.pdf
   (accessed October 10, 2012)
3. Timonen, M.: Categorization of very short documents. In: Internation Conference on Knowledge Discovery and Information Retrieval (KDIR 2012), Spain, pp. 5–16 (2012)
4. Forman, G.: An extensive empirical study of feature selection metrics for text classification. Journal of Machine Learning Research 3, 1289–1305 (2003)
5. Rennie, J.D.M., Jaakkola, T.: Using term informativeness for named entity detection. In: Proceedings of the 28th Annual International ACM SIGIR Conference on Research and Development in Information Retrieval (SIGIR 2005), Brazil, pp. 353–360 (2005)
6. Clark, K., Gale, W.: Inverse Document Frequency (IDF): A measure of deviation from Poisson. In: Third Workshop on Very Large Corpora, pp. 121–130. Massachusetts Institute of Technology, Cambridge (1995)
7. Mladenic, D., Grobelnik, M.: Feature selection for unbalanced class distribution and Naive Bayes. In: Proceedings of the Sixteenth International Conference on Machine Learning (ICML 1999), Slovenia, pp. 258–267 (1999)
8. Forman, G.: BNS feature scaling: an improved representation over TF-IDF for SVM text classification. In: Proceedings of the 17th ACM Conference on Information and Knowledge Management (CIKM 2008), USA, pp. 263–270 (2008)

9. Yang, Y., Pedersen, J.: Feature selection in statistical learning of text categorization. In: Proceedings of the Fourteenth International Conference on Machine Learning (ICML 1997), USA, pp. 412–420 (1997)
10. Yang, Y.: An evaluation of statistical approaches to text categorization. Information Retrieval 1, 69–90 (1999)
11. Yang, Y., Liu, X.: A re-examination of text categorization methods. In: Proceedings of the 22nd Annual International ACM SIGIR Conference on Research and Development in Information Retrieval (SIGIR 1999), USA, pp. 42–49 (1999)
12. Krishnakumar, A.: Text categorization building a kNN classifier for the Reuters-21578 collection (2006), http://citeseerx.ist.psu.edu/viewdoc/summary?doi=10.1.1.135.9946 (accessed October 10, 2012)
13. Joachims, T.: Text categorization with Support Vector Machines: Learning with many relevant features. In: Nédellec, C., Rouveirol, C. (eds.) ECML 1998. LNCS, vol. 1398, pp. 137–142. Springer, Heidelberg (1998)
14. Rennie, J.D., Shih, L., Teevan, J., Karger, D.R.: Tackling the poor assumptions of Naive Bayes text classifiers. In: Proceedings of the Twentieth International Conference on Machine Learning (ICML 2003), USA, pp. 616–623 (2003)
15. Kibriya, A.M., Frank, E., Pfahringer, B., Holmes, G.: Multinomial Naive Bayes for text categorization revisited. In: Webb, G.I., Yu, X. (eds.) AI 2004. LNCS (LNAI), vol. 3339, pp. 488–499. Springer, Heidelberg (2004)
16. Pak, A., Paroubek, P.: Twitter as a corpus for sentiment analysis and opinion mining. In: Proceedings of the International Conference on Language Resources and Evaluation (LREC 2010), Malta (2010)
17. Irani, D., Webb, S., Pu, C., Li, K.: Study of trend-stuffing on Twitter through text classification. In: Seventh Annual Collaboration, Electronic Messaging, Anti-Abuse and Spam Conference (CEAS 2010), USA (2010), http://ceas.cc/2010/papers/Paper%2013.pdf (accessed October 10, 2012)
18. Benevenuto, F., Mango, G., Rodrigues, T., Almeida, V.: Detecting spammers on Twitter. In: Seventh Annual Collaboration, Electronic Messaging, Anti-Abuse and Spam Conference (CEAS 2010), USA (2010), http://ceas.cc/2010/papers/Paper%2021.pdf (accessed October 10, 2012)
19. Joachims, T.: Making large-Scale SVM Learning Practical. In: Advances in Kernel Methods - Support Vector Learning, pp. 41–56. MIT Press (1999)

An Approach to Model Selection in Spectral Clustering with Application to the Writing Style Determination Problem

Renata Avros, Avi Soffer, Zeev Volkovich, and Orly Yahalom

Department of Software Engineering, ORT-Braude College of Engineering, Karmiel, Israel
{r_avros,asoffer,vlvolkov,oyahalom}@braude.ac.il

Abstract. An open problem in spectral clustering concerning automatically finding the number of clusters is studied. We generalize the method for selecting the scale parameter offered in the Ng-Jordan-Weiss (NJW) algorithm and reveal a connection with the distance learning methodology. Values of the scaling parameter estimated via clustering of samples drawn are considered as a cluster stability indicator such that the clusters quantity corresponding to the most concentrated distribution is accepted as the "correct" number of clusters. Several numerical experiments have been conducted in order to establish the proposed spectral clustering approach and its application towards the style determination problem. The results reported here demonstrate high potential ability of the proposed method.

Keywords: Spectral Clustering, Model Selection, Text Mining.

1 Introduction

Cluster analysis is a significant component in information retrieval and intelligent processing, aimed at identifying meaningful, homogenous data groups, named clusters. Dissimilarity among the data items in a cluster is typically measured by a distortion function, whose value is desired to be minimal.

Spectral clustering has emerged as one of the main clustering approaches following Shi and Malik [46] and Ng, Jordan and Weiss [42]. Over the last decade, various spectral clustering algorithms have been developed and applied to computer vision [42], [46], [55], network science [16], [53], biometrics [52], text mining [32], natural language processing [6] and other areas. We note that spectral clustering methods have been found equivalent to kernel k-means [8], [29] as well as to nonnegative matrix factorization [10]. Several surveys review the use of spectral clustering methods [41], [34], [13].

The main idea in spectral clustering is to use eigenvectors of the Laplacian matrix, based on an affinity (similarity) function over the data. The Laplacian is a positive semi-definite matrix whose eigenvalues are nonnegative reals. It is well-known that the smallest eigenvalue of the Laplacian is 0, and it corresponds to an eigenvector with all entries equal. Moreover, viewing the data similarity function as an adjacency

A. Fred et al. (Eds.): IC3K 2012, CCIS 415, pp. 19–36, 2013.

matrix of a graph, the multiplicity of the 0 eigenvalue is the number of connected components [39]. While in clustering problems the corresponding graph is typically connected, we partition the data into k clusters using the k eigenvectors corresponding to the k smallest eigenvalues. These would either be the k smallest eigenvectors or the k largest eigenvectors, depending on the Laplacian version being used. For example, a simple way of partitioning the data into two clusters would be considering the second eigenvector as an indicator vector, assigning items with positive coordinate values into one cluster, and items with negative coordinate values to another cluster.

Spectral clustering algorithms have several significant advantages. First, they do not make any assumptions on the clusters, which allows flexibility in discovering various partitions (unlike the k-means algorithm, for example, which assumes that the clusters are spherical). Second, they rely on basic linear algebra operations. And finally, while spectral clustering methods can be costly for large and dense data sets, they are particularly efficient when the Laplacian matrix is sparse (i.e., when many pairs of points are of zero affinity). Spectral methods can also serve in dimensionality reduction for high-dimensional data sets (the new dimension being the number of clusters k).

Note, that the problem to determine the optimal ("correct") number of groups for a given data set is very crucial in cluster analysis. This need, known as cluster validation, arises in many applications. As usual, the clustering solutions obtained for several numbers of clusters are compared according to the chosen criteria. The sought number yields the optimal quality in accordance with the chosen rule. The problem may have more than one solution and is known as an "ill posed" [21] and [18]. For instance, an answer here can depend on the scale in which the data is measured. Many approaches were proposed to solve this problem, yet none has been accepted as superior so far.

From a geometrical point of view, cluster validation has been studied in the following papers: Dunn [12], Hubert and Schultz [20], Calinski-Harabasz [4], Hartigan [19], Krzanowski-Lai [28], Sugar-James [49], Gordon [17], Milligan and Cooper [38] and Tibshirani, Walter and Hastie [48] (the Gap Statistic method). Here, the "elbow" criterion plays a central role in the indication of the correct number of clusters.

Several methods using the goodness of fit concepts are suggested by Volkovich, Barzily and Morozensky [51], Barzily, Volkovich, Akteke-Ozturk and Weber [1], Toledano-Kitai, Avros and Volkovich [49]. Here, the source cluster distributions are constructed based on a model designed to represent well-mixed samples within the clusters.

Another common approach in this area employs the cluster stability concept. Apparently, Jain and Moreau [22] were the first to propose such a point of view in the cluster validation thematic and used the dispersions of empirical distributions of the cluster object function as a stability measure. Following this perception, differences between solutions obtained via rerunning a clustering algorithm on the same datum evaluate the partitions stability. Hence, the number of clusters minimizing partitions' changeability is used to assess the "correct" number of clusters. In papers of Levine and Domany [31], Ben-Hur, Elisseeff and Guyon [2], Ben-Hur and Guyon [3] and Dudoit and Fridlyand [11] (the CLEST method), stability criteria are understood to be the fraction of times that pairs of elements maintain the same membership under reruns of the clustering algorithm. Mufti, Bertrand, and El Moubarki [40] exploit Loevinger's measure of isolation to determine a stability function.

In this paper we propose a new approach to an open problem in spectral clustering which is aimed at automatically finding the number of clusters. Our approach is based on the cluster stability concept. We generalize the method for selecting the scale parameter suggested by the Ng-Jordan-Weiss algorithm (NJW) [42], and reveal a connection with the distance learning methodology. Values of the scaling parameter, estimated via clustering of the drawn samples for the number of clusters allocated in a given area, are considered as a cluster stability indictor such that the preferred number of clusters corresponds to the most concentrated empirical distribution of the parameter. The numerical experiments that we conducted demonstrate high potential ability of the proposed method. In addition, we applied this enhanced clustering method to several sets of texts, attempting to address the style determination challenge through our proposed spectral clustering methodology.

The rest of the paper is organized in the following way: Section 2 is devoted to establishing the basic statements of cluster analysis used in this work, and to a discussion of the scale parameter selection approaches. In section 3 we propose an application of the proposed new methodology to the cluster validation problem. Section 4 include a description of the numerical experiments conducted in order to test the ability of our methodology to reveal the "true" number of clusters. Applications of this methodology to the author's style determination problem are studied and presented in section 5, while section 6 includes the conclusion.

2   Clustering

We consider a finite subset $X = \{x_1, \dots, x_n\}$ of the Euclidean space R^d. A partition of the set X into k clusters is a collection of k non-empty disjoint its subsets $\Pi_k = \{\pi_1, \dots, \pi_k\}$ satisfying the condition:

$$\bigcup_{i=1}^{k} \pi_i = X.$$

The partition's elements are named *clusters*. Two partitions are identical if and only if every cluster in the first partition is also presented in the second one and vice versa. In cluster analysis a partition is chosen so that a given quality

$$Q(\Pi_k) = \sum_{i=1}^{k} q(\pi_i)$$

is optimized for some real valued function q whose domain is the set of subsets of X. The function q is a distance-like function and, commonly, it is not required to be positive or to satisfy the triangle inequality. In case of the hard clustering the underlying distribution of X is assumed to be represented in the form

$$\mu_X = \sum_{i=1}^{k} p_i \eta_i,$$

where p_i, $i = 1,..,k$ are the clusters' probabilities and η_i, $i = 1,..,k$ are the clusters' distributions. Particularly, the most prevalent Gaussian Mixture Model considers distributions η_i having densities

$$f_i(x) = \phi(x \mid m_i, \Gamma_i), i = 1,...,k,$$

where $\phi(x \mid m_i, \Gamma_i)$ denotes the Gaussian density with mean vector m_i and covariance matrix Γ_i. Usually, the mixture parameters

$$\theta = (p_i, m_i, \Gamma_i), i = 1,...,k$$

are estimated in this case by maximizing the likelihood

$$L(\theta \mid x_1,...,x_n) = \sum_{j=1}^{n} \ln \left(\sum_{i=1}^{k} p_i \phi(x_j \mid m_i, \Gamma_i) \right). \tag{1}$$

The most common procedure for maximum likelihood clustering solution is the EM algorithm (see, for example [36]). The EM algorithm provides, in many cases, meaningful results. However, the algorithm often converges slowly and has a strong dependence on its starting position. One of the important EM related algorithms is a Classification EM algorithm (CEM) introduced by Celeux and Govaert in [5]. CEM maximizes the Classification Likelihood criterion which is different from the Maximum Likelihood criterion (1).

2.1 Spectral Clustering

The k-means approach has been introduced in [15] and in [35]. It provides the clusters which approximately minimize the sum of the items' squared Euclidean distances from cluster centers, which are called *centroids*. The algorithm generates linear boundaries among clusters. Celeux and Govaert [5] showed that, in the case of the Gaussian Mixture Model, this procedure actually assumes that all mixture proportions are equal, and   the covariance matrixes of the components are identical sperical ones.

In the recent decade, spectral clustering has emerged as one of the main approaches to data clustering, as it enables discovery of clusters of various forms, in addition to being very efficient for sparse data. As in other clustering methods, one of the difficulties in spectral clustering is the need to determine various parameter values. In our research, we focus on the scaling parameter σ as defined by Ng, Jordan and Weiss for their popular spectral clustering algorihm (NJW) [42].

Spectral clustering capability commonly leverage the spectrum of a given similarity matrix in order to perform dimensionality reduction for clustering in fewer dimensions. Here, we concentrate on a relatively simple technique offered in [42] in order to demonstrate the ability of our proposed approach.

Algorithm 2.1. *Spectral Clustering(* X , k, σ *)* *(NJW).*

> **Input**
> - X - the data to be clustered;
> - k - number of clusters;
> - σ - the scaling parameter.
>
> **Output**
> $\Pi_{k,\sigma}(X)$ - a partition of X into k clusters depending on σ.

========================

- Construct the affinity matrix $A(\sigma^2)$

$$\{a_{ij}(\sigma^2)\} = \begin{cases} \exp\left(\dfrac{-\|x_i - x_j\|^2}{2\sigma^2}\right) & \text{if } i \neq j, \\ 0 & \text{otherwise} \end{cases}$$

- Introduce $L = D^{-\frac{1}{2}}A(\sigma^2)D^{-\frac{1}{2}}$ where D is the diagonal matrix whose (i,i)-element is the sum of $A's$ i-th row.
  (Note, that the acceptable point of view proposes to deal with the Laplacian $I-L$. However, the authors Ng et al. [42] prefer to work with L and only to change the eigenvalues (from A to $I-A$) without any changing of the eigenvectors.)
- Compute $z_1, z_2, ..., z_k$, the k largest eigenvectors of L (chosen to be orthogonal to each other in the case of repeated eigenvalues);
- Create the matrix $Z = \{z_1, z_2, ..., z_k\} \in R^{n \times k}$ by joining the eigenvectors as consequent columns;
- Compute the matrix Y from Z by normalizing each of Z's rows to have a unit length;
- Cluster the rows of Y into k clusters via K-means or any other algorithm (that attempts to minimize distortion) to obtain a partition $\Pi_{k,\sigma}(Y)$;
- Assign each point x_i according to the cluster that was assigned to the row i in the obtained partition.

Note, that there is a one to one correspondence between the partitions $\Pi_{k,\sigma}(X)$ and $\Pi_{k,\sigma}(Y)$. The magnitude parameter σ represents the increasing rate of the affinity of the distance function. This parameter plays a very important role in the clustering process and can be naturally reached as the outcome of an optimization problem intended to find the best possible partition configuration. An appropriate meta algorithm could be presented in the following form.

Algorithm 2.2. *Self-Learning Spectral Clustering* (X,k,F).

Input
- X - the data to be clustered;
- k - number of clusters;
- F - cluster quality function to be minimized.

Output
- σ^* - an optimal value of the the scaling parameter;
- $\Pi_{k,\sigma^*}(X)$ – a partition of X into k clusters corresponding to σ^*.

=====================
    Return

$$\sigma^* = arg\ \min_{\sigma}(\ F(\ \Pi_{k,\sigma}(\ X\ )=$$

$$= SpectralClustering(\ X,k,\sigma\ )\ )).$$

When σ is described as a human-specified parameter which is selected to form the ``tight" k clusters on the surface of the k -sphere. Consequently, it is recommended in the source algorithm to search over σ and to take the value that gives the tightest (smallest distortion) clusters of the set Y . Here

$$F_1(\ \Pi_{k,\sigma}(\ X\ )) = \frac{1}{|Y|}\sum_{i=1}^{k}\sum_{y\in\pi_i}\|y-r_i\|^2, \tag{2}$$

where r_i , $i=1,...,k$ are cluster's centroids. This procedure can be generalized other partition quality functions. Functions of this kind can be found in the framework of the distance learning methodology. Let us presume that the degree of similarity between pairs of elements of data collection is known:

S : $\{(x_i,x_j)$; *if* x_i *and* x_j *are similar: belong to the same cluster}*

and

D : $\{(x_i,x_j)$; *if* x_i *and* x_j *are not similar: belong to different clusters}*

the goal is to learn a distance metric $d(x,y)$ such that all "similar" data points are kept in the same cluster, (i.e., close to each other) while distinguishing the "dissimilar" data points. To this end, we define a distance metric in the form:

$$d_C^2(\ x,y\ ) = \|x-y\|_A^2 = (x-y)^T\cdot C\cdot(x-y),$$

where C is a positive semi-definite matrix, $C\succ 0$ which is learned. We can formulate a constrained optimization problem where we aim to minimize the sum of similar distances concerning pairs in S while maximizing the sum of dissimilar distances related to pairs in D in the following way:

$$\min_{C} \sum_{(x_i, x_j) \in S} \|x_i - x_j\|_C^2$$

s.t.

$$\sum_{(x_i, x_j) \in D} \|x_i - x_j\|_C^2 \geq 1, C \succ 0$$

If we suppose that the purported metric matrix is diagonal then minimizing the function is equivalent to solving the stated optimization problem [54] up to a multiplication of C by a positive constant. So, we take this function as the second quality function:

$$F_2(\Pi_{k,\sigma}(X)) = \sum_{(x_i, x_j) \in S, i \neq j} \|y_i - y_j\|^2 - \log \left(\sum_{(x_i, x_j) \in D} \|y_i - y_j\| \right). \tag{3}$$

Finally, in the spirit of the Fisher's linear discriminant analysis we can consider the function:

$$F_3(\Pi_{k,\sigma}(X)) = \frac{\sum_{(x_i, x_j) \in S, i \neq j} \|y_i - y_j\|^2}{\sum_{(x_i, x_j) \in D} \|y_i - y_j\|^2}. \tag{4}$$

3   An Application to the Cluster Validation Problem

In this section we discuss an application of the offered methodology to the cluster validation problem. We suggest that these values should be learned from samples clustered for several clusters quantities such that the most stable behaviour of the parameter is exhibited when the cluster structure is the most stable. In our case, it means that the number of clusters is chosen by the best possible way. The drawbacks of the used algorithm together with the complexity of the dataset structure add to the uncertainty of the process outcome. To overcome this ambiguity, a sufficient amount of data has to be involved. This is achieved by drawing many samples and constructing an empirical distribution of the scaling parameter values. The most concentrated distribution corresponds to the appropriate number of clusters.

Algorithm 3.1. *Spectral Clustering Validation* (X , K , F , J , m , Ind).

    Input

-   X - the data to be clustered;
-   K - maximal number of clusters to be tested;
-   F - cluster quality function to be minimized;
-   J - number of samples to be drawn;
-   m - size of samples to be to be drawn;
-   Ind - concentration index.

    Output

-   k^* − an estimated number of clusters in the dataset.

========================

- For $k = 2$ to K do
-   For $j = 1$ to J do
-     $S = sample(X, m)$;
-     σ_j = **Self-Learning Spectral Clustering**(X, k, F);
-   end For j
-     Compute $C_k = Ind\{\sigma_1, ..., \sigma_J\}$
- end For k

The "true" number of clusters - k^* is chosen according to the most concentrated distribution indicated by an appropriate value of C_k , $k = 2, ..., K$.

3.1   Remarks Concerning the Algorithm

Here, $sample(X, m)$ denotes a procedure of drawing a sample of size m from the population X without repetitions. Concentration indexes can be provided in several ways. The most widespread instrument used for the evaluation of a distribution's concentration is the standard deviation. However, it is sensitive to outliers and can be principally dependent, in our situation, on the number of clusters examined. To counterbalance this reliance, the values have be normalized. Unfortunately, it has been specified in the clustering literature that the standard " correct" strategy, for normalization and scaling, does not exist (see, for example [45] and [48]). We use the coefficient of variation (*CV*) which is defined as the ratio of the sample standard deviation to the sample mean. For comparison between arrays with different units this value is preferred to the standard deviation because it is a dimensionless number.

    Several experiments have been conducted in order to establish the improved stability of the proposed spectral clustering approach and exemplify its application towards the style determination problem. The experiments are divided into 2 parts, according to 2 objectives:

1. Validate the cluster stability achieved by algorithm 3.1 - presented in section 4.
2. Demonstrate the application of the proposed clustering approach to determination of writing style – described in section 5.

4 Numerical Experiments on Cluster Stability Discovering

Several experiments have been conducted in order to validate the improved stability of the proposed spectral clustering approach defined by algorithm 3.1. We demonstrate and test the proposed approach by means of various numerical experiments on synthetic and real datasets provided for the three functions $F1$, $F2$, and $F3$ presented in (2)-(4) (in section 2). We choose $K=7$ in all tests and perform 10 trials for each experiment. The results are presented via the error-bar plots of the coefficient of variation within the trials. The presentation is followed by an internal and external comparison discussion.

4.1 Synthetic Data

The first example consists of a mixture of 5 two-dimensional Gaussian distributions with independent coordinates with the same standard deviation $\sigma = 0.25$. The components means are placed on the unit circle with the angular neighboring distance $2\pi / 5$. The dataset contains (denoted as $G5$) 4000 items.

We set here J (number of samples) = 100 and m (size of samples) = 400.

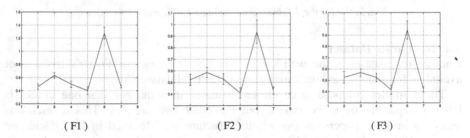

(F1)   (F2)   (F3)

Fig. 1. CV for the $G5$ dataset using $F1$, $F2$, and $F3$ functions

The CV index demonstrates approximately the same performance for all object functions hinting to a 5 or 7 clusters structure. However, the bars do not overlap only in the first case where a 5 cluster partition is properly indicated.

4.2 Real-World Data

We now exemplify and analyze the proposed algorithm by application to several known real-world datasets.

Three Texts Collection

The first real dataset is chosen from the text collection available at: http://ftp.cs.cornell.edu/pub/smart

This set (denoted as $T3$) includes the following three text collections:

- DC0--Medlars Collection (1033 medical abstracts);
- DC1--CISI Collection (1460 information science abstracts);
- DC2--Cranfield Collection (1400 aerodynamics abstracts).

This dataset was considered in many works [9], [25], [26], [27], and [50]). Usually, following the well-known ``bag of words" approach, 600 ``best" terms were selected (see, [7] for term selection details). So, the dataset was mapped into Euclidean spaces with dimensions 600. A dimension reduction is provided by the Principal Component Analysis (PCA). The considered dataset is recognized to be well-separated by means of the two leading principal components. We use this data representation in our experiments. The results presented in Fig. 2 for $m=J=100$ show that the number of clusters was properly determined for all functions F.

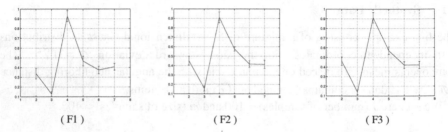

(F1)   (F2)   (F3)

Fig. 2. CV for the $T3$, 600 terms, using $F1$, $F2$, and $F3$ functions

The Iris Flower Dataset

Another real dataset is the well-known Iris flower dataset or Fisher's Iris dataset available, for example, at http://archive.ics.uci.edu/ml/datasets/Iris

These species compose three clusters situated in a manner that one cluster is linearly separable from the others, but the other two are not. This dataset was analyzed in many papers. A two cluster structure was detected in [44]. Here, we selected 100 samples of size 140 for each tested number of clusters. As it can be seen, the "correct" number of clusters has been successfully found for the $F2$ and $F3$ objective functions. The experiments with $F1$ offer a two clusters configuration.

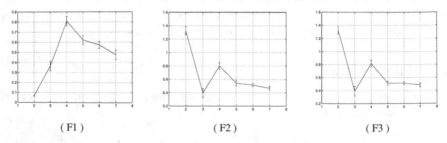

(F1)   (F2)   (F3)

Fig. 3. CV for the Iris dataset using $F1$, $F2$, and $F3$ functions

The Wine Recognition Dataset

This real dataset contains 178 results of a chemical analysis of three different types (cultivates) of wine given by their 13 ingredients. This collection is available at: *http://archive.ics.uci.edu/ml/machine-learning-databases/wine*

This data collection is relatively small however it exhibits a high dimension. The parameters in use were J=m=100. Fig. 4 demonstrates undoubtedly that for $F2$ and $F3$ the true number of clusters is revealed, however $F1$ detects a wrong structure.

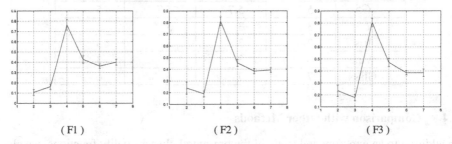

(F1)   (F2)   (F3)

Fig. 4. *CV* for the Wine dataset using $F1$, $F2$, and $F3$ functions

The Glass Dataset

This dataset is taken from the UC Irvine Machine Learning Repository collection. (http://archive.ics.uci.edu/ml/index.html). The study of classification of glass types was motivated by criminology investigation.The glass found at the place of a crime, can be used as evidence. Number of Instances: 214. Number of Attributes: 9.

Fig. 5 demonstrates outcomes obtained for J=100. Note, that this relatively small dataset possess a comparatively large dimension and a significantly larger, in comparison with previous collection, suggested number of clusters. To eliminate the influence of the sample size on the clustering solutions we draw samples with growing sizes $m = max((k-1)*40, 214)$. The minimal value depicted in the graph corresponding to the $F3$ function is 6, however the bars of "2" and "6" overlap. Since the index behavior is more stable once the number of clusters is 6, this value is accepted as the true number of clusters. Other function do not success in determining the true number of clusters.

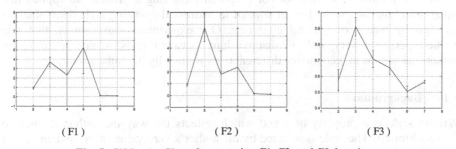

(F1)   (F2)   (F3)

Fig. 5. *CV* for the Glass dataset using $F1$, $F2$, and $F3$ functions

4.3   Comparison of the Partition Quality Function Used

Table 1 summarizes the results of the numerical experiments provided. As can be seen, the functions $F2$ and $F3$, introduced in (3)-(4) (in section 2), subsume the previously offered function $F1$.

Table 1. Comparison of the partition quality function used

Dataset	$F1$	$F2$	$F3$	TRUE
$G5$	5	5,7	5,7	5
$T3$	3	3	3	3
Iris	2	3	3	3
Wine	2	3	3	3
Glass	7	7	6	6

4.4   Comparison with Other Methods

In addition to an experimental study of the presented cluster quality functions, we also provide a comparison of our method with several other cluster validation approaches. In particular, we evaluate the results obtained by the Calinski and Harabasz index (CH) [4], the Krzanowski and Lai index (KL) [28], the Sugar and James index (SJ) [47], the GAP-index [48], and the Clest-index [11]. Our method succeeds quite well in the comparison in case ones an appropriate quality function was chosen.

Table 2. Comparison with other methods

Data-	CH	KL	SJ	Gap	$Clest$
$G5$	5	5	5	3	6
$T3$	3	3	1	3	2
Iris	2	2	4	7	7
Wine	3	2	3	6	1
Glass	2	2	2	6	3

5   Writing Style Determination Experiments

In order to reinforce the value of our proposed clustering approach, we applied this enhanced clustering method in an attempt to address the writing style determination problem by means of spectral clustering. Our experiments have shown that this method performs very well in distingushing between texts written by different authors, as well as in recognizing different texts written by the same author.

5.1   Background

Writing style is a property of a text which reflects the way the author chooses to express himself. The style is affected by the author's perception of his audience, and the way he wishes to convey his thoughts and messages to them. Attributes of the writing style are the choice of words, the syntactical structure (simple/complex), choice of prose (simple/formal wording), tone (casual/proper/formal), use of figurative language (e.g. use of metaphors), and overall organization of the thought. The writing style is one of the fundamental character of a text, thus it may be useful as a key to identifying an unknown author.

5.2   Experiments

The experiments in this category were done in two stages:

First, we exemplify the algorithm by application to several texts written by known authors. The objective is to test how well the clustering algorithm is capable of distinguishing between the various books and group them according to their author, as determined by the writing style. In the second experiment we used the algorithm in analyzing a set of books whose authors are unknown, with a goal to strengthen or weaken the hypothesis that some books were written by a certain author.

In the prepocessing step of each experiment we divided all the documents into equal sized chunks of 10KBites. Then, the Bag of Words method has been applied in order to represent each chunk as a vector. As a result, each book is represented by means of several vectors. The folowing sub-sections include a detailed description of the two experiments and their results.

1. Text Classification by Writing Style Determination
This experiment's data set (CAR) includes 5 books by Arthur C. Clark, 4 books by Isaac. Asimov ("Foundation" series), and 4 books by J. K. Rowling (Harry Potter series). At the first step we have tried to determine the "true" number of clusters in this collection by means of our method. (Note that it is evidently expected to be equal to 3). We choose $K = 7$ repeat all tests 10 times for 100 samples of size 50 drawn from 67 subdocuments which are obtained here as vectors having 16694 coordinates. The results presented in Fig. 6 clearly indicate 3 a clusters' structure.

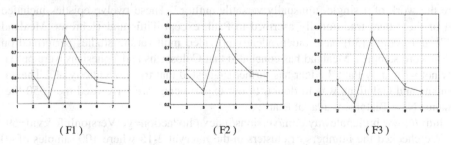

(F1)   (F2)   (F3)

Fig. 6. CV for the CAR dataset using $F1$, $F2$, and $F3$ functions

The clustering solution attained for k=3 demonstrate perfect separation between authors achieved by our method as can be seen in the following contingency table:

Table 3. Contingency table of the authors' partition amid the clusters

cluster/author	Clark	Asimov	Rowling
1	0	18	0
2	18	0	0
3	0	0	27

2. Authorship of the Deutro-Pauline Epistles

In this section, we use our clustering method in order to examine the Deutero-Pauline Epistles authorship question. The Pauline Epistles are the fourteen letters in the New Testament traditionally attributed to Paul the Apostle, although the Epistle to the Hebrews does not bear his name. The Pauline Epistles are classified into several groups:

- The Undisputed Epistles: Romans, 1 Corinthians, 2 Corinthians, 1 Thessalonians, Galatians, Philippians, and Philemon, are considered genuine work of Paul by most scholars [43].
- The Pastoral Epistles: 1 Timothy, 2 Timothy and Titus, are thought by most modern scholars to be pseudepigraphic, that is, written by an imposter of Paul in order to justify ecclesiastical innovations [24].
- The Deutero-Pauline Epistles: Ephesians, Colossians and 2 Thessalonians, have no consensus on whether or not they are authentic letters of Paul. Thus, they are the focus of our experiments.
- Almost all scholars agree that the Epistle to the Hebrews was not written by Paul [14].

The Pauline Epistles were among the first documents whose authorship was examined using statistical tools, when, in 1851, Augustus de Morgan conducted a study on word length statistics on them (see [33]). In the recent decades, various studies have been carried out on the Epistles, often reaching to conflicting conclusions. We mention the comprehensive work of Kenny [23], who has conducted a univariate study and then combined features in order to show the relationships between different epistles. He concluded: "I see no reason to reject the hypothesis that twelve of the Pauline Epistles are the work of a single, unusually versatile author." These twelve epistles included all of the letters traditionally ascribed to Paul, except Titus and Hebrews. Mealand [37] and Ledger [30] have independently provided multivariate studies performed on 1000 word samples. Mealand has concluded that Colossians and Ephesians were probably not written by Paul. Ledger has further suggested that the authenticity of 1 Thessalonians and Galatians was in doubt. In our experiments, we have used the 1550 Stephanus New Testament available at:

http://www.biblegateway.com/versions/index.php?action=getVersionInfo&vid=69.

We checked the numbers of clusters in the interval 2-15 where 100 samples of 40 documents were drawn from 49 for each number of clusters. The subdocuments have been obtained as vectors with 7192 features.

Fig. 7 shows the values of the functions $F1$, $F2$ and $F3$, respectively, used for determining the number of clusters in the Pauline data set. As can be seen, all the functions clearly indicate that the optimal number of clusters is three.

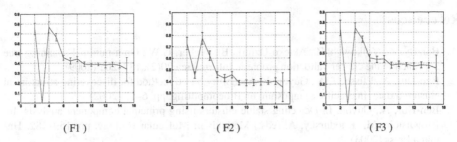

(F1)   (F2)   (F3)

Fig. 7. *CV* for the Deutero-Pauline Epistles using *F*1, *F*2, and *F*3 functions

Applying the adaptive spectral clustering algorithm with k = 3 has resulted in the following partition:

- <u>Cluster 1</u>: 1 Thessalonians, Philippians, 2 Corinthians , 2 Thessalonians, Colossians, Ephesians.
- <u>Cluster 2</u>: 1 Corinthians, Galatians, Romans.
- <u>Cluster 3</u>: 1 Timothy, 2Timothy, Titus, Hebrews.

The Deutero-Pauline Epistles, marked in boldface, all belong to Cluster 1, together with Undisputed Epistles. Cluster 2 consists entirely of Undisputed Epistles, whereas Cluster 3 consists of the Pastoral Epistles and Hebrews. According to the partition above, some of the Undisputed Epistles are more similar to the Deutero-Pauline Epistles than to other Undisputed Epistles. Thus, our experiments seem to reinforce the hypothesis that the Disputed Epistles were in fact written by Paul.

6   Conclusions and Future Work

In this paper a new approach to determine the number of the groups in spectral clustering was presented. An empirical distribution of the scaling parameter, found resting upon samples clusterization, is considered as a new cluster stability feature. In order to establish the proposed spectral clustering method we conducted several experiments, in which we analyzed three cost functions which can be used in a self-tuning version of a spectral clustering algorithm. Our experiments have shown that this method performs very well in identifying and distingushing between texts written by different authors, as well as different texts written by the same author.

As continued research we plan to generalize our method to the Local Scaling methodology [56] and compare the obtained outcomes. Another direction which will be investigated in our futur work consist of a study of the model behavior when the number of clusters is suggested to be relatively big. An essential ingredient of each resampling cluster validation approach is the selection of the parameters values in an implementation. It is difficult to treat this task from a theoretical point of view (as indicated, for example, in [11], [45]   and [31]).

References

1. Barzily, Z., Volkovich, Z., Akteko-Ozturk, B., Weber, G.W.: On a minimal spanning tree approach in the cluster validation problem. Informatica 20(2), 187–202 (2009)
2. Ben-Hur, A., Elisseeff, A., Guyon, I.: A stability based method for discovering structure in clustered data. In: Pacific Symposium on Biocomputing, pp. 6–17 (2002)
3. Ben-Hur, A., Guyon, I.: Detecting stable clusters using principal component analysis. In: Brownstein, M., Khodursky, A. (eds.) Methods in Molecular Biology, pp. 159–182. Humana Press (2003)
4. Calinski, R., Harabasz, J.: A dendrite method for cluster analysis. Communications in Statistics 3, 1–27 (1974)
5. Celeux, G., Govaert, G.: A classification EM algorithm for clustering and two stochastic versions. Computational Statistics and Data Analysis 14, 315–332 (1992)
6. Dasgupta, S., Ng, V.: Mine the easy, classify the hard: a semi-supervised approach to automatic sentiment classification. In: ACL-IJCNLP 2009: Proceedings of the Main Conference, pp. 701–709 (2009)
7. Dhillon, I., Kogan, J., Nicholas, C.: Feature selection and document clustering. In: Berry, M. (ed.) A Comprehensive Survey of Text Mining, pp. 73–100. Springer, Berlin (2003)
8. Dhillon, I.S., Guan, Y., Kulis, B.: Kernel k-means, spectral clustering and normalized cuts. In: Proceedings of the Tenth ACM SIGKDD International Conference on Knowledge Discovery and Data Mining (KDD), pp. 551–556 (2004)
9. Dhillon, I.S., Modha, D.S.: Concept decompositions for large sparse text data using clustering. Machine Learning 42(1), 143–175 (2001), Also appears as IBM Research Report RJ 10147 (July 1999)
10. Ding, C., He, X., Simon, H.D.: On the equivalence of nonnegative matrix factorization and spectral clustering. In: Proceedings of the Fifth SIAM International Conference on Data Mining, vol. 4, pp. 606–610 (2005)
11. Dudoit, S., Fridlyand, J.: A prediction-based resampling method for estimating the number of clusters in a dataset. Genome Biol. 3(7) (2002)
12. Dunn, J.C.: Well Separated Clusters and Optimal Fuzzy Partitions. Journal on Cybernetics 4, 95–104 (1974)
13. Filippone, M., Camastra, F., Masulli, F., Rovetta, S.: A survey of kernel and spectral methods for clustering. Pattern Recognition 41(1), 176–190 (2008)
14. Fonck, L.: Epistle to the Hebrews, The Catholic Encyclopedia, vol. 7. Robert Appleton Company, New York (1910)
15. Forgy, E.W.: Cluster analysis of multivariate data - efficiency vs interpretability of classifications. Biometrics 21(3), 768–769 (1965)
16. Fortunato, S.: Community detection in graphs. Phys. Rep. 486(3-5), 75–174 (2010)
17. Gordon, A.D.: Identifying genuine clusters in a classification. Computational Statistics and Data Analysis 18, 561–581 (1994)
18. Gordon, A.D.: Classification. Chapman and Hall, CRC, Boca Raton, FL (1999)
19. Hartigan, J.A.: Statistical theory in clustering. J. Classification 2, 63–76 (1985)
20. Hubert, L., Schultz, J.: Quadratic assignment as a general data-analysis strategy. Br. J. Math. Statist. Psychol. 76, 190–241 (1974)
21. Jain, A., Dubes, R.: Algorithms for Clustering Data. Prentice-Hall, Englewood Cliffs (1988)
22. Jain, A.K., Moreau, J.V.: Bootstrap technique in cluster analysis. Pattern Recognition 20(5), 547–568 (1987)

23. Kenny, A., Stylometric, A.: Study of the New Testament. Oxford University Press, USA (1986)
24. Knight III, G.W.: The Pastoral Epistles: A Commentary on the Greek Text. Eerdmans Publishing Company (1992)
25. Kogan, J., Nicholas, C., Volkovich, V.: Text mining with hybrid clustering schemes. In: Berry, M.W., Pottenger, W. (eds.) Proceedings of the Workshop on Text Mining (held in conjunction with the Third SIAM International Conference on Data Mining), pp. 5–16 (2003)
26. Kogan, J., Nicholas, C., Volkovich, V.: Text mining with information– theoretical clustering. Computing in Science & Engineering, pp. 52–59 (November/December 2003)
27. Kogan, J., Teboulle, M., Nicholas, C.: Optimization approach to generating families of k–means like algorithms. In: Dhillon, I., Kogan, J. (eds.) Proceedings of the Workshop on Clustering High Dimensional Data and its Applications (held in conjunction with the Third SIAM International Conference on Data Mining) (2003)
28. Krzanowski, W., Lai, Y.: A criterion for determining the number of groups in a dataset using sum of squares clustering. Biometrics 44, 23–34 (1985)
29. Kulis, B., Basu, S., Dhillon, I., Mooney, R.J.: Semi-supervised graph clustering: A kernel approach. In: Proceedings of the 22nd International Conference on Machine Learning, Bonn, Germany, pp. 457–464 (2005)
30. Ledger, G.: An Exploration of Dierences in the Pauline Epistles using Multivariate Statistical Analysis. Oxford University Press (1995)
31. Levine, E., Domany, E.: Resampling method for unsupervised estimation of cluster validity. Neural Computation 13, 2573–2593 (2001)
32. Liu, X., Yu, S., Moreau, Y., Moor, B.D., Glanzel, W., Janssens, F.A.L.: Hybrid clustering of text mining and bibliometrics applied to journal sets. In: SDM 2009, pp. 49–60 (2009)
33. Lord, R.: De Morgan and the Statistical Study of Literary Style. Biometrica, 282 (1958)
34. Luxburg, U.V.: A tutorial on spectral clustering. Statistics and Computing 17(4), 395–416 (2007)
35. MacQueen, J.B.: Some methods for classification and analysis of multivariate observations. In: Proceedings of 5th Berkeley Symposium on Mathematical Statistics and Probability, vol. 1, pp. 281–297. University of California Press, Berkeley (1967)
36. McLachlan, G.J., Peel, D.: Finite Mixture Models. Wiley (2000)
37. Mealand, D.L.: The Extent of the Pauline Corpus: A Multivariate Approach. JSNT 59 (1995)
38. Milligan, G., Cooper, M.: An examination of procedures for determining the number of clusters in a data set. Psychometrika 50, 159–179 (1985)
39. Mohar, B.: Some applications of Laplace eigenvalues of graphs. In: Hahn, G., Sabidussi, G. (eds.) Graph Symmetry: Algebraic Methods and Applications. Springer (1997)
40. Mufti, G.B., Bertrand, P., Moubarki, E.: Determining the number of groups from measures of cluster validity. In: Proceedings of ASMDA 2005, pp. 404–414 (2005)
41. Nascimento, M., Carvalho, A.D.: Spectral methods for graph clustering – a survey. European Journal of Operational Research 2116(2), 221–231 (2011)
42. Ng, A.Y., Jordan, M.I., Weiss, Y.: On spectral clustering: analysis and an algorithm. In: Advances in Neural Information Processing Systems 14 (NIPS 2001), pp. 849–856 (2001)
43. Perkins, P.: Reading the New Testament: An Introduction, p. 47. Paulist Press (1988)
44. Roth, V., Lange, V., Braun, M., Buhmann, J.: A resampling approach to cluster validation. In: COMPSTAT (2002), http://www.cs.uni-bonn.De/~braunm
45. Roth, V., Lange, V., Braun, M., Buhmann, J.: Stability-based validation of clustering solutions. Neural Computation 16(6), 1299–1323 (2004)

46. Shi, J., Malik, J.: Normalized cuts and image segmentation. IEEE Transactions on Pattern Analysis and Machine Intelligence 22(8), 888–905 (2000)
47. Sugar, C., James, G.: Finding the number of clusters in a data set: An information theoretic approach. J. of the American Statistical Association 98, 750–763 (2003)
48. Tibshirani, R., Walther, G., Hastie, T.: Estimating the number of clusters via the gap statistic. J. Royal Statist. Soc. B 63(2), 411–423 (2001)
49. Toledano-Kitai, D., Avros, R., Volkovich, Z.: A fractal dimension standpoint to the cluster validation problem. International Journal of Pure and Applied Mathematics 68(2), 233–252 (2011)
50. Volkovich, V., Kogan, J., Nicholas, C.: k–means initialization by sampling large datasets. In: Dhillon, I., Kogan, J. (eds.) Proceedings of the Workshop on Clustering High Dimensional Data and its Applications (held in conjunction with SDM), pp. 17–22 (2004)
51. Volkovich, Z., Barzily, Z., Morozensky, L.: A statistical model of cluster stability. Pattern Recognition 41(7), 2174–2188 (2008)
52. Wechsler, H.: Intelligent biometric information management. Intelligent Information Management 2, 499–511 (2010)
53. White, S., Smyth, P.: A spectral clustering approach to finding communities in graphs. In: Proceedings of the Fifth SIAM International Conference on Data Mining, vol. 119, pp. 274–285. Society for Industrial Mathematics (2005)
54. Xing, E.P., Ng, A.Y., Jordan, M.I., Russell, S.: Distance metric learning, with application to clustering with side-information. In: Advances in Neural Information Processing Systems 15 (NIPS 2002), pp. 505–512 (2002)
55. Yu, S.X., Shi, J.: Multiclass spectral clustering. In: Proceedings of the Ninth IEEE International Conference on Computer Vision, vol. 1, pp. 313–319 (2003)
56. Zelnik-manor, L., Perona, P.: Self-tuning spectral clustering. In: Advances in Neural Information Processing Systems 17, pp. 1601–1608. MIT Press (2004)

A Seed-Based Inter-Domain Supervised Framework to Cluster Mixed Data Types

Artur Abdullin and Olfa Nasraoui

Knowledge Discovery & Web Mining Lab,
Department of Computer Engineering and Computer Science, University of Louisville,
Louisville, KY, U.S.A.

Abstract. We propose a Seed-based Inter-Domain Supervised (IDS) framework to handle possibly diverse data formats, mixed-type attributes and different sources of data. This approach can be used for combining diverse representations of the data, in particular where data comes from different sources, some of which may be unreliable or uncertain, or for exploiting optional external concept set labels to guide the clustering of the main data set in its original domain. Unlike semi-supervised clustering, our approach exploits the synergy between different domains instead of external labels. Also ensemble clustering and our proposed IDS clustering are different mechanisms with distinct goals: (i) the former aims to combine independent clustering results from many independent subsets sampled from the same data set or from different clustering algorithm results on the same data set, into one consensus, (ii) while the latter (IDS) aims to exploit a mutual synergy between different data domains or sources to guide the clustering in these different domains or sources. Our preliminary results in clustering data with mixed numerical and categorical attributes show that the proposed IDS framework gives better clustering results in the categorical domain. Thus the seeds or the constraints obtained from clustering the numerical domain give an additional knowledge to the categorical clustering algorithm. Additional results show that our approach outperforms clustering either domain on its own or clustering both domains converted to the same target domain.

Keywords: Semi-supervised Clustering, Mixed Data Type Clustering.

1 Introduction

Many algorithms exist for clustering. However most of them have been designed to optimally handle specific types of data, e.g the spherical k-means was proposed to cluster text data [1]. The algorithm in [2] has been designed for data with directional distributions that lie on the unit hypersphere such as text data. The k-modes has been designed specifically for categorical data [3]. Special data types and domains have been also handled using specialized dissimilarity or distance measures. For example, the k-means, using the Euclidean distance, is optimal for compact globular clusters with numerical attributes.

Suppose that a data set comprises multiple types of data that can each be best clustered with a different specialized clustering algorithm or with a specialized dissimilarity measure. In this case, the most common approach has been to either convert all data

A. Fred et al. (Eds.): IC3K 2012, CCIS 415, pp. 37–52, 2013.

types to the same type (e.g: from categorical to numerical or vice-versa) and then cluster the data with a standard clustering algorithm in that target domain; or to use a different dissimilarity measure for each domain, then combine them into one dissimilarity measure and cluster this dissimilarity matrix with an $O(N^2)$ algorithm.

To handle possibly diverse data formats and different sources of data, we propose a new approach for combining diverse representations or types of data. Examples of data with mixed attributes include network activity data (e.g. the KDD cup data), most existing census or demographic data, environmental data, and other scientific data. We propose a new methodology for clustering data comprising multiple domains or parts, in such a way that the separate domains mutually supervise each other within a semi-supervised learning framework. Unlike existing uses of semi-supervised learning, our methodology does not assume the presence of labels from part of the data, but rather, each of the different domains of the data separately undergoes an unsupervised learning process, while sending and receiving supervised information in the form of data seeds to/from the other domains. The entire process is an alternation of semi-supervised learning stages in the different data domains. The same approach can also be used for multi-source data, since each source of data can be considered a domain.

The rest of this chapter is organized as follows. Section 2 gives an overview of related work. Section 3 presents our proposed framework to cluster mixed and multi-source data. Section 4 evaluates the proposed approach and Section 5 presents the conclusions of our preliminary investigation.

2   Related Work

Most successful clustering algorithms are specialized for specific types of attributes. For instance, categorical attributes have been handled using specialized algorithms such as k-modes, ROCK or CACTUS. The main idea of the k-modes is to select k initial modes, followed by allocating every object to the nearest mode [3]. The k-modes algorithm uses the match dissimilarity measure to measure the distance between categorical objects [4]. ROCK is an adaptation of an agglomerative hierarchical clustering algorithm, which heuristically optimizes a criterion function defined in terms of the number of "links" between transactions or tuples, defined as the number of common neighbors between them. Starting with each tuple in its own cluster, they repeatedly merge the two closest clusters until the required number of clusters remain [5]. The central idea behind CACTUS is that a summary of the entire data set is sufficient to compute a set of "candidate" clusters which can then be validated to determine the actual set of clusters. The CACTUS algorithm consists of three phases: computing the summary information from the data set, using this summary information to discover a set of candidate clusters, and then determining the actual set of clusters from the set of candidate clusters [6]. The spherical k-means algorithm is a variant of the k-means algorithm that uses the cosine similarity instead of the Euclidean distance. The algorithm computes a disjoint partition of the document vectors and for each partition, computes a centroids that is then normalized to have unit Euclidean norm [1]. This algorithm was successfully used for clustering transactional or text (text documents are often represented as sparse high-dimensional vectors) data. Numerical data has been clustered using k-means, DBSCAN and many other algorithms. The k-means algorithm is a partitional or non-hierarchical clustering method, designed to cluster numerical data in which each cluster has a center called mean. The k-means algorithm operates as follow: starting with a specified number k of initial cluster centers, the remaining data is reallocated or assigned , such that each data point is assigned to the nearest cluster. This is continued with repeatedly recomputing the new centers of the data assigned to each cluster and changing the membership assignments of the data points to belong to the nearest cluster until the objective function (which is the sum of distance values between the data and the assigned cluster's centroids), centroids or membership of the data points converge [7]. DBSCAN is a density-based clustering algorithm designed to discover arbitrarily shaped clusters. A point x is directly density reachable from a point y if it is not farther than a given distance ϵ (i.e., it is part of its ϵ-neighborhood), and if the ϵ-neighborhood of y has more points than an input parameter N_{min} such that one may consider y and x to be part of a cluster [8].

The above approaches have the following limitations:

– Specialized clustering algorithms can fall short when they must handle different data types.
– Data type conversion can result in the loss of information or the creation of artifacts in the data.
– Different data sources may be hard to combine for the purpose of clustering because of the problem of duplication of data and the problem of missing data from one of the sources, in addition to the problem of heterogeneous types of data from multiple sources.

Algorithms for mixed data attributes exist, for instance the k-prototypes [9] and IN-CONCO algorithms [10]. The k-prototypes algorithm integrates the k-means and the k-modes algorithms to allow for clustering objects described by mixed numerical and categorical attributes. The k-prototypes works by simply combining the Euclidean distance and the categorical (matching) distance measures in a weighted sum. The choice of the weight parameter and the weighting contribution of the categorical versus numerical domains cannot vary from one cluster to another, and this can be a limitation for some data sets. The INCONCO algorithm extends the Cholesky decomposition to model dependencies in heterogeneous data and, relying on the principle of Minimum Description Length, integrates numerical and categorical information in clustering. The limitations of INCONCO include that it assumes a known probability distribution model for each domain, and it assumes that the number of clusters is identical in both the categorical and the numerical domains. It is also limited to two domains.

Many semi-supervised algorithms have been proposed [11] including co-training [12], the transductive support vector machine [13], entropy minimization [14], semi-supervised Expectation Maximization [15], graph-based approaches [16,17], and clustering-based approaches [18]. In semi-supervised clustering, labeled data can be used in the form of *(1) initial seeds* [19], *(2) constraints* [20], or *(3) feedback* [21]. All these existing approaches are based on model-based clustering [22] where each cluster is represented by its centroid. *Seed-based* approaches use labeled data *only to help initialize* cluster centroids, while *constrained* approaches keep the grouping of labeled data unchanged throughout the clustering process, and *feedback-based* approaches start by running a regular clustering process and finally adjusting the resulting clusters based on labeled data.

Our proposed approach is inspired from and yet distinct from semi-supervised learning, because it is an alternation of semi-supervised learning stages in the different data domains, where the clustering of each domain serves as a guide to the other domain instead of externally provided labels. Our approach is also reminiscent of ensemble-based clustering [23,24]. However, one main distinction is that our approach enables the different algorithms running in each domain to reinforce or supervise each other during the intermediate stages, until the final clustering is obtained. In other words, our approach is more collaborative. Ensemble-based methods, on the other hand, were not intended to provide a collaborative exchange of knowledge between different data "domains," *while* the individual algorithms are still running, but rather to combine the *end* results of several runs, several algorithms, and in fact they benefit more when the individual clustering results are as independent and decoupled as possible from each other. Indeed ensemble clustering and our proposed IDS clustering are different mechanisms with distinct goals: (i) the former aims to combine independent clustering results from many independent subsets sampled from the same data set or from different clustering algorithm results on the same data set, into one consensus, (ii) while the latter (IDS) aims to exploit a mutual synergy between different data domains or sources to guide the clustering in these different domains or sources.

3   Inter-Domain Supervised (IDS) Framework for Clustering Mixed Data Types

Our proposed IDS framework can use specifically designed clustering algorithms which can be distinct and specialized for the following different types of data, however all the algorithms are bound together within a collaborative scheme:

1. For categorical data types, the algorithms k-modes [3], ROCK [5], CACTUS [6], etc, can be used.
2. For transactional or text data , the spherical k-means algorithm [1], or any specialized algorithm can be used
3. For numerical data types, one can use the K-means [7], DBSCAN [8], etc.
4. For graph data, one can use KMETIS [25], spectral clustering [26], etc.

In the following sub-sections, we distinguish between two cases depending on whether the number of clusters is the same across the different domains of the data.

3.1   The Case of an Equal Number of Clusters in Each Data Type or Domain

Our initial implementation reported in this chapter, can handle data records composed of two parts: numerical and categorical, within an IDS framework that consists of the following stages, as shown in Figure 1:

1. **Splitting Across Domains.** The first stage consists of dividing the set of attributes into two subsets: one subset, called domain T_1, with only attributes of numerical type (age, income, etc), and another subset, called domain T_2, with attributes of categorical type (eyes color, gender, etc).

2. **Baseline Clustering per Domain.** The next stage is to cluster each subset using a specifically designed algorithm for that particular data type. In our experiments, we used k-means [7] for numerical type attributes T_1, and k-modes [3] for categorical type attributes T_2. Both algorithms start from the same random initial seeds and run for a small number of iterations (t_n and t_c for k-means and k-modes, respectively), yielding (data-cluster) membership matrices M_{T_1} and M_{T_2}, respectively.

3. **Best Cluster Selection from All Domains.** In the third stage, we compare the cluster centroids obtained in the first domain, T_1, and the second domain, T_2, and find the best combination of both for each of the domains.

   (a) **Cluster Matching.** First, we solve a cluster correspondence problem between the two domains using the Hungarian matching method [27,28] using as weight matrix, the entry-wise reciprocal of the Jaccard coefficient matrix, which is computed using the cluster memberships M_{T_1} and M_{T_2} of the T_1 and T_2 domains respectively.

   (b) **Cluster Validation across Domains.** Then using the membership matrices M_{T_1} and M_{T_2}, we compute the Davies-Bouldin (DB) indices $db_{M_{T_1}}^{T_1}$ and $db_{M_{T_2}}^{T_1}$ [29] in data domain T_1 for each cluster centroid obtained respectively, from clustering the data in domain T_1 and from clustering the data in domain T_2 from the previous stage (2). Similarly, we also compute the DB indices $db_{M_{T_1}}^{T_2}$ and $db_{M_{T_2}}^{T_2}$ in data domain T_2 for each cluster centroid obtained respectively, from clustering the data in domain T_1 and from clustering the data in domain T_2. Note that computing a DB index for every cluster centroid is essentially the same as computing the original overall DB index but without taking the sum over all centroids.

   (c) **Best Cluster Selection across Domains.** To find the best combination of centroids for domain T_1, we compare $db_{M_{T_1}}^{T_1}$ and $db_{M_{T_2}}^{T_1}$ for each centroid resulting from clustering the data in domain T_1 and resulting from clustering the data in domain T_2, and then take only those centroids which score a lower value in the DB index, thus forming better clusters in one domain compared to the other. We then perform a similar operation for domain T_2. The outputs of this stage are two sets, each consisting of the best combination of cluster centroids or prototypes for each of the data domains T_1 and T_2, respectively.

4. **Inter-Domain Supervised Clustering in Domain 1.** In this stage, we use the best seeds obtained from stage 3 to recompute the cluster centroids in the first domain by running k-means for a small number (t_n) of iterations; then compare these re-computed centroids against the cluster centroids that were computed in the second domain in the previous iteration (as explained in detail in stage 3) and find the best cluster centroids' combination for the second domain (T_2).

5. **Inter-Domain Supervised Clustering in Domain 2.** In this stage, we use the best seeds obtained from stage 4 to initialize the k-modes algorithm in domain T_2, and run it for t_c iterations. Then again, we compare these recomputed centroids against the cluster centroids computed in the first domain in the previous iteration (as explained in detail in stage 3) and find the best cluster centroids' combination for the first domain (T_1).

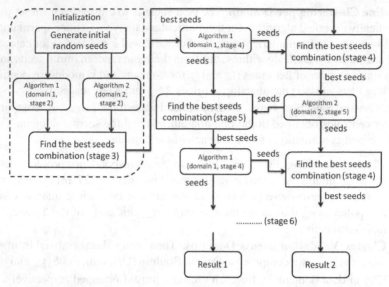

Fig. 1. Overview of the semi-supervised seeding clustering approach

6. We repeat stages 4 and 5 until both algorithms converge or the number of exchange iterations exceeds a maximum number.

We compared the proposed mixed-type clustering approach with the following two classical baseline approaches for clustering mixed numerical and categorical data.

Baseline 1: Conversion. The first baseline approach is to convert all data to the same attribute type and cluster it. We call this method the *conversion* algorithm. Since we have attributes of two types, there are two options to perform this algorithm:

1. Converting all numerical type attributes to categorical type attributes and run k-modes.
2. Converting all categorical type attributes to numerical type attributes and run k-means.

Baseline 2: Splitting. The second classical baseline approach is to run k-means and k-modes independently on the numerical and categorical subsets of attributes, respectively. We call this method the *splitting* algorithm.

The conversion algorithm requires data type conversion: from numerical to categorical and from categorical to numerical. There are several ways to convert a numerical type attribute z, ranging in $[z_{min}, z_{max}]$, to a categorical type attribute y, also known as "discretization" [30]:

(i) by mapping the n numerical values, z_i, to N categorical values y_i using direct categorization. Then the categorical value is defined as $y_i = \lfloor \frac{N(z_i - z_{min})}{(z_{max} - z_{min})} \rfloor + 1$, where $\lfloor \ \rfloor$ denotes the largest integer less than or equal to z. Obviously, if $z_i = z_{max}$, we get $y_i = N + 1$, and we should set $y_i = N$.

(ii) by mapping the n numerical values to N categorical values using a histogram binning based method.

(iii) by clustering the n numerical values into N clusters using any numerical clustering algorithm (e.g. k-means). The optimal number of clusters N can be chosen based on some validation criterion.

In the current baseline implementation, we use cluster-based conversion (iii) with the Silhouette index as a validity measure because this results in the best conversion. There are also several methods to convert categorical type attributes to numerical type attributes:

(i) by mapping the n values of a nominal attribute to binary values using 1-of-n encoding, resulting into transactional-like data, with each nominal value becoming a distinct binary attribute

(ii) by mapping the n values of an ordinal nominal attribute to integer values in the range of 1 to n, resulting in numerical data with n values

(iii) without any mapping, but instead modifying our distance measure so that for those nominal attributes, a suitable deviation is computed (e.g. a match between two categories results in a distance of 0, while a non-match results in a maximal distance of 1). In this case, we need a clustering algorithm that works on the distance matrix instead of the input data vectors (also called relational clustering).

In our baseline implementation, we used the first method (1-of-n encoding).

Computational Complexity. The complexity of the proposed approach is mainly determined based on the complexity of the embedded base algorithms used in each domain. In addition, there is the overhead complexity resulting from the coordination and alternating seed exchange process between the different domains during the mutual supervision process. The main overhead computation in the latter step is the cluster matching, validity scoring and comparison performed in stage 3 (which is then repeatedly invoked at the end of the subsequent stages 4 and 5). Stage 3 involves the following computations: first, the computation of the Jaccard coefficient matrix using the cluster memberships of the domains in time $O(k^2 N)$ (assuming the number of clusters to be of similar order k), then, solving the correspondence problem between the two domains using the Hungarian method in time $O(k^3)$, and finally, computing the DB indices for each cluster centroid in both domains in time $O(k^2 N)$. Thus, the total overhead complexity of stage 3 is $O(k^2 N)$ since $k \ll N$. With the k-means and k-modes as the base algorithms, the total computational complexity of the proposed approach is $O(N)$.

3.2 The Case of a Different Number of Clusters or Different Cluster Partitions in Each Data Type or Domain

In our preliminary design above, the number of clusters is assumed to be the same in each domain. This can be considered as the default approach, and has the advantage of being easier to design. However, in real life data, there are two challenges:

– Case 1. The first challenge is when each data domain naturally gives rise to a different number of clusters, which is simple to understand.

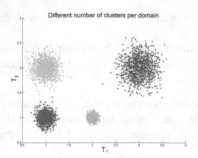

Fig. 2. An example with a different number of clusters in each domain

- Case 2. The second challenge is when regardless of whether the number of clusters are similar or different in the different domains, their nature is actually completely different, and this will be illustrated with the following example.

How do we combine the results of clustering in different domains if the numbers of clusters are different? Let us look at the example shown in Figure 2, which for visualization purposes, artificially splits the two numerical features into two distinct domains, thus illustrating the difficulties with mixed domains. Here we have two domains or (artificially different) data types T_1 and T_2. In total, taking into account both data domains or types T_1 and T_2, we have four distinct clusters, however if we cluster each domain separately, we see that in T_1, we have three clusters, while in T_2, we have only two clusters. This illustrates Case 2 and gives rise to the problem of judiciously combining the clustering results emerging from each domain into a coherent clustering result with correct cluster labelings for all the data points.

We propose the following algorithm to cluster such a data set, that we emphasize, *actually targets completely different data domains or types that cannot be compared using traditional attribute-based distance measures.*

1. First, cluster T_1 with k_{T_1} number of clusters and cluster T_2 with k_{T_2} number of clusters. Let M_{T_1} be the cluster membership matrix of domain T_1 and M_{T_2} be the cluster membership matrix of domain T_2. Therefore, M_{T_1} is an $n \times k_{T_1}$ matrix and M_{T_2} is an $n \times k_{T_2}$ matrix, where n is the number of data records. The membership matrix M_T is such that entry $M_T[i,j]$ is 1 or 0 depending on whether or not point i belongs to cluster j in the current domain T.
2. Next, match each cluster j_1 in T_1 to the corresponding cluster j_2 in T_2. For this purpose, we compute the Jaccard coefficient matrix J of size $k_{T_1} \times k_{T_2}$ in which entry $J[j_1, j_2]$ is defined as follows:

$$J[j_1, j_2] = \frac{|C_{T_1, j_1} \cap C_{T_2, j_2}|}{|C_{T_1, j_1} \cup C_{T_2, j_2}|},$$

where $C_{T,j}$ is the set of points that belong to cluster j in domain T, i.e.,

$$C_{T,j} = \{x_i | M_T(i,j) > 0\}.$$

Then we compute the optimal cluster correspondence using the Hungarian method with the weight matrix W in which every entry, $W[j_1, j_2] = 1/J[j_1, j_2]$ [28,27].

3. Finally, merge the clustering results of domains T_1 and T_2 using Algorithm 1. Let T_{max} be the domain with the highest number of clusters $k_{T_{max}} = max\{k_{T_1}, k_{T_2}\}$ and $M_{T_{max}}$ be the membership matrix of that domain. Let T_{other} be the other domain with a number of clusters $k_{T_{other}}$ ($k_{T_{other}} \leq k_{T_{max}}$) and let its membership matrix be $M_{T_{other}}$.

Algorithm 1. IDS merging algorithm for domains differing in the number of clusters.

---

input: $M_{T_{max}}, M_{T_{other}}, k_{T_{max}}, th$
output: M_{merge}
  {for all clusters in the domain with more clusters}
  for $j_1 = 1$ to $k_{T_{max}}$ **do**
    $j_2 = argmax_{j_2}[J_k(j_1, j_2)]$ {find nearest cluster from the other domain, i.e. with most data overlap compared to cluster j_1 in this domain}
    $C_{T_{max},j_1} = \{x_i | M_{T_{max}}(i, j_1) > 0\}$ {find points in cluster j_1 in this domain}
    $C_{T_{other},j_2} = \{x_i | M_{T_{other}}(i, j_2) > 0\}$ {find points in nearest cluster j_2 from other domain}
    $C_{merge} = C_{T_{max},j_1} \cap C'_{T_{other},j_2}$ {find points in intersection between these two clusters}
    {if intersection ratio is high}
    if $\frac{|C_{merge}|}{|C_{T_{max},j_1}|} > th$ **then**
      $C_{new} = \{x_i | x_i \in C_{merge}\}$ {then assign intersection points to a new cluster}
      $C_{T_{max},j_1} = C_{T_{max},j_1} - C_{new}$ {remove intersection points from first cluster in this domain}
    end if
  end for

Computational Complexity. The complexity of the IDS merging algorithm is mainly determined by stage 2, where we compute the Jaccard coefficient matrix J using the cluster memberships of the domains in time $O(k_{T_1} k_{T_2} N)$. The complexity of Algorithm 1 itself is $O(k_{T_{max}})$. Thus, the total complexity of the merging algorithm is $O(N)$.

4   Experimental Results

4.1   Clustering Evaluation

The proposed IDS framework was evaluated using several internal and external clustering validity metrics. Note that in calculating all internal indices that require a distance measure, we used the square of the Euclidean distance for numerical data types and the simple matching distance [4] for categorical data types.

- Internal validity metrics
  - The Davies-Bouldin (DB) index is a function of the ratio of the sum of within-cluster scatter to between-cluster separation [29]. Hence the ratio is small if the clusters are compact and far from each other. That is, the DB index will have a small value for a good clustering.
  - The Silhouette index is calculated based on the average silhouette width for each sample, average silhouette width for each cluster and overall silhouette width for the entire data set [31]. Using this approach, each cluster can be

represented by its silhouette, which is based on the comparison of its tightness and separation. A silhouette value close to 1 means that the data sample is well-clustered and assigned to an appropriate cluster. A silhouette value close to zero means that the data sample could be assigned to another cluster, and the data sample lies halfway between both clusters. A silhouette value close to −1 means that the data sample is misclassified and is located somewhere in between the clusters.

- The Dunn index is based on the concept of cluster sets that are compact and well separated [32]. The main goal of this measure is to maximize the inter-cluster distances and minimize the intra-cluster distances. A higher value of the Dunn index mean a better clustering.

– External validity metrics

- Purity is a simple evaluation measure that assumes that an external class label is available to evaluate the clustering results. First, each cluster is assigned to the class which is most frequent in that cluster, then the accuracy of this assignment is measured by the ratio of the number of correctly assigned data samples to the number of data points. A bad clustering has a purity close to 0, and a perfect clustering has a purity of 1. Purity is very sensitive to the number of clusters, in particular, purity is 1 if each point gets its own cluster [33].

- Entropy is a commonly used information theoretic external validation measure that measures the purity of the clusters with respect to given external class labels [34]. A perfect clustering has an entropy close to 0 which means that every cluster consists of points with only one class label. A bad clustering has an entropy close to 1.

- Normalized mutual information (NMI) estimates the quality of the clustering with respect to a groundtruth class membership [35]. It measures how closely a clustering algorithm could reconstruct the underlying label distribution in the data. The minimum NMI is 0 if the clustering assignment is random with respect to true class membership. The maximum NMI is 1 if the clustering algorithm could perfectly recreate the true class memberships.

4.2 Equal Number of Clusters in Each Data Type or Domain

Real Data Sets. We experimented with three real-life data sets with the characteristics shown in Table 1. All three data sets were obtained from the UCI Machine Learning Repository [36].

- *Adult Data.* The adult data set was extracted by Barry Becker from the 1994 Census database. The data set has two classes: People who make over $50K a year and people who make less than $50K. The original data set consists of 48, 842 instances. After deleting instances with missing and duplicate attributes we obtained 45, 179 instances.
- *Heart Disease Data.* The heart disease data, generated at the Cleveland Clinic, contains a mixture of categorical and numerical features. The data comes from two classes: people with no heart disease and people with different degrees of heart disease.

Table 1. Data sets properties

Data set	No. of Records	No. of Numerical Attributes	No. of Categorical Attributes	Missing Values	No. of Classes
Adult	45179	6	8	Yes	2
Heart Disease Data	303	6	7	Yes	2
Credit Approval Data	690	6	9	Yes	2

Table 2. Clustering results for the adult data set (10 runs, $k = 2$ clusters per domain)

Data type	Numerical		
Algorithm	Inter-domain supervision	Conversion	Splitting
DB Index	$\mathbf{0.43 \pm 0.01}[0.42, 0.42, 0.45]$	$12.02 \pm 5.78[2.54, 12.31, 26.72]$	$3.05 \pm 0.83[0.42, 3.29, 3.29]$
Silhouette Index	$\mathbf{0.71 \pm 0.01}[0.70, 0.71, 0.71]$	$0.17 \pm 0.21[0.07, 0.07, 0.64]$	$0.19 \pm 0.08[0.21, 0.21, 0.71]$
Dunn Index	$\mathbf{1.1e-3 \pm 7e-4}[1e-4, 1.1e-3, 2.8e-3]$	$3e-4 \pm 8e-4[0.00, 0.00, 2.8e-3]$	$1e-4 \pm 3e-4[0, 0, 1.1e-3]$
Purity	$\mathbf{0.75 \pm 0.00}[0.75, 0.75, 0.75]$	$0.61 \pm 0.19[0.25, 0.71, 0.71]$	$0.65 \pm 0.03[0.64, 0.64, 0.75]$
Entropy	$0.79 \pm 0.00[0.79, 0.80, 0.80]$	$\mathbf{0.72 \pm 0.05}[0.69, 0.69, 0.81]$	$0.72 \pm 0.02[0.71, 0.71, 0.80]$
NMI	$0.03 \pm 0.01[0.02, 0.03, 0.03]$	$\mathbf{0.10 \pm 0.05}[2.2e-4, 0.13, 0.13]$	$0.09 \pm 0.02[0.02, 0.10, 0.11]$

Data type	Categorical		
Algorithm	Inter-domain supervision	Conversion	Splitting
DB Index	$\mathbf{1.11 \pm 0.01}[1.10, 1.12, 1.12]$	$1.79 \pm 0.36[1.16, 1.86, 2.38]$	$1.87 \pm 0.88[1.11, 1.36, 3.10]$
Silhouette Index	$\mathbf{0.25 \pm 0.01}[0.23, 0.24, 0.26]$	$0.16 \pm 0.03[0.11, 0.16, 0.20]$	$0.19 \pm 0.08[0.06, 0.24, 0.27]$
Dunn Index	$\mathbf{0.125 \pm 0.0}[0.125, 0.125, 0.125]$	$0.07 \pm 0.00[0.07, 0.07, 0.07]$	$0.125 \pm 0[0.125, 0.125, 0.125]$
Purity	$0.58 \pm 0.04[0.55, 0.55, 0.63]$	$0.64 \pm 0.08[0.51, 0.66, 0.74]$	$\mathbf{0.65 \pm 0.06}[0.55, 0.67, 0.72]$
Entropy	$0.73 \pm 0.01[0.72, 0.73, 0.74]$	$\mathbf{0.73 \pm 0.02}[0.68, 0.72, 0.76]$	$0.74 \pm 0.04[0.71, 0.73, 0.80]$
NMI	$0.09 \pm 0.01[0.08, 0.09, 0.09]$	$\mathbf{0.09 \pm 0.03}[0.05, 0.10, 0.13]$	$0.07 \pm 0.04[0.01, 0.09, 0.10]$

- *Credit Approval Data.* The data set has 690 instances, which were classified in two classes: approved and rejected.

Results with the Real Data Sets. Since all three data sets have two classes, we clustered them in two clusters.[1] We repeated each experiment 50 times (10 times for the larger adult data set), and report the mean, standard deviation, minimum, median, and maximum values for each validation metric (in the format of mean±std [min, median, max]).

- Adult Data. Table 2 shows the results of the adult data set using the IDS framework, the conversion algorithm, and the splitting algorithm, with the best results in a bold font. As the table illustrates, the IDS framework performs better in both domains: showing significant improvements in all internal indices. The conversion algorithm shows better clustering results for the numerical and categorical domains based on the entropy and NMI indices.
- Heart Disease Data. Table 3 shows the results of clustering the heart disease data set using the three approaches. The IDS framework yielded better clustering results for the categorical domain for all validation indices. The conversion algorithm shows better clustering results for the numerical domain based on all external and Dunn indices. The splitting algorithm shows better clustering result for the numerical domain based on DB and Silhouette indices.
- Credit Approval Data. Table 4 shows the results of clustering the credit approval data set. Again, the IDS approach outperforms the traditional algorithms for the

---

[1] We realize the possibility of more than one cluster per class, however we defer such an analysis to the future.

Table 3. Clustering results for the heart disease data set (50 runs, $k = 2$ clusters per domain)

Data type	Numerical		
Algorithm	Inter-domain supervision	Conversion	Splitting
DB Index	$1.73 \pm 0.05[1.64, 1.71, 1.90]$	$2.97 \pm 0.56[\mathbf{0.21}, 2.95, 5.16]$	$\mathbf{1.65 \pm 0.003}[1.65, 1.65, 1.65]$
Silhouette Index	$0.35 \pm 0.02[0.26, 0.35, 0.39]$	$0.26 \pm 0.07[0.16, 0.25, 0.75]$	$\mathbf{0.36 \pm 0.005}[0.36, 0.36, 0.36]$
Dunn Index	$3.5e-3 \pm 1.2e-3[1.6e-3, 3.3e-3, 6.5e-3]$	$0.04 \pm 0.14[0.015, 0.015, 0.98]$	$4.6e-3 \pm 0[4.6e-3, 4.6e-3, 4.6e-3]$
Purity	$0.73 \pm 0.02[0.66, 0.73, 0.77]$	$\mathbf{0.77 \pm 0.11}[0.47, 0.82, 0.82]$	$0.75 \pm 0.003[0.75, 0.75, 0.75]$
Entropy	$0.83 \pm 0.03[0.77, 0.83, 0.89]$	$\mathbf{0.72 \pm 0.11}[0.67, 0.67, 0.99]$	$0.80 \pm 0.003[0.80, 0.80, 0.81]$
NMI	$0.16 \pm 0.03[0.10, 0.16, 0.22]$	$\mathbf{0.28 \pm 0.11}[2.1e-4, 0.32, 0.32]$	$0.19 \pm 0.004[0.18, 0.19, 0.19]$
Data type	Categorical		
Algorithm	Inter-domain supervision	Conversion	Splitting
DB Index	$\mathbf{0.76 \pm 0.01}[0.75, 0.75, 0.77]$	$1.40 \pm 0.37[0.97, 1.31, 2.97]$	$0.97 \pm 0.32[0.75, 0.78, 2.31]$
Silhouette Index	$\mathbf{0.30 \pm 0.00}[0.30, 0.30, 0.30]$	$0.14 \pm 0.04[0.06, 0.15, 0.18]$	$0.25 \pm 0.06[0.13, 0.28, 0.30]$
Dunn Index	$\mathbf{0.14 \pm 0.00}[0.14, 0.14, 0.14]$	$0.10 \pm 0.04[0.07, 0.07, 0.23]$	$0.14 \pm 0.00[0.14, 0.14, 0.14]$
Purity	$\mathbf{0.79 \pm 0.01}[0.77, 0.81, 0.81]$	$0.72 \pm 0.08[0.53, 0.75, 0.82]$	$0.73 \pm 0.09[0.49, 0.78, 0.81]$
Entropy	$\mathbf{0.72 \pm 0.03}[0.70, 0.70, 0.75]$	$0.82 \pm 0.09[0.66, 0.79, 0.99]$	$0.80 \pm 0.11[0.70, 0.76, 0.98]$
NMI	$\mathbf{0.27 \pm 0.04}[0.13, 0.24, 0.30]$	$0.17 \pm 0.10[1.4e-3, 0.20, 0.34]$	$0.19 \pm 0.11[0.001, 0.23, 0.29]$

Table 4. Clustering results for the credit card data set (50 runs, $k = 2$ clusters per domain)

Data type	Numerical		
Algorithm	Inter-domain supervision	Conversion	Splitting
DB Index	$1.84 \pm 0.33[0.99, 1.79, 3.00]$	$5.63 \pm 2.03[0.29, 5.06, 11.65]$	$\mathbf{1.88 \pm 0.36}[0.18, 1.97, 1.97]$
Silhouette Index	$0.58 \pm 0.07[0.36, 0.59, 0.73]$	$0.30 \pm 0.23[0.09, 0.29, 0.92]$	$\mathbf{0.63 \pm 0.06}[0.62, 0.62, 0.95]$
Dunn Index	$3e-4 \pm 3e-4[e-4, 3e-4, 9e-4]$	$\mathbf{0.03 \pm 0.10}[1.1e-3, 0.01, 0.77]$	$0.003 \pm 0.012[1.1e-4, 1.1e-4, 0.06]$
Purity	$0.64 \pm 0.02[0.62, 0.64, 0.69]$	$0.65 \pm 0.12[0.48, 0.56, 0.81]$	$\mathbf{0.64 \pm 0.02}[0.56, 0.64, 0.64]$
Entropy	$0.92 \pm 0.02[0.87, 0.91, 0.95]$	$\mathbf{0.86 \pm 0.13}[0.68, 0.97, 0.99]$	$0.93 \pm 0.01[0.93, 0.93, 0.98]$
NMI	$0.08 \pm 0.02[0.04, 0.08, 0.14]$	$\mathbf{0.13 \pm 0.13}[1.2e-4, 0.03, 0.31]$	$0.07 \pm 0.01[0.02, 0.08, 0.08]$
Data type	Categorical		
Algorithm	Inter-domain supervision	Conversion	Splitting
DB Index	$\mathbf{1.78 \pm 0.14}[1.37, 1.82, 1.86]$	$1.98 \pm 0.33[1.23, 1.98, 2.57]$	$2.01 \pm 0.67[0.95, 1.90, 3.32]$
Silhouette Index	$0.23 \pm 0.01[0.19, 0.23, 0.23]$	$0.15 \pm 0.03[0.06, 0.16, 0.22]$	$\mathbf{0.23 \pm 0.06}[0.19, 0.23, 0.36]$
Dunn Index	$0.12 \pm 0.01[0.11, 0.12, 0.12]$	$0.07 \pm 0.002[0.07, 0.07, 0.08]$	$\mathbf{0.13 \pm 0.04}[0.11, 0.11, 0.22]$
Purity	$\mathbf{0.79 \pm 0.01}[0.77, 0.79, 0.79]$	$0.73 \pm 0.08[0.47, 0.76, 0.82]$	$0.68 \pm 0.12[0.50, 0.75, 0.83]$
Entropy	$\mathbf{0.72 \pm 0.01}[0.72, 0.72, 0.78]$	$0.80 \pm 0.09[0.66, 0.78, 0.99]$	$0.83 \pm 0.12[0.64, 0.77, 0.99]$
NMI	$\mathbf{0.26 \pm 0.01}[0.22, 0.27, 0.27]$	$0.20 \pm 0.09[0.01, 0.21, 0.34]$	$0.17 \pm 0.13[0.01, 0.22, 0.36]$

categorical type attributes based on the external indices but concedes to the splitting algorithm in terms of internal indices The splitting algorithms yielded better clustering results for the numerical domain based on the DB and Silhouette indices but conceded to the splitting algorithm based on entropy and NMI.

4.3 Different Number of Clusters or Different Cluster Partitions in Each Data Type or Domain

Data Sets. In the following experiments, as we have done with our illustrating example above, we validate our second algorithm on the data sets satisfying Case 2 (described in Section 3.2) of non-coherent cluster partitions across the different domains. Although in the Iris data set, the attributes are of the same type, we artificially split them into two domains so that we can validate the method and visualize the input data and the results. The validity of this example can generalize for different domains, because we do not exploit any attribute-based distance measure between the different data domains (as would be the case for really different domains). The Iris data set is a benchmark set that contains 3 classes of 50 data instances each, where each class refers to a type of iris plant: iris Setosa, iris Versicolour, iris Virginica.

Table 5. Basic characteristics of the Iris data set

Attribute	Min	Max	Mean	Standard deviation	Class Correlation (Pearson's CC)
Sepal length	4.3	7.9	5.84	0.83	0.7826
Sepal width	2.0	4.4	3.05	0.43	−0.4194
Petal length	1.0	6.9	3.76	1.76	0.9490
Petal width	0.1	2.5	1.20	0.76	0.9565

Table 6. Overview of the experiments

Experiment Number	Data set	T_1	T_2	k_{T_1}	k_{T_2}	K	Threshold
1	Iris	$\{1\}$	$\{3\}$	2	3	3	0.6
2	Iris	$\{2\}$	$\{4\}$	2	3	3	0.6
3	Iris	$\{1,3\}$	$\{2,4\}$	2	3	3	0.6

Experiments. All three experiments were performed on the Iris data set. We repeated each experiment 10 times and report only the best results. For validation purposes, we used the class labels. For class - cluster assignment we used the Jaccard coefficient, meaning that a class will be assigned to the cluster with highest Jaccard coefficient.

Experiment 1. We first take only the first and third features of the IRIS data set. Let T_1 be the sepal length (first feature) and T_2 be the petal length (third feature). We chose those two features because the first feature has a low class correlation index while the third feature has a high class correlation index. The first experiment was performed in the following steps:

1. Cluster the data in domain T_1 using K-means ($k = 2$)
2. Cluster the data in domain T_2 using K-means ($k = 3$)
3. Merge the two clustering results using the merging algorithm described in Algorithm 1. Notice that one of the output parameters is K, which is the number of clusters *after* merging. We set the overlap threshold parameter to $th = 0.6$.
4. Compare this result with the clustering result of K-means using $k = K$ clusters, performed on domains T_1 and T_2 together.

Table 7 shows the results in terms of classification accuracy, precision, recall, F-measure, purity, entropy and NMI for each algorithm (the results in **bold** are better).

Experiment 2. We next performed a similar experiment to the above but this time, domain T_1 is the 2nd feature - sepal width, while T_2 is the 4th feature - petal width. The overlap threshold was set to $th = 0.6$. In Table 8 we show the results for the second experiment in terms of the same validity metrics as before.

Experiment 3. We repeated the same experiment as above, but this time, domain T_1 consists of the 1st and 3rd features, while T_2 consists of the 2nd and 4th features. The overlap threshold was set to $th = 0.6$. The results of the second experiment are shown in Table 9.

Table 7. Experiment 1: Evaluation measures for the IDS merging algorithm and k-means (bold results are best)

Algorithm	Class	Precision	Recall	F-measure	Accuracy	Purity	Entropy	NMI
K-means	Setosa	0.9804	1.0	0.9901	0.9933	0.8800	0.2967	0.7065
	Versicolour	0.7758	0.9	0.8333	0.88			
	Virginica	0.9224	0.74	0.8132	0.8868			
IDS merging	Setosa	**1.0**	**1.0**	**1.0**	**1.0**	**0.9467**	**0.1642**	**0.8366**
	Versicolour	**0.8889**	**0.96**	**0.9231**	**0.9467**			
	Virginica	**0.9565**	**0.88**	**0.9167**	**0.9467**			

Table 8. Experiment 2: Evaluation measures for the IDS merging algorithm and k-means (bold results are best)

Algorithm	Class	Precision	Recall	F-measure	Accuracy	Purity	Entropy	NMI
K-means	Setosa	1.0	0.98	0.9899	0.9933	0.9268	0.2265	0.7738
	Versicolour	0.8679	0.92	0.8932	0.9267			
	Virginica	0.9167	0.88	0.8979	0.9333			
IDS merging	Setosa	1.0	**1.0**	**1.0**	**1.0**	**0.9600**	**0.1360**	**0.8642**
	Versicolour	**0.9231**	**0.96**	**0.9412**	**0.96**			
	Virginica	**0.9583**	**0.92**	**0.9388**	**0.96**			

Table 9. Experiment 3: Evaluation measures for the IDS merging algorithm and k-means (bold results are best)

Algorithm	Class	Precision	Recall	F-measure	Accuracy	Purity	Entropy	NMI
K-means	1 - Setosa	1.0	**1.0**	**1.0**	1	0.8933	0.2485	0.7582
	2 - Versicolour	0.7742	**0.96**	0.8574	0.8933			
	3 - Virginica	**0.9474**	0.72	0.8182	0.8933			
IDS merging	1 - Setosa	1.0	0.98	0.9899	0.9933	**0.9267**	0.2265	0.7738
	2 - Versicolour	**0.86792**	0.92	**0.8932**	**0.92667**			
	3 - Virginica	0.9167	**0.88**	**0.8980**	**0.9333**			

Results with the IRIS Data Set. In the first and second experiments, the IDS merging algorithm outperformed the K-means algorithm. In the third experiment, the merging algorithm obtained similar results to the K-means algorithm, although it still outperformed it in terms of the purity of the results and giving a lower entropy of the clustering overall.

5   Conclusions

Our preliminary results show that the proposed semi-supervised framework tends to yield better clustering results in the categorical domain. Thus the seeds obtained from clustering the numerical domain tend to provide additional helpful knowledge to the categorical clustering algorithm (in this case, the K-modes algorithm). This information may in turn be used to avoid local minima and obtain a better clustering in the categorical domain. We are currently completing our study by (1) extending our experiments and methodology to mixed data involving transactional information (particularly text and clickstreams) as one of the types, and (2) devising a suitable method for further combining the results of the multiple clusterings performed on each data type or domain. This is because, although each one of the data type-specific algorithms receives some guidance from the algorithm that clustered the other data types, the final results are currently not combined, but are rather still being evaluated in each domain separately. One promising direction is to combine the multiple clustering results such that the best clustering decisions are selected (or merged) from each result. A challenging issue is whether to merge decisions at the cluster prototype/parameter level or at the data partitioning/labeling level, or both.

Acknowledgements. This work was supported by US National Science Foundation *Data Intensive Computation* Grant IIS-0916489.

References

1. Dhillon, I.S., Modha, D.S.: Concept decompositions for large sparse text data using clustering. Mach. Learn. 42, 143–175 (2001)
2. Banerjee, A., Dhillon, I.S., Ghosh, J., Sra, S., Ridgeway, G.: Clustering on the unit hypersphere using von mises-fisher distributions. Journal of ML Research 6 (2005)
3. Huang, Z.: A fast clustering algorithm to cluster very large categorical data sets in data mining. In: Research Issues on KDD, pp. 1–8 (1997)
4. Kaufman, L., Rousseeuw, P.: Finding Groups in Data An Introduction to Cluster Analysis (1990)
5. Guha, S., Rastogi, R., Shim, K.: Rock: A robust clustering algorithm for categorical attributes. Information Systems 25, 345–366 (2000)
6. Ganti, V., Gehrke, J., Ramakrishnan, R.: Cactus - clustering categorical data using summaries. In: Proc. of the 5th ACM SIGKDD International Conference on KDD, pp. 73–83 (1999)
7. MacQueen, J.B.: Some methods for classification and analysis of multivariate observations. In: Proc. of the 5th Berkeley Symposium on Math. Statistics and Probability, vol. 1, pp. 281–297 (1967)
8. Ester, M., Kriegel, H.P., Sander, J., Xu, X.: A density-based algorithm for discovering clusters in large spatial databases with noise. In: Proc. of the Second International Conference on KDD, pp. 226–231 (1996)
9. Huang, Z.: Extensions to the k-means algorithm for clustering large data sets with categorical values. In: Data Mining and Knowledge Discovery, vol. 2, pp. 283–304 (1998)
10. Plant, C., Böhm, C.: Inconco: interpretable clustering of numerical and categorical objects. In: Proc. of the 17th ACM SIGKDD International Conference on KDD, pp. 1127–1135 (2011)
11. Zhong, S.: Semi-supervised model-based document clustering: A comparative study. Mach. Learn. 65, 3–29 (2006)
12. Blum, A., Mitchell, T.: Combining labeled and unlabeled data with co-training. In: Proc. of the 11th Annual Conference on CL Theory, pp. 92–100 (1998)
13. Joachims, T.: Transductive inference for text classification using support vector machines. In: Proc. of 16th ICML, Bled, SL, pp. 200–209 (1999)

14. Guerrero-Curieses, A., Cid-Sueiro, J.: An entropy minimization principle for semi-supervised terrain classification. In: 2000 International Conference on Image Processing, vol. 3, pp. 312–315 (2000)
15. Nigam, K., McCallum, A.K., Thrun, S., Mitchell, T.: Text classification from labeled and unlabeled documents using em. Mach. Learn. 39, 103–134 (2000)
16. Blum, A., Chawla, S.: Learning from Labeled and Unlabeled Data Using Graph Mincuts. In: Proc. 18th ICML, pp. 19–26 (2001)
17. Zhu, X., Ghahramani, Z., Lafferty, J.: Semi-supervised learning using gaussian fields and harmonic functions. In: Proc. 20th International Conf. on ML, pp. 912–919 (2003)
18. Zeng, H.J., Wang, X.H., Chen, Z., Lu, H., Ma, W.Y.: Cbc: clustering based text classification requiring minimal labeled data. In: Third IEEE Data Mining, ICDM, pp. 443–450 (2003)
19. Basu, S., Banerjee, A., Mooney, R.: Semi-supervised clustering by seeding. In: Proc. of 19th ICML (2002)
20. Wagstaff, K., Cardie, C., Rogers, S., Schrödl, S.: Constrained k-means clustering with background knowledge. In: Proc. of the 18th ICML, pp. 577–584 (2001)
21. Cohn, D., Caruana, R., Mccallum, A.: Semi-supervised clustering with user feedback. Tech. rep. (2003)
22. Zhong, S., Ghosh, J.: A unified framework for model-based clustering. Journal of ML Research 4, 1001–1037 (2003)
23. Al-Razgan, M., Domeniconi, C.: Weighted clustering ensembles. In: Proc. of the 6th SIAM ICML (2006)
24. Ghaemi, R., Sulaiman, M.N., Ibrahim, H., Mustapha, N.: A survey: Clustering ensembles techniques (2009)
25. Karypis, G., Kumar, V.: A fast and high quality multilevel scheme for partitioning irregular graphs. SIAM Journal on Scientific Comp. 20, 359–392 (1998)
26. Shi, J., Malik, J.: Normalized cuts and image segmentation. IEEE Transactions on Pattern Analysis and Machine Intelligence 22, 888–905 (2000)
27. Frank, A.: On kuhn's hungarian method - a tribute from hungary. Naval Research Logistics (NRL) 52, 2–5 (2005)
28. Kuhn, H.W.: The hungarian method for the assignment problem. Naval Research Logistic Quarterly 2, 83–97 (1955)
29. Davies, D.L., Bouldin, D.W.: A cluster separation measure. Pattern Analysis and Machine Intelligence, 224–227 (1979)
30. Gan, G., Ma, C., Wu, J.: Data Clustering: Theory, Algorithms, and Applications. Society for Industrial and Applied Mathematics (2007)
31. Rousseeuw, P.: Silhouettes: a graphical aid to the interpretation and validation of cluster analysis. J. Comput. Appl. Math. 20, 53–65 (1987)
32. Dunn, J.C.: Well separated clusters and optimal fuzzy partitions. J. Cybern. 4, 95–104 (1974)
33. Manning, C.D., Raghavan, P., Schtze, H.: Introduction to Information Retrieval (2008)
34. Xiong, H., Wu, J., Chen, J.: K-means clustering versus validation measures: a data distribution perspective. In: Proc. of the 12th ACM SIGKDD International Conference on KDD, pp. 779–784 (2006)
35. Strehl, A., Strehl, E., Ghosh, J., Mooney, R.: Impact of similarity measures on web-page clustering. In: Workshop on AI for Web Search, pp. 58–64 (2000)
36. Frank, A., Asuncion, A.: UCI machine learning repository (2010)

Strategies for Guided Exploratory Search
on the Mobile Web

Günter Neumann[1] and Sven Schmeier[2]

[1] DFKI - German Research Center for Artificial Intelligence,
Stuhlsatzenhausweg 3, 66123 Saarbrücken, Germany
[2] DFKI - German Research Center for Artificial Intelligence,
Alt–Moabit 91c, 10559 Berlin, Germany
{guenter.neumann,sven.schmeier}@dfki.de
http://www.dfki.de/~neumann
http://www.dfki.de/~schmeier

Abstract. We propose and develop new innovative methods for *guided* exploratory search on the mobile web. The approach has been fully implemented in a system called *MobEx* on a tablet, i.e. an Apple iPad, and on a mobile device/phone, i.e. Apple iPhone or iPod. Starting from a user's search query a set of web snippets is collected by a standard search engine in a first step. After that the snippets are collected into one document from which the topic graph is computed. This topic graph is presented to the user in different touchable and interactive graphical representations depending on the screensize of the mobile device. However due to possible semantic ambiguities in the search queries the snippets may cover different thematic areas and so the topic graph may contain associated topics for different semantic entities of the original query. This may lead the user to wrong directions while exploring the solution space. Hence we present our approach for an interactive disambiguation of the search query and so we provide assistance for the users towards a *guided* exploratory search.

Keywords: Exploratory Search, Unsupervised Topic Extraction, Search Query Disambiguation.

1 Introduction

The World Wide Web is a huge set of hyperlinked and semantically correlated documents. When searching the Web using standard search engines users get presented just some nodes of this Web in form of ranked lists of (text snippets and pointers to) documents. The underlying link structure is more or less hidden from the users' perspective.

This kind of Web lookup search has been shown to be quite successful if the user is mainly interested in retrieving facts or answers for her query [15]. Important reasons are:

- It is hard to find any alternative successful competing way of searching the Web for ordinary users. Hence people got used to that way of searching the Web.

A. Fred et al. (Eds.): IC3K 2012, CCIS 415, pp. 53–67, 2013.

- On ordinary computers human-computer interactions are mainly done by typing on the keyboard. It is not hard to reformulate search queries in case the desired results are not in the best n documents [10] .

So it seems that the simplicity and easiness of the interactions with current search engines are strongly correlated with the still keyboard-dominated human-computer interfaces.

Nowadays, the mobile Web and mobile touchable devices, like smartphones and tablet computers, are getting more and more prominent and widespread. For such devices the most convenient way to interact with is by tapping on buttons, swiping the screen, squeezing it with two or more fingers etc. It is reasonable to assume that the current success and popularity of such mobile devices is also due to the fact that ordinary users have vastly accepted these kind of touchable interfaces as a very convenient way of interacting with them.

Furthermore, we are convinced that touchable devices and interfaces also support the development and breeding of alternative Web search strategies like *exploratory search*. In such a search activity the user only has a vague idea of the information in question and just wants to explore the information space in order to develop new knowledge about the topic in question which usually involves multiple iterations of search [15].

In [16] and [17] we have shown that mobile touchable devices can be a very convenient way for realizing simple and intuitive exploratory search strategies and to provide an usable mobile device searches to "find out about something". The core idea of the underlying search strategy is:

1. A user query is considered as a specification of a topic that the user wants to know and learn more about. Hence, the search result is basically a graphical structure of that topic and associated topics that are found.
2. The user can interactively further explore this topic graph using a simple and intuitive touchable interface in order to either learn more about the content of a topic or to interactively expand a topic with newly computed related topics.

However the success of working with such a system heavily depends on the quality of the topics presented in the topic graph. One possible source of insufficient quality is the uncovered, implicit ambiguity of a search query (which usually the user is not aware of or at least not of all possible readings, e.g., natural entities). For example, if the user looks for information about the person *Jim Clark* she might only have in mind either the racing driver or the Netscape founder[1]. As the retrieved search results may contain information about both entities or all of them, the topic graph will show associated topics that might lead the user into wrong directions while further exploring the search space (Fig. 1).

Consequently, the search strategy should be able to detect and uncover this sort of ambiguity and should explicitly use it for guiding the user's further exploratory search into the direction of the selected preferred reading. Hence, the goal and major contribution of the work presented in this paper is twofold: 1) to extend the above mentioned search strategy to a *guided exploratory search* by proposing a method for interactive disambiguation, and 2) to propose an automatic method for its evaluation.

---

[1] ... or the baseball player, the football player, the bank robber, the film editor, the war hero,...

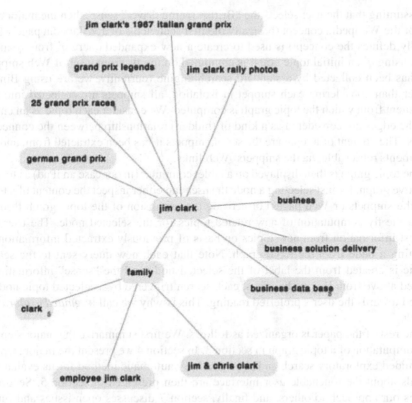

Fig. 1. The associated topics refer to three different entities

2   A Strategy for Guided Exploratory Search

To begin with we will use the topic "Jim Clark" as a running example to briefly describe our approach of guided exploratory search before we present and discuss its details in the next sections.

A user starts her exploratory search by entering a query q consisting of one or more keywords used to represent the topic in question (in our example, just the two words "Jim" and "Clark"). Instead of directly computing and presenting a topic graph for q (as done in the previous mentioned non–guided exploratory search approach), possible senses of q are identified and enumerated by referring to an external knowledge base, Wikipedia in our case. Beside the fact that Wikipedia is known to cover a huge number of possible senses for a very large number of topics, we also consider Wikipedia as a suitable means of a human–computer interface in the sense that both, a human and a computer, can directly communicate in natural language (NL). Continuing our running example, this means that the search strategy determines all possible senses (i.e., Wikipedia pages) that entail q as part of the Wikipedia title (i.e., the NL name of the concept described in the Wikipedia page). All found readings are then sorted and presented to the user and the user is asked to select her preferred one.

Assuming that the user selects the "British racing driver" sense, then the major content of the Wikipedia concept (basically the first sentence s of a Wikipedia page which usually defines the concept) is used to create a new expanded query q' from q and s. Now, using q' an initial topic graph is computed on the fly from a set of Web snippets that has been collected by a standard search engine (currently, we are using Bing[2]). Rather than considering each snippet in isolation, all snippets are collected into one document from which the topic graph is computed. We consider each topic as an entity, and the edges are considered as a kind of (hidden) relationship between the connected topics. The content of a topic are the set of snippets it has been extracted from, and the documents retrievable via the snippets' Web links.

The topic graph is then displayed on a tablet computer (in our case an iPad) as touch-sensitive graph. By just selecting a node the user can either inspect the content of a topic (i.e, the snippets or Web pages) or activate the expansion of the topic graph through an on the fly computation of new related topics for the selected node. The user can request information from new topics on basis of previously extracted information by selecting a node from the topic graph. Note that each new query sent to the search engine is created from the label of the selected node and the "sense"-information s created above from Wikipedia. Thus, each search triggered by a selected topic node is guided towards the user's preferred reading. This is why we call it *guided exploratory search*.

The rest of the paper is organized as follows. We first summarize the major steps of the computation of a topic graph in section 3. In section 4 we present the major steps of the guided exploratory search and present a fully automatic method for its evaluation. Details about the touchable user interface are then presented in section 5. Section 6 relates our approach to others, and finally, section 7 discusses open issues and future plans.

3   Unsupervised Topic Graph Construction

The representation of results in a topic graph provides alternative possibilities to perform search on a mobile device. The process is as follows:

- Show main topics that are generated from snippets retrieved by an ordinary search engine instead of documents in a first step.
- Present topics as interactive graphical structures.
- Let the user interact with the system by different interaction methods.
- Presenting a complete document is the last step in the search process.

We consider the extraction of topics as (1) a specific empirical collocation extraction task where collocations are extracted between chunks combined with (2) the cluster descriptions of an online clustering algorithm. (1) and (2) are computed in parallel for efficiency reasons.

The collocation extraction (step (1)) is done by using a special measure of pointwise mutual information (PMI c.f. [26]) that explicitly takes distance information into

---

[2] http://www.bing.com/

account. For this we first tag the snippets with Part–of–Speech (PoS) information using the SVMTagger [8] and chunk the PoS-tagged text in the next step. The chunker recognizes two types of word chains. Each chain consists of longest matching sequences of words with the same PoS class, namely noun chains or verb chains, where an element of a noun chain belongs to one of the extended noun tags[3], and elements of a verb chain only contains verb tags. We finally apply a kind of "phrasal head test" on each identified chunk to guarantee that the right–most element only belongs to a proper noun or verb tag. For example, the chunk "a/DT british/NNP formula/NNP one/NN racing/VBG driver/NN from/IN scotland/NNP" would be accepted as proper NP chunk, where "compelling/VBG power/NN of/IN" is not.

We compute the chunk–pair–distance model CPD_M using the frequencies of each chunk, each chunk pair, and each chunk pair distance. CPD_M is used for constructing the topic graph in the final step. Formally, a topic graph $TG = (V, E, A)$ consists of a set V of nodes, a set E of edges, and a set A of node actions. Each node $v \in V$ represents a chunk and is labeled with the corresponding PoS–tagged word group. The nodes and edges are computed from the chunk–pair–distance elements. Since the number of these elements is quite large (up to several thousands), the elements are ranked according to a weighting scheme which takes into account the frequency information of the chunks and their collocations. More precisely, the weight of a chunk–pair–distance element $cpd = (c_i, c_j, D_{ij})$, with $D_{i,j} = \{(freq_1, dist_1), (freq_2, dist_2), ..., (freq_n, dist_n)\}$, is computed based on point–wise mutual information (PMI, cf. [26]) as follows:

$$PMI(cpd) = log_2((p(c_i, c_j)/(p(c_i) * p(c_j))))$$

$$= log_2(p(c_i, c_j)) - log_2(p(c_i) * p(c_j))$$

where relative frequency is used for approximating the probabilities $p(c_i)$ and $p(c_j)$. For $log_2(p(c_i, c_j))$ we take the (unsigned) polynomials of the corresponding Taylor series using $(freq_k, dist_k)$ in the k-th Taylor polynomial and adding them up:

$$PMI(cpd) = (\sum_{k=1}^{n} \frac{(x_k)^k}{k}) - log_2(p(c_i) * p(c_j))$$

$$, where \; x_k = \frac{freq_k}{\sum_{k=1}^{n} freq_k}$$

For step (2), we use the online clustering system *Carrot2* [19] to cluster the snippets and to generate sensible cluster descriptions. *Carrot2* is based on the Lingo [18] algorithm. It firstly extracts frequent terms from the input documents and produces a term–document matrix. Secondly, it performs a reduction of this matrix using Singular Value Decomposition (SVD) for the identification of latent structure in the search results.

---

[3] Concerning the English PoS tags, "word/PoS" expressions that match the following regular expression are considered as extended noun tag: "/(N(N|P))|/VB(N|G)|/IN|/DT". The English Verbs are those whose PoS tag start with VB. We are using the tag sets from the Penn treebank (English) and the Negra treebank (German).

Finally, we combine the results of both methods (1) and (2), such that the cluster labels are used to filter out the collocation results using simple fuzzy matching methods. The visualized part of the topic graph is then computed from a subset of the filtered CPD_M using the m highest ranked chunk–pair–distance elements for fixed c_i. In other words, we restrict the complexity of a topic graph by restricting the number of edges connected to a node.

4   Guiding Text Exploration by Enumerating Senses

We already mentioned in sec.1 that the topic extraction process may suffer from possible ambiguities of the search query. Suppose, for example, the search query has two prominent senses then the set of retrieved snippets will quite likely also cover two different thematic areas and so the set of extracted topics, too. If the user performs an investigative search (see section 6) she will then possibly end up with confusion more than solution. In [21] it is reported that between 7% and 23% of frequent queries in the logs of two search engines are ambiguous. This not only includes ambiguous queries (e.g., caused by homonym keywords like "bank" or "jaguar") but also queries that may lead to a different solution space of a search engine's document pool. Hence in this context the disambiguation task is strongly correlated to the automatic determination of the user's intension or goals. For todays search engines these tasks become very tricky to solve as often enough both problems are correlated and occur at the same time for users' search queries.

Regarding our solution for exploratory search on mobile devices the disambiguation part is less hard to solve. As our system supports the idea and paradigm of exploratory search we also let the user decide in which thematic area her exploration should go. Hence our solution divides the above mentioned problems, query disambiguation and determination of the user goal, in a natural way by presenting to the user the possible directions before actually presenting the topic graph. The difficult part is to filter out topics and to gather new topics in case too many nodes in the current topic graph do not fit the chosen context.

In order to detect possible ambiguities and to present them in an appropriate way we are focussing on a knowledge–based method by making use of Wikipedia (cf. sec. 2 for our motivation of using Wikipedia). The idea is to first match the user query with entries in Wikipedia. If we find more than one match we trigger the disambiguation process. As a starting point we indexed a snapshot of Wikipedia into a *structured* Lucene[4] index containing the title and the abstract of each article in separate fields. The index contains 2.999.597 articles with 4.320.497 different terms and has a size of 7.63 GB on a disc. Using this knowledge base, our query disambiguation algorithm works as follows:

```
10 let Q=user's query;
20 let TG=produce_TG(Q); // initial topic graph TG
30 let LI=Lucene Index;
40 let q[]=SA(tokenize(Q));
50 let query=(title:+q[1] ... +q[n]);
60 let results[]=search(LI ,query);
```

---

[4] http://lucene.apache.org/core/

```
70  if (num(results[]) > 1) {
80    let ass[]=SA(associated_topics(TG));
90    let Qexp=(title:+q[1] ... +q[n]) AND
               (body:+ass[1] ... +ass[m]);
100   let docs[]=search(LI, Qexp);
110   if (user chooses docs[i]) {
120     let s=definition_sentences(docs[i]);
130     let TGnew=produce_TG(Q + s);
140     return TGnew;
140   }} else {
150     return TG;} // return initial TG
```

We start to compute an initial Topic Graph TG with the original user query (20) using the TG construction process described in section 3. The steps (30) to (60) then compute the degree of sense ambiguity using Wikipedia in the following way. Firstly (40), we tokenize the query and apply Lucene's SimpleAnalyzer SA which lowercases all words in the query and deletes numbers. In a next step Lucene retrieves all documents that entail all tokens of the query in the titles of the articles (50+60). In this way it is guaranteed that we find all instances for an entity. The title of an article uniquely identifies each instance because it typically describes the entity in the article and is further qualified by parenthetical expressions. For example, the query for "Jim Clark" also matches "James (Jim) Clark", "Jim Clark (sheriff)", "Jim Clark (film editor)", etc. If only a single title matches or if there is no match at all, we return the initial topic graph TG (150). Otherwise (70) we know that the query matches different Wikipedia articles, and hence, that the query is potentially ambiguous.

In principle, we could now present the different concepts to the user just in the order determined by Lucene. However, the problem is that this ordering actually ignores the information already expressed in the topic graph TG. It could happen that the higher ranked elements in the ranked list are unrelated with the information used by the search engine and covered in TG. On the other hand, TG already expresses some interesting latent semantic information computed via the use of PMI, e.g., expressing that neighboring nodes of a node n are semantically more related to n than nodes with larger distance. Thus in order to achieve a more user query and TG related ordering we perform the following steps (80) to (140). Firstly, we perform a query expansion by adding topics from TG that are determined by a $1NN$ strategy (80) to the original query, i.e. we use only the directly associated topics. In the next steps (90 ff) we again formulate a query against our Wikipedia index. This time we use the associated topics to also search in the articles' body. The result is an ordered list according to the main topics in TG where the most probable meaning is listed first. The abstracts of the articles are presented to the user to chose from. We extract the most important terms (using the function definition_sentences() defined more precisely in the next listing) from the chosen article (120) and produce the final TG using the combination of the terms and the original query (130).

```
10 let first=article.firstSentence
20 let first_pos=POS_Tagging(first)
30 let sep=first_pos.indexOf(((is|was)(a|the)));
40 let isa_part=substr(first_pos,sep);
50 return filter_pos("N",isa_part);
```

According to Wikipedia article guidelines[5] usually an article contains a definition in the first sentence (10). Therefore we first tag the sentence with PoS information (20). If we find the definition phrases "is a", "is the", "was a" or "was the" we choose its right adjacent substring (30+40). If the definition phrase cannot be found, we choose the whole sentence. We filter out all tokens that are not tagged as nouns and return the remaining list (50).

4.1 Experimental Evaluation

In the experimental evaluation we present an automatic way of how to determine the accuracy of the knowledge–based disambiguation algorithm. In a first step we use the above mentioned algorithm. Please note we evaluate real ambiguous queries only. Then we alter the original algorithm in the following way:

```
110 let right=0; all=0;
120 foreach(doc in docs) {
130   let s=definition_sentences(doc);
140   let TGnew=produce_TG(Q + s);
150   let ass[]=SA(associated_topics(TGnew));
160   let Qexp=(title:+q[1] ... +q[n]) AND
             (body:+ass[1] ... ass[m]);
170   let articles[]=search(LI,Qexp);
180   if(doc==articles[0]) {
190       right++;
200   }
210   all++;
220 }
230 final_accuracy=right/all ;
```

The idea behind this automatic evaluation is as follows: the topic graph produced, starting from a disambiguated document, results in a new Topic Graph $TGnew$. A search against the Wikipedia index using the original query for the title–field and the $1NN$ associated topics from $TGnew$ should have the disambiguated document as its best result.

In our experiments we took the entries of 'List of celebrity guest stars on Sesame Street'[6] ($Set1$) and the "List of film and television directors"[7] ($Set2$). Furthermore we evaluated both kinds of the topic graph construction process described above in sec. 3: Topic retrieval based on collocations only ($TopCol$) and its combination with the cluster descriptions ($TopClus$). Table 1 shows the results on the two datasets and the two different TG construction approaches (The first column says: 1:Set1; 2:Set2; A:TopCol; B:TopClus).

---

[5] http://en.wikipedia.org/wiki/Wikipedia:Lead_
section#Introductory_text

[6] http://en.wikipedia.org/wiki/List_of_celebrity_guest_stars_
on_Sesame_Street

[7] http://en.wikipedia.org/wiki/List_of_film_and_
television_directors

Table 1. Accuracy of disambiguation

Set	All	Ambig	Good	Bad	Acc
1+A	406	209	375	54	87.41%
1+B	406	209	378	51	88.11%
2+A	1028	229	472	28	94.4%
2+B	1028	229	481	19	96.2%

Table 2. Manaual evaluation

Set	All	Topics	Good	Bad	Guidance
A	20	167	132	35	ca. 95%
B	20	145	129	16	ca. 95%
A	20	167	108	59	> 97%
B	20	145	105	40	> 97%

4.2 Manual Evaluation

To doublecheck the results of the previous section we also did manual evaluations on datasets by randomly picking results from several test runs and let two independent human judges (not the authors) check the correctness and usefulness of the topics for the chosen senses. This kind of evaluation is often used to evaluate unsupervised methods, cf. [6]. The general setup was to count the number of correct vs. incorrect topics for a given sense. We furthermore gave the judges the chance to intuitively decide whether they would have followed right paths while exploring the solution space. i.e. the task of guiding the exploratory search would have been successful. Table 2 shows the results: the first column denotes the kind of topic retrieval like in the automatic evaluation, A for *TopCol* and B for *TopClus*. The next column shows the number of examples or senses that have been checked[8]. Column 3 shows the total number of extracted topics. The combined retrieval delivers less topics but as you can see in column 4 the quality seems to be improved as the ratio between correct and incorrect topics decreases for both testers. The last column shows whether the guidance towards topics for the chosen sence has been successful, i.e. the percentage of followed paths that are appropriate for the given sense. Please note the values in the columns 3–5 are highly subjective. So for example, for the second judge lots of tokens do not make sense in her opinion but on the other hand she would not have followed them during exploration anyway. Hence although she generally judged more topics not to fit, she rated the algorithms original sense, i.e. guiding the search towards the right direction, as more successful than the first judge.

However we see that the manual evaluations seem to proove the results and the method of the automatic evaluation.

5 Visualisation on Mobile Devices

In this section we briefly introduce the *guiding* part as it is implemented on the mobile device. Whenever the system finds any possible ambiguities in the search query the user receives a list of cells containing short expressive context information for the search term. In our example (Fig. 2) the search query has been "Jim Clark" and the user gets presented all possible found meanings. After selecting one of the cells by simply tapping on it the list-view flips back and the related topic graph is shown. In our example Fig. 3 shows the associated topics for Jim Clark the racing driver, Fig. 4 shows

---

[8] Each judge checked the same examples independently.

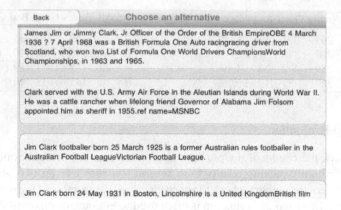

Back   Choose an alternative

James Jim or Jimmy Clark, Jr Officer of the Order of the British EmpireOBE 4 March 1936 ? 7 April 1968 was a British Formula One Auto racingracing driver from Scotland, who won two List of Formula One World Drivers ChampionsWorld Championships, in 1963 and 1965.

Clark served with the U.S. Army Air Force in the Aleutian Islands during World War II. He was a cattle rancher when lifelong friend Governor of Alabama Jim Folsom appointed him as sheriff in 1955.ref name=MSNBC

Jim Clark footballer born 25 March 1925 is a former Australian rules footballer in the Australian Football LeagueVictorian Football League.

Jim Clark born 24 May 1931 in Boston, Lincolnshire is a United KingdomBritish film

Fig. 2. The alternatives to choose from (part)

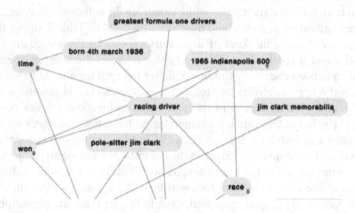

Fig. 3. Excerpt of the topics for the British racing driver Jim Clark

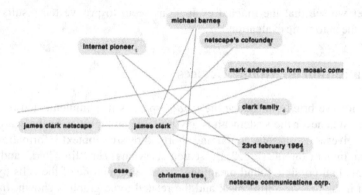

Fig. 4. Excerpt of the topics for the Netscape founder James "Jim" Clark

the results for Jim Clark the Netscape founder. The user now may interact with the graph by a single tap on a node - shows new associated topics for the node; squeezing with two fingers - zooms the view; sliding around - moves the topic graph; double tap - brings in a new view showing the snippets containing the topic of the node. The cells are interactive and by tapping on a cell the corresponding Web page will be shown.

In this way the user is able to explore the solution space by simple and well known interaction patterns.

6   Related Work

6.1   Exploratory Search

Nowadays information has become more and more ubiquitous and the demands of searchers on search engines have been growing, i.e. is a growing need for systems that support search behaviors beyond document oriented simple "one-shot" lookup. The research field *Exploratory Search* embedded in the field of Human Computer Interaction *HCI* explores the process of information seeking and tries to find solutions to support it. Exploratory search systems should for example discover new associations and kinds of knowledge, resolve complex information problems, or develop an understanding of terminology and information space structure. The general aim of this research is to come to a next generation of search interfaces to support users to find information even if the goal is vague, to learn from the information, and to investigate solutions for complex information problems. "Exploratory search can be used to describe an information-seeking problem context that is open-ended, persistent, and multi-faceted; and to describe information-seeking processes that are opportunistic, iterative, and multi-tactical" [27].

Exploratory searches are driven by curiosity or a desire to learn about or investigate something. According to Marchionini [15] a more detailed view on search is:

1. Lookup: Fact retrieval, Known-item search, Navigation, Transaction, Verification, and Question answering[9]
2. Learn: Knowledge acquisition, Interpretation, Comparison, Integration, and Socialize
3. Investigate: Accretion, Analysis, Exclusion, Synthesis, Evaluation, Discovery, Planning, and Transformation

The still dominating ranked list approach is well suited for lookup up search strategies, but probably less suited for a learn search strategy. For investigative search strategies it is too simple and does not support a discourse of questions and answers. Furthermore it is also known that information placed at the end of a ranked list will perhaps never be accessed [23].

The clustering interface *Grouper* [28] has been originally implemented for the HuskySearch engine and it has been compared with the ranked list interface of the same. A clustering algorithm called SuffixTree Clustering (STC) groups the search results into coherent groups. Through the analysis of behavior logs of the search engine

---

[9] Answering specific question like: when, who, where, how much - in contrast to: how, why, . . .

with and without clustering it could be proven that finding specific documents that are ranked very high in the result set of the engine without clustering could be used more efficiently. After some time working with the system people enjoyed the clustering system more although not in all cases.

Findex [13] again used clustering to organize search results. An automatic computation of labelled categories/clusters based on the search results by Google is shown to the left side of the web interface. The clusters may be clicked to filter the overall search result set. The evaluation of the system has been based on an analysis of Web-logs and by a final questionnaires for the testers. The results pretty much confirmed the findings by Zamir and Etzioni: specific searches show less improvement than vague searchers concerning user's performance. Also users need to get used to the new kind of result presentation but they accept and even like it more after a very short time.

WordBars [11] provides active user interaction during the search process in contrast to the previous systems. It visualizes an ordered list of terms that occur in the titles and snippets of the first 100 documents gathered by Google. The user has the possibility to add or remove terms from his query and thereby resorting the search results. In fact WordBars helps the user to refine her query and supports result exploration for specific and vague initial queries. They report that one fundamental design of their system is to create the right balance between computer automation and human control. Hence WordBars does not simply expand the original query but instead actually waits for user interaction before activating next steps. The crucial part is to present the possible choices as good as possible in order to create a real interactive Web information retrieval system.

[1] apply the results of a clustering algorithm on the representation like a fractal tree. [5] organize search results of Web clustering engines. The *WhatsOnWeb*–system uses graphs instead of trees to present the clusters and sub–clusters of the result document set for a query.

6.2 Web Query Disambiguation

There are several approaches for Web query disambiguation. The goal is not only to detect ambiguities in the words of the query but also to decide the right direction in the solution space and present it to the user. Some approaches like [20] or [3] try to automatically learn a user's interest based on the click history. In order to achieve this they provide a three step algorithm: 1) a model representing a user's interest based on the click history; 2) a process that estimates the user's hidden interest based on the click history; 3) a ranking mechanism that reranks the search engine result on the basis of 1) and 2). Other approaches like [22] or [7] follow the same principle but with different learning and ranking algorithms.

Another approach is based on hyperlink structures of the Web and aims for a personal PageRank that modifies the search engines' PageRanks. Examples for this approach are [9] and [12].

A more generalizing approach consists of collaborative filtering methods. Here the search history of groups with similar interests are used to refine the search. This method has been used in [24] where users' profiles are constructed using a collaborative filtering algorithm [2], or [25] where the correlation among users, queries, and clicked Web pages is analyzed. The advantage for the user is by the increased completeness of the search results because the knowledge base for the filtering process is already filled by other users - provided there are users with similar interests.

In contrast there has also been much research trying to post process the search results using clustering algorithms. [14] propose a very promising approach for disambiguation of person names. This approach does not require user models or a learning and personalization phase. The results from a search process are clustered by taking different document properties into account: Title, URL, metadata, snippet, context window (around the original query), context sentence, and bag of words of the whole document. The main property of this algorithm is the robustness and speed and hence the disambiguation performance. However it lacks - as reported in this paper - the labeling or definition of the clusters. So again the user has to check by reading at least some snippets inside a cluster [4].

7 Conclusions and Outlook

We presented an approach of guided interactive topic graph extraction for exploration of web content. The initial information request is issued online by a user to the system in the form of a query topic description. Instead of directly computing and presenting a topic graph for the user query, possible senses of the query are identified and enumerated by referring to an external knowledge base, Wikipedia in our case. All found readings are then sorted and presented to the user and the user is asked to select her preferred one. The user–selected sense is then used for constructing an initial topic graph from a set of web snippets returned by a standard search engine. At this point, the topic graph already represents a graph of strongly correlated relevant entities and terms. The topic graph is then displayed on a tablet computer (in our case an iPad) as touch–sensitive graph. The user can then request further detailed information through multiple iterations.

Experimental results achieved by means of an automatic evaluation procedure demonstrate the benefit of the disambiguation method for exploratory search strategies. The automatic evaluation has been approved by another human evaluation. Currently, the main problem of our approach arises when an ambiguous query cannot be found in Wikipedia using our strategy. For example, the query "Famous Jim Clark" would not be found as we require that all words of the query occur in an Wikipedia article's title. Even if we could cope with this using a modified fuzzy search strategy we still would not find out ambiguities in queries that simply are not present in Wikipedia. However, in the running system we plan to give some feedback to the user by changing the color of the search entry. Then the user knows that there may be more then just one meaning for her query. Another open question is whether an improvement of our rather simple way of expanding the query using Wikipedia abstracts will lead to significant improvements of the disambiguation results. We are planning to do some research on this.

References

1. Akhavi, M.S., Rahmati, M., Amini, N.N.: 3d visualization of hierarchical clustered web search results. In: Proceedings of the Computer Graphics, Imaging and Visualisation, CGIV 2007, pp. 441–446. IEEE Computer Society, Washington, DC (2007)
2. Breese, J.S., Heckerman, D., Kadie, C.: Empirical analysis of predictive algorithms for collaborative filtering, pp. 43–52. Morgan Kaufmann (1998)
3. Chirita, P.A., Nejdl, W., Paiu, R., Kohlschütter, C.: Using odp metadata to personalize search. In: Proceedings of the 28th Annual International ACM SIGIR Conference on Research and Development in Information Retrieval, SIGIR 2005, pp. 178–185. ACM, New York (2005)
4. Cucerzan, S.: Large-scale named entity disambiguation based on wikipedia data. In: Proc. 2007 Joint Conference on EMNLP and CNLL, pp. 708–716 (2007)
5. Di Giacomo, E., Didimo, W., Grilli, L., Liotta, G.: Graph visualization techniques for web clustering engines. IEEE Transactions on Visualization and Computer Graphics 13, 294–304 (2007)
6. Fader, A., Soderland, S., Etzioni, O.: Identifying relations for open information extraction. In: Proceedings of the 2011 Conference on Empirical Methods in Natural Language Processing, EMNLP 2011, pp. 1535–1545 (2011)
7. Gauch, S., Chaffee, J., Pretschner, A.: Ontology-based personalized search and browsing. Web Intelli. and Agent Sys. 1, 219–234 (2003)
8. Gimenez, J., Marquez., L.: Svmtool: A general pos tagger generator based on support vector machines. In: Proceedings of the 4th International Conference on Language Resources and Evaluation (LREC 2004), Lisbon, Portugal, vol. I, pp. 43–46 (2004) ISBN 2-9517408-1-6
9. Haveliwala, T.H.: Topic-sensitive pagerank. In: Proceedings of the 11th International Conference on World Wide Web, WWW 2002, pp. 517–526. ACM, New York (2002)
10. Hearst, M.A.: Search User Interfaces. Cambridge University Press (2009)
11. Hoeber, O., Yang, X.D.: Interactive web information retrieval using wordbars. In: Proceedings of the 2006 IEEE/WIC/ACM International Conference on Web Intelligence, WI 2006, pp. 875–882. IEEE Computer Society, Washington, DC (2006)
12. Jeh, G., Widom, J.: Scaling personalized web search. In: Proceedings of the 12th International Conference on World Wide Web, WWW 2003, pp. 271–279. ACM, New York (2003)
13. Käki, M.: Findex: search result categories help users when document ranking fails. In: Proceedings of the SIGCHI Conference on Human Factors in Computing Systems, CHI 2005, pp. 131–140. ACM, New York (2005)
14. Liu, Z., Lu, Q.: High performance clustering for web person name disambiguation using topic capturing. Ratio (2011)
15. Marchionini, G.: Exploratory search: from finding to understanding. Commun. ACM 49, 41–46 (2006)
16. Neumann, G., Schmeier, S.: A mobile touchable application for online topic graph extraction and exploration of web content. In: Proceedings of the ACL 2011 System Demonstrations. ACL (2011)
17. Neumann, G., Schmeier, S.: Exploratory search on the mobile web. In: 4th International Conference on Agents and Artificial Intelligence (ICAART 2012), pp. 110–119. SciTePress (2012)
18. Osinski, S., Stefanowski, J.: WeissOsinski, D.: Lingo: Search results clustering algorithm based on singular value decomposition. In: Proceedings of the International IIS: Intelligent Information Processing and Web Mining Conference, Zakopane, Poland. Advances in Soft Computing, pp. 359–368. Springer (2004)
19. Osinski, S., Weiss, D.: Carrot2: Making sense of the haystack. In: ERCIM News (2008)

20. Qiu, F., Cho, J.: Automatic identification of user interest for personalized search. In: Proceedings of the 15th International Conference on World Wide Web, WWW 2006, pp. 727–736. ACM, New York (2006)
21. Sanderson, M.: Ambiguous queries: test collections need more sense. In: Proceedings of the 31st Annual international ACM SIGIR Conference on Research and Development in Information Retrieval, SIGIR 2008, pp. 499–506. ACM, New York (2008)
22. Shen, X., Tan, B., Zhai, C.: Implicit user modeling for personalized search. In: Proceedings of the 14th ACM International Conference on Information and Knowledge Management, CIKM 2005, pp. 824–831. ACM, New York (2005)
23. Sping, A., Wolfram, D., Jansen, M., Saracevic, T.: Searching the web: The public and their queries. Journal of the American Society for Information Science and Technology, 226–334 (2001)
24. Sugiyama, K., Hatano, K., Yoshikawa, M.: Adaptive web search based on user profile constructed without any effort from users. In: Proceedings of the 13th International Conference on World Wide Web, WWW 2004, pp. 675–684. ACM, New York (2004)
25. Sun, J.-T., Zeng, H.-J., Liu, H., Lu, Y., Chen, Z.: Cubesvd: a novel approach to personalized web search. In: Proceedings of the 14th International Conference on World Wide Web, WWW 2005, pp. 382–390. ACM, New York (2005)
26. Turney, P.D.: Mining the web for synonyms: Pmi-ir versus lsa on toefl. In: Proceedings of the Twelfth European Conference on Machine Learning (2001)
27. White, R.W., Roth, R.A.: Exploratory search: Beyond the query-response paradigm. Synthesis Lectures on Information Concepts, Retrieval, and Services, vol. 1(1), pp. 1–98 (January 2009)
28. Zamir, O., Etzioni, O.: Grouper: a dynamic clustering interface to web search results. In: Proceedings of the Eighth International Conference on World Wide Web, WWW 1999, pp. 1361–1374. Elsevier North-Holland, Inc., New York (1999)

Towards a Unified Thematic Model for Recommending Context-Sensitive Content

Mihaela Dinsoreanu and Rodica Potolea

Computer Science Department, Technical University of Cluj-Napoca, Romania
{mihaela.dinsoreanu,rodica.potolea}@cs.utcluj.ro

Abstract. The objective of our work is to identify the most relevant content given unstructured, text-based context. In this respect, we propose a unified model that includes a generic context model and the similarity metrics in order to provide context-sensitive content. The context model relies on the underlying thematic structure of the context by means of lexical and semantic analysis. Moreover, we analyse both the static characteristics and dynamic evolution. The model has a high degree of generality by not being committed to a certain domain, nor a constrained context structure. Based on the model, we have implemented a system dedicated to contextual advertisements for which the content is the set of relevant ads while the context is represented by a web page visited by a given user. The dynamic component refers to the changes of the user's interest over time. From all the composite criteria the system could accept for assessing the quality of the result, we have considered relevance and diversity. The design of the model and its ensemble underlines our original view on the problem. From the conceptual point of view, the unified thematic model and its category based organization are original concepts together with the implementation.

Keywords: (Dynamic) Content Context Match, Classification, Topic Model, Parallelization, Text Mining, Taxonomy, Design and Implementation, Evaluation.

1   Problem Statement

Since we live in an information centric society, we are flooded with data (the so called "deluge of data" [17] yet we still struggle to filter the information that we need. In this context, an automatic identification of relevant pieces of information for each specific user is paramount. The goal is not only to expose to the user only meaningful information (relevant and of interest), but more important, at the right time and in the right context [8].

Considering the above mentioned challenges we propose a generic model capable of providing context-sensitive content based on the underlying thematic similarity between an analysed context and the recommended content. We present our approach based on a unified topic model that extracts the topics describing both the context and the content. This process is fuelled by the portions of the analysed entities having the highest descriptive value. Once these entities are annotated with thematic information,

A. Fred et al. (Eds.): IC3K 2012, CCIS 415, pp. 68–83, 2013.

their reciprocal affinity, within the unified topic model, can be measured. Using such values, a topic based coverage aims to improve diversity and achieve serendipitous recommendations.

To summarize, the main steps of our process are:

- the extraction of the highest descriptive valued n-grams among the analysed entities;
- attaching thematic information to the analysed entities;
- maximizing diversity of the recommended content.

We apply our approach, as proof of concept, in the Online Advertising problem of the Best Match (OABM) [6] between a web page (active context), advertisements (suitable content) and the user that is currently interacting with that context (dynamic context). We claim that this problem can be mapped on our model by using a double instantiation of the context-to-content similarity relation. One instantiation describes the relation between a web page and an advertisement. The second has the same mapping for the content but describes the context as being the user, moreover his/her historical information. The combined, triple recommendation between an active context, the content and the dynamic context is constructed by further processing the two instantiations.

2   State of the Art

Most of the reported approaches start by describing the matching content with relevant keywords. These keywords are compared with the descriptors of the ads (bid phrase) hence obtaining a lexical similarity [12], [19]. Such an approach follows a pipeline with a few, well-defined stages. A pre-processing stage is [19] to prepare the content by sanitizing, removing stop words, stemming and extracting some keyword candidates (words from the context, annotated with some descriptive features). Then the annotated keyword candidates are processed, in a Monolithic Combinedi approach, by a binary classifier. This is how the keywords are selected and the keyword selection step is completed.

Such an approach is generally enhanced with additional models that sustain the semantic similarity between web pages and advertisements [5], [20], [15]. This association generates a semantic score which, combined with the lexical, consolidates the match. This semantic information can be embedded in a taxonomy [5] and used to score the similarity based on the distance to the least common ancestor, if both the context and the advertisement can be mapped on it.

The third aspect to be considered in such a model consists of the particularities of an actual user. The associated historical information, if present, will influence the final match [2], [7]. User information can be attached to the advertisement or to the page [7] but, recent research explores the idea of user interest and behavioural trend [2]. Such a model can extract the dynamics of behaviour and make better recommendations.

The concept of a "topic" is described using a specialized mixed membership model called topic model. Such a model describes the hidden thematic structure [4] in large collections of documents. The Latent Dirichlet Allocation (LDA) [4] is such a topic model. Its distinguishing characteristic is that all documents in the collection share the same set of topics, but each document exhibits those topics with different proportions [3]. The topics are defined by a distribution over the whole set of available words (within the document corpus). The documents are described by a distribution over topics based on the used words are just sampled from the topic's word distribution. In a model like that, the only observable data are the document's words. The topics, their distribution in documents and the distribution of words between topics need to be inferred. Direct inference is not tractable so approximation techniques are used [10].

Many systems produce highly accurate recommendations, with reasonable coverage, yet with limited benefit for practical purposes due to their "obviousness" hence lack of novelty, therefore new dimensions that consider the "non-obviousness" should be considered. Such dimensions are coverage (percentage of items part of the problem domain for which predictions are made), novelty and serendipity, dimensions for which also [21] advocates. Since serendipity is a measure of the degree to which the recommendations are presenting items that are attractive and surprising at the same time, it is rather difficult to quantify. "A good serendipity metric would look at the way the recommendations are broadening the user's interests over time" [11], so, again the need for introducing timing and sequence of items analysis for RS.

3   The Unified Model for Structured Content Representation

3.1   Concepts and Terminology

In the following we define the notions used throughout the article.

- A *topic* $\beta_t = \{p(w|t) \mid w \in V\}$ is a probability distribution over a finite vocabulary V of words w where $\sum_{i \in \beta_t} i = 1$. Let T be the set of all topics;
- A *document* $d = \{w_1, w_2, ..., w_N\}$ is a sequence of N words. Each document has an associated distribution over topics $\theta_d = \{p(t|d)|t \in T\}$ where $\sum_{i \in \theta_d} i = 1$;
- A *keyword* k_d is an element of a document with high descriptive value for that document. Let $K_d = keywords(d)$ represent all the keywords of a document;
- A *context* C_x is a document with no specific structure;
- A *dynamic context* ΔC_x is a specialized context that evolves over time;
- An *active context* $C_x A$ is the current specific context;
- A *content* C_n is a specialized document that is to be associated with a context;
- A *corpus* $C = \{d_1, d_2, ..., d_D\}$ is a collection of D documents;
- A *unified topic model* is a 5-tuple $UTM = \langle V, T, \{\beta_t | t \in T\}, C, \{\theta_d | d \in C\}\rangle$ describes the set of all the topics T, their distribution over words β_t, the underlying vocabulary V, all the documents C and their distribution over topics θ_d;

- A *category* ψ is a named subset of topics with similar probability distributions over the topic set. The associated topic distribution of the category ψ is denoted by θ_ψ and is organized in a *category taxonomy* Ψ;
- A *contextual relevance* $rel(C_x A, C_n)$ measures the similarity between the context and content;
- A *dynamic relevance* $rel(\Delta C_x, C_n)$ measures the similarity between the dynamic context and content.

3.2   The Unified Thematic Model

We propose an approach for extracting the thematic structure from a corpus of documents, with no restriction to the type (content/context) or structure (actually no specific structure is considered for the underlined documents). For doing this, the Latent Dirichlet Allocation (LDA) generative model proposed in [4] is considered

The LDA is formally described in Figure 1. This model describes a corpus of D documents on which a number of K topics are defined with a β_k topic distribution for topic k.

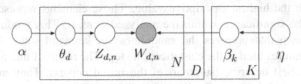

Fig. 1. LDA model representation

Each document has a θ_d distribution over topics from which, for each of the N words in the document, a topic $Z_{d,n}$ is sampled followed by the sampling of the word $W_{d,n}$ from $\beta_{Zd,n}$. The α parameter controls the sparsity [10] of θ while parameter η influences β. For smaller α values, the topics space is narrowed, with higher weights for the selected topics. In the same direction, smaller values of the η parameter would allow for a limited number of words to be selected.   In Fig. 1 shadowed nodes define an observable variable hence all the information we have are the words $W_{d,n}$.

Direct inference of the latent (unobservable) variables is not tractable [10]. Therefore approximation techniques have to be considered. We chose to employ a *collapsed Gibbs sampling* approximation technique [9].

We adopted, as a starting point, the solution proposed in [14] and extended it to a *parallel Gibbs sampling LDA*. We employed an input decomposition technique by dividing the documents analyzed during each of the Gibbs iteration in evenly distributed work packages for the parallel processes.

3.3   Our Proposed Approach

Our approach is based on the issue of selection rather than generation. The context is generaly unstructured while the content might respect a certain format hence the solution should be able to handle the context's lack of structure and select the most

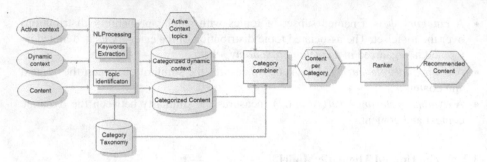

Fig. 2. Conceptual Architecture

relevant content from a repository. Moreover, it should be able to bridge the lexical and semantic context –to-content gap by employing a common representation. Our proposed solution that addresses the above mentioned challenges and consists of four main modules that interact in a pipeline manner to generate recommendations (Figure 2).

The first module, *Keyword Extractor* (KE), identifies and extracts the elements of the context with the highest descriptive value. Those elements represent keywords, which outline the significance within the analysed context. This module performs a pre-processing step that prepares the candidates for keyword status by annotating them with the features used in the classification step. The result of this module is a set of n-grams that best describe (summarize) the initial context. They are used as an input by the next module. Formally, KE is described as follows:

$$KE(C_X A) = K_{C_X A}. \tag{1}$$

The *Topic Identifier* (TI) is responsible for associating topic information to the analysed context ($C_X A$) based on the keywords that describe it ($K_{C_X A}$). The association is accomplished using the TI's underlying *topic model*. At this point the topic level unification takes place by associating to $C_X A$ a distribution over topics $\theta_{C_X A}$. From now on, all the analysed entities are modelled by a distribution over topics within the *unified topic model*. Formally, TI is described as follows:

$$TI(K_{C_X A}) = \theta_{C_X A}. \tag{2}$$

The *Category Combiner* (CC) is responsible for computing the similarity between the topic distribution generated by TI ($\theta_{C_X A}$) and the distribution associated to the managed content (θ_{C_n}) or dynamic context ($\theta_{\Delta C_x}$). The main limitation of the topics discovered by TI is anonymity (topics have no semantic information). To overcome this shortcoming we added an abstraction layer above the topics called category. Such categories are nodes in a taxonomy having a pre-computed topic distribution. In CC we also analyse the dynamic context that describes the evolution of the interaction based on previously acquired data. The output of this module is a set of advertisements (Υ) with two associated relevance values. One is computed from the perspective of the active context ($rel(C_x A, C_n)$) and the other, from the perspective of the dynamic context ($rel(\Delta C_x, C_n)$). Formally, CC is described as follows:'

$$CC(\theta_{C_xA}, \theta_{\Delta C_x}) = \Upsilon,\tag{3}$$

Where

$$\Upsilon = \left\{ \left\{ \begin{matrix} C_n, \\ \langle rel(C_xA, C_n),\rangle \\ rel(\Delta C_x, C_n) \end{matrix} \right\} \Big| C_n \in \psi \right\}.\tag{4}$$

The *Ranker* (R) is responsible for filtering the output according to the actual performance criteria (Γ). Such criteria can range from relevance to diversity or trustworthiness. In the case of diversity, we aim a low thematic overlap between recommendations while maintaining their relevance to the considered context (whether it is C_xA or ΔC_x). This reduced overlap induces an increased context thematic coverage that is more likely to produce serendipitous recommendations. Formally, R is described as follows:

$$R(\Upsilon) = \left\{ C_n | C_n \in \text{argmax}_{C_n \in \Upsilon} \Gamma \right\}\tag{5}$$

4 Tailoring the Model on Contextual Advertisement

The components of the generic solution described above are not committed to any domain of application. For experiments we instantiated our generic solution to address the contextual advertisement problem. To do so we considered the triple <web page, user history, advertisement> and the similarities between the elements of the triple.

The OABM problem can be decomposed in two *context-to-content* relations defined based on the same underlying *unified topic model*. Due to this common reference, the values describing the context/content similarity can be extrapolated at model level. We formally specialize the generic terms to OABM specific concepts:

Table 1. Mapping of generic model to the OABM problem

Generic	Specific
Active context	*web page* C_xW
Content	*advertisement* C_nA
Dynamic context	*user interaction* described by its *overall interest* I
Contextual relevance function	$rel(C_xW, C_nA)$;
Behavioral relevance function	$rel(I, C_nA)$

The active context is represented by a web page that can be described by a set of keywords. The advertisements have associated bid phrases that are considered keywords. User history reflects the dynamic context, namely the set of <web page, past ads> pairs the user previously dealt with. Thus, the keywords describing the active context can also reflect the user interests. From the conceptual point of view, we employ a unified technique for the recommendation of relevant advertisements. Thus, the

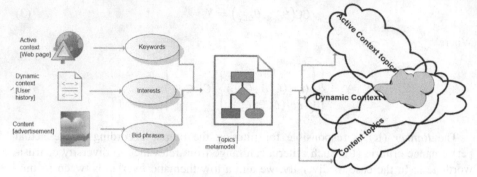

Fig. 3. The Unified Topic Model

topic concept describes all the three components of OABM. The advertisement is described by a single, targeted topic. From the active context a static set of topics is extracted. The user is described by a dynamic set of topics to reflect the evolution of interests over time. Thus, the tree components can be defined, at topic level, based on a common reference i.e. the unified topic model in Figure 3.

Let $D(\theta_e)$ be a probability distribution over a set of topics describing entity e. Given an entity e we define LDA_e a function from the set of keywords describing the entity e to its probability distribution over topics θ_e as follows:

$LDA_e: \{\vartheta | \vartheta \in keywords(e)\} \to D(\theta_e)$.

Accordingly, *active context* topic distribution is:

$$LDA_{C_xA}: \{\vartheta | \vartheta \in keywords(C_xA)\} \to D(\theta_{C_xA}). \tag{6}$$

and the *content* topic distribution is:

$$LDA_{C_n}: \{\vartheta | \vartheta \in keywords(C_n)\} \to D(\theta_{C_n}). \tag{7}$$

We claim that $\exists\, t \in T$ s.t. $\operatorname{argmax}_t(p(t|C_n)) = \{t\}$ where $p(t|C_n) \in \theta_{C_n}$ hence the content is described by a dominant topic.

The user history, instance of the dynamic context, is defined based on the hierarchy levels:short (I_s), medium (I_m) and long (I_l) term [2]. We define the user's overall interest (I) based on a convex combination between the three sub-interests. Let $\kappa \in (0,1)$ and $K = \kappa + \kappa^2 + \kappa^3$ such that:

$$I = \frac{\kappa}{K} * I_s + \frac{\kappa^2}{K} * I_m + \frac{\kappa^3}{K} * I_l \tag{8}$$

Let I_i, $i \in \{s, m, l\}$ be one of the three sub-interests and $C_i = \{\varepsilon | visit(\varepsilon) \in interval(i)\}$ the set of all accessed contexts during the interval associated with the sub-interest. At this level, we employ the following definition for $I_i, i \in \{s, m, l\}$:

$$LDA_{C_i}: \bigcup_{c \in C_i} keywords(c) \to D(I_i) \tag{9}$$

and describe the sub-interest as the distribution over topics of a pseudo-context described by the union of the keywords associated to a context accessed during the sub-interest interval.

4.1   Keyword Extractor

To find D(θ_e) for each

e∈ {*active context* [*web page*], *dynamic context* [*user history*], *content*[*ad*]}

a keyword extraction step is performed. The internal structure of the module is presented in Figure 4.

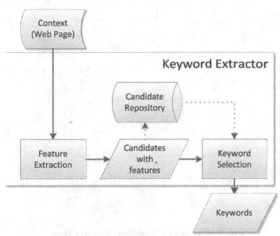

Fig. 4. Keyword Extractor detailed view

The **Feature Extraction** (FE) sub-module is responsible for the initial preprocessing of the analysed context and assumes stop-words removal, stemming and associating additional information in order to bring the context's elements in an annotated candidate state. We associate to each candidate its statistic information occurrences together with characteristics dependent on the context's nature like the candidate's localization within the context, its styling information or its inner structure. The *candidates with features* generated by the FE sub-module are persisted in a repository.

The **Keyword Selection** (*KS*) sub-module uses a binary classifier to identify the actual keywords among the extracted candidates. The classifier is chosen based on the specific criteria required by CA. We have to maximise precision allowing for moderate degradation of recall as from the business point of view an advertisement recommendation that is out of context is worse then no advertisement at all.

For the classification step the features that best differentiate between the components of the text are needed to efficiently describe the classification category to which the membership relation is in question. Such features can range from statistical descriptors (of the occurrences of a word based on both its local and global statistics) to localization markers or unique style definitions.

4.2   Category Combiner (Cc)

The CC module is responsible for computing the triple <web page, user history, adverisement> similarity based on their distribution over topics {$D(\theta_{C_xA})$, $D(\theta_{C_n})$}.

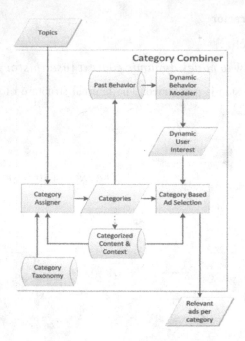

Fig. 5. Category Combiner detailed view

$D(I_i)$} within the unified topic model. The internal structure of CC is presented in Figure 5. The Category Assigner (CA) receives as input the distribution of topics for the given context and a taxonomy of categories (with their topic distribution) to construct the mapping between a category and a topic. In order to quantify the similarity between two probability distributions we use the Hellinger distance [13]. The module returns the list of candidate categories ordered by their relevance. A subset of these categories is processed by the next module.

These categories are further refined by the *dynamic context* to select a final subset of advertisements that qualify for the next processing step.

The *dynamic context* generation is a process enforced by the **Dynamic Behaviour Modeller** (DBM). This sub-module aggregates the user information (in the form of interests that is further used to enable behavioural recommendations.

The **Category-Based Ad Selector** analyses ads associated to categories, in a reduced search space due to the category set cardinality reduction yielded by CA and DBM. For each advertisement we further compute two similarity scores (the contextual relevance and the behavioral relevance) based on the Hellinger distance between their probability distributions. Both scores will be further integrated in the ranking module.

4.3   Ranker

The ads are recommended from two perspectives: (1) contextual relevance based on <web page, ad> similarity and (2) behavioral relevance basedon <user history, ad>

similarity. The page topic coverage and user interest coverage may be antagonistic by nature as they compete for the same page advertisement slots. Balancing them by choosing the ones with the maximal relevance in the given overall context is required. The value of the *correlation coefficient* λ determines the proportion/weight of the two perspectives. Hence we represent the overall recommendation as a weighted sum (convex combination) of the two previously defined relevance functions:

$$\text{similarity}_\lambda(C_x W, I, C_n A) = rel(C_x W, C_n A) * \lambda + rel(I, C_n A) * (1 - \lambda). \quad (10)$$

Moreover, the behavioural recommendation maps on $\text{similarity}_0(C_x W, I, C_n A)$ and the contextual recommendation maps on $\text{similarity}_1(C_x W, I, C_n A)$.

5   Results and Evaluation

For the classification problems, we performed several evaluations with classifier candidates from Weka [18]. The experiments were conducted on a dataset with 1333 instances with 75%/25% class distribution. The Percentage Split technique, with 66% for train and 10 repetitions has been considered. The results in Table 2 show that the best performing classifiers are the Multilayer Perceptron (MLP), J48 and a Bagged Predictor with underlying J48 (BJ48). We discarded MLP as candidate due to the large training time required (an order of magnitude compared to the others). The remaining candidates have the same underlying classifier but BJ48 performs additional replications and voting with a minimal improvement of the target measurements. Thus our final choice is the J48 classifier

Table 2. Classifier comparison for the Keyword Selection step

Name	Correct Classification	Precision	False Positive Rate	F-measure
Bayes Network	0.90	0.96	0.12	0.93
Naïve Bayes	0.86	0.92	0.24	0.91
Multilayer Perceptron	0.91	0.95	0.14	0.94
SMO	0.86	0.90	0.31	0.91
Bagging (J48)	0.92	0.96	0.13	0.94
Decision Table	0.89	0.93	0.22	0.93
J48	**0.92**	**0.95**	**0.14**	**0.94**
Decision Stump	0.79	1.0	0.0	0.84
AdaBoost M1	0.88	0.93	0.22	0.92
SPegasos	0.88	0.92	0.24	0.92

The largest computational effort appears during the computation of the topic distribution. This process is dependent on the number of words that describe the topic model. We chose to adopt a parallel implementation for this critical area of the flow. We measured the relative speedup (ΔS) while varying the dimension of input parameters like the number of topics to be discovered (#T), the number of iterations to approach convergence (#I) and the number of analyzed documents (#D) for the estimation and inference (inf.) use-cases. For the experimental results covered in Table 3 we used two processing elements. Further investigation showed proportional growth of ΔS as the number of processing elements increases.

We can observe that the growth of a single measure of interest with a controlled increment will generate a proportional growth in both sequential and parallel results by maintaining the relative speedup in a constant range. But when we increase multiple measures of interest with significant increments we observe a spike in the relative speedup hence favoring the parallel implementation.

Table 3. Improvement of parallel LDA

Use case	#T	#I	#D	Sequential [s]	Parallel [s]	ΔS
Estimation	30	20	2246	5.52	3.05	1.81
	50	20	2246	8.02	4.40	1.82
	50	40	2246	16.26	8.74	1.86
	100	40	2246	30.39	16.46	1.83
	100	100	2246	77.72	40.47	1.92
	50	40	1123	8.23	4.40	1.87
Inf.	100	100	1123	25.38	14.03	1.81

Two processing elements prove to bring a significant boost for the specific needs. We further considered a fixed workload scenario where we varied the processing units. Our findings are summarized in Figure 6. We observe 82% efficiency by the time we consider four processing elements and 70% as we get to eight, our available maximum. In a highly parallelized environment, intensive topic model interactions will generate contention on our critical section that makes us slowly converge to our Amdahl limit. For the parallel implementation of LDA we used a shared memory model. Using the technique an average 1.85 relative speed-up for 2 processing elements was obtained.

Another aspect of interest is represented by the benefit introduced by the usage of categories as an additional abstraction layer above topics. Figure 7 shows the evolution of execution time as the number of analysed contexts grows both with and without the usage of categories. This behaviour appears in the CA sub-module of the CC. Our category taxonomy has 100 nodes organized on 6 levels. We can see that even if

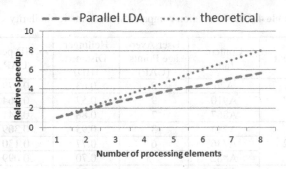

Fig. 6. Relative speedup evolution

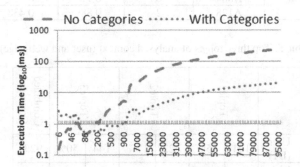

Fig. 7. Evolution of category-based selection

the category based approach starts with a default overhead, as the number of advertisements grows, the two curves will intersect when the total number of advertisements equals the number of categories combined with the number of ads in the top categories and from that point on, their growth patterns differ with almost a *decade*.

Another important aspect is the evaluation of the provided recommendations. Due to the subjective nature of the underlying problem and to the lack of annotated benchmark datasets we chose to employ an end user evaluation of the recommended advertisements. We considered the manual inspection performed by a group of users that were asked to assess, on 1-10 rating scale, the similarity of a content with a designated context. The representative results are presented in Table 4.

As we aim to minimize Hellinger distance lower values of the HD per UAP is better. We can see good recommendations like A26 and A910 for C27 having an 8 and 9 UAP with smaller than 0.1 HD per UAP. This shows a good correlation between the user evaluation and the results of the Hellinger distance.

Moreover we explored the influence of the *correlation coefficient* λ on the recommendations. We assess the marginal cases for which we have fully contextual or fully behavioral recommendations and the case in which the two are combined. We consider a user and a web page having their top three topics illustrated in Table 5.

The topics in Table 5 cover, in different proportions, the context but their combined coverage is sometimes enough to describe them. We consider a set of recommended advertisements for which we compute the degree of coverage and the combined similarity, using different versions of the correlation coefficient (λ).

Table 4. User evaluation compared with thematic similarity

Context	Content (ad)	User Average Points (UAP)	Hellinger Distance (HD)	HD per UAP
C27	A26	8	0.63	0.079
	A910	9	0.57	0.064
	A867	2	0.88	0.441
C42	A283	3	0.93	0.309
	A736	6	0.77	0.130
	A882	7.5	0.70	0.099
C54	A801	2	0.86	0.427
	A884	6	0.77	0.127
	A128	2	0.88	0.438

Table 5. Top three topics of analysed context (user and web page)

	Topic1		Topic2		Topic3		Combined Coverage
User (U1)	T21	25%	T22	17%	T44	50%	92%
Web Page (C27)	T5	35%	T27	18%	T48	15%	68%

Table 6. Full behavioural ($\lambda=0$) recommendation comparison

Content	T21 (%)	T22 (%)	T44 (%)	Coverage (%)	HD to User	Rank
A26	4	12	21	13.05	0.88	3
A910	8.8	80	8.8	22.05	0.68	1
A867	32	14	9.4	15.40	0.73	2

Table 7. Full contextual ($\lambda=1$) recommendation comparison

Content	T5 (%)	T27 (%)	T48 (%)	Coverage (%)	HD to Web Page	Rank
A26	11.2	16.7	35	12.17	0.63	2
A910	26.4	44.4	8.8	18.5	0.57	1
A867	4.7	40	4.7	9.52	0.88	3

Table 8. Combined (λ=0.5) recommendation comparison

Content	HD2User	HD 2 WP	HD2Overall	Rank
A26	0.88	0.63	0.75	2
A910	0.78	0.57	0.71	1
A867	0.86	0.88	0.83	3

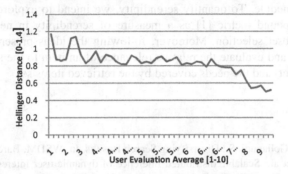

Fig. 8. Hellinger evolution with user evaluation

We observe that there is a correlation between the low values of the Hellinger distance and the associated coverage from both perspectives. We also considered a manual user evaluation for exploring the correlation with the Hellinger distance. The relation between the Hellinger distance and the user's evaluation is represented in the curve depicted in Figure 8.

At the lowest end of the evaluation interval (ranging between 0 and 4) one can observe a strong variation between the actual distance and the score. But at this level is difficult to assess the correlation because users see content within this band from different perspectives. The important aspect is that content in this range is not desirable. On the other hand, at the other extremity of the interval one can observe an evolution of the distance to ever-decreasing values. The area of confusion is between 4 and 7. Here, the distance varies with small increments making it harder to discriminate.

We concluded that Hellinger values above 0.9 have a higher chance of being bad content and values below 0.7 of being good content. If we consider that this measure is theoretically bound between 0 and $\sqrt{2}$ this results seems promising.

6   Conclusions

This paper proposes a model capable of providing context-sensitive content based on the similarity relation between the analyzed context and the recommended content.

The similarity is measured at thematic level by mapping both the context and the content within a common reference system, the proposed unified thematic model. In order to achieve topic information, we analyze the underlying thematic structure induced by the keywords of the considered entities (contexts, both static and dynamic, and content) using a parallel topic extraction approach. The content search space is reduced using the category abstraction and the quality is measured based on relevance and diversity.

We applied the model on the OABM problem considering as context a web page, as content an advertisement and as dynamic context the user interaction within the system. Our current focus is on feature selection, approximation techniques to deal with the LDA intractability, and the integration of serendipity as performance metric in the ranking module. To quantify serendipity, we intend to explore the use of the NDGG-IA compound metric [1] as a measure of serendipity in new, original approaches for subset selection. Moreover, following the ideas presented in [16] we intend to define and evaluate the relative relevance of an item to the set of the user's needs (optimal set) and the needs covered by the retrieved items set.

References

1. Agrawal, R., Gollapudi, S.: Diversifying Search Results. In: WSDM, Barcelona (2009)
2. Ahmed, A., et al.: Scalable distributed inference of dynamic user interests for behavioral targeting, San Diego, California, USA, pp. 114–122. ACM (2011)
3. Blei, D.M.: Introduction to probabilistic topic models. Communications of the ACM 54(12), 77–78 (2011)
4. Blei, D.M., Ng, A.Y., Jordan, M.I.: Latent dirichlet allocation. The Journal of Machine Learning Research 3, 993–1022 (2003)
5. Broder, A., et al.: A semantic approach to contextual advertising, Amsterdam, The Netherlands, pp. 559–566. ACM (2007)
6. Broder, A., Josifovki, V.: Introduction to computational advertising (MS&E 239). Stanford University, Stanford(California) (2011)
7. Chakrabarti, D., Agarwal, D., Josifovski, V.: Contextual advertising by combining relevance with click feedback, Beijing, China. ACM (2008)
8. Garcia-Molina, H., Koutrika, G., Parameswaran, A.: Information seeking: convergence of search, recommendations and advertising. Communications of the ACM 54(11), 121–130 (2011)
9. Griffiths, T.L., Steyvers, M.: A probabilistic approach to semantic representation, Fairfax, Virginia, s.n., pp. 381–386 (2002)
10. Heinrich, G.: Parameter estimation for text analysis. Fraunhofer IGD, Darmstadt (2009)
11. Herlocker, J.L., Konstan, J.A., Terveen, L.G., Riedl, J.T.: Evaluating collaborative filtering recommender systems. ACM Transactions on Information Systems 1(22), 5–53 (2004)
12. Manning, C.D., Raghavan, P., Schtze, H.: Introduction to Information Retrieval. Cambridge University Press, New York (2008)
13. Nikulin, M.: Hellinger distance - Encyclopedia of Mathematics (2011)
14. Phan, X.-H., Nguyen, C.-T.: JGibbLDA: A Java implementation of latent Dirichlet allocation, LDA (2008)
15. Ribeiro-Neto, B., Cristo, M., Golgher, P.B., Silva de Moura, E.: Impedance coupling in content-targeted advertising, Salvador, Brazil, pp. 496–503. ACM (2005)
16. Santos, R.L., Macdonald, C., Ounis, I.: Exploiting query reformulations for web search result diversification, Raleigh, North Carolina, USA, pp. 881–890. ACM (2010)
17. The Economist: The data deluge (2010)

18. Witten, I.H., Frank, E., Hall, M.A.: Data Mining: Practical Machine Learning Tools and Techniques, 3rd edn. Morgan Kaufmann Publishers Inc., San Francisco(CA) (2011)
19. Yih, W.-T., Goodman, J., Carvalho, V.R.: Finding advertising keywords on web pages, Edinburgh, Scotland, pp. 213–222. ACM (2006)
20. Zhang, Y., Surendran, A.C., Platt, J.C., Narasimhan, M.: Learning from multi-topic web documents for contextual advertisement, Las Vegas, Nevada, USA, pp. 1051–1059. ACM (2008)
21. Ziegler, C.-N., McNee, S.M., Konstan, J.A., Lausen, G.: Improving recommendation lists through topic diversification, Chiba, Japan, pp. 22–32. ACM (2005)

ILP Characterization of 3D Protein-Binding Sites and FCA-Based Interpretation

Emmanuel Bresso[1,2], Renaud Grisoni[2], Marie-Dominique Devignes[1,2,3],
Amedeo Napoli[1,2,3], and Malika Smail-Tabbone[1,2]

[1] Université de Lorraine, LORIA, UMR 7503, Vandoeuvre-les-Nancy, F-54506, France
[2] Inria, Villers-lès-Nancy, F-54600, France
[3] CNRS, LORIA, UMR 7503, Vandoeuvre-les-Nancy, F-54506, France
{bressoem,rgrisoni,devignes,napoli,malika}@loria.fr

Abstract. Life sciences are continuously producing large amounts of complex data that require relational learning to facilitate knowledge discovery. Inductive Logic Programming (ILP) is a powerful method which allows expressive representation of the data and produces explicit knowledge. However, ILP systems return variable theories depending on heuristic user-choices of various parameters and may miss potentially relevant rules. Accordingly, we propose an original approach based on post-ILP propositionalization of the examples and Formal Concept Analysis for effective interpretation of reached rules with the possibility of adding domain knowledge. Our approach is applied to the characterization of three-dimensional (3D) protein-binding sites which are protein portions on which interactions with other proteins take place. We define a relational representation of protein 3D patches and formalize the problem as a concept learning problem using ILP. We report here the results we obtained on particular protein-binding sites namely phosphorylation sites using ILP followed by FCA-based interpretation.

Keywords: Inductive Logic Programming, Formal Concept Analysis, Knowledge Discovery, Propositionalization, 3D Protein Binding Sites.

1   Introduction and Motivation

Relational or logical learning is a well established method in knowledge discovery, especially for complex application domains [1]. Life sciences provide a wide variety of such applications. In our work we investigate how relational learning can contribute to the understanding of protein-protein interactions which are important for most cellular processes. Great effort has been put into both experimental and computational methods to identify or predict protein-protein interactions. In protein docking, geometric and steric considerations are used to fit two protein structures into a bound complex [2]. Alternative computational methods predict bindings between pairs of proteins based either on their homology with known binding pairs of proteins or on integrated data from a wide variety of sources [3-5]. However, despite the large number of reported computational methods, precise characterization of protein-protein

A. Fred et al. (Eds.): IC3K 2012, CCIS 415, pp. 84–100, 2013.
© Springer-Verlag Berlin Heidelberg 2013

interaction sites is still challenging. Such sites are called hereafter Protein-Binding Sites (PBS) to make a clear distinction from ligand-binding sites on protein surface.

In fact most reported methods for structure-based prediction of protein-protein interactions apply on a single data table where each PBS is described by a set of descriptors or attributes including diverse physico-chemical properties aggregated on the whole binding site such as the residue composition, hydrophobicity, accessible surface area [6,7]. However, this data model prevents from representing individual properties of the PBS components (accessible surface of a particular residue) or spatial relations between components (e.g., distance between two residues). Hence, more expressive languages are necessary to represent the structural aspect of 3D interaction sites. Moreover, most current methods do not provide explicit characterization of PBS along with the prediction model. For instance, methods based on Support Vector Machines (SVM) act as black-boxes returning predictions with respect to inputs without explanation [7]. Explicit characterization of PBS would obviously provide good insights of the underlying biological phenomena.

In this context our aim is to exploit the growing set of available protein 3D structures for characterizing PBS and go beyond the limitations of the most current approaches qualified as black-box and single-table. We propose, as a first contribution, to apply Inductive Logic Programming (ILP) on a logical representation of protein 3D patches corresponding to positive or negative examples of PBS in order to induce a general definition of the PBS concept. Although ILP is a powerful method which allows expressive representation of the data and produces explicit knowledge, ILP systems return variable theories depending on heuristic user-choices of various parameters and may miss potentially relevant rules. Consequently, we propose as a second contribution an approach using Formal Concept Analysis for effective interpretation of reached ILP rules with the possibility of adding domain knowledge.

For a first validation of our approach, we choose a specific group of PBS, the phosphorylation sites. Indeed, phosphorylation is an important biological process and phosphorylation sites are exhaustively listed in a unique data source [14]. The rest of the paper is organized as follows. Section 2 introduces the methods we use (ILP and FCA). Section 3 describes the knowledge discovery problem. Section 4 describes our proposal for theory interpretation using FCA and domain knowledge. Section 5 summarizes the results obtained with our application on PBS. We discuss our results and describe related work in Section 6.

2   Methods

2.1   Inductive Logic Programming (ILP)

ILP allows learning a concept definition from observations, i.e., a set of positive examples (E+) and a set of negative examples (E-), and background knowledge (B) [8]. Given E+, E-, and B, the goal of ILP is to induce a set of rules or a theory T that is consistent (T∪B covers or explains each positive example in E+), and complete (T∪B does not cover any negative example in E-).

In most ILP systems both B and T are represented as definite clauses (or prolog programs) in First-Order Logic (FOL), i.e., a disjunction of literals with one positive literal. A rule has the form "head :-body" and is interpreted as: if the conditions in the body are true then the head is true as a logical consequence. The background knowledge B includes (i) the relational description of the examples using a set of relevant n-ary predicates and (ii) a priori domain knowledge, i.e., a set of rules and facts which don't refer to any example but express what is known about the elements which describe the examples. The theory T is a set of rules which cover as many of the positive examples as possible and the fewest negative examples. The head of each rule is the concept to learn whereas the body is a conjunction of literals and represents the induced description of the concept by generalization of examples. The rule search is performed in a clause space where the clause subsumption allows building generalizations or specializations of the clauses [9]. As the clause space is too large to be exhaustively explored, heuristic mechanisms exist to reduce its size and make the induction process feasible. These mechanisms allow the user to define which kind of rules she wants to get. This learning bias is defined by setting some parameters which orientate the search and often lead to a specific set of rules (theory). Hence, albeit ILP systems are relevant for knowledge discovery from complex data, they can return variable theories depending on heuristic user-choices of various parameters and may miss potentially relevant rules [10].

2.2   Aleph Program and Parameters

The experiments reported in this paper were conducted with the Aleph ILP program whose basic algorithm is described in four steps [18]:

1. Select a seed example to be generalized. If none exists, stop.
2. Construct the most specific clause that entails the example selected, and is compliant with the language restrictions provided. This is usually a definite clause with many literals, and is called the "bottom clause".
3. Find a clause more general than the bottom clause. This is done by searching for some subset of the literals in the bottom clause that has the "best" evaluation score.
4. The clause with the best score is added as a rule to the current theory, and all examples made redundant are removed. Return to Step 1.

Many parameters can be set for tuning some aspect of the theory construction with Aleph and reducing the size of space search. For instance, the rule evaluation function can be chosen and the default one is based on the difference between the number of covered positive examples and the number of covered negative examples. The *noise* parameter is the maximum negative examples that an acceptable rule may cover (default value is 0). This parameter can be set to higher values in case of noisy data. The *min-pos* parameter is the minimal number of positive examples that a rule must cover (default value is 1). Aleph requires other learning *bias* to be defined as (i) a set of determinations, defining the predicate to learn and the predicates which can appear in the rules; (ii) a set of modes, defining the types of predicate arguments and the way they can be chained in a rule.

As the above algorithm suggests, Aleph iterates on the positive examples of the learning set for building the most specific clause of a chosen seed example which is compliant with the defined *bias* and covers the maximum number of positive examples and the minimum number of negative ones. When finding the best rule, the examples covered by the best rule may be removed or not from the seed set and/or from the learning set (used for the rule evaluation). Hence with regard to this removal step, a *induce-type* parameter defines three ways of theory construction, (i) *induce* (covered examples are removed from both the seed and learning sets), (ii) *induce-cover* (covered examples are removed from the seed set and not from the learning set), and (iii) *induce-max* (covered examples are removed neither from the seed set nor from the learning set). Consequently, both *induce* and *induce-cover* are sensitive to the order in which the seed examples are presented contrasting with *induce-max* (each example is generalized). Both *induce-cover* and *induce-max* ways of theory construction produce more overlapping rules than the *induce* one.

2.3   Formal Concept Analysis (FCA)

The framework of FCA is fully detailed in [12]. FCA starts with a formal context (G,M,I) where G denotes a set of objects, M a set of attributes, and $I \subseteq G \times M$ a binary relation between G and M. The statement $(g,m) \in I$ is interpreted as "the object g has attribute m" (also noted gIm). Two operators (.)' define a Galois connection between the powersets (2G,$\subseteq$) and (2M, $\subseteq$), with $A \subseteq G$ and $B \subseteq M$:

$$A' = \{m \in M \mid \forall g \in A : gIm\} \text{ and}$$
$$B' = \{g \in G \mid \forall m \in B : gIm\}$$

For $A \subseteq G$, $B \subseteq M$, a pair (A,B), such that $A' = B$ and $B' = A$, is called a formal concept. In (A,B), the set A is called the extent and the set B the intent of the concept (A,B). Formal concepts are partially ordered by the concept subsumption:

$$(A1, B1) \leq (A2,B2) \Leftrightarrow A1 \subseteq A2 \Leftrightarrow B2 \subseteq B1$$

With respect to this partial order, the set of all formal concepts forms a full lattice called the concept lattice of (G,M,I).The FCA implementation used in our experiments was the one of the Coron data mining platform [19].

3   Characterizing 3D Protein-Binding Sites as ILP Concept Learning

3.1   Protein 3D Patch Definition

Protein surface patches were first introduced by Jones and Thornton [6] who defined a surface patch as a central surface accessible residue (amino acid) along with the nearest surface accessible neighbours. For our part, we define a protein 3D patch as a spherical fragment of a protein 3D structure centered on a selected residue of the protein similarly to [13]. A 3D patch has a radius *r* corresponding to the sphere radius and the residues composing the patch are those having an atom whose distance to the central residue does not exceed r. The RCSB PDB database (www.pdb.org) stores the resolved 3D structures of proteins [15].

3.2 Protein 3D Patch Descriptors

We propose to describe a 3D patch at two levels: (i) the patch is characterized as a whole by a set of global descriptors such as patch solvent accessible surface area (ASA), the number of carbon atoms occurring in the patch, and the number of residues in the patch; (ii) the patch composition and structure are characterized by a set of descriptors describing secondary and tertiary structure information on the patch residues. At the latter level, each residue of the patch is described by its name and its relative position in the primary sequence of the protein with respect to the central residue of the patch. The ASA value of each residue is used as a local descriptor of the residue in the patch. Two descriptors indicate if a given residue is on a helix, respectively on a sheet. Finally, one descriptor represents the spatial distance between each patch residue and the central residue. This distance information may play an important role in the interaction building. Figure 1 shows an example of a protein 3D patch. The variety of relationships between the elements of a patch clearly requires a relational or logical representation language as a feature-based language could only represent the global descriptors of the patches.

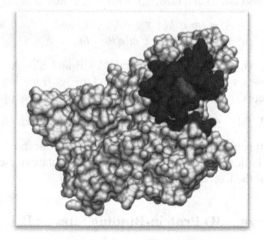

Fig. 1. Visualization of a protein (PDB ID: 1opk) 3D patch. The central residue of the patch is represented in red. The patch surface accessible to the solvent is represented in blue.

3.3 Characterization of 3D PBS as a Relational Learning Problem

Our objective is to learn with ILP a definition of the Protein-Binding Site concept, given relational descriptions of positive and negative examples of this concept.

We define a set of first-order logic predicates relevant for the ILP problem (Table 1). The unary predicate "pbs" is the concept to learn and has a 3D patch identifier as argument. Several binary predicates represent the global descriptors of the patch such as the predicates "p_asa" and "p_c" which correspond respectively to the ASA value of a patch and the number of carbon atoms. A set of ternary and 4-ary predicates represent the structural descriptors of the 3D patches. One 4-ary predicate is "p_r_distance" which represents the distance value of a patch residue to the central residue. One ternary predicate is "p_r_helix" which expresses that a patch residue belongs to a helix. A supplementary ternary predicate is "p_r_surface" which is derived from the "p_r_asa" predicate. Finally, a prolog rule allows to infer that a residue is on the patch surface if its ASA value is greater than 10Å^2:

```
p_r_surface (pid,res,pos) :-  p_r_asa (pid,res,pos,v),
                  greater_than (v, 10).
```

As mentioned before, ILP allows using domain knowledge during the induction process. Classifications of residues in several classes exist, reflecting shared physicochemical properties which may play a role in protein-protein interfaces [6]. We choose to use two classifications as a priori domain knowledge (Table 2). Consequently, we define a unary predicate for each residue class (e.g., *acidic*, *basic*) whose interpretation is defined according to the class membership (e.g., *acidic*(asp), *basic*(arg)).

To sum up, global and structural descriptors are computed for a set of 3D patches and represented with respect to the defined FOL predicates, forming a learning set. A ILP program can then be used to learn FOL rules characterizing or covering subsets of positive patches.

Table 1. First order-logic predicates to describe protein 3D patches

Predicate	Interpretation
pbs (pid)	The patch identified by pid is a protein-binding site
p_asa (pid,v)	v is the solvent accessible surface value of patch 'pid'
p_c (pid,n), p_o (pid,n), p_n (pid,n), p_s (pid,n)	n is the number of carbon/oxygen/nitrogen/sulphur atoms in patch 'pid'
p_r (pid,res,pos)	res is the name of the residue at relative position pos in patch 'pid'
p_r_asa (pid,res,pos,v)	v is the ASA value of the residue named res at the relative position pos (in the primary sequence) in patch 'pid'
p_r_distance (pid,res,pos,v)	v is the spatial distance between the residue (res, pos) and the central residue in patch 'pid'
p_r_helix (pid,res,pos), p_r_sheet (pid,res,pos)	the residue (res, pos) is on a helix/sheet in patch 'pid'

Table 2. Physico-chemical classes of residues defined in [16, 17]

Class Name	Residues in the Class
Acidic	asp, glu
Basic	arg., hais, lys
Aromatic	phe, trp, yr
Amide	asn, gln
small hydroxyl	ser, thr
sulphur containing	cys, met
aliphatic1	ala, gly, pro
aliphatic2	ile, leu, val
Aliphatic	ala, gly, pro, ile, leu, val
Small	gly,ala,ser,cys,thr,pro,asp
Medium	asn, val, glu, gln, ile, leu
Large	met, his, lys, phe, arg, tyr, trp
low polarizability	gly, ala, ser, asp, thr
medium polarizability	cys, pro, asn, val, glu, gln, ile, leu
high polarizability	lys, met, his, phe, arg, tyr, trp
hydrophobic	cys, val, leu, ile, met, phe, trp

4   FCA-Based Interpretation of a ILP Theory

Different theories are reached by the ILP program depending on the values of the program parameters set by the user. It is then important to allow the domain expert to explore each theory. Each theory is globally evaluated by the learning set coverage, *i.e.,* the proportion of positive examples that are covered by at least one rule of the theory (theory coverage). Another global criterion is the coverage of the best rule. Furthermore, we propose a FCA-based analysis of the ILP learning results including or not domain knowledge to help the expert in the interpretation task. This can possibly lead to one or more preferred theories.

Prior to further analyses, we perform a post-ILP propositionalization [29] for describing our 514 positive examples by the rules of a theory. Each rule is considered here as a boolean feature (in fact, only the body part of a rule is necessary). Table 3 shows an example of a formal context with 8 patches and 3 rules (for instance, patch_1 satisfies two rules, rule_1 and rule_3). Formal Concept Analysis can then be applied on the binary *patch x rule* matrix for effective interpretation of the theory rules.

A first interpretation mechanism of a theory is defined on the concept lattice issued from FCA applied to the propositional representation of positive examples. In our case, a formal concept gathers a subgroup of protein 3D patches which are covered by a conjunction of a set of rules. Figure 2 shows the concept lattice obtained from the context presented in Table 3. The concept lattice forms a good navigation support in the ILP results allowing the expert to move from a set of patches covered by one rule

Table 3. Formal context example

	rule_1	rule_2	rule_3
patch_1	x		x
patch_2			
patch_3	x	x	
patch_4	x	x	x
patch_5		x	
patch_6	x		x
patch_7		x	x
patch_8			

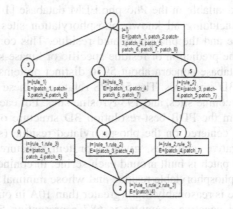

Fig. 2. The concept lattice obtained by FCA on Table 3. Each numbered concept has intent (I) and an extent (E). Edges correspond to subsumption relationships.

to more specific concepts containing subsets of the patches covered by two rules, three rules and so on. It is relevant, for each theory, to count the number of concepts having more than two rules (multiple-rule concepts) and a significant extent size (number of patches). Furthermore, multiple-rule concepts can be examined by the domain expert to see if any concept intent (rule conjunction) provides a relevant description of the corresponding patch subgroup.

A second interpretation mechanism of a theory is defined on the results of FCA applied to the propositional representation of positive examples enriched with supplementary domain properties of the examples that were not used in the learning process. This allows the expert to analyze how the theory rules are related to known properties of the examples. More precisely, interesting concepts to look at are those whose intent includes specific properties along with rules and whose extent has a significant size. These formal concepts gather 3D patches which share domain properties and covered by the same rule(s). These rule(s) can thus be associated with the domain properties found in the same formal concept.

5   Results

As a first validation of our approach, we apply it to a particular category of protein-binding sites, i.e., phosphorylation sites. We first present the learning set we established (Section 5.1) and then the results of the FCA-based interpretation of 18 theories produced with the Aleph program (Section 5.2).

5.1   The Learning Dataset of the Case Study

Phosphorylation is a reversible post-translation modification of a protein due to a protein kinase adding a phosphate group to a serine, threonine, or tyrosine residue (so called phosphorylated residue). A set of experimentally verified eukaryotic Phosphorylation Sites (PS) is available in the Phospho.ELM database [14]. In this study, we build a learning set including all known phosphorylation sites irrespective of the phosphorylating kinase and the phosphorylated residue. This contrasts with the systems which perform the prediction of residue-specific or kinase-specific PS [20, 21]. The Phospho.ELM database concern about 5976 distinct proteins among which only 286 have exploitable 3D structures. The 514 PS relative to these proteins divide into 231 serine sites, 90 threonine sites, and 193 tyrosine sites. For each PS we extract one positive 3D patch from the PDB best-resolution 3D structure of the corresponding protein. This patch is centered on the phosphorylated residue (serine, threonine, or tyrosine). As for negative examples, given a protein structure for which PS are known, a negative 3D patch is built around each serine, threonine or tyrosine residue which is not listed as phosphorylable residue and whose minimal spatial distance to a phosphorylable residue is reasonably high (greater than 10Å in our experiments). The total number of found negative examples is 687 representing 57% of the learning dataset. The proposed descriptors for protein patches are computed on the basis of the 3D coordinates of the patch atoms extracted from the PDB structures. The numeric descriptor values were discretized by equal-frequency binning.

Our objective is to characterize 3D PBS rather than predicting them, especially since there exist for phosphorylation sites sequence-based prediction programs which exhibit good performances [20, 21]. Besides, the amount of available protein 3D structures is fairly smaller than the available protein sequences making it difficult to reach the prediction accuracy of these programs. Nevertheless, the prediction accuracy of a ILP theory (defined as the number of true positives plus true negatives over the total number of examples) is worth considering when comparing various learning-set proportions in positive and negative examples, and comparing theories produced with different sets of parameters.

Hence theory prediction accuracy was computed by ten-fold cross-validation for variable percentages of negative examples in the learning set (induce-type is set to induce-cover and min-pos to 13). The results are given in Table 4 and confirm that the prediction accuracy continuously increases with the percentage of negative example. It is thus relevant to use the whole set of found negative 3D patches (i.e., 57% negative examples and 43% positive examples). To sum up, the dataset is composed of 514 positive examples and 687 negative examples of PBS which are described by their FOL descriptors as defined in Table 1.

Table 4. Prediction accuracy values (ten-fold cross validated) obtained when varying the negative example percentage in the learning set for two *noise* values (0 or 1)

Perc. neg. examples	*Noise* parameter	Prediction accuracy
35	0	52
35	1	54
40	0	54
40	1	55
45	0	54
45	1	55
50	0	55
50	1	57
55	0	58
55	1	58
57	0	59
57	1	61

5.2   FCA-Based Interpretation of ILP Theories

In our experiments, we consider three parameters which alter the way the rule search is seeded and the rules are selected in the Aleph program: induce-type (induce, induce-cover, induce-max), min-pos (from 8 to 12), and noise (0 or 1). The rest of the parameters were set to their default value. In order to reduce the number of theories to interpret, we used as theory ranking criterion the prediction accuracy value (ten-fold cross validation). We selected the 18 theories having a prediction value greater than 60%. In fact the accuracy values vary from 54 to 63%. As mentioned before the relatively weak values of prediction accuracy can be explained by the small size of the training set due to the small number of currently available 3D structures and its heterogeneity with respect to the phosphorylated residue and the phosphorylating kinase. The global criteria of the 18 selected theories are reported in Table 5 with the quantitative results of FCA-based interpretation.

We notice that the three ILP parameters have a combined effect on the theory size, the theory coverage and the best rule coverage as well as on the concept lattice composition. The most obvious trend is the continuous decrease of theory size, theory coverage and the number of multiple-rule concepts when increasing min-pos for a given induce-type and noise value. This makes the choice of an optimal configuration difficult because some trade-off must be found between these criteria. For instance, a small size is expected for a valuable theory but this should not be at the expense of its coverage.

To illustrate the effect of the *induce-type* parameter on the clause search, the theory reached with the induce/10/1 configuration is provided (Table 6.). One rule found in a

Table 5. Quantitative results of FCA applied on the *patch x rule* matrices for the 18 selected theories

Theory parameters (*induce-type/ min-po*s)	Theory size / best rule cov.	Theory coverage (%)	# multiple-rule concepts/ best concept extent size
noise=0			
ind-cov/9	64/ 15	56	106/ 9
ind-cov/10	37/ 16	39	71/ 14
ind-max /9	64/ 15	56	106/ 9
ind-max/10	39/ 16	39	88/ 14
noise=1			
induce/8	26/ 17	48	8/ 3
induce/9	19/ 17	39	5/ 3
induce /10	14/ 16	32	3/ 3
induce /11	10/ 15	24	0/0
ind-cov/10	85/ 18	73	215/ 13
ind-cov/11	58/ 18	61	144/ 16
ind-cov/12	38/ 18	48	91/ 13
ind-cov/13	20/ 18	30	50/ 13
ind-cov/14	15/ 18	23	31/ 16
ind-max/10	101/ 18	73	386/ 16
ind-max/11	69/ 18	61	254/ 16
ind-max/12	43/ 18	48	136/ 16
ind-max/13	24/ 18	30	80/ 16
ind-max/14	15/ 18	23	31/ 16

different configuration (induce-max/10/1) and missing in the previous theory is the following:

```
pbs(A) :- p_r_helix(A,B,3), high_polarizability(B), p_r(A,pro,1).
```

This rule covers 17 positive examples and expresses the fact that the residue at position +3 from central residue is on a helix and is highly polarizable, and also that a proline residue is present at position +1.

Browsing through the concept lattice is useful for exploring multiple-rule concepts in a post-ILP lattice. For instance, the theory with induce-max/12/1 produces a formal concept of 14 patches sharing the two following rules:

```
pbs(A) :- p_r_surface(A,B,0), p_r_helix(A,C,21), p_r_helix(A,D,28).
pbs(A) :- p_r_surface(A,B,0), pr_helix(A,C,17),p_r_helix(A,D,28).
```

When browsing up in the lattice, we find a subgroup composed of 11 patches which share the two previous rules besides the following third one:

Table 6. Rules of the theory *induce*/10/1

Rule n°	Rule body	Covered positive /negative examples
1.	p_o(A,29-inf), p_r_distance(A,B,-2,6-8), p_r_distance(A,C,-1,0-6), aliphatic(C).	16/1
2.	p_r_surface(A,B,0), p_r_helix(A,C,21), high_polarizability(C).	16/1
3.	p_r_surface(A,ser,0), p_r(A,pro,1), p_r_distance(A,B,-2,0-6), p_r_distance(A,C,2,0-6).	15/1
4.	p_r_helix(A,B,4), p_r_distance(A,C,6,8-10), p_c(A,105-inf).	15/0
5.	p_r_helix(A,ser,0), p_r(A,r,-3), p_r_distance(A,B,1,0-6), medium_polarizability(B).	15/0
6.	p_r(A,pro,1), p_r_distance(A,B,-3,8-10), p_r_distance(A,C,-1,0-6), medium_polarizability(C).	15/1
7.	p_r_helix(A,B,-2), p_r_distance(A,C,7,10-11), aliphatic(C).	14/1
8.	p_n(A,22-26), p_c(A,63-75).	14/1
9.	p_r(A,val,-1), p_r_distance(A,B,-4,0-6), p_r_distance(A,C,8,0-6).	14/1
10.	p_r_distance(A,B,-3,10-11), p_res(A,46-inf).	13/1
11.	p_r_helix(A,B,13), small(B), p_r_helix(A,C,17).	13/1
12.	p_r(A,arg,-2), p_r_distance(A,B,-1,0-6), basic(B).	13/1
13.	p_r_distance(A,B,2,6-8), p_r_distance(A,t,1,0-6).	13/1
14.	p_r(A,arg,-3), p_r_distance(A,B,3,8-10), medium_polarizability(B).	13/1
15.	p_n(A,26-inf), p_r_sheet(A,B,-4), p_res(A,46-inf). [12,0]	12/0
16.	p_r_helix(A,B,27), p_r_distance(A,C,9,13-inf).	12/1
17.	p_r_surface(A,B,0), p_r_helix(A,ser,-6), p_r_distance(A,C,2,0-6).	12/1
18.	p_r_distance(A,B,-10,8-10), high_polarizability(B), p_r_sheet(A,B,-10).	11/1

```
pbs(A) :- p_r_surface(A,B,0), p_r_helix(A,C,20), p_r_helix(A,D,28).
```

Hence, this subgroup of 11 patches is better characterized by the conjunction of the three rule bodies (after variable renaming and removal of redundant literals):

```
p_r_surface(A,B,0), p_r_helix(A,C,21), p_r_helix(A,D,28),
p_r_helix(A,E,17), p_r_helix(A,F,20).
```

In more general terms, selecting formal concepts with several rules constitutes an interesting way to achieve longer rules than in the theory issued by Aleph. Indeed, the compression heuristic used during the clause search leads most ILP programs to favor shorter clauses to longer ones. Finn et al. [22] proposed a solution to this drawback suitable to the pharmacophore search case.

An additional way to explore the concept lattices is to count how many multiple-rule concepts are enriched in one or the other type of patches, with respect to the nature of the phosphorylated residue (serine or threonine versus tyrosine). Over the 18 analysed lattices we observed that more than 75% of the multiple-rule concepts are either tyrosine specific or serine-or-threonine specific. This provides the expert with relevant descriptions for each type of phosphorylation sites as well as descriptions common to both types.

5.3   FCA-Based Interpretation of ILP Theories Including Domain Knowledge

We then applied FCA on the *patch x rule* table enriched with supplementary domain properties of the examples not used in the learning process, namely the phosphorylating kinase and the functional domain on which the 3D patch is located [23]. ILP produces a set of rules covering 3D patches and our interpretation procedure helps the expert to analyze how the rules are related to known properties of the examples. About 50 kinases phosphorylate more than 2 patches of the learning set whereas 40 Pfam domains are associated with more than 2 patches. The *patch x rule* table is thus updated with these kinases and functional domains as supplementary properties of the patches. Some formal concepts of the resulting lattice include at least one rule along with a kinase (respectively a Pfam domain) and display an extent containing more than 1/3 of the patches associated with the kinase (respectively the Pfam domain). The rules involved in such concepts can be examined by the expert in order to check their consistency with previous knowledge and whether they reveal novel relevant knowledge units regarding the kinase (respectively Pfam domain) concerned. For example the following rule was found associated with PKB kinases in concepts grouping the majority of 3D patches phosphorylated by this type of kinases:

```
pbs(A) :-   p_r_helix(A,B,-4), p_r_helix(A,arg,-3),
p_r_helix(A,ser,0).
```

In this rule the expert recognizes in particular that an arginine residue is present at position -3 which is a well-established observation for PKB kinases [23]. More precisely the second literal expresses that the arginine residue at -3 position should be on a helix, which reveals to be interesting for characterizing those 3D patches.

Quantitative analysis of domain-knowledge FCA was performed in order to help the expert assessing the relative relevance of a theory with respect to another one. To this aim, the number of kinases (respectively Pfam domains) appearing at least once in a concept with an extent containing at least 1/3 of the concerned patches was computed for each of the 18 theories selected in this study. The results are presented in Table 7. Interestingly, it appears that, in particular with the *induce-max* value of the *induce-type* parameter, the number of kinases does not continuously decrease with increasing values of the *min-pos* parameter. This contrasts with the continuous decrease of the theory coverage and size (Table 5). Thus even at low coverage values, the theory rules still exhibit descriptive ability correlated with domain knowledge.

Table 7. Number of kinases (respectively Pfam domains) appearing in concepts whose extent contains more than 1/3 of the patches associated to the kinase (respectively Pfam domain)

Theory parameters (*induce-type/ min-pos*)	Kinase number	Pfam domain number
noise=0		
induce-cover/9	10	10
induce-cover/10	6	8
induce-max /9	10	9
induce-max/10	6	10
noise=1		
induce/8	8	8
induce/9	9	9
induce /10	8	8
induce /11	4	6
induce-cover/10	12	11
induce-cover/11	7	10
induce-cover/12	7	7
induce-cover/13	8	4
induce-cover/14	8	2
induce-max/10	12	11
induce-max/11	11	7
induce-max/12	7	7
induce-max/13	8	4
induce-max/14	10	3

One can notice that such low-coverage theories would be discarded if the objective was prediction instead of characterization.

6   Related Work and Discussion

Because biological data exhibit rich relational structures as a consequence of the progress in integrative biology and high-throughput technologies, relational data mining methods were naturally applied on these data [25]. On the one hand, many feature-based methods were upgraded to handle relational data such as the search of frequent itemsets, association rule extraction, and Bayesian networks [28]. On the other hand, Inductive Logic Programming (ILP) was proposed by Muggleton as a logic-based approach for learning concepts from relational data [8]. ILP was successfully applied on various biological problems since then.

Turcotte et al. used ILP for predicting protein 3D structure [27]. Each protein domain is described by global features (e.g., length, number of helices), adjacency relationships between two consecutive 2D structure elements, and some local properties of the 2D structure elements (e.g., length, presence of some residue).

Another application was aimed at pharmacophore design for virtual screening purposes [22]. A pharmacophore is defined as an abstract 3D structure of a molecule that interacts with a protein target. In this case, pairwise distances between atoms or atom groups (e.g., hydrogen donors) of a set of interacting (respectively not interacting) molecules with a specific target are considered for learning.

More recently, 3D information on molecules was exploited for searching structurally diverse molecules (drugs) which share a biological activity [26]. Finally, ILP was applied on genomic annotations of proteins coming from public databases (e.g., Pfam, InterPro, PROSITE) in order to predict protein-protein interactions for one specific species [5].

To our knowledge, our study is the first which aims to characterize 3D proteinbinding sites using ILP except the very recently published work about ILP-based extraction of protein-ligand interaction features [29].

As for post-ILP analysis, our approach is innovative and represents a step forward in the interpretation of ILP results in the frame of a knowledge discovery process. It is worth noticing that our transformation of the examples representation is a propositionalization procedure as defined in [30] where body rules are used as boolean features of the examples. Analyzing this propositional representation of the examples offers to the expert an effective assistance when exploring the ILP results including confrontation with domain knowledge. By facilitating theory interpretation, our approach puts the tricky problem of heuristic parameter selection into perspective. Otherwise, investigation into the effect of numerous parameters on discriminative selection criteria is required as reported in [27].

Our approach can be extended in two ways. The first perspective concerns the scaling-up of ILP programs. Indeed, theories can be produced on distinct descriptors subsets reflecting distinct views on the examples, in the same spirit as Berthold's socalled parallel universes [31]. FCA-based interpretation of the resulting theories can then enable the discovery of concepts involving first-order features from distinct universes. The second extension is, beside FCA, to perform supervised classification (e.g. decision trees, classification rules) on the propositionalized data. In this case, one can run ILP on one or several universes with loose constraints (small value of *min-pos* and relatively high value of *noise* parameters) to get rules forming local patterns in the examples. The final supervised classification step is then comparable to global model construction as in the Lego framework [32].

References

1. De Raedt L.: Logical and Relational Learning. Springer (2008)
2. Smith, G., Sternberg, M.: Prediction of protein-protein interactions by docking methods. Current Opinion in Structural Biology 12(1), 28–35 (2002)
3. Aloy, P., Russell, R.: InterPreTS: Protein Interaction Prediction through Tertiary Structure. Bioinformatics Applications Note 19(1), 161–162 (2003)
4. Jansen, R., Yu, H., Greenbaum, D., Kluger, Y., Krogan, N.J., Chung, S., Emili, A., Snyder, M., Greenblatt, J.F., Gerstein, M.: A Bayesian networks approach for predicting proteinprotein interactions from genomic data. Science 302(5644), 449–453 (2003)

5. Tran, T.N., Satou, K., Ho, T.B.: Using Inductive Logic Programming for Predicting Protein-Protein Interactions from Multiple Genomic Data. In: Jorge, A.M., Torgo, L., Brazdil, P.B., Camacho, R., Gama, J. (eds.) PKDD 2005. LNCS (LNAI), vol. 3721, pp. 321–330. Springer, Heidelberg (2005)
6. Jones, S., Thornton, J.: Analysis of protein-protein interaction sites using surface patches. J. Mol. Biol. 272, 121–132 (1997)
7. Zhu, H., Domingues, F.S., Sommer, I., Lengauer, T.: NOXclass: prediction of protein-protein interaction types. BMC Bioinformatics 7, 27 (2006)
8. Muggleton, S.: Inductive Logic Programming. New Generation Computing 8(4), 295–318 (1991)
9. Muggleton, S., De Raedt, L.: Inductive Logic Programming: Theory And Methods. Journal of Logic Programming 19(20), 629–679 (1994)
10. Page, D., Srinivasan, A.: ILP: A Short Look Back and a Longer Look Forward. Journal of Machine Learning Research 4, 415–430 (2003)
11. King, R.: Logic, Automation, and the Future of Biology. In: Proceedings of the Spring School on Modelling Complex Biological Systems, Sophia-Antipolis, France (2011)
12. Ganter, B., Wille, R.: Formal concept analysis: Mathematical foundations. Springer, Heidelberg (1999)
13. Guharoy, M., Chakrabarti, P.: Conservation and relative importance of residues across protein-protein interfaces. PNAS 102(43), 15447–15452 (2005)
14. Diella, F., Gould, C.M., Chica, C., Via, A., Gibson, T.J.: Phospho.ELM: a database of phosphorylation sites. Nucleic Acids Res. 36(Database issue), D240-D244 (2008)
15. Berman, H.M., Westbrook, J., Feng, Z., Gilliland, G., Bhat, T.N., Weissig, H., Shindyalov, I.N., Bourne, P.E.: The Protein Data Bank. Nucleic Acids Research 28, 235–242 (2000)
16. Yu, C.S., Chen, Y.C., Lu, C.H., Hwang, J.K.: Prediction of protein subcellular localization. Proteins 64, 643–651 (2006)
17. Dubchak, I., Muchnik, I., Mayor, C., Dralyuk, I., Kim, S.-H.: Recognition of a protein fold in the context of the SCOP classification. Proteins: Structure, Function, and Genetics 35(4), 401–407 (1999)
18. Srinivasan, A.: The Aleph Manual (2007), http://www.comlab.ox.ac.uk/oucl/research/areas/machlearn/Aleph/
19. Szathmary, L.: Symbolic Data Mining Methods with the Coron Platform. PhD Thesis in Computer Science, Univ. Henri Poincaré – Nancy 1, France (2006)
20. Wong, Y., et al.: Kinasephos 2.0: A Web Server For Identifying Protein Kinase-Specific Phosphorylation Sites Based on Sequences and Coupling Patterns. Nucleic Acids Res. 35(Web Server issue), W588–W594 (2007)
21. Durek, P., Schudoma, C., Weckwerth, W., Selbig, J., Walther, D.: Detection and characterization of 3D-signature phosphorylation site motifs and their contribution towards improved phosphorylation site prediction in proteins. BMC Bioinformatics 10, 117 (2009)
22. Finn, P., Muggleton, S., Page, D., Srinivasan, A.: Pharmacophore Discovery Using the Inductive Logic Programming System PROGOL. Machine Learning 30(2-3), 241–273 (1998)
23. Punta, M., et al.: The Pfam protein families database. Nucleic Acids Research 40(Database Issue), D290–D301 (2012)
24. Obata, T., Yaffe, M.B., Leparc, G.G., Piro, E.T., Maegawa, H., Kashiwagi, A., Kikkawa, R., Cantley, L.C.: Peptide and protein library screening defines optimal substrate motifs for AKT/PKB. J. Biol. Chem. 275, 36108–36115 (2000)
25. Page, D., Craven, M.: Biological applications of multi-relational data mining. SIGKDD Explorations 5(1), 69–79 (2003)

26. Tsunoyama, K., Ata Amini, A., Sternberg, M., Muggleton, S.: Scaffold Hopping in Drug Discovery Using Inductive Logic Programming. Journal of Chemical Information and Modeling 48(5), 949–957 (2008)
27. Turcotte, M., Muggleton, S., Sternberg, M.: Automated discovery of structural signatures of protein fold and function. Journal of Molecular Biology 306(3), 591–605 (2001)
28. Dzeroski, S., Lavrac, N.: Relational Data Mining. Springer (2001)
29. Santos, J., Nassif, H., Page, D., Muggleton, S., Sternberg, M.: Automated identification of protein-ligand interaction features using Inductive Logic Programming: a hexose binding case study. BMC Bioinformatics 13, 162 (2012)
30. Kramer, S., Lavrac, N., Flach, P.: Propositionalization Approaches to Relational data Mining. In: Dzeroski, S., Lavrac, N. (eds.) Relational Data Mining. Springer (2001)
31. Berthold, M.R., Morik, K., Siebes, A. (eds.): Parallel universes and local patterns. Dagstuhl Seminar No. 07181 (2007)
32. Knobbe, A., Crémilleux, B., Fürnkranz, J., Scholz, M.: From Local Patterns to Global Models: The LeGo Approach to Data Mining. In: Proc. of the Int. Workshop From Local Patterns to Global Models co-located with ECML/PKDD 2008, Antwerp, Belgium, pp. 1–16 (2008)

Combination of Lexical and Structure-Based Similarity Measures to Match Ontologies Automatically

Thi Thuy Anh Nguyen and Stefan Conrad

Heinrich-Heine-University Düsseldorf, Institute of Computer Science,
Universitätsstr. 1, D-40225 Düsseldorf, Germany
{thuyanh,conrad}@cs.uni-duesseldorf.de
http://dbs.cs.uni-duesseldorf.de

Abstract. The great development of semantic web in the distributed environment leads to the different forms of ontologies. Therefore, ontology matching is an important task in order to share knowledge among applications more easily. In this paper, we propose an automatic ontology matching method by combining lexical and structure-based measures. A basic lexical similarity measure is applied to all pairs of concepts of two ontologies to achieve an initial matrix. With this matrix, we calculate the similarity between concepts based on a new structural similarity measure. Additionally, the structure-based matching method is improved by using a set of centroid concepts to reduce the computation time. We use I^3CON 2004 benchmark to evaluate the proposed method. The experimental results show that our measure has some prominent features for ontology matching.

Keywords: Ontology Matching, Structure-based Similarity, Centroid Concept.

1 Introduction

The speedy development of the web technology leads to an increase in knowledge sharing among web applications. However, these applications are difficult to interact because of using different tools and knowledge in a distributed system. Therefore, ontologies have been developed to express knowledge bases improving the understanding between applications. An ontology usually includes some kinds of entities, that are classes, relations, instances, data types and data values [7] to describe a specific domain.

Nowadays ontologies can be used to represent and store knowledge in many different application domains such as peer-to-peer information sharing, information integration, e-commerce, web service composition and so on; there are also a number of ontologies within the same subject. In fact, such ontologies are about the same area but they may use different concepts because they are developed by different communities independently. Therefore, ontology matching plays a crucial role of sharing knowledge and data. Ontology matching can be described as follows. Given two ontologies O_1 and O_2. Ontology matching is the process using parameters, external resources and a predefined alignment to find the correspondences between entities of one ontology to entities of another ontology.

Because of the different types of heterogeneity, several kinds of ontology matching strategies are distinguished. A detailed classification of ontology matching can be found

A. Fred et al. (Eds.): IC3K 2012, CCIS 415, pp. 101–112, 2013.

in [7]. In this survey, Euzenat J. and Shvaiko P. discuss two main techniques to match ontologies, which are the element-level techniques and structure-level techniques. The former may be based on strings, languages, constraint or external techniques to compute correspondences. In these techniques, considering the elements such as entities and instances of those entities is independent from their structure. Besides, these techniques ignore the relations to other entities. In string-based techniques, strings are considered chains of letters in an alphabet. The more similar the strings, the more likely they are to denote the same concept [7]. Language-based techniques determine the similarities of concepts based on natural language processing techniques. They consider the labels of concepts as words of a natural language. Constraint-based techniques consider internal constraints of entities, for example data types, properties and so on. External techniques use extrinsic resources such as Wordnet dictionary, predefined alignments and lexicons to find how the similarities between concepts are. In contrast, the latter considers the structures and relations among the entities of the ontologies to detect correspondences. Particularly, entities and their instances are considered together in a structure. These techniques can be divided into three categories named graph-based, taxonomy-based and model-based [21]. Graph-based techniques consider given ontologies as labeled graph structures. They use graph algorithms to compute similarities of pairs of nodes. Taxonomy-based techniques use also graph structures but they analyze only "is-a" relationships. Model-based techniques exploit the semantic interpretation of ontologies and apply techniques such as propositional satisfiability or description logics reasoning methods.

In this paper, we use a edit-distance measure which is a basic element-level technique, and a new structure similarity measure to find correspondences between the concepts of source and target ontologies. In particular, we use Levenshtein measure to begin ontology matching processing. Firstly, Levenshtein measure is applied to each pair of concepts of the ontologies to obtain a lexical similarity matrix representing their similarity. Secondly, we use the lexical matrix as an initial value for our structural similarity matrix which is based on a new similarity measure. Next, by using a threshold value, we get a set of matched pairs of concepts. Finally, our structure-based ontology matching technique is improved by using centroid concepts which were mentioned in [24].

The remainder of this paper is organized as follows. In Section 2, we present related work in the area. A description of our suggested measure is given in Section 3. In Section 4 we provide an example in order to illustrate our structure-based method. In Section 5 we describe the experimental results of applying our method. Finally, some conclusions are given in Section 6.

2   Related Work

Many ontology matching methods have been proposed to obtain alignments among entities of given ontologies. In this section, we briefly present some approaches related to lexical and structure-based techniques.

Lexical approaches are string-based methods for finding the correspondences in the input ontologies. In other words, the lexical methods consider the names of concepts as strings in order to align concepts. The lexical methods can be classified as prefix, suffix, n-gram, and edit distance. [21]

Prefix and suffix methods check two given strings whether the first string is the prefix or suffix of the second one. The approaches are useful for comparing strings and similar abbreviations. The n-gram measure is also useful for comparing strings. It calculates the number of common n-grams between two given strings. An edit distance between two objects is the minimal cost of operations to be applied to one of the objects in order to obtain the other one [7].

The structural techniques consider the position of the concepts and the relationships among ontologies. The intuition of these approaches is that two concepts of the ontologies are similar if their structures are similar.

In [19], the authors consider the matching of two ontologies as a maximum likelihood problem and resolve it by using the expectation-maximization technique. In particular, ontology schemas are modeled as directed graphs and then the structural, lexical and instance methods are applied to take a map between these graphs.

OLA [8] constructs graphs from the input ontologies and combines two techniques, that are concept-based and structure-based techniques, to match these graphs. However, in the alignment process, OLA only considers contributions of the neighbors in the same type.

The approach used in ASMOV [11] integrates lexical, structural and extensional methods for the calculation of similarities. It is similar to OLA, but it has more flexibility to the calculations for different features.

To match ontologies, RiMOM [23] applies some approaches such as lexical-based, constraint-based and linguistic-based approaches. Additional, it also uses a taxonomic technique. However, in this technique, it only considers the direct super-concepts and sub-concepts of a node.

Furthermore, structure-based methods usually calculate the similarity of two concepts in the ontologies based on parents, neighbors, children or leaves.

[4], [13] map structural similarities based on children and leaves. The authors assume two objects which are not leaves can be considered similar if their children and leaves are similar.

Moreover, the similarity between nodes can be found their relationships (object characters) as we can see in [5], [14].

Similarity flooding algorithm is established in [15]. The authors build PCGs and use structural characteristics for propagating similarities between elements on a directed labeled graph. This approach finds similarities in the graph structure. It spreads the similarity from similar nodes to their neighbors and back based on propagation coefficients.

VBOM approach [6] transfers the relationships and entities of the given ontologies into the space of vectors in N dimensions. Particularly, each entity contains a vector with the weights of its ancestors and successors classes. The similarity score between two concepts is computed by the cosine of the pair of these concept vectors.

[20] represents ontologies and schemas as graphs. Using these graphs can help to calculate the weighted value for each node on the graph using the lexical similarity of ancestors. However, this method only calculates the values up to the grandparent level.

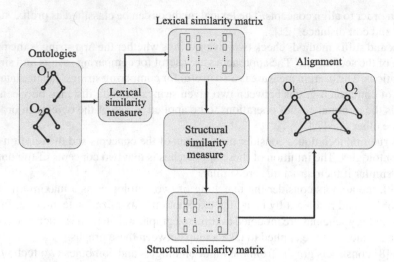

Fig. 1. A new structural measure

MLMA+ algorithm [3] and its improvement [1] are approaches using neighbor searching techniques to find the best method for matching.

[2] uses a structural approach for creating a grid around each node in the ontologies by linking it with its neighbors and the neighbor of its neighbors. After that, the similarity can be computed based on these grids.

In [22] the structural similarities of nodes are calculated by the similarities of pairs of parents as well as these of siblings. The authors proposed the DSI method which allows ancestors of a concept to identify that concept. We know that different ontologies have different characteristics. In case two ontologies do not have the same depth they can omit some pairs of nodes similarities because they only take care of pairs of ancestors at the same generation.

In [18] the authors analyze similar paths between a set of anchor matches to obtain new concept mapping. In other words, if two pairs of terms from the input ontologies are similar and there are paths connecting the terms then the entities in those paths are similar. This measure tries to find relationships between entities based on the primary relations recognized before.

In [24] the local similarities of nodes in the ontologies are established by the centroid concepts of two given ontologies through a partition technique. The similarity spreading method is used to get the similarities between concepts.

3   A New Structural Measure

In this section, we present a new structure-based similarity measure for matching two nodes of the given ontologies based on combining lexical and structural similarities. Firstly, two input ontologies are matched by using a basic lexical similarity measure (edit-distance) to obtain initial mappings. Based on that, we will use our structure-based similarity measure to find correspondences among the concepts of the ontologies.

In Figure 1 the input ontologies are described as graphs. The nodes in the graphs correspond to entities in ontologies and the edges represent relationships between nodes which are connected. The process of similarity calculation gives values between all pairs of concepts between c_i and c'_j, where c_i and c'_j belong to the ontologies O_1 and O_2, respectively. All these values are stored in a matrix. In this step an edit-distance similarity measure is used. Next, a new structure-based similarity metric is applied for calculating similarity of each pair of nodes. At that time, we obtain the similarity degrees between each c_i in O_1 and all the nodes c'_j in O_2. Based on these results and a threshold th, the alignment is finally obtained. This structure-based method allows us to determine the similarity degree of two concepts based on the combinations of similarity measures of these concepts and their ancestors. The details of this calculation are explained in the subsequent subsections.

3.1   Lexical Similarity

The lexical matching techniques consider similarities of the entities of the given ontologies by computing string similarities of the entities. There have been many lexical similarity methods proposed so far. However, in this paper we only focus on using Levenshtein measure to calculate the lexical similarity of two concepts.

In computer science, the Levenshtein measure [12] is used to determine the amount of differences between two strings. In particular, it computes the minimum number of operations needed to transform one string into another. Let c_i and c'_j are two arbitrary strings. Three types of operations are used including the substitution of a character of c_i by a character of c'_j, the deletion of a character of c_i or the insertion of a character of c'_j. The total cost of the operations used is equal to the sum of the costs of each of the operations.

The similarity measure for two strings (c_i, c'_j) is defined as follows:

$$Lex\_sim(c_i, c'_j) = \max \left(0, \frac{\min(|c_i|, |c'_j|) - ed(c_i, c'_j)}{\min(|c_i|, |c'_j|)} \right) . \qquad (1)$$

where $|c_i|$, $|c'_j|$ are lengths of strings c_i and c'_j, respectively; ed(c_i, c'_j) is Levenshtein measure.

Consider the following example.

Given two strings c_i='kitten' and c'_j='kitchen'. There are two edits transforming c_i into c'_j:

– the substitution of 't' with 'c'
– insert 'h' follow 'c'

Therefore, the Levenshtein distance between two strings 'kitten' and 'kitchen' is 2.

Applying the Eq. (1), the similarity between two strings 'kitten' and 'kitchen' is:

$$Lex\_sim(kitten, kitchen) = \max \left(0, \frac{\min(|kitten|, |kitchen|) - ed(kitten, kitchen)}{\min(|kitten|, |kitchen|)} \right)$$

$$= \max \left(0, \frac{6 - 2}{6} \right)$$

$$= 0.67$$

Now, we use Levenshtein measure for computing the similarities between any two concepts with one from each ontology. The similarity matrix is obtained representing the lexical similarities of all pairs of nodes in the given ontologies. With this similarity matrix, we compute the similarity between concepts based on our structural similarity measure as discussed in the following subsection.

3.2   Structural Similarity

In this section, we propose Structural_sim metric for calculating the similarities between concepts of the input ontologies with the lexical similarity measure introduced in the previous section.

Consider the two example ontologies in Figure 2.

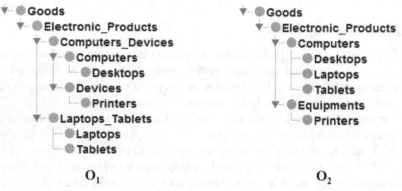

Fig. 2. The example ontologies

Intuitively, the nodes Goods, Electronic_Products, Desktops and Printers from ontology O_1 are similar to the nodes Goods, Electronic_Products, Desktops and Printers from ontology O_2, respectively.

Our measure is based on the idea in [22] about the similarity between two concepts. However, the proposed method is different from the DSI method in [22]. In the DSI method (Descendant's Similarity Inheritance), the authors concentrate on automatic structure-based technique to enhance the alignment results that were obtained from using the base similarity method, which is concept-based. Therefore, they point out not only the labels of the concepts, but also the positions of the concepts in the hierarchy. The main characteristic of the DSI method is that it allows for the parent and in general for any ancestor to play a role in the identification of the concept [22]. It means the DSI method is based on a structure-based technique in which it uses the relations between ancestors. However, these relations are considered at the same levels. The idea of our new metric is as follows: the similarity measure of any two concepts comes from the similarity of themselves and the contribution of their ancestors with their levels in the graphs can be different.

Relating the computations of the structural similarities, we can easily recognize the important point with our intuitions that the more similar the structure is, the more likely nodes are similar. Besides we assume that two concepts are similar if their ancestors are similar where the generations of ancestors can different.

Let m, n be the lengths of paths from the concepts c_i and c'_j in the ontologies O_1 and O_2 to the roots, respectively (where $c_i \in O_1$ and $c'_j \in O_2$); $lex\_sim(c_i, c'_j)$ stands for the lexical similarity of two concepts c_i and c'_j; and $parent_k(c_i)$, $parent_l(c'_j)$ for the kth, l^{th} concepts from the concepts c_i, c'_j to the roots of the ontologies O_1 and O_2, respectively. The similarity between the concept c_i in the source ontology O_1 and the concept c'_j in the target ontology O_2 is defined by the following equation:

$$Structural\_sim(c_i, c'_j) = \alpha * lex\_sim(c_i, c'_j)$$

$$+\beta * \left(\frac{\sum\limits_{k=1}^{n} (n+1-k)*\max\left(lex\_sim(parent_k(c_i),parent_{l=1..m}(c'_j))\right)}{n*(n+1)} + \frac{\sum\limits_{l=1}^{m} (m+1-l)*\max\left(lex\_sim(parent_l(c'_j),parent_{k=1..n}(c_i))\right)}{m*(m+1)} \right). \quad (2)$$

where the values of α and β satisfy the relationship: $\alpha + \beta = 1$.

Applying Eq. (2), when the algorithm finishes, we obtain a structural similarity matrix using the lexical similarity matrix as an initial matrix.

From Eq. (2) our similarity metric takes values from the interval [0, 1] which can be found in the literature.

The main feature of the approach is that the similarity of two concepts in the source and target ontologies is not only the similarity of themselves but also the contribution of their ancestors. However, the closer to the root of the ancestor, the smaller role the ancestor has. The factors (n+1-k) and (m+1-l) perform this rule. Moreover, the measure here is applied for the maximum similarity of pairs of ancestors in order to contribute the similarities.

3.3   Improving the Structure-Based Similarity Measure

In this study, we present the efficiency of the structure-based matching method by using a set of centroid concepts proposed in [24]. In this approach, the authors chose a set of centroid concepts from two input ontologies to partition these ontologies. In particular, each concept in the ontologies is represented by its descriptive information including its name and comment. After that, the authors apply the element-level techniques and the vector space model method to achieve the similarities between names and comments, respectively. Finally, the overall similarity measures between concepts result from the combination of component similarities. In case each entity in an ontology matches perfectly to one in another ontology, that entity is picked as a centroid concept. We used the same set of centroid concepts as in [24].

We calculate the structural similarities between nodes in ontologies as well as the ones of ancestors. The advantage of this improvement is that we do not need to consider all pairs of nodes from these nodes to roots. That means if there is any pair of centroid nodes on the path from the considered nodes to the roots, we only calculate the similarities of these nodes and their ancestors with the nearest pair of centroid concepts.

When applying Eq. (2), m and n are the lengths of the paths from c_i to the centroid node e_i of the ontology O_1 and c'_j to the centroid node e'_j of the ontology O_2, and $parent_k(c_i)$, $parent_l(c'_j)$ for the kth, l^{th} concepts from the concepts c_i, c'_j to the centroid concepts e_i, e'_j of the ontologies O_1 and O_2, respectively.

4 Illustrative Example

In this section, we describe an example to illustrate our similarity matching method proposed in the previous section.

Consider the two input ontologies shown in Figure 2.

By applying the lexical matching method for all pairs of nodes in the source and target ontologies with one from each of the two ontologies, we obtain the similarity values between them. The result of the lexical measure is shown below:

$$
Lex\_sim(O_1, O_2) = \begin{pmatrix}
1 & 0 & 0 & 0 & 0 & 0 & 0 & 0 \\
0 & 1 & 0 & 0 & 0 & 0 & 0 & 0 \\
0 & 0.06 & 0.11 & 0 & 0 & 0 & 0 & 0 \\
0 & 0 & 1 & 0.13 & 0.14 & 0 & 0.11 & 0.38 \\
0 & 0 & 0.13 & 1 & 0.43 & 0 & 0 & 0.25 \\
0 & 0 & 0 & 0.29 & 0.14 & 0.14 & 0 & 0.14 \\
0 & 0 & 0.38 & 0.25 & 0.14 & 0.14 & 0.13 & 1 \\
0 & 0.07 & 0 & 0 & 0 & 0 & 0 & 0 \\
0 & 0 & 0.14 & 0.43 & 1 & 0.29 & 0 & 0.14 \\
0 & 0 & 0 & 0 & 0.29 & 1 & 0 & 0.14
\end{pmatrix}
$$

At this time, the lexical similarity matrix is used to calculate the similarity values of all pairs of concepts based on their structures. Once applying the structure matching measure with Eq. (2), we get the similarities between concepts of the ontologies represented as follows:

$$
Structural\_sim(O_1, O_2) = \begin{pmatrix}
1 & 0 & 0 & 0 & 0 & 0 & 0 & 0 \\
0 & 1 & 0.23 & 0.20 & 0.20 & 0.20 & 0.23 & 0.20 \\
0 & 0.27 & 0.42 & 0.26 & 0.26 & 0.26 & 0.35 & 0.26 \\
0 & 0.20 & 0.92 & 0.28 & 0.29 & 0.19 & 0.34 & 0.43 \\
0 & 0.19 & 0.32 & 0.95 & 0.58 & 0.30 & 0.23 & 0.40 \\
0 & 0.20 & 0.27 & 0.38 & 0.29 & 0.29 & 0.27 & 0.27 \\
0 & 0.19 & 0.48 & 0.32 & 0.25 & 0.25 & 0.32 & 0.79 \\
0 & 0.28 & 0.35 & 0.26 & 0.26 & 0.26 & 0.35 & 0.26 \\
0 & 0.20 & 0.36 & 0.46 & 0.83 & 0.37 & 0.27 & 0.27 \\
0 & 0.20 & 0.27 & 0.18 & 0.37 & 0.83 & 0.27 & 0.27
\end{pmatrix}
$$

In case we choose the threshold value th to be 0.7, the matched pairs of concepts are obtained and given in Table 1.

By applying the centroid concepts algorithm, we selected a set of centroid concepts from both ontologies including Goods and Electronic_Products. Since Electronic_Products node is on the paths from all nodes to roots, we only need to calculate similarities of pairs of nodes and their ancestors to the node Electronic_Products.

The results are the same when applying both of the methods. However, omitting the roots "Goods" in the matching algorithm leads to reducing the computation time.

Table 1. The matched pairs of concepts applying the threshold value $th = 0.7$

No.	The matched concept in ontology O_1	The matched concept in ontology O_2
1	Goods	Goods
2	Electronic_Products	Electronic_Products
3	Computers	Computers
4	Desktops	Desktops
5	Printers	Printers
6	Laptops	Laptops
7	Tablets	Tablets

5 Experiments and Results

To test the performance of our structural similarity measure, we use five pairs of ontologies taken from the I^3CON 2004 [21] data set, that are: People and pets (without instances), Russia, Weapons, Networks, and CS. The parameter α in Eq. (2) is also chosen as in [17].

We compare our similarity with the DSI method and Similarity Flooding algorithm based on these four pairs, that are People and pets (without instances), Russia, Weapons, and Networks. We also give a comparison of our results with Avg. F-Measures which are the F-Measure average values of five participants executed four pairs of ontologies (People and pets (without instances), Russia, CS, and Networks) in the I^3CON 2004 benchmark. These participants include a research program of Lockheed Martin ATL, an algorithm from INRIA, Lexicon-based Ontology Mapping (LOM) from Teknowledge, Intelligent Agent Systems from AT&T, and Institut AIFB from University of Karlsruhe [10].

We evaluate the implementation of the structural similarity using the following classical metrics: precision, recall and F-measure.

$$Precision = \frac{No.\_correct\_found\_correspondences}{No.\_found\_correspondences},$$

$$Recall = \frac{No.\_correct\_found\_correspondences}{No.\_existing\_correspondences}, \quad (3)$$

$$F - measure = \frac{2 * Precision * Recall}{Precision + Recall}.$$

According to the results represented in [17], our measure has a better F-measure value than original DSI method generally. It means it is more effective than the DSI method.

In order to compare the performance of our measure with the other measures, we also show F-Measure values applying our method, the DSI measure, Similarity Flooding algorithm and measures of five participants based on five pairs of ontologies as in Table 2 [17,10].

Consider the results in Table 2. For the test case People and pets the ontologies contain a number of properties so the F-measure value of our measure is less than the

Table 2. F-Measure values of different methods for five pairs of ontologies

Ontologies	F-Measures			
	Avg.	DSI	Structural_sim	Similarity Flooding
People and pets	0.73	0.63	0.68	0.63
Weapons	N/A	0.87	0.95	0.86
Networks	0.66	0.45	0.49	0.69
Russia	0.39	0.57	0.57	0.54
CS	0.27	N/A	0.45	N/A

average F-measure of five participants. However, our result is better than these of two other - the DSI and Similarity Flooding methods. The test case Weapons has good structural and lexical characteristics so applying our metric yields a good result compared to other ones. Since the labels of the nodes in the test case Networks contain some omitted words so our method is not so good in comparison with the Similarity Flooding method as well the average F-measure. Matching pairs of ontologies CS and Russia is difficult because these ontologies have big differences in their structures. In particular, the CS ontologies include 109 concepts and 52 edges, 20 concepts and 7 edges respectively, while the Russia ontologies contain internal structures and the labels of the concepts are very different [19]. That leads to not good implementations of the approaches. Moreover, the test case Russia includes a lot of properties and instances so we have not got good result. However, even in this case, the F-measure applying our method is one of the best values. The ontology pairs CS are not available for the DSI and Similarity Flooding methods. The F-measures of the other approaches are compared to our F-measure value. The results show that our measure is also the best one.

6   Conclusions

In this paper, we proposed a new measure to find correspondences of concepts of two input ontologies based on the combination of lexical and structural similarities. We used a basic lexical measure for computing the lexical similarities among nodes. These similarities are used as initial values for estimating structural similarities. The structure-based similarity value of two nodes is the contribution of the lexical similarity of the concept names and similarities of their descendants. With the proposed similarity measure, a structural matrix describing the similarities of all pairs of nodes is received.

The other methods based on structure usually focus on the similarities of neighbor nodes such as parents, grandparents, children and siblings. The important aspect of our structural approach is that the similarity of a pair of nodes depends not only on their similarity but also on the similarity of all possible pairs of their ancestors. Unlike the approach in [22], our method does not omit any pairs in order to deal with the situation that the ontologies do not have the same structure.

For the improvement of the structural matching, we calculate the similarities between two nodes based on similarities of their descendant to the nearest centroid nodes instead of roots. This implementation leads to reducing the calculation.

We performed our metric on the I^3CON 2004 [21] data set. We compared our similarity measure to the DSI method and Similarity Flooding algorithm as well as measures of five participants. The experimental results showed that our approach obtains some prominent features comparing to ontology matching algorithms and is effective in case the given ontologies are very different in their structures.

In the future research, we are going to integrate our approach with linguistic matching using WordNet dictionary in order to increase the accuracy of the lexical matching and to match the relations of the given ontologies. With this method we can match ontologies based on weighted graphs using the properties as the weights.

References

1. Akbari, I., Fathian, M., Badie, K.: An Improved MLMA+ Algorithm and its Application in Ontology Matching. In: Innovative Technologies in Intelligent Systems and Industrial Applications (CITISIA), pp. 56–60. IEEE Xplore (2009)
2. Akbari, I., Fathian, M.: A Novel Algorithm for Ontology Matching. Journal of Information Science 36(3), 324–334 (2010)
3. Alasoud, A., Haarslev, V., Shiri, N.: An Empirical Comparison of Ontology Matching Techniques. Journal of Information Science 35(4), 379–397 (2009)
4. Do, H.H., Rahm, E.: COMA - A System for Flexible Combination of Schema Matching Approaches. In: 28th International Conference on Very Large Data Bases, pp. 610–621. ACM (2002)
5. Dong, X., Madhavan, J., Halevy, A.Y.: Mining Structures for Semantics. SIGKDD Explorations Newsletter 6(2), 53–60 (2004)
6. Eidoon, Z., Yazdani, N., Oroumchian, F.: A Vector based Method of Ontology Matching. In: 3rd International Conference on Semantics, Knowledge and Grid (SKG), pp. 378–381. IEEE Xplore (2007)
7. Euzenat, J., Shvaiko, P.: Ontology Matching. Springer, Leipzig (2007)
8. Euzenat, J., Valtchev, P.: Similarity-Based Ontology Alignment in OWL-Lite. In: European Conference on Artificial Intelligence (ECAI), pp. 333–337. IOS Press (2004)
9. Hanif Seddiqui, M., Aono, M.: An Efficient and Scalable Algorithm for Segmented Alignment of Ontologies of Arbitrary Size. Journal of Web Semantics: Science, Services and Agents on the World Wide Web, 344–356 (2009)
10. Hughes, T., Ashpole, B.: Information Interpretation and Integration Conference (2004), http://www.atl.external.lmco.com/projects/ontology/i3con.html
11. Jean-Mary, Y.R., et al.: Ontology Matching with Semantic Verification. Journal on Web Semantics: Science, Services and Agents on the World Wide Web 7(3), 235–251 (2009)
12. Levenshtein, V.I.: Binary Codes Capable of Correcting Deletions, Insertions, and Reversals. Soviet Physics Doklady 10, 707–710 (1966)
13. Madhavan, J., Bernstein, P., Rahm, E.: Generic Schema Matching with Cupid. In: 27th International Conference on Very Large Data Bases, pp. 49–58. Morgan Kaufmann, San Francisco (2001)
14. Maedche, A., Staab, S.: Measuring Similarity between Ontologies. In: 13th International Conference on Knowledge Engineering and Knowledge Management: Ontologies and the Semantic Web, pp. 251–263. Springer, London (2002)
15. Melnik, S., Garcia-Molina, H., Rahm, E.: Similarity Flooding: A Versatile Graph Matching Algorithm and its Application to Schema Matching. In: 18th International Conference on Data Engineering (ICDE), pp. 117–128. IEEE Xplore (2002)
16. Miller, G.A.: WordNet: A Lexical Database for English. Communications of the ACM 38, 39–41 (1995)

17. Nguyen, T.T.A., Conrad, S.: A New Structure-based Similarity Measure for Automatic Ontology Matching. In: 4th International Conference on Knowledge Discovery and Information Retrieval, pp. 443–449. SciTePress (2012)

18. Noy, N.F., Musen, M.A.: Anchor-PROMPT: Using Non-local Context for Semantic Matching. In: Workshop on Ontologies and Information Sharing at the 17th International Joint Conference on Artificial Intelligence, IJCAI (2001)

19. Prashant, D., Ravikanth, K., Christopher, T.: Inexact Matching of Ontology Graphs Using Expectation-Maximization. Journal on Web Semantics: Science, Services and Agents on the World Wide Web 7(2), 90–106 (2009)

20. Sharma, A.: Ontology Matching Using Weighted Graphs. In: 1st International Conference on Digital Information Management (ICDIM), pp. 121–124. IEEE Xplore (2006)

21. Shvaiko, P., Euzenat, J.: A Survey of Schema-Based Matching Approaches. In: Spaccapietra, S. (ed.) Journal on Data Semantics IV. LNCS, vol. 3730, pp. 146–171. Springer, Heidelberg (2005)

22. Sunna, W., Cruz, I.F.: Structure-Based Methods to Enhance Geospatial Ontology Alignment. In: Fonseca, F., Rodríguez, M.A., Levashkin, S. (eds.) GeoS 2007. LNCS, vol. 4853, pp. 82–97. Springer, Heidelberg (2007)

23. Tang, J., Li, J., Liang, B., Huang, X., Li, Y., Wang, K.: Using Bayesian Decision for Ontology Mapping. Journal on Web Semantics: Science, Services and Agents on the World Wide Web 4(4), 243–262 (2006)

24. Wang, Y., Liu, W., Bell, D.A.: A Structure-Based Similarity Spreading Approach for Ontology Matching. In: Deshpande, A., Hunter, A. (eds.) SUM 2010. LNCS, vol. 6379, pp. 361–374. Springer, Heidelberg (2010)

Toponym Extraction and Disambiguation Enhancement Using Loops of Feedback

Mena B. Habib and Maurice van Keulen

Faculty of EEMCS, University of Twente, Enschede, The Netherlands
{m.b.habib,m.vankeulen}@ewi.utwente.nl

Abstract. Toponym extraction and disambiguation have received much attention in recent years. Typical fields addressing these topics are information retrieval, natural language processing, and semantic web. This paper addresses two problems with toponym extraction and disambiguation. First, almost no existing works examine the extraction and disambiguation interdependency. Second, existing disambiguation techniques mostly take as input extracted named entities without considering the uncertainty and imperfection of the extraction process. In this paper we aim to investigate both avenues and to show that explicit handling of the uncertainty of annotation has much potential for making both extraction and disambiguation more robust. We conducted experiments with a set of holiday home descriptions with the aim to extract and disambiguate toponyms. We show that the extraction confidence probabilities are useful in enhancing the effectiveness of disambiguation. Reciprocally, retraining the extraction models with information automatically derived from the disambiguation results, improves the extraction models. This mutual reinforcement is shown to even have an effect after several automatic iterations.

Keywords: Toponyms Extraction, Toponym Disambiguation, Uncertain Annotations.

1 Introduction

Named entities are atomic elements in text belonging to predefined categories such as the names of persons, organizations, locations, expressions of times, quantities, monetary values, percentages, etc. Named entity extraction (a.k.a. named entity recognition) is a subtask of information extraction that seeks to locate and classify those elements in text. This process has become a basic step of many systems like Information Retrieval (*IR*), Question Answering (*QA*), and systems combining these, such as [1].

One major type of named entities is the toponym. In natural language, *toponyms* are names used to refer to locations without having to mention the actual geographic coordinates. The process of *toponym extraction* (a.k.a. toponym recognition) aims to identify location names in natural text. The extraction techniques fall into two categories: rule-based or based on supervised-learning.

Toponym disambiguation (a.k.a. toponym resolution) is the task of determining which real location is referred to by a certain instance of a name. Toponyms, as with named

A. Fred et al. (Eds.): IC3K 2012, CCIS 415, pp. 113–129, 2013.

entities in general, are highly ambiguous. For example, according to GeoNames[1], the toponym "Paris" refers to more than sixty different geographic places around the world besides the capital of France.

A general principle in our work is our conviction that Named entity extraction (NEE) and disambiguation (NED) are highly dependent. In previous work [2], we studied not only the positive and negative effect of the extraction process on the disambiguation process, but also the potential of using the result of disambiguation to improve extraction. We called this potential for mutual improvement, the *reinforcement effect*.

To examine the reinforcement effect, we conducted experiments on a collection of holiday home descriptions from the EuroCottage[2] portal. These descriptions contain general information about the holiday home including its location and its neighborhood (See Figure 3 for examples). As a representative example of toponym extraction and disambiguation, we focused on the task of extracting toponyms from the description and using them to infer the country where the holiday property is located.

In general, we concluded that many of the observed problems are caused by an improper treatment of the inherent ambiguities. Natural language has the innate property that it is multiply interpretable. Therefore, none of the processes in information extraction should be 'all-or-nothing'. In other words, all steps, including entity recognition, should produce *possible* alternatives with associated likelihoods and dependencies.

In this paper, we focus on this principle. We turned to statistical approaches for toponym extraction. The advantage of statistical techniques for extraction is that they provide alternatives for annotations along with confidence probabilities (confidence for short). Instead of discarding these, as is commonly done by selecting the top-most likely candidate, we use them to enrich the knowledge for disambiguation. The probabilities proved to be useful in enhancing the disambiguation process. We believe that there is much potential in making the inherent uncertainty in information extraction explicit in this way. For example, phrases like "Lake Como" and "Como" can be both extracted with different confidence. This restricts the negative effect of differences in naming conventions of the gazetteer on the disambiguation process.

Second, extraction models are inherently imperfect and generate imprecise confidence. We were able to use the disambiguation result to enhance the confidence of true toponyms and reduce the confidence of false positives. This enhancement of extraction improves as a consequence the disambiguation (the aforementioned reinforcement effect). This process can be repeated iteratively, without any human interference, as long as there is improvement in the extraction and disambiguation.

The rest of the paper is organized as follows. Section 2 presents related work on NEE and NED. Section 3 presents a problem analysis and our general approach to iterative improvement of toponym extraction and disambiguation based on uncertain annotations. The adaptations we made to toponym extraction and disambiguation techniques are described in Section 4. In Section 5, we describe the experimental setup, present its results, and discuss some observations and their consequences. Finally, conclusions and future work are presented in Section 6.

---

[1] www.geonames.org

[2] http://www.eurocottage.com

2   Related Work

NEE and NED are two areas of research that are well-covered in literature. Many approaches were developed for each. NEE research focuses on improving the quality of recognizing entity names in unstructured natural text. NED research focuses on improving the effectiveness of determining the actual entities these names refer to. As mentioned earlier, we focus on toponyms as a subcategory of named entities. Is this section, we briefly survey a few major approaches for toponym extraction and disambiguation.

2.1   Named Entity Extraction

NEE is a subtask of Information Extraction (IE) that aims to annotate phrases in text with its entity type such as names (e.g., person, organization or location name), or numeric expressions (e.g., time, date, money or percentage). The vast majority of proposed approaches for NEE fall in two categories: handmade rule-based systems and supervised learning-based systems.

One of the earliest rule-based system is FASTUS [3]. It is a nondeterministic finite state automaton text understanding system used for IE. In the first stage of its processing, names and other fixed form expressions are recognized by employing specialized microgrammars for short, multi-word fixed phrases and proper names. The idea behind supervised learning is to discover discriminative features of named entities by applying machine learning on positive and negative examples taken from large collections of annotated texts. The aim is to automatically generate rules that recognize instances of a certain category entity type based on their features. Supervised learning techniques applied in NEE include Hidden Markov Models (HMM) [4], Decision Trees [5], Maximum Entropy Models [6], Support Vector Machines [7], and Conditional Random Fields (CRF) [8][9].

Imprecision in information extraction is expected, especially in unstructured text where a lot of noise exists. Imprecision in information extraction can be represented by associating each extracted field with a probability value. Other methods extend this approach to output multiple possible extractions instead of a single extraction. It is easy to extend probabilistic models like HMM and CRF to return the k highest probability extractions instead of a single most likely one and store them in a probabilistic database [10].

2.2   Toponym Disambiguation

According to [11], there are different kinds of toponym ambiguity. One type is structural ambiguity, where the structure of the tokens forming the name are ambiguous (e.g., is the word "Lake" part of the toponym "Lake Como" or not?). Another type of ambiguity is semantic ambiguity, where the type of the entity being referred to is ambiguous (e.g., is "Paris" a toponym or a girl's name?). A third form of toponym ambiguity is reference ambiguity, where it is unclear to which of several alternatives the toponym actually refers (e.g., does "London" refer to "London, UK" or to "London, Ontario, Canada"?). In this work, we focus on the structural and the reference ambiguities.

Toponym reference disambiguation or resolution is a form of Word Sense Disambiguation (WSD). According to [12], existing methods for toponym disambiguation can be classified into three categories: (i) map-based: methods that use an explicit representation of places on a map; (ii) knowledge-based: methods that use external knowledge sources such as gazetteers, ontologies, or Wikipedia; and (iii) data-driven or supervised: methods that are based on machine learning techniques. An example of a map-based approach is [13], which aggregates all references for all toponyms in the text onto a grid with weights representing the number of times they appear.

Knowledge-based approaches are based on the hypothesis that toponyms appearing together in text are related to each other, and that this relation can be extracted from gazetteers and knowledge bases like Wikipedia.

Supervised learning approaches use machine learning techniques for disambiguation. [14] trained a naive Bayes classifier on toponyms with disambiguating cues such as "Nashville, Tennessee" or "Springfield, Massachusetts", and tested it on texts without these clues. Similarly, [15] used Hidden Markov Models to annotate toponyms and then applied Support Vector Machines to rank possible disambiguations.

In this paper, we chose to use HMM and CRF to build statistical models for extraction. We developed a clustering-based approach for the toponym disambiguation task. This is described in Section 4.

3   Problem Analysis and General Approach

The task we focus on is to extract toponyms from EuroCottage holiday home descriptions and use them to infer the country where the holiday property is located. We use this country inference task as a representative example of disambiguating extracted toponyms.

Our initial results from our previous work, where we developed a set of hand-coded grammar rules to extract toponyms, showed that effectiveness of disambiguation is affected by the effectiveness of extraction. We also proved the feasibility of a reverse influence, namely how the disambiguation result can be used to improve extraction by filtering out terms found to be highly ambiguous during disambiguation.

One major problem with the hand-coded grammar rules is its "All-or-nothing" behavior. One can only annotate either "Lake Como" or "Como", but not both. Furthermore, hand-coded rules don't provide extraction confidences which we believe to be useful for the disambiguation process. We therefore propose an entity extraction and disambiguation approach based on uncertain annotations. The general approach illustrated in Figure 3 has the following steps:

1. Prepare training data by manually annotating named entities (in our case toponyms) appearing in a subset of documents of sufficient size.
2. Use the training data to build a statistical extraction model.
3. Apply the extraction model on test data and training data. Note that we explicitly allow uncertain and alternative annotations with probabilities.
4. Match the extracted named entities against one or more gazetteers.
5. Use the toponym entity candidates for the disambiguation process (in our case we try to disambiguate the country of the holiday home description).

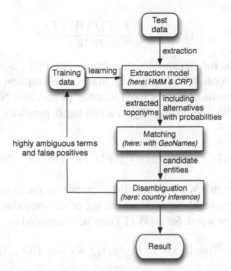

Fig. 1. General approach

6. Evaluate the extraction and disambiguation results for the training data and determine a list of highly ambiguous named entities and false positives that affect the disambiguation results. Use them to re-train the extraction model.
7. The steps from 2 to 6 are repeated automatically until there is no improvement any more in either the extraction or the disambiguation.

Note that the reason for including the training data in the process, is to be able to determine false positives in the result. From test data one cannot determine a term to be a false positive, but only to be highly ambiguous.

4   Our Approaches

In this section we illustrate the selected techniques for the extraction and disambiguation processes. We also present our adaptations to enhance the disambiguation by handling uncertainty and the imperfection in the extraction process, and how the extraction and disambiguation processes can reinforce each other iteratively.

4.1   Toponym Extraction

For toponym extraction, we trained two statistical named entity extraction modules[3], one based on Hidden Markov Models (HMM) and one based on Conditional Ramdom Fields (CRF).

HMM Extraction Module. The goal of HMM is to find the optimal tag sequence $T = t_1, t_2, ..., t_n$ for a given word sequence $W = w_1, w_2, ..., w_n$ that maximizes:

---

[3] We made use of the *lingpipe* toolkit for development: http://alias-i.com/lingpipe

$$P(T \mid W) = \frac{P(T)P(W \mid T)}{P(W)} \tag{1}$$

where $P(W)$ is the same for all candidate tag sequences. $P(T)$ is the probability of the named entity (NE) tag. It can be calculated by Markov assumption which states that the probability of a tag depends only on a fixed number of previous NE tags. Here, in this work, we used $n = 4$. So, the probability of a NE tag depends on three previous tags, and then we have,

$$P(T) = P(t_1) \times P(t_2|t_1) \times P(t_3|t_1, t_2) \times P(t_4|t_1, t_2, t_3) \times \ldots \times P(t_n|t_{n-3}, t_{n-2}, t_{n-1}) \tag{2}$$

As the relation between a word and its tag depends on the context of the word, the probability of the current word depends on the tag of the previous word and the tag to be assigned to the current word. So $P(W|T)$ can be calculated as:

$$P(W|T) = P(w_1|t_1) \times P(w_2|t_1, t_2) \times \ldots \times P(w_n|t_{n-1}, t_n) \tag{3}$$

The prior probability $P(t_i|t_{i-3}, t_{i-2}, t_{i-1})$ and the likelihood probability $P(w_i|t_i)$ can be estimated from training data. The optimal sequence of tags can be efficiently found using the Viterbi dynamic programming algorithm [16].

CRF Extraction Module. HMMs have difficulty with modeling overlapped, non-independent features of the output part-of-speech tag of the word, the surrounding words, and capitalization patterns. Conditional Random Fields (CRF) can model these overlapping, non-independent features [17]. Here we used a linear chain CRF, the simplest model of CRF.

A linear chain Conditional Random Field defines the conditional probability:

$$P(T \mid W) = \frac{\exp\left(\sum_{i=1}^{n}\sum_{j=1}^{m}\lambda_j f_j\left(t_{i-1}, t_i, W, i\right)\right)}{\sum_{t,w}\exp\left(\sum_{i=1}^{n}\sum_{j=1}^{m}\lambda_j f_j\left(t_{i-1}, t_i, W, i\right)\right)} \tag{4}$$

where f is set of m feature functions, λ_j is the weight for feature function f_j, and the denominator is a normalization factor that ensures the distribution p sums to 1. This normalization factor is called the *partition function*. The outer summation of the *partition function* is over the exponentially many possible assignments to t and w. For this reason, computing the *partition function* is intractable in general, but much work exists on how to approximate it [18].

The feature functions are the main components of CRF. The general form of a feature function is $f_j\left(t_{i-1}, t_i, W, i\right)$, which looks at tag sequence T, the input sequence W, and the current location in the sequence (i).

We used the following set of features for the previous w_{i-1}, the current w_i, and the next word w_{i+1}:

- The tag of the word.
- The position of the word in the sentence.
- The normalization of the word.

- The part of speech tag of the word.
- The shape of the word (Capitalization/Small state, Digits/Characters, etc.).
- The suffix and the prefix of the word.

An example for a feature function which produces a binary value for the current word shape is *Capitalized*:

$$f_i(t_{i-1}, t_i, W, i) = \begin{cases} 1 & \text{if } w_i \text{ is } Capitalized \\ 0 & otherwise \end{cases} \qquad (5)$$

The training process involves finding the optimal values for the parameters λ_j that maximize the conditional probability $P(T \mid W)$. The standard parameter learning approach is to compute the stochastic gradient descent of the *log* of the objective function:

$$\frac{\partial}{\partial \lambda_k} \sum_{i=1}^{n} \log p(t_i|w_i)) - \sum_{j=1}^{m} \frac{\lambda_j^2}{2\sigma^2} \qquad (6)$$

where the term $\sum_{j=1}^{m} \frac{\lambda_j^2}{2\sigma^2}$ is a Gaussian prior on λ to regularize the training. In our experiments we used the prior variance $\sigma^2 = 4$. The rest of the derivation for the gradient descent of the objective function can be found in [17].

Extraction Modes of Operation. We used the extraction models to retrieve sets of annotations in two ways:

- **First-Best.** In this method, we only consider the first most likely set of annotations that maximizes the probability $P(T \mid W)$ for the whole text. This method does not assign a probability for each individual annotation, but only to the whole retrieved set of annotations.
- **N-Best.** This method returns a top-N of possible alternative hypotheses in order of their estimated likelihoods $p(t_i|w_i)$. The confidence scores are assumed to be conditional probabilities of the annotation given an input token. A very low cut-off probability is additionally applied as well. In our experiments, we retrieved the top-25 possible annotations for each document with a cut-off probability of 0.1.

4.2 Toponym Disambiguation

For the toponym disambiguation task, we only select those toponyms annotated by the extraction models that match a reference in GeoNames. We furthermore use a clustering-based approach to disambiguate to which entity an extracted toponym actually refers.

The Clustering Approach. The clustering approach is an unsupervised disambiguation approach based on the assumption that toponyms appearing in same document are likely to refer to locations close to each other *distance-wise*. For our holiday home descriptions, it appears quite safe to assume this. For each toponym t_i, we have, in general, multiple entity candidates. Let $R(t_i) = \{r_{ix} \in \text{GeoNames gazetteer}\}$ be the set of reference candidates for toponym t_i. Additionally each reference r_{ix} in GeoNames belongs to a country $Country_j$. By taking one entity candidate for each toponym, we form a cluster. A cluster, hence, is a possible combination of entity candidates, or in other words, one possible entity candidate of the toponyms in the text. In this approach, we consider all possible clusters, compute the average distance between the candidate locations in the cluster, and choose the cluster $Cluster_{min}$ with the lowest average distance. We choose the most often occurring country in $Cluster_{min}$ for disambiguating the country of the document. In effect the abovementioned assumption states that the entities that belong to $Cluster_{min}$ are the true representative entities for the corresponding toponyms as they appeared in the text. Equations 7 through 11 show the steps of the described disambiguation procedure.

$$Clusters = \{\{r_{1x}, r_{2x}, \ldots, r_{mx}\} \mid \forall t_i \in d \bullet r_{ix} \in R(t_i)\} \qquad (7)$$

$$Cluster_{min} = \underset{Cluster_k \in Clusters}{\arg\min} \quad \text{average distance of } Cluster_k \qquad (8)$$

$$Countries_{min} = \{Country_j \mid r_{ix} \in Cluster_{min} \wedge r_{ix} \in Country_j\} \qquad (9)$$

$$Country_{winner} = \underset{Country_j \in Countries_{min}}{\arg\max} \quad freq(Country_j) \qquad (10)$$

where

$$freq(Country_j) = \sum_{i=1}^{n} \begin{cases} 1 & \text{if } r_{ix} \in Country_j \\ 0 & \text{otherwise} \end{cases} \qquad (11)$$

Illustrative Example. To illustrate our clustering approach, we plot the toponyms' entity candidates of the holiday property description shown in figure 3(b). Figures 2(a) and 2(b) show the entity candidates of each toponym with a different color. For example, the candidates of the toponym "**Steinbach**" have red color. The correct candidates of the mentioned toponyms are characterized with a dotted icon. The cluster $Cluster_{min}$ is shown with an oval in figure 2(b). We can see that $Cluster_{min}$ contains all the correct representatives of the mentioned toponyms. Given the candidates belonging to $Cluster_{min}$, we could easily infer "**Belgium**" to be the $Country_{winner}$ of that property.

Handling Uncertainty of Annotations. Equation 11 gives equal weights to all toponyms. The countries of toponyms with a very low extraction confidence probability are treated equally to toponyms with high confidence; both count fully. We can take the uncertainty in the extraction process into account by adapting Equation 11 to include the confidence of the extracted toponyms.

$$freq(Country_j) = \sum_{i=1}^{n} \begin{cases} p(t_i|w_i) & \text{if } r_{ix} \in Country_j \\ 0 & \text{otherwise} \end{cases} \qquad (12)$$

In this way terms which are more likely to be toponyms have a higher contribution in determining the country of the document than less likely ones.

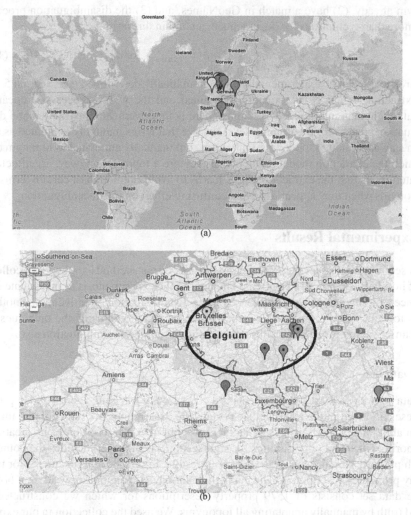

Fig. 2. Map plot of candidate entities for toponym of property description shown in figure 3(b)

4.3 Improving Certainty of Extraction

In the abovementioned improvement, we make use of the extraction confidence to help the disambiguation to be more robust. However, those probabilities are not accurate and reliable all the time. Some extraction models (like HMM in our experiments) retrieve some false positive toponyms with high confidence probabilities. Moreover, some of these false positives have many entity candidates in many countries according to GeoNames (e.g., the term "Bar" refers to 58 different locations in GeoNames in 25 different countries; see Figure 6). These false positives affect the disambiguation process.

This is where we take advantage of the reinforcement effect. To be more precise, we introduce another class in the extraction model called 'highly ambiguous' and annotate those terms in the training set with this class that (1) arc not manually annotated as a toponym already, (2) have a match in GeoNames, and (3) the disambiguation process finds more than τ countries for documents that contain this term, i.e.,

$$\left|\{c \mid \exists d \bullet t_i \in d \wedge c = Country_{winner} \text{ for } d\}\right| \geq \tau \tag{13}$$

The threshold τ can be experimentally and automatically determined (see Section 5.3). The extraction model is subsequently re-trained and the whole process is repeated without any human interference as long as there is improvement in extraction and disambiguation process for the training set. Observe that terms manually annotated as toponym stay annotated as toponyms. Only terms not manually annotated as toponym but for which the extraction model predicts that they are a toponym anyway, are affected. The intention is that the extraction model learns to avoid prediction of certain terms to be toponyms when they appear to have a confusing effect on the disambiguation.

5   Experimental Results

In this section, we present the experimental results of our methods applied to a collection of holiday properties descriptions. The goal of the experiments is to investigate the influence of using annotation confidence on the disambiguation effectiveness. Another goal is to show how to improve the imperfect extraction model using the outcomes of the disambiguation process and subsequently improving the disambiguation also.

5.1   Data Set

The data set we use for our experiments is a collection of traveling agent holiday property descriptions from the EuroCottage portal. The descriptions not only contain information about the property itself and its facilities, but also a description of its location, neighboring cities and opportunities for sightseeing. The data set includes the country of each property which we use to validate our results. Figure 3 shows examples for two holiday properties descriptions. The manually annotated toponyms are written in bold.

The data set consists of 1579 property descriptions for which we constructed a ground truth by manually annotating all toponyms. We used the collection in our experiments in two ways:

- **Train_Test Set.** We split the data set into a training set and a validation test set with ratio 2 : 1, and used the training set for building the extraction models and finding the highly ambiguous toponyms, and the test set for a validation of extraction and disambiguation effectiveness against "new and unseen" data.
- **All_Train Set.** We used the whole collection as a training and test set for validating the extraction and the disambiguation results.

The reason behind using the **All_Train set** for traing and testing is that the size of the collection is considered small for NLP tasks. We want to show that the results of the **Train_Test set** can be better if there is enough training data.

2-room apartment 55 m2: living/dining room with 1 sofa bed and satellite-TV, exit to the balcony. 1 room with 2 beds (90 cm, length 190 cm). Open kitchen (4 hotplates, freezer). Bath/bidet/WC. Electric heating. Balcony 8 m2. Facilities: telephone, safe (extra). Terrace Club: Holiday complex, 3 storeys, built in 1995 2.5 km from the centre of **Armacao de Pera**, in a quiet position. For shared use: garden, swimming pool (25 x 12 m, 01.04.-30.09.), paddling pool, children's playground. In the house: reception, restaurant. Laundry (extra). Linen change weekly. Room cleaning 4 times per week. Public parking on the road. Railway station "**Alcantarilha**" 10 km. Please note: There are more similar properties for rent in this same residence. Reception is open 16 hours (0800-2400 hrs). Lounge and reading room, games room. Daily entertainment for adults and children. Bar-swimming pool open in summer. Restaurant with Take Away service. Breakfast buffet, lunch and dinner(to be paid for separately, on site). Trips arranged, entrance to water parks. Car hire. Electric cafetiere to be requested in adavance. Beach football pitch. IMPORTANT: access to the internet in the computer room (extra). The closest beach (350 m) is the "**Sehora da Rocha**", **Playa de Armacao de Pera** 2.5 km. Please note: the urbanisation comprises of eight 4 storey buildings, no lift, with a total of 185 apartments. Bus station in **Armacao de Pera** 4 km.

(a) Example 1.

Le Doyen cottage is the oldest house in the village of **Steinbach** (built in 1674). Very pleasant to live in, it is situated right in the heart of the **Ardennes**. Close to **Robertville** and **Butchembach**, five minutes from the ski slopes and several lakes.

(b) Example 2.

Fig. 3. Examples of EuroCottage holiday home descriptions (toponyms in bold)

5.2   Experiment 1: Effect of Extraction with Confidence Probabilities

The goal of this experiment is to evaluate the effect of allowing uncertainty in the extracted toponyms on the disambiguation results. Both a HMM and a CRF extraction model were trained and evaluated in the two aforementioned ways. Both modes of operation (**First-Best** and **N-Best**) were used for inferring the country of the holiday descriptions as described in Section 4.2. We used the unmodified version of the clustering approach (Equation 11) with the output of **First-Best** method, while we used the modified version (Equation 12) with the output of **N-Best** method to make use of the confidence probabilities assigned to the extracted toponyms.

Results are shown in Table 1. It shows the percentage of holiday home descriptions for which the correct country was successfully inferred. We can clearly see that the **N-Best** method outperforms the **First-Best** method for both the HMM and the CRF models. This supports our claim that dealing with alternatives along with their confidences yields better results.

5.3   Experiment 2: Effect of Extraction Certainty Enhancement

While examining the results of extraction for both HMM and CRF, we discovered that there were many false positives among the extracted toponyms, i.e., words extracted as a toponym and having a reference in GeoNames, that are in fact not toponyms. Samples of such words are shown in Figures 4(a) and 4(b). These words affect the disambiguation result, if the matching entities in GeoNames belong to many different countries.

bath	shop	terrace	shower	at
house	the	all	in	as
they	here	to	table	garage
parking	and	oven	air	gallery
each	a	farm	sauna	sandy

(a) Sample of false positive toponyms extracted by HMM.

north	zoo	west	well	travel
tram	town	tower	sun	sport

(b) Sample of false positive toponyms extracted by CRF.

Fig. 4. False positive extracted toponyms

Table 1. Effectiveness of the disambiguation process for First-Best and N-Best methods in the extraction phase

(a) On Train_Test set

	HMM	CRF
First-Best	62.59%	62.84%
N-Best	68.95%	68.19%

(b) On All_Train set

	HMM	CRF
First-Best	70.7%	70.53%
N-Best	74.68%	73.32%

Table 2. Effectiveness of the disambiguation process using manual annotations

Train_Test set	All_Train set
79.28%	78.03%

We applied the proposed technique introduced in Section 4.3 to reinforce the extraction confidence of true toponyms and to reduce them for highly ambiguous false positive ones. We used the N-Best method for extraction and the modified clustering approach for disambiguation. The best threshold τ for annotating terms as highly ambiguous has been experimentally determined (see section 5.3).

Table 2 shows the results of the disambiguation process using the manually annotated toponyms. Table 4 show the extraction results using the state of the art Stanford named entity recognition model[4]. Stanford is a NEE system based on CRF model which incorporates long-distance information [9]. It achieves good performance consistently across different domains. Tables 3 and 5 show the effectiveness of the disambiguation and the extraction processes respectively along iterations of refinement. The "No Filtering" rows show the initial results of disambiguation and extraction before any refinements have been done.

We can see an improvement in HMM extraction and disambiguation results. It starts with lower extraction effectiveness than Stanford model but it outperforms after retraining the model. This support our claim that the reinforcement effect can help imperfect extraction models iteratively. Further analysis and discussion shown in Section 5.5.

---

[4] http://nlp.stanford.edu/software/CRF-NER.shtml

Table 3. Effectiveness of the disambiguation process after iterative refinement

(a) On Train_Test set

	HMM	CRF
No Filtering	68.95%	68.19%
1st Iteration	73.28%	68.44%
2nd Iteration	73.53%	68.44%
3rd Iteration	73.53%	-

(b) On All_Train set

	HMM	CRF
No Filtering	74.68%	73.32%
1st Iteration	77.56%	73.32%
2nd Iteration	78.57%	-
3rd Iteration	77.55%	-

Table 4. Effectiveness of the extraction using Stanford NER

(a) On Train_Test set

	Pre.	Rec.	F1
Stanford NER	0.8385	0.4374	0.5749

(b) On All_Train set

	Pre.	Rec.	F1
Stanford NER	0.8622	0.4365	0.5796

Table 5. Effectiveness of the extraction process after iterative refinement

(a) On Train_Test set

	HMM		
	Pre.	Rec.	F1
No Filtering	0.3584	0.8517	0.5045
1st Iteration	0.7667	0.5987	0.6724
2nd Iteration	0.7733	0.5961	0.6732
3rd Iteration	0.7736	0.5958	0.6732

(b) On All_Train set

	HMM		
	Pre.	Rec.	F1
No Filtering	0.3751	0.9640	0.5400
1st Iteration	0.7808	0.7979	0.7893
2nd Iteration	0.7915	0.7937	0.7926
3rd Iteration	0.8389	0.7742	0.8053

	CRF		
No Filtering	0.6969	0.7136	0.7051
1st Iteration	0.6989	0.7131	0.7059
2nd Iteration	0.6989	0.7131	0.7059
3rd Iteration	-	-	-

	CRF		
No Filtering	0.7496	0.7444	0.7470
1st Iteration	0.7496	0.7444	0.7470
2nd Iteration	-	-	-
3rd Iteration	-	-	-

5.4   Experiment 3: Optimal Cutting Threshold

Figures 5(a), 5(b), 5(c) and 5(d) show the effectiveness of the HMM and CRF extraction models at first iteration in terms of Precision, Recall, and F1 measures versus the possible thresholds τ. Note that the graphs need to be read from right to left; a lower threshold means more terms being annotated as highly ambiguous. At the far right, no terms are annotated as such anymore, hence this is equivalent to no filtering.

indent We select the threshold with the highest F1 value. For example, the best threshold value is 3 in figure 5(a). Observe that for HMM, the F1 measure (from right to left) increases, hence a threshold is chosen that improves the extraction effectiveness. It does not do so for CRF, which is prominent cause for the poor improvements we saw earlier for CRF.

5.5   Further Analysis and Discussion

For deep analysis of results, we present in Table 6 detailed results for the property description shown in Figure 3(a). We have the following observations and thoughts:

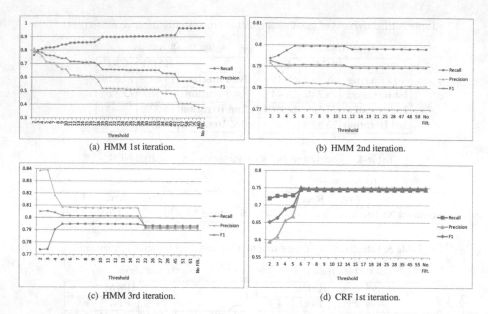

(a) HMM 1st iteration.

(b) HMM 2nd iteration.

(c) HMM 3rd iteration.

(d) CRF 1st iteration.

Fig. 5. The filtering threshold effect on the extraction effectiveness (On All_Train set)[5]

– From table 1, we can observe that both HMM and CRF initial models were improved by considering confidence of the extracted toponyms (see Section 5.2). However, for HMM, still many false positives were extracted with high confidence scores in the initial extraction model.
– The initial HMM results showed a very high recall rate with a very low precision. In spite of this our approach managed to improve precision significantly through iterations of refinement. The refinement process is based on removing highly ambiguous toponyms resulting in a slight decrease in recall and an increase in precision. In contrast, CRF started with high precision which could not be improved by the refinement process. Apparently, the CRF approach already aims at achieving high precision at the expense of some recall (see Table 5).
– In table 5 we can see that the precision of the HMM outperforms the precision of CRF after iterations of refinement. This results in achieving better disambiguation results for the HMM over the CRF (see Table 3)
– It can be observed that the highest improvement is achieved on the first iteration. This where most of the false positives and highly ambiguous toponyms are detected and filtered out. In the subsequent iterations, only few new highly ambiguous toponyms appeared and were filtered out (see Table 5).
– It can be seen in Table 6 that initially non-toponym phrases like ".-30.09.)" and "IMPORTANT" were falsely extracted by HMM. These don't have a GeoNames reference, so were not considered in the disambiguation step, nor in the subsequent re-training. Nevertheless they disappeared from the top-N annotations. The reason

---

[5] These graphs are supposed to be discrete, but we present it like this to show the trend of extraction effectiveness against different possible cutting thresholds.

Table 6. Deep analysis for the extraction process of the property shown in Figure 3(a) (∈: present in GeoNames; #refs: number of references; #ctrs: number of countries)

	Extracted Toponyms	GeoNames lookup ∈	#refs	#ctrs	Confidence probability	Disambiguation result
Manually annotated toponyms	Armacao de Pera	√	1	1	-	
	Alcantarilha	√	1	1	-	
	Sehora da Rocha	×	-	-	-	Correctly Classified
	Playa de Armacao de Pera	×	-	-	-	
	Armacao de Pera	√	1	1	-	
Initial HMM model with First-Best extraction method	Balcony 8 m2	×	-	-	-	
	Terrace Club	√	1	1	-	
	Armacao de Pera	√	1	1	-	
	.-30.09.)	×	-	-	-	
	Alcantarilha	√	1	1	-	
	Lounge	√	2	2	-	
	Bar	√	58	25	-	Misclassified
	Car hire	×	-	-	-	
	IMPORTANT	×	-	-	-	
	Sehora da Rocha	×	-	-	-	
	Playa de Armacao de Pera	×	-	-	-	
	Bus	√	15	9	-	
	Armacao de Pera	√	1	1	-	
Initial HMM model with N-Best extraction method	Alcantarilha	√	1	1	1	
	Sehora da Rocha	×	-	-	1	
	Armacao de Pera	√	1	1	1	
	Playa de Armacao de Pera	×	-	-	0.999849891	
	Bar	√	58	25	0.993387918	
	Bus	√	15	9	0.989665883	
	Armacao de Pera	√	1	1	0.96097006	
	IMPORTANT	×	-	-	0.957129986	
	Lounge	√	2	2	0.916074183	Correctly Classified
	Balcony 8 m2	×	-	-	0.877332628	
	Car hire	×	-	-	0.797357377	
	Terrace Club	√	1	1	0.760384949	
	In	√	11	9	0.455276943	
	.-30.09.)	×	-	-	0.397836259	
	.-30.09.	×	-	-	0.368135755	
		×	-	-	0.358238066	
	. Car hire	×	-	-	0.165877044	
	adavance.	×	-	-	0.161051997	
HMM model after 1st iteration with N-Best extraction method	Alcantarilha	√	1	1	0.999999999	
	Sehora da Rocha	×	-	-	0.999999914	Correctly Classified
	Armacao de Pera	√	1	1	0.999998522	
	Playa de Armacao de Pera	×	-	-	0.999932808	
Initial CRF model with First-Best extraction method	Armacao	×	-	-	-	
	Pera	√	2	1	-	
	Alcantarilha	√	1	1	-	
	Sehora da Rocha	×	-	-	-	Correctly Classified
	Playa de Armacao de Pera	×	-	-	-	
	Armacao de Pera	√	1	1	-	
Initial CRF model with N-Best extraction method	Alcantarilha	√	1	1	0.999312439	
	Armacao	×	-	-	0.962067016	
	Pera	√	2	1	0.602834683	
	Trips	√	3	2	0.305478198	Correctly Classified
	Bus	√	15	9	0.167311005	
	Lounge	√	2	2	0.133111374	
	Reception	√	1	1	0.105567287	

for this behavior is that initially the extraction models were trained on annotating for only one type (toponym), whereas in subsequent iterations they were trained on two types (toponym and 'highly ambiguous non-toponym'). Even though the aforementioned phrases were not included in the re-training, their confidences still fell below the 0.1 cut-off threshold after the 1st iteration. Furthermore, after one iteration the top-25 annotations contained 4 toponym and 21 highly ambiguous annotations.

6   Conclusions and Future Work

NEE and NED are inherently imperfect processes that moreover depend on each other. The aim of this paper is to examine and make use of this dependency for the purpose of improving the disambiguation by iteratively enhancing the effectiveness of extraction, and vice versa. We call this mutual improvement, the *re-inforcement effect*. Experiments were conducted with a set of holiday home descriptions with the aim to extract and disambiguate toponyms as a representative example of named entities. HMM and CRF statistical approaches were applied for extraction. We compared extraction in two modes, First-Best and N-Best. A clustering approach for disambiguation was applied with the purpose to infer the country of the holiday home from the description.

We examined how handling the uncertainty of extraction influences the effectiveness of disambiguation, and reciprocally, how the result of disambiguation can be used to improve the effectiveness of extraction. The extraction models are automatically retrained after discovering highly ambiguous false positives among the extracted toponyms. This iterative process improves the precision of the extraction. We argue that our approach that is based on uncertain annotation has much potential for making information extraction more robust against ambiguous situations and allowing it to gradually learn. We provide insight into how and why the approach works by means of an in-depth analysis of what happens to individual cases during the process.

We claim that this approach can be adapted to suit any kind of named entities. It is just required to develop a mechanism to find highly ambiguous false positives among the extracted named entities. Coherency measures can be used to find highly ambiguous named entities. For future research, we plan to apply and enhance our approach for other types of named entities and other domains. Furthermore, the approach appears to be fully language independent, therefore we like to prove that this is the case and investigate its effect on texts in multiple and mixed languages.

References

1. Habib, M.B.: Neogeography: The challenge of channelling large and ill-behaved data streams. In: Workshops Proc. of the 27th ICDE 2011, pp. 284–287 (2011)
2. Habib, M.B., van Keulen, M.: Named entity extraction and disambiguation: The reinforcement effect. In: Proc. of MUD 2011, Seattle, USA, pp. 9–16 (2011)
3. Hobbs, J., Appelt, D., Bear, J., Israel, D., Kameyama, M., Stickel, M., Tyson, M.: Fastus: A system for extracting information from text. In: Proc. of Human Language Technology, pp. 133–137 (1993)

4. Zhou, G., Su, J.: Named entity recognition using an hmm-based chunk tagger. In: Proc. ACL 2002, pp. 473–480 (2002)
5. Sekine, S.: NYU: Description of the Japanese NE system used for MET-2. In: Proc. of MUC-7 (1998)
6. Borthwick, A., Sterling, J., Agichtein, E., Grishman, R.: NYU: Description of the MENE named entity system as used in MUC-7. In: Proc. of MUC-7 (1998)
7. Isozaki, H., Kazawa, H.: Efficient support vector classifiers for named entity recognition. In: Proc. of COLING, pp. 1–7 (2002)
8. McCallum, A., Li, W.: Early results for named entity recognition with conditional random fields, feature induction and web-enhanced lexicons. In: Proc. of CoNLL 2003, pp. 188–191 (2003)
9. Finkel, J.R., Grenager, T., Manning, C.: Incorporating non-local information into information extraction systems by gibbs sampling. In: Proc. of the 43nd Annual Meeting of the Association for Computational Linguistics, ACL 2005, pp. 363–370 (2005)
10. Michelakis, E., Krishnamurthy, R., Haas, P.J., Vaithyanathan, S.: Uncertainty management in rule-based information extraction systems. In: Proc. of the 35th SIGMOD International Conference on Management of Data, SIGMOD 2009, pp. 101–114. ACM, New York (2009)
11. Wacholder, N., Ravin, Y., Choi, M.: Disambiguation of proper names in text. In: Proc. of ANLC 1997, pp. 202–208 (1997)
12. Buscaldi, D., Rosso, P.: A conceptual density-based approach for the disambiguation of toponyms. Int'l Journal of Geographical Information Science 22, 301–313 (2008)
13. Smith, D.A., Crane, G.: Disambiguating geographic names in a historical digital library. In: Constantopoulos, P., Sølvberg, I.T. (eds.) ECDL 2001. LNCS, vol. 2163, pp. 127–136. Springer, Heidelberg (2001)
14. Smith, D., Mann, G.: Bootstrapping toponym classifiers. In: Workshop Proc. of HLT-NAACL 2003, pp. 45–49 (2003)
15. Martins, B., Anastácio, I., Calado, P.: A machine learning approach for resolving place references in text. In: Proc. of AGILE 2010 (2010)
16. Viterbi, A.: Error bounds for convolutional codes and an asymptotically optimum decoding algorithm. IEEE Transactions on Information Theory 13, 260–269 (1967)
17. Wallach, H.: Conditional random fields: An introduction. Technical Report MS-CIS-04-21, Department of Computer and Information Science, University of Pennsylvania (2004)
18. Sutton, C., McCallum, A.: An introduction to conditional random fields. Foundations and Trends in Machine Learning (to appear, 2011)

Keyword Extraction from Short Documents
Using Three Levels of Word Evaluation

Mika Timonen[1], Timo Toivanen[1], Melissa Kasari[2], Yue Teng[3],
Chao Cheng[3], and Liang He[3]

[1] VTT Technical Research Centre of Finland, PO 1000, FI-02044 Espoo, Finland
[2] Department of Computer Science, PO 68, FI-00014 University of Helsinki, Finland
[3] East China Normal University, Institute of Computer Applications,
No.500 Dongchuan Road, 200241 Shanghai, China

Abstract. In this paper we propose a novel approach for keyword extraction from short documents where each document is assessed on three levels: corpus level, cluster level and document level. We focus our efforts on documents that contain less than 100 words. The main challenge we are facing comes from the main characteristic of short documents: each word occurs usually only once within the document. Therefore, the traditional approaches based on term frequency do not perform well with short documents. To tackle this challenge we propose a novel unsupervised keyword extraction approach called Informativeness-based Keyword Extraction (IKE). We compare the performance of the proposed approach is against other keyword extraction methods, such as CollabRank, KeyGraph, Chi-squared, and TF-IDF. In the experimental evaluation IKE shows promising results by out-performing the competition.

Keywords: Keyword Extraction, Machine Learning, Short Documents, Term Weighting, Text Mining.

1 Introduction

In this paper we propose a novel approach for keyword extraction from short documents that evaluates words on three levels. We focus on short documents as there is a rising number of sources that produce only documents that are short. For example, Twitter messages, product descriptions at several online stores, and TV-show description in most TV-guides are all short; in most cases they contain at most two or three sentences.

We focus our efforts on keyword extraction which is an area of text mining that aims to identify the most informative and important words, also called terms, of the document. Keyphrase extraction, even though sometimes used as a synonym to keyword extraction, may differ from keyword extraction by including steps and heuristics to identify collections of words instead of just single words. Both of them have uses in several different domains, including text summarization, text categorization, document tagging and recommendation systems. Our motivation comes from the need to utilize the extracted keywords as tags in user modeling.

In our previous work with short documents [1–3] we have identified differences between categorization of short and long documents. These differences are relevant with

A. Fred et al. (Eds.): IC3K 2012, CCIS 415, pp. 130–146, 2013.

keyword extraction also. The most notable difference is due to the number of words in each document: when documents are short, a word appears usually only once within a single document. This is also known as *hapax legomenon*, and we call it *Term Frequency=1* (*TF=1*) challenge. Because of this challenge, approaches that rely on the term frequency within a document, such as *Term Frequency - Inverse Document Frequency* (*TF-IDF*) [2] become less effective.

TF=1 challenge is very relevant as many of the traditional keyword extraction approaches use term frequency. In addition to TF-IDF, *KeyGraph* by Ohsawa et al. [4], and a Chi-Squared based approach by Matsuo and Ishizuka [5] rely on word co-occurrence and word frequencies. Both of these studies have focused on longer documents such as news articles or scientific articles.

An example of a naive approach for keyword extraction from short documents is to extract all the nouns. However, this is rarely beneficial as it results extracting words that are significantly less informative than others. In addition, if the documents are longer than 30 words, the number of nouns would often be too large. In Section 4 of this paper we compare the results of our approach against the naive approach and conclude that the benefits of identifying the most relevant words and extracting only them has significant impact on performance.

The motivation of our work is to use the extracted keywords in user models and in a recommendation system. The idea is to extract the keywords from the documents user has seen (i.e., clicked open) and use the keywords as tags. We have based our work on CollabRank keyword extraction approach by Wan and Xiao [6], and our previous approach for feature weighting in short documents [1]. The idea behind our work is to extract words that are important in two levels: corpus level where the words are more general, and cluster level where the words are important among a smaller set of documents with similar topic. The words are extracted from a single document at a time by taking some document level information into account. We call this approach *Informativeness-based Keyword Extraction* (*IKE*). The approach presented in this paper was originally proposed in our previous paper [7]. In this paper, we extend the evaluation of the method with a new experimental case from a real world setting where our dataset consists of TV-show descriptions.

We conduct three experimental evaluation cases using four different datasets: movie descriptions, event descriptions, tv-show descriptions, and company descriptions. These datasets are in three different languages: English, Finnish and Chinese. In the first experiment we manually annotated approximately 300 documents from movie, event and company datasets and measured the precision and recall of each approach. In the second experiment we used the extracted keywords for user modeling and evaluated the recommendation precision with the models. As the keywords are used as tags, this evaluation demonstrates the utilization of the extracted keywords. In these two test cases we compared the results against several other keyword extraction methods such as CollabRank, KeyGraph, TF-IDF, Matsuo's Chi-Squared, and the naive approach. In all of these experiments our approach out-performed the competition.

In the final experiment we test the use of our approach in a website called *skimm.tv*[1] that offers a personalized TV-guide (in Finnish). The aim is to extract the keywords

---

[1] http://www.skimm.tv/

from the descriptions user has seen and use them as tags for the site's personalization and recommendation. In this paper, we evaluate the keywords extracted from the descriptions manually and report the results.

This paper is organized as follows: in Section 2 we discuss related keyword extraction approaches. In Section 3 we describe our approach. In Section 4 we report the results of the experimental evaluation and compare the results against other keyword extraction approaches. We conclude the paper in Section 5.

2   Related Work

Several authors have presented keyword extraction approaches in recent years. The methods often use supervised learning. In these cases the idea is to use a predefined seed set of documents as a training set and learn the features for keywords. The training set is built by manually tagging the documents for keywords.

One approach that uses supervised learning is called Kea [8, 9]. It uses Naive Bayes learning with Term Frequency - Inverse Document Frequency (TF-IDF) and normalized term positions as the features. The approach was further developed by Turney [10] who included keyphrase cohesion as a new feature. One of the latest updates to Kea is to include linguistic information, such as section information, as features [11].

Before developing the Kea approach, Turney experimented with two other approaches: decision tree algorithm C4.5 and an algorithm called GenEx [12]. GenEx is an algorithm that has two components: hybrid genetic algorithm Genitor, and Extractor. The latter is the keyword extractor and it needs twelve parameters to be tuned. Genitor is used for finding these optimal parameters from the training data.

Hulth et al. [13] describe a supervised approach that utilizes domain knowledge found from Thesaurus, and TF-IDF statistics. Later, Hulth added linguistic knowledge and several other models to improve the performance of the extraction process [14, 15]. The models use four different attributes: term frequency, collection frequency, relative position of the first occurrence, and Part-of-Speech tags.

Ercan and Cicekli [16] describe a supervised learning approach that uses lexical chains for extraction. The idea is to find semantically similar terms, i.e., lexical chains, from text and utilize them for keyword extraction.

There are also approaches that do not use supervised learning but rely on term statistics instead. KeyGraph is an approach described by Ohsawa et al. [4] that does not use POS-tags, large corpus, nor supervised learning. It is based on term co-occurrence, graph segmentation, and clustering. The idea is to find important clusters from a document and to assume that each cluster holds keywords. Matsuo and Ishizuka [5] describe an approach that also uses a single document as its corpus. The idea is to use the co-occurrences of frequent terms to evaluate if a candidate keyword is important for a document. The evaluation is done using Chi-squared (χ^2) measure. Paukkeri et al. [17] describe a language independent approach that ranks the document's terms according to their occurrence rank in the document and in the corpus, and picks the top n terms as the keywords.

Mihalcea and Tarau [18] describe an unsupervised learning approach called Text-Rank. It is based on PageRank [19] which is a graph-based ranking algorithm. The idea is to create a network where the vertices corresponds to terms of the document and edges are links between co-occurring terms. A term pair is co-occurring if they are within 2 to 10 words of each other in the document. The edges hold a weight that is assigned using the PageRank algorithm. The edges are undirected and symmetric. The keywords are extracted by ranking the vertices and picking the top n ones. This approach produced improved results over the approach described by Hulth.

There are some approaches developed that extract keywords from abstracts. These abstracts often contain 200-400 words making them considerably longer than documents in our corpus. One approach, which is presented by HaCohen-Kerner [20], uses term frequencies and importance of sentences for extracting the keywords. Later, HaCohen-Kerner et al. [21] continue the work and include other statistics as well. Andrade and Valencia [22] use Medline abstracts for extracting protein functions and other biological keywords. The previously mentioned work done by Ercan and Cicekli [16] also uses abstracts as the corpus. Finally, SemEval-2010 had a task where the objective was to extract keywords from scientific articles. Kim et al. [23] presents the findings of the task.

Keyword extraction is often used as a synonym to keyphrase extraction even though the latter aims to extract *n-grams*, i.e., word groups that are in the form of phrases (e.g., "digital camera"). Yih et al. [24] present a keyphrase extraction approach for finding keyphrases from web pages. Their approach is based on document structure and word locations. This approach is designed to identify keyphrases, i.e., word groups, instead of just words. A logistic regression is used for training the classifier for the extraction process. Tomokiyo and Hurst [25] present an approach for keyphrase extraction based on language models. They use pointwise Kullback-Leibler –divergence between language models to assess the informativeness and "phraseness" of the keyphrases.

Wan and Xiao [6] introduce an unsupervised approach called CollabRank that clusters the documents and extracts the keywords within each cluster. The assumption is that documents with similar topics contain similar keywords. The keywords are extracted in two levels. First, the words are evaluated in the cluster level using graph-based ranking algorithm similar to PageRank [19]. Next, the words are scored on the document level by summing the cluster level saliency scores. In the cluster level evaluation POS-tags are used to identify suitable candidate keywords; the POS-tags are also used when determining if the candidate keyphrases are suitable. Wan and Xiao use news articles as their corpus.

Assessing the term informativeness is an important part of keyword extraction. Rennie and Jaakkola [26] have survey term informativeness measures in the context of named entity recognition and conclude that Residual IDF produced the best results for their case. Residual IDF is based on the idea of comparing the word's observed Inverse Document Frequency (IDF) against predicted IDF ($\widehat{IDF}$) [27]. Predicted IDF is assessed using the term frequency and assuming a random distribution of the term in the documents (Poisson model). If the difference between the two IDF measures is large, the word is informative.

3   Keyword Extraction from Short Documents

In this work we utilize the previous works by Wan and Xiao [6], and our previous work on text categorization [1]. From the former we use the idea of multi-level word assessment through clustering, and from the latter we use the term weighting approach. The term weighting approach is designed for short documents which makes it relevant for this case also.

The extraction process has two steps: (1) preprocessing that includes document clustering, and (2) word informativeness evaluation. The latter is divided into three levels of evaluation: corpus level, cluster level, and document level, and it aims to identify and extract the most informative words of each level. The input for the process is the set of documents (corpus). The process produces a set of keywords for each document as its output.

3.1   Problem Description

We define a short document as a document that contains no more than 100 words, which is equal to a very short scientific abstract. Depending on the dataset, the documents are often much shorter: for example, we have used a Twitter dataset that holds 15 words on average [2]. In this paper we concentrate on event and movie descriptions that have 30 to 60 words per description. These word counts are considerably less than in corpora previously used in keyword extraction. We consider this type of data relevant for two reasons: 1) a short description often contains the same information than the longer one but in a concise form, and 2) when we want to model user's interests we need to extract the information from the text the user has seen. In many cases, this means a short description read in a mobile application.

It is usually important to extract keywords of different abstraction levels. Consider the following example: "The White Stripes is performing at the tonight's rock concert at the stadium." If we only extract words of lower abstraction level such as "The White Stripes" and do not map it to any other information, people who do not know who or what "The White Stripes" is, would not be interested. On the other hand, more general words such as "Rock" would not include enough information. By combining these two levels we get more information and can make better user profiles and document summaries.

3.2   Preprocessing and Clustering

Before the keywords are extracted the text is preprocessed by stemming the text and removing stop words. In this step we use a filter that removes words with less than 3 characters, and a standard stop word list. In addition, as we do not focus on keyphrase extraction we use a heuristic approach for identifying some important noun phrases, such as names. That is, we identify names by their spelling: if two or more consecutive words have a capital letter as its first letter, the words are tagged as a proper noun group (e.g., "Jack White"). In addition, we include some connecting words like 'and' to the heuristics to identify phrases such as "Rock and Roll". For Chinese, we do not use noun phrase tagging.

The term evaluation approach that we will describe in the next section requires clustering of the documents. We use Agglomerative (CompleteLink) clustering, which produced the best results for [6]. CompleteLink is a bottom-up clustering approach where at the beginning, each document forms its own cluster. The most similar clusters are joined in each iteration until there are at most k clusters. The similarity between the clusters c_n and c_m is the minimum similarity between two documents d_n $(d_n \in c_n)$ and d_m $(d_m \in c_m)$ and the similarity is assessed using cosine similarity between the documents. Instead of a predefined number of clusters we use a predefined minimum similarity. That is, if the similarity between the clusters is below a threshold, the clusters are not joined.

3.3 Word Informativeness Evaluation

We focus on keywords of two different levels: ones that are more abstract and are therefore more common, and the ones that are more expressive and therefore more rare. The more common keywords usually describe the text as a whole; for example, terms like "Rock and Roll" define the content of the document in a more abstract level. Words like "Aerosmith" and "Rambo" give a more detailed description of the topic of the document. In this section we give a short description of the word informativeness evaluation. More complete description can be found from our previous paper on the topic [7].

Corpus Level Word Evaluation. The aim of the corpus level evaluation is to find words that define the document in a more abstract level. These words tend to be more common than the more expressive words but they should not be too common either. For example, we want to find terms like "Rock and Roll" instead of just "event" or "music". Our hypothesis is that the most informative words in the corpus level are those that are neither too common nor too rare in the corpus; however, an informative word will more likely be rare than common.

In order to find these types of words we rely on word frequency in the corpus level (tf_c). As in most cases when using a corpus of short documents, the term frequency within a document (tf_d) for each term is often 1. In this step, we use the following idea: we want to find words that have an IDF close to the assumed optimal value as this will emphasize words that are not too common nor too rare. Here, the greater the difference between the observed IDF and the assumed optimal IDF is, the less informative the word is in the corpus level.

To get the corpus level score $s_{corpus}(t)$ we use an approach we call *Frequency Weighted IDF* (IDF_{FW}). It is based on the idea of updating the observed IDF using *Frequency Weight* (FW):

$$IDF_{FW}(t) = IDF(t) - FW(t), \tag{1}$$

where *FW(t)* is the assumed optimal IDF described below.

The intuition behind $FW(t)$ is to penalize words when the corpus level term frequency does not equal the estimated optimal frequency n_o. Equation 2 shows how *FW* is calculated:

$$FW(t) = \alpha \times | \log_2 \frac{tf_c}{n_o} | . \tag{2}$$

We use α to emphasize words that are more rare than common; $\alpha = 1.1$ is used in our experiments.

An important part of the equation is the selection of n_o. After studying the data and experimental results we decided to use a value related to the corpus size: $n_o = 0.03 \times |D|$. That is, we consider that a word is optimally important in the corpus level when it occurs in 3 % of documents. Even though this has performed well in all of our experiments, it should be noted that it may be beneficial to change this value when using different datasets. However, we believe this value is a good starting point for experimentation.

Cluster Level Word Evaluation. In the second step, the word's informativeness is assessed in the cluster level. The idea is to group documents together that may share a topic and find words that are important for the group: if the word w appears often in the cluster c and seldom in other clusters, the word is informative in c. Assessing the cluster level informativeness is done by using the same word weighting approach we have used previously [1].

The first step of the process is to break the documents into smaller pieces called fragments. The text fragments are extracted from sentences using breaks such as question mark, comma, semicolon, and other similar characters. In addition, words such as 'and', 'or', and 'both' are also used as breaks. For example, the description "Photo display and contemporary art exhibition at the central museum" consists of two fragments, "photo display" and "contemporary art exhibition at the central library".

The cluster level weight consists of two components: *TCoR* and *TCaR*; relevance of the word in the corpus, and in the cluster (i.e., category in our previous work), respectively. There are two features used for assessing *TCoR*: *inverse average fragment length (ifl(t))* and *inverse cluster count (ic(t))*. The average fragment length for word t is simply the average number of words in the set of fragments the word appears in:

$$ifl(t) = \frac{1}{\frac{1}{|f_t|} \sum l_f(t)} , \tag{3}$$

where l_f is the length of the fragment f, and $|f_t|$ is the number of fragments t appears in. Inverse cluster count ($ic(t)$) is the inverse count of categories for the word t:

$$ic(t) = \frac{1}{c_t} , \tag{4}$$

where c_t is the number of different categories word t appears in. The first component of the score, *TCoR* is the sum of $ifl(t)$ and $ic(t)$:

$$TCoR(t) = (ifl(t) + ic(t)) . \tag{5}$$

The second component, cluster relevance (*TCaR(t)*), evaluates the word's informativeness among clusters. The idea is to identify words that occur often within the cluster and rarely in other clusters. *TCaR* consists of two probabilities: $P(c|d)$, the probability that the document d with word t occurs within the cluster c, and $P(d|c)$, the probability for a document d with word t within the cluster c.

The conditional probability $P(c|d)$ is estimated simply by taking the number of documents in the cluster that contain the word t ($|\{d : t \in d, d \in c\}|$) and dividing it by the total number of documents that contain word t ($|\{d : t \in d\}|$):

$$P(c|d) = \frac{|\{d : t \in d, d \in c\}|}{|\{d : t \in d\}|} . \tag{6}$$

The probability that a document in cluster c contains word t is the final component of the weight:

$$P(d|c) = \frac{|\{d : t \in d, d \in c\}|}{|\{d : d \in c\}|} , \tag{7}$$

where $|\{d : d \in c\}|$ is the number of documents in cluster c. $TCaR(t,c)$ for the word t in the cluster c is calculated as follows:

$$TCaR(t, c) = (P(c|d) + P(d|c)) . \tag{8}$$

The two scores $TCoR$ and $TCaR$ are combined when the final cluster level score is calculated:

$$s_{cluster}(t, c) = TCoR(t) \times TCaR(t, c) . \tag{9}$$

The result of the cluster level evaluation is a score for each word and for each of the clusters it appears in. If the word appears only in a single cluster, the weight will be considerably higher than if it would appear in two or more clusters. More information and the intuition behind components can be found from our previous publications [1, 3].

Document Level Word Evaluation. For finding the most important words of the document, we use the corpus and cluster level word scores. We aim to extract words that get a high score on either of the levels. As the extraction process is done in the document level, i.e., words are extracted from a single document at a time, the final step of the process is called document level word evaluation.

In order to make the corpus and the cluster level scores comparable, the corpus level scores are normalized to fall between [0,1]. The normalization is done by taking the maximum corpus level word score found in the document and dividing each score with the maximum value. That is, the word with the highest corpus level score gets a normalized corpus level score of 1.

The document level score s_{doc} for word t in document d, which belongs to cluster c, is calculated by taking the weighted average of the cluster level score $s_{cluster}(t, c)$ and the normalized corpus level score $s_{n,corpus}(t)$. This is shown in Equation 10:

$$s_{doc}(t, d) = \frac{\beta \times s_{cluster}(t, c) + (1 - \beta) \times s_{n,corpus}(t)}{2} , \tag{10}$$

where β indicates a weight used for giving more emphasis to either cluster or the corpus level score. To give more emphasis for the cluster level scores, we should use $\beta > 0.5$, and vice versa.

We included a *distance* factor $d(t)$ to the assessment as we want to emphasize word locations. Distance is based on the same idea used in Kea [8] and it indicates the location of the word's first occurrence in the document. That is, it is the number of words that precede the first occurrence of t in the document divided by the length of the document:

$$d(t) = 1 - \frac{i(t)}{|d|} , \tag{11}$$

where $|d|$ is the number of words in the document and $i(t)$ is the number of words before the word's first occurrence in the document; $i(t) = 0$ for the first word of the document.

To filter unwanted words and give emphasis on others, we use the Part Of Speech-tags of the words as an optional component for the weight: different tags get a different POS-weight (w_{POS}) in the final score. The simplest approach is to give weight 1.0 to all tags that are accepted, such as NP and JJ (nouns and adjectives), and 0.0 to all others. To emphasize some tags over the others, $w_{POS}(tag_1) > w_{POS}(tag_2)$ can be used. If POS-tags are not available, $w_{POS} = 1.0$ is used for all words.

We get the final score $s(t, d)$ for each word t in the document d by combining the three components $s_{doc}(t, d)$, $w_{POS}(t)$ and $d(t)$:

$$s(t, d) = s_{doc}(t, d) \times d(t) \times w_{POS}(t) , \tag{12}$$

where $w_{POS}(t)$ is the POS weight for the word t. If t has several POS-tags, the one with the largest weight is used.

Each word t in the document d now has a score $s(t, d)$ that indicates its informativeness for the document. The top k most informative words are then assigned as keywords for the document. As the number of informative words per document may vary, we pick only words that have high enough score. That is, only words that get the score of $s(t, d) > \max s(t, d) \times 0.5$ are accepted as keywords (i.e., a word should have a score of at least half of the highest score). The number 0.5 was selected after evaluating the performance of the keyword extraction. If there are more than nine word that fulfill this criterion, we pick the top-9 words as keywords.

4   Experimental Evaluation

Our experimental evaluation consisted of three different cases using four different datasets. The datasets are described in Section 4.1. In two of the experiments we compared the results between IKE and the following methods: CollabRank, KeyGraph, Matsuo's χ^2 measure, TF-IDF and Chi-squared feature weighting. In the first experiment we utilized manually annotated keywords as the gold standard to see which approach produces the best results. In the second experiment the keywords were used as tags to create user models and see which of the models produces the best recommendations. We consider the latter experiment the most indicative of performance as it is the most objective. In the third experiment, we manually evaluated the relevancy of the extracted keywords for user modeling.

Even though IKE uses several parameters that need to be selected, we leave the evaluation of the different parameters and their selection outside the scope of this paper.

The results reported here are obtained with the parameters that produced highest scores in our experiments. In our experiments, the impact of the parameter selection was quite small as they only fine tune the results.

4.1 Data

We used data of three different languages: Finnish, English and Chinese. Finnish is a complex language with lots of suffixes, Chinese is a simpler language without prefixes and suffixes but a complex language due to its different character set and writing system. English is the language often used in the experimental evaluation of this area.

For Finnish, we used two datasets. First, we use an event dataset that consists of approximately 5,000 events from the Helsinki Metropolitan area. After preprocessing, the event data held 32 words per document on average. The average term frequency per document in this dataset was 1.04. The second dataset consisted of TV-show descriptions from skimm.tv. The descriptions were from the actual TV-shows shown in television at the time. The type of TV-shows varied from documentaries and sports to drama and comedies. The average number of words per description was 34 and the average term frequency per document was 1.05.

For Chinese, we used data received from Velo coupon machines from Shanghai. The dataset contained 1,000 descriptions of companies and their products stored in the Velo databases. The descriptions held 80 words on average; even though longer than most of our data this was short enough to be used in our experiments. One of the challenges with Chinese is to tokenize the text into words; for this, paodingjieniu, a Chinese Word segmentation tool was used to divide the descriptions into words separated by blank space.

For English, we use approximately 7,000 movie abstracts from Wikipedia[2]. We used MovieLens dataset[3] when selecting the movies: if a movie was found from MovieLens dataset, we downloaded its Wikipedia page. We used only the abstract of the Wikipedia page. If the abstract was longer than 100 words, we removed the last full sentences to shorten the document under the given limit. The average word frequency per document in this dataset was 1.07 and the average length of a document was 63 words.

For POS-tagging in English and Chinese we used the Stanford's Log-Linear Part-of-Speech tagger[4]. For POS-tagging in Finnish we used LingSoft's commercial FinTWOL tagger.

4.2 Evaluation of Keyword Precision

We evaluated the feasibility of the extracted keywords using a set of manually annotated keywords. We used Event, Movie and Velo datasets for evaluation. We created the test set by randomly selecting 300 documents from each of the test sets and annotated the keywords. Event and Wikipedia data was tagged by two research scientist from VTT Technical Research Centre of Finland, and Velo data was tagged by two students from

---

[2] http://www.wikipedia.org/
[3] http://www.grouplens.org/node/12
[4] http://nlp.stanford.edu/software/tagger.shtml

East China Normal University. At most nine keywords were chosen per document. The agreement rate among annotators was 69 % for the Event data, 64 % for the Wikipedia data, and 70 % for the Velo data. The test set was updated after disagreements were resolved.

Evaluation Setup. We implemented CollabRank algorithm as described by Wan and Xiao [6]. For Term Frequency - Inverse Document Frequency (TF-IDF) and Chi-Squared we used the standard implementation of the approach. For Ohsawa's KeyGraph [4] and Matsuo's χ^2 keyword extraction approach [5] we used Knime[5] and its implementation of the two algorithms. We used the default parameters. To improve the results we used stop word lists and N char filter ($N = 3$) to remove uninformative words and characters from the documents.

After experimenting with parameters, we selected the following values for IKE: $\beta = 0.3$, and POS-tag weights $w_{POS}(N) = 1.0$, $w_{POS}(JJ) = 1.0$, $w_{POS}(V) = 0$, $w_{POS}(Others) = 0$. However, with event data, we used $w_{POS}(N) = 3.0$ and $\beta = 0.6$.

Results. For the baseline, we used the naive method that extracts all nouns and adjectives. We included adjectives as they are relevant in some domains; for example, adjective *explosive* can be considered relevant in the description "explosive action movie". Therefore, the words with the following POS-tags are extracted: N, A, ADJ, AD, AD-A, -, JJ, NN, NNS, NNP, and NNPS. Some of these tags are used in FinTWOL and some in Stanford POS-tagger. The tag "-" indicates unknown, which is usually a name not recognized by the tagger. Therefore, we treated the tag "-" as NP in our experiments. This resulted the following baselines: Event 0.39, Wikipedia 0.22, and Velo 0.15.

Table 1 shows the results of our experiments. The best F-scores for both Event and Wikipedia data were obtained using IKE. For the Chinese Velo data the difference between IKE and CollabRank, and IKE and the baseline is notable. However, we can see that TF-IDF and χ^2 performs similarly in this case. After careful review of the results, we conclude that most of the keywords extracted by IKE were feasible even though not originally picked by the annotators. This applies to all of the datasets. The keywords picked by both χ^2 and TF-IDF were in most cases the uncommon words in the dataset, such as the name of the movie. Keywords extracted by IKE were both uncommon and common; for example, name of the movie, actors, and the genre were all extracted.

Table 1. F-scores for each of the method in keyword precision experiment. Chi-squared is the traditional feature weighting approach, and χ^2 KE is the keyword extraction approach presented by [5]. Due to the Chinese character set, we were unable to evaluate KeyGraph and Matsuo's χ^2 keyword extraction approach on Velo data using Knime.

	IKE	CollabRank	Chi-squared	TF-IDF	KeyGraph	χ^2 KE	Baseline
Wikipedia	0.57	0.29	0.35	0.35	0.22	0.21	0.22
Events	0.56	0.46	0.49	0.49	0.36	0.35	0.39
Velo	0.31	0.18	0.26	0.22	-	-	0.15

---

[5] Konstanz Information Miner: http://www.knime.org/

The rest of the approaches did not produce very good results. The reason for poor performance is most likely the same for all the compared approaches: as most words occur only once in the document, the approaches based on term frequency within the document cannot get enough information about the words in order to distinguish the informative words from the uninformative ones.

4.3 Evaluation of Keyword Utilization for User Modeling

As the manually annotated keywords tend to be subjective, we conducted an experiment we consider more objective. We compared the recommendation precision of the keywords extracted by each of the approaches. The recommendation precision was assessed by recommending a top-n list of movies and comparing how many of them the user has liked.

Evaluation Setup. We used Wikipedia data for keyword extraction and the user ratings from MovieLens data for user modeling and recommendation. To test the recommendation precision we created a simple user model: first, we randomly selected 10,000 users from the MovieLens dataset. Then for each user, all the movies they rated were retrieved. This set was divided into a training set and a test set with 75 % - 25 % ratio. However, only movies with a strong positive rating (rating 4.5 or 5) were included to the test set. The rating scale in MovieLens is $0 - 5$.

For each of the movies in the training set, the keywords were extracted from the Wikipedia page. Each of the keywords were then used as a tag in the user model. The tags were weighted using the user's rating for the movie: for example, if the rating was 3, the tag was assigned a weight of 0, if the rating was 0, the weight was -1.0, and if the rating was 5, the weight was 1.0. If the same keyword is found from several movies, we use the user's average rating among the movies.

To evaluate the model's precision, we randomly included $k \times 5$ movies to each user's test set from the set of all movies, where k is the initial size of the test set of that user. That is, if we have 5 movies in the test set, we take randomly 25 movies among all movies the user hasn't seen to make the total size of the test set 30 movies. The recommendation is done by assigning a score to each of the movies. The score of a movie is the summed weight of the matching movie–user -tags: for each of the movie's keywords found from the user model, the weight of the tag from the user model is added to the score. The top k scoring movies are then selected as the recommendations. The precision of the user model is the number of user-rated movies in the recommendation list. That is, if the list consists solely of original k movies, the precision is 1.0.

Table 2. Comparison of the recommendation scores for each of the keyword extraction methods

	IKE	CollabRank	Chi-squared	TF-IDF	KeyGraph	χ^2 KE	Baseline
Precision	0.55	0.41	0.84	0.86	0.30	0.33	0.39
Coverage	0.75	0.59	0.27	0.29	0.86	0.85	0.89
Total Score	**0.41**	**0.24**	**0.23**	**0.25**	**0.26**	**0.28**	**0.35**

Results. The baseline method used here is the same as before, i.e., all nouns and adjectives are selected as keywords. To assess the precision of the model, we skipped the movies that did not have any matching tags in the user model. This was done to simulate an actual recommendation system: when assessing the precision, we are only interested in movies that can be linked to the user model.

In some cases, for example, with KeyGraph, there was a problem of overspecialization as the approach produced too specific models. In these cases the model was able to produce only a very limited number of recommendations. An example of this was James Bond movies: the model consisted solely of keywords like agent, and James Bond. When using this model, the recommendation precision of James Bond movies was high but it could not recommend any other movies. With users that only had rated James Bond movies, the recommendation precision was high but for other users the models were unusable. As we wanted to experiment the use of the keywords for recommendation, we wanted to emphasize broader models. Therefore, in addition to precision we included *coverage* to the assessment of performance. Coverage, in this case, measures the percentage of users with broad enough models that can produce recommendations. That is, recommendation coverage is the percentage of users that receive recommendations using the user model. The *recommendation score* for each of the approach is then calculated as *recommendation precision* × *recommendation coverage*.

Table 2 shows the results of our experiment. When all the nouns and adjectives are used as tags in the user model, the average precision was 0.39, i.e., approximately 2 movies out of 5 were found from the top-5 list. The recommendation coverage for the baseline, i.e., the percentage of users that receive recommendations, was 89 %. This produced the recommendation score of 0.35. When using IKE, the precision was 0.55 with the recommendation coverage of 75 %, making the score of 0.41. We consider this result better as the precision is considerably higher and the coverage is good. Finally, CollabRank produced the score of 0.24, TF-IDF 0.25, and χ^2 0.23.

The problem with Chi-squared and TF-IDF was that the models were too narrow as they extract only uncommon words. KeyGraph and χ^2 KE produced models that contained only words that are common. Only the baseline and IKE were able to produce models that contained both uncommon and common words. However, as the baseline contained words that were not informative, its precision was poor.

These results indicate that the extraction of the informative words is more feasible than extraction of all nouns and adjectives. In addition, we can see that IKE can extract more useful words for recommendation than CollabRank, TF-IDF and χ^2. However, if we are able to combine the precision of Chi-squared and/or TF-IDF with IKE, the results might improve.

4.4 Manual Evaluation of TV-Show Keywords for User Profiling

In this experiment the keywords were extracted from TV-show descriptions using IKE. Currently, we are using the keyword extraction and user modeling in a personalized TV-guide. As this is a real world system, evaluation of the quality of the extracted keywords is important. In this experiment, we describe the results of the evaluation. The keywords assessed here were extracted from approximately 100 TV-show descriptions found from skimm.tv. The descriptions were in Finnish.

Table 3. Keyword scoring guidelines. The examples of the different levels are only suggestive as the relevancy depends heavily on the context.

Score	Guideline	Examples
1.0	Very relevant keywords; name of the show, type of the show, contents of the show	Matlock, Documentary, Africa
0.75	Relevant keywords; Contents of the show (not as descriptive as above)	Wildlife, Dog
0.50	Quite Relevant keywords; Contents of the show (relevant in the context)	Animal, Nature, Steel, Ocean
0.25	Quite irrelevant but may be useful in the context; name of the characters	Ben, Arthur, Joey
0.0	Useless; Noise, non-descriptive words	TV-Show, Season, Episode

As the different TV-shows form naturally their own clusters, instead of agglomerative clustering we cluster the documents based on the title of the TV-shows. In addition, we included a list of words the approach should always extract. This list was created by the experts at skimm.tv and it contains such words as "Documentary", "Sports", and "News". This was done to guarantee that these keywords are always extracted. We rated these words also during the evaluation.

We used the following method for evaluating the results: for each extracted keyword we assess its relevancy in the context; i.e., how well the keywords describe the content. If the keyword was very relevant, we gave it a score of 1.0. If it was relevant/somewhat useful, we gave it either 0.75, 0.5 or 0.25 depending on the level of relevance. If the keyword was not useful at all, we gave it 0.0. Table 3 shows examples of the types of keywords and their scores. There were two evaluators and we report the average relevancy scores from both of them. In addition, we counted the number of very relevant keywords (score 1.0) that the method did not extract from the descriptions. It should be noted that this is a very subjective experiment and gives only some indication about the feasibility of the keywords.

In this experiment, the relevancy scores for the extracted keywords were: for the first evaluator 0.66, and the second evaluator 0.58, making the average of 0.62. That is, both evaluated that most of the keywords were at least quite relevant. In addition, the evaluators gave the score 0.0 on average to 1.1 and 1.4 keywords in the keyword list of usually nine keywords (15 %). There were only 0.32 keywords per document on average that would have been very relevant (score 1.0) but were not extracted from the description.

Even though we can see that this experiment is quite subjective, the results indicate that the overall quality of the extracted keywords was quite. There were several extracted keywords where the evaluators disagreed on the relevancy but overall the high average score of 0.62 indicates that the keywords are useful.

The list of predefined keywords had a small impact on the results. The average relevancy score for the evaluated keywords from the predefined keyword set was 0.82 (evaluator 1) and 0.71 (evaluator 2). The relevancy score for the rest of the keywords was 0.62 (evaluator 1) and 0.52 (evaluator 2). However, after a careful review of the results we see that half of the keywords extracted due to their presence in the predefined list would have been extracted in any case due to their high score. In addition, only 10 % of the extracted keywords are from the predefined list.

After reviewing these results we consider our approach feasible. However, using the extracted keywords from TV-series like soap operas is not feasible, as the description concentrates on the plot of the episode. For documentaries and other shows where the description is more relevant, the extracted keywords are very useful. In our opinion, the quality of the extracted keywords is good and they can be used in user models. However, we also believe that the utilization of the keywords is not trivial. We may need to include other information to the model and use keyword pairs so that the context of the keywords would be utilized. For example, combining words like "Romantic,Comedy" instead of using the them alone as "Romantic" and "Comedy" would be very useful for recommendation. At the moment we are integrating a semantic analyzer to the recommender that enriches the keywords with semantic information.

5   Conclusions

In this paper, we have described the challenge of keyword extraction from short documents. We consider a document short when it contains at most 100 words, which is equal to a short abstract in a research article. We have proposed Informativeness-based Keyword Extraction (IKE) approach for extracting keywords from short documents. It is based on word evaluation that is done in three levels: corpus level, cluster level, and document level. We use text clustering to find documents of similar topic and use the clusters in the cluster level word evaluation.

We performed three experiments and compared the results against several other keyword extraction approaches. The results are promising as IKE out-performed the competition. When comparing the effectiveness of the extracted keywords for user modeling and recommendations, we demonstrated that by extracting the most relevant common and uncommon words, IKE produced considerably better results than any other approach. This is encouraging as it shows the feasibility of Informativeness-based Keyword Extraction for user modeling and recommendation.

The biggest challenge with this approach is the number of parameters to be tuned. In this paper we have provided values for the parameters that we feel are a good starting point when using the approach. In addition, these parameters have been designed such that their impact is small; i.e., they fine tune and improve the results but are not the main component of the results. We believe that IKE can produce good results even if the parameters are not used. However, we leave a more comprehensive evaluation of each of the parameters and their impact for future work.

In the future, we aim to include a more sophisticated noun phrase identification as it may benefit summarization and user modeling. In addition, we plan to study the parameters and each of the components of the proposed approach to see if it can be simplified and made more effective. As this is a new area of keyword extraction, we hope that our work can used as stepping stones in the research of keyword extraction and text mining from short documents.

Acknowledgements. Authors wish to thank the Finnish Funding Agency for Technology and Innovation (TEKES) for funding a part of this research. In addition, the authors wish to thank Prof. Hannu Toivonen and the anonymous reviewers for their valuable comments.

References

1. Timonen, M., Silvonen, P., Kasari, M.: Classification of short documents to categorize consumer opinions. In: Online Proceedings of 7th International Conference on Advanced Data Mining and Applications (ADMA 2011), China (2011), http://aminer.org/PDF/adma2011/session3D/adma11_conf_32.pdf (accessed October 10, 2012)
2. Timonen, M.: Categorization of very short documents. In: International Conference on Knowledge Discovery and Information Retrieval (KDIR 2012), Spain, pp. 5–16 (2012)
3. Timonen, M.: Term Weighting in Short Documents for Document Categorization, Keyword Extraction and Query Expansion. PhD thesis, University of Helsinki, Faculty of Science, Department of Computer Science (2013)
4. Ohsawa, Y., Benson, N.E., Yachida, M.: KeyGraph: Automatic indexing by co-occurrence graph based on building construction metaphor. In: Proceedings of IEEE International Forum on Research and Technology Advances in Digital Libraries (ADL 1998), pp. 12–18 (1998)
5. Matsuo, Y., Ishizuka, M.: Keyword extraction from a single document using word co-occurrence statistical information. In: Proceedings of the Sixteenth International Florida Artificial Intelligence Research Society Conference (FLAIR 2003), USA, pp. 392–396 (2003)
6. Wan, X., Xiao, J.: CollabRank: Towards a collaborative approach to single-document keyphrase extraction. In: Proceedings of 22nd International Conference on Computational Linguistics (COLING 2008), United Kingdom, pp. 969–976 (2008)
7. Timonen, M., Toivanen, T., Teng, Y., Chen, C., He, L.: Informativeness-based keyword extraction from short documents. In: International Conference on Knowledge Discovery and Information Retrieval (KDIR 2012), Spain, pp. 411–421 (2012)
8. Frank, E., Paynter, G.W., Witten, I.H., Gutwin, C., Nevill-Manning, C.G.: Domain-specific keyphrase extraction. In: Proceedings of the Sixteenth International Joint Conference on Artificial Intelligence (IJCAI 1999), Sweden, pp. 668–673 (1999)
9. Witten, I.H., Paynter, G.W., Frank, E., Gutwin, C., Nevill-Manning, C.G.: KEA: Practical automatic keyphrase extraction. In: Proceedings of the Fourth ACM Conference on Digital Libraries (DL 1999), USA, pp. 254–255 (1999)
10. Turney, P.D.: Coherent keyphrase extraction via web mining. In: Proceedings of the Eighteenth International Joint Conference on Artificial Intelligence (IJCAI 2003), Mexico, pp. 434–442 (2003)
11. Nguyen, T.D., Kan, M.-Y.: Keyphrase extraction in scientific publications. In: Goh, D.H.-L., Cao, T.H., Sølvberg, I.T., Rasmussen, E. (eds.) ICADL 2007. LNCS, vol. 4822, pp. 317–326. Springer, Heidelberg (2007)
12. Turney, P.D.: Learning algorithms for keyphrase extraction. Information Retrieval 2, 303–336 (2000)
13. Hulth, A., Karlgren, J., Jonsson, A., Boström, H., Asker, L.: Automatic keyword extraction using domain knowledge. In: Gelbukh, A. (ed.) CICLing 2001. LNCS, vol. 2004, pp. 472–482. Springer, Heidelberg (2001)
14. Hulth, A.: Improved automatic keyword extraction given more linguistic knowledge. In: Conference on Empirical Methods in Natural Language Processing (EMNLP 2003), Japan, pp. 216–223 (2003)
15. Hulth, A.: Enhancing linguistically oriented automatic keyword extraction. In: Proceedings of the Human Language Technology Conference - North American Chapter of the Association for Computational Linguistics (HLT-NAACL 2004), USA, pp. 17–20 (2004)
16. Ercan, G., Cicekli, I.: Using lexical chains for keyword extraction. Information Processing and Management 43, 1705–1714 (2007)

17. Paukkeri, M., Nieminen, I.T., Pöllä, M., Honkela, T.: A language-independent approach to keyphrase extraction and evaluation. In: Posters Proceedings of 22nd International Conference on Computational Linguistics (COLING 2008), United Kingdom, pp. 83–86 (2008)

18. Mihalcea, R., Tarau, P.: Textrank: Bringing order into text. In: Proceedings of the 2004 Conference on Empirical Methods in Natural Language Processing (EMNLP 2004), A meeting of SIGDAT, a Special Interest Group of the ACL, held in conjunction with ACL 2004, Spain, July 25-26, pp. 404–411 (2004)

19. Page, L., Brin, S., Motwani, R., Winograd, T.: The PageRank citation ranking: Bringing order to the web. Technical report, Stanford (1998)

20. HaCohen-Kerner, Y.: Automatic extraction of keywords from abstracts. In: Palade, V., Howlett, R.J., Jain, L. (eds.) KES 2003. LNCS, vol. 2773, pp. 843–849. Springer, Heidelberg (2003)

21. HaCohen-Kerner, Y., Gross, Z., Masa, A.: Automatic extraction and learning of keyphrases from scientific articles. In: Gelbukh, A. (ed.) CICLing 2005. LNCS, vol. 3406, pp. 657–669. Springer, Heidelberg (2005)

22. Andrade, M., Valencia, A.: Automatic extraction of keywords from scientific text: Application to the knowledge domain of protein families. Bioinformatics 14, 600–607 (1998)

23. Kim, S., Medelyan, O., Kan, M., Baldwin, T.: Semeval-2010 task 5: Automatic keyphrase extraction from scientific articles. In: Proceedings of the 5th International Workshop on Semantic Evaluation (ACL 2010), pp. 21–26 (2010)

24. Yih, W., Goodman, J., Carvalho, V.R.: Finding advertising keywords on web pages. In: Proceedings of the 15th international conference on World Wide Web (WWW 2006), Scotland, May 23-26, pp. 213–222 (2006)

25. Tomokiyo, T., Hurst, M.: A language model approach to keyphrase extraction. In: Proceedings of ACL Workshop on Multiword Expressions (2003)

26. Rennie, J.D.M., Jaakkola, T.: Using term informativeness for named entity detection. In: Proceedings of the 28th Annual International ACM SIGIR Conference on Research and Development in Information Retrieval (SIGIR 2005), Brazil, pp. 353–360 (2005)

27. Clark, K., Gale, W.: Inverse Document Frequency (IDF): A measure of deviation from Poisson. In: Third Workshop on Very Large Corpora, pp. 121–130. Massachusetts Institute of Technology, Cambridge (1995)

Part II

Knowledge Engineering
and Ontology Development

Applying Simple Ontology Relations for Receiving Better Recommendations

Lamiaa Abdelazziz and Khaled Nagi

Dept. of Computer and Systems Engineering, Faculty of Engineering,
Alexandria University, Egypt
{lamiaa.abdelazziz,khaled.nagi}@alexu.edu.eg

Abstract. Huge number of documents must be shared in virtual organizations; which render artifact recommender systems indispensable. Recommender systems use several information retrieval techniques to enhance the quality of their results. However, the problem arises when a user tries to search for some information in his/her peers' exposed data due to the difference in classification systems in use. The seeker categories must be matched with the responders categories. In this work, we purpose a way to enhance the recommendation process based on using simple implicit ontology relations. This helps in recognizing better matched categories in the exposed data. We show that this approach improves the quality of the results using two different real-life datasets.

Keywords: Recommender Systems, Ontology Mapping, Quality of Recommendation, Performance Analysis.

1   Introduction

Due to the autonomous nature of the participants in virtual organizations [1], knowledge sharing cannot be done in a structured and centralized way. Peers in a virtual organization possess huge amount of documents and each peer or group of peers have their own way of classifying them. This creates great challenges to any document recommender system. Recommender Systems (RS) acting in this environment must be distributed and autonomous in order to match the nature of virtual organizations. Typical RS use several Information Retrieval (IR) techniques to generate good results. Moreover, RS should also match the category structure of the seeker with that of the responder. For this to work, the search engines within the RS should be extended.

We base our work on KARe; which stands for Knowledgeable Agent for Recommendations [2] and [3]. It is a multi-agent recommender system that supports nomadic users sharing knowledge in a peer-to-peer environment with the support of a nomadic service. We extend this system in order to enhance the quality of the results coming from search component of the RS. A way of doing this is by enriching the search query and enhancing the ranking process of the result set. We employ extra ontological information provided by the peers. In our distributed and autonomous scenario, we

A. Fred et al. (Eds.): IC3K 2012, CCIS 415, pp. 149–164, 2013.

restrict ourselves to using simple implicit ontological relations since we do not want to burden the peer with defining their own elaborate ontology (or else they will simply not do it) or force them to use a centralized ontology (since it is not applicable in such a heterogeneous environment).

However, improving the quality of results often involves more computation. For this reason, we test our extension against the original system using the same dataset to quantify the increase in quality versus the increase in computation cost during indexing and searching. We also use a second dataset to verify the generality of our solution.

The rest of the paper is organized as follows. Section 2 provides a background on recommender systems. Our proposed system is presented in Section 3. Section 4 contains an assessment of our proposed system and its implementation while Section 5 concludes the paper.

2   Background

2.1   Recommender Systems

[4] provides a good taxonomy for distinguishing between recommender systems. Recommender Systems can be categorized in the following classes: *content-based*, *collaborative filtering*, *demographic*, *knowledge-based*, *community-based*, and *hybrid recommender systems*.

In *content-based* RS, the system learns to recommend items that are similar to the ones that the user liked in the past. The similarity of items is calculated based on the features associated with the compared items. Content-based RS can be even found in early standard literature as in [5].

Collaborative filtering RS are also called "people-to-people correlation". In their simplest form, implementations of this approach recommend to the active user the items that other users with similar tastes liked in the past [6]. The similarity in taste of two users is calculated based on the similarity in the rating history of the users.

Demographic RS recommend items based on the demographic profile of the user. Many web sites dispatch their users to particular pages based on their language or country. Other criteria include age, gender, etc., if this information is collected in the user profile.

Knowledge-based RS recommend items based on specific domain knowledge about how certain item features meet user needs and preferences. Most knowledge based recommender systems are constraint based or case-based [7]. In these systems, a similarity function estimates how much the user needs match the recommendations. Here the similarity score can be directly interpreted as the utility of the recommendation for the user.

Community-based RS recommend items based on the preferences of the user friends. The emergence of social networks gave rise to this type of systems. Social networks contain billions of records holding user behavioral patterns and combining them with a mapping of their social relationships.

Hybrid RS are based on the combination of the above mentioned techniques. Collaborative filtering methods suffer from new item problems, i.e., they cannot recommend items that have no ratings. This does not limit content-based approaches since the prediction for new items is based on their description (features) that are typically available. Given two (or more) basic RS techniques, several ways have been proposed for combining them to create a new hybrid system. Four different recommendation techniques and seven different hybridization strategies are compared in [4].

Complementary Role of Information Retrieval. Information Retrieval (IR) assists users in storing and searching various forms of content, such as text, images and videos [8]. IR generally focuses on developing global retrieval techniques, often neglecting the individual needs and preferences of users.

Nevertheless, both IR and RS are faced with similar filtering and ranking problems. That's why at the heart of RS usually lies a search engine, such as the open source Lucene [9]. Queries submitted to the search engine are enriched with RS-relevant attributes collected by the RS and associated to the resultset.

Nowadays, various search engines also apply some form of personalization by generating results to a user query that are not only relevant to the query terms but are also tailored to the user context (e.g., location, language), and his/her search history. Clearly, both RS and IR will eventually converge to one intelligent user assistant agent.

Complementary Role of Taxonomies and Ontologies. Taxonomy is a hierarchical grouping of entities. Ontology is a machine readable set of definitions that create a taxonomy of classes and subclasses and relationships between them [10].

Both taxonomies and ontologies are used to enhance the quality of results suggested by RS. The RS can use the taxonomy structures and ontology to refine the filtering and adjust the ranking of the results sent by the IR internal component. Since most of our knowledge is not hierarchical, it is intuitive to assume that an ontology-based approach would lead to better results. Yet, there is an overhead in defining ontologies by the user and creating a match for the nodes of different ontologies in case of peer-based autonomous systems, such as multi-agent environments.

Clearly, a trade-off would be using *implicit* ontology relations which can easily defined by the user and then matched by the system. The *simplicity* of the definition of the ontology is critical factor in convincing the autonomous peer to define it.

2.2   Example of Artifact Recommendation Systems

KARe (Knowledgeable Agent for Recommendations) [2-3] is a multi-agent recommender system that supports nomadic users sharing knowledge in a peer-to-peer environment. Supporting social interaction, KARe allows users to share knowledge through questions and answers. Furthermore, it is assumed that nearby users are more suitable for answering user questions in some scenarios and it uses this information for choosing the answering partners during the recommendation request process.

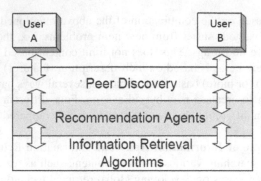

Fig. 1. KARe architecture

The first goal of KARe is to develop a distributed system for artifact recommendation. It aims at increasing the precision of current recommendation algorithms. KARe mainly consists of the following components, illustrated in Fig. 1 [11].

The *information retrieval* component is divided into two parts. The first part is the process where the user *creates an index* of the knowledge artifacts and concepts. The second is the recommendation mechanism which consists of a *searching* process for the knowledge artifacts. KARe includes the user context in the searching mechanism, providing semantics to the artifacts (i.e., relating it to the concept it is associated with). Similar documents are grouped by the user under the same concept in the context tree. Before submitting the query, the user assigns it to a specific concept. By doing this, the user gives the system extra information on the query content leading to more accurate results.

The *recommendation agent* component simulates the natural social process involved in knowledge sharing by exchanging requests (*questions*) and recommendations (*answers*). Social interaction involved in the recommendation process is modeled as agent interaction.

The *peer discovery* component finds potential peers based on proximity information. The system scans the neighborhood for other devices mainly using Bluetooth. The KARe scanner prepares a message and sends it to the *peer assistant* agent. If the peer assistant finds the agent representing the device in the KARe platform, then it sends a message back to the KARe scanner.

We choose KARe as a base for our work due to the following reasons:

- Its multi-agent nature suits the environment of autonomous virtual organizations.
- Documents fit perfectly in the artifact concept of KARe.
- It comprises a standard search engine: Lucene [9].
- It originally uses taxonomies in structuring its recommendations.
- It is extendible due to its origins in research labs.
- It belongs to the most general class of RS; which is the *hybrid* family. It combines knowledge-based, location (similar to demographic), community-based, and content-based approaches.

3   Proposed Solution

One challenge in KARe is gaining user acceptance to spend more time in feeding the system with documents and classifying them. Since the way each user classifies his/her own knowledge is particular, we cannot impose a common classification for their artifacts.

Each user is allowed to define and use his/her own taxonomy represented in OWL format [12] to classify artifacts. Fig. 2 (left) illustrates a user-defined ontology that holds the artifacts; computer science research papers in this case. This ontology expresses the peer's point of view and does not typically match the standard classification system of the Association of Computing Machinery [13] used as a reference base in KARe and illustrated in Fig. 2 (right).

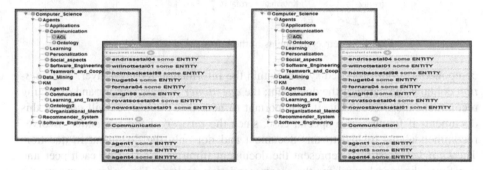

Fig. 2. User-defined vs. ACM ontology

We integrate the Protégé Ontology Editor [14] in our proposed system to enable the peer editing his/her OWL ontology file. Through the editor, the user can extend the basic taxonomy trees with other implicit relations such as *sibling*, *parent* and *related-to* between the different tree nodes. The more relations are added, the more support is given for the recommender system in detecting the best matched categories. Fig. 3 illustrates a sample ontology relation added to the ACM taxonomy.

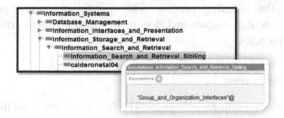

Fig. 3. Sample ontology relation added to ACM

3.1   Component Architecture

The ontology files defined in the previous section are fed to the adapted information retrieval component first proposed in [15]. Fig. 4 illustrates the integration of new

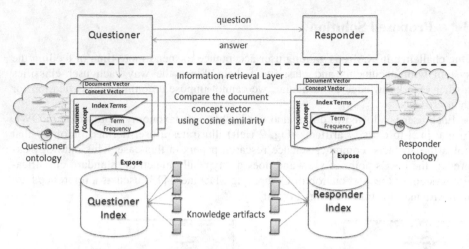

Fig. 4. System components of the information retrieval layer

components of our proposed system within the information retrieval layer. The *question* contains the name of the document; the *questioner* is searching for and expecting recommendations around it. It is important to note that the answer will contain this document- if found - as a special case, since the *answer* of the *responder* involves recommendations related to this document and not only the document in question. The *knowledge artifacts* represent the document libraries exposed by each peer and constitute the search pool for the recommendation seekers. The *index terms* are the words contained in each document after performing text preprocessing steps. The *term frequency* vector defines the frequency of occurrence of each specific term in a document. The *document vector* represents the terms in the documents together with their frequencies. It is used to formulate the *document matrix* where the documents are represented as rows and the terms as columns and the term frequencies as cell values. The *concept vectors* represent concepts or nodes in the ontology. The dimension size is that of the vocabulary (i.e., number of indexed terms). In order to create the vectors it is necessary to read all indexed terms and count their occurrences in the documents during the indexing process. The *questioner* and *responder indices* represent the store for the concept and document vectors representing the indexed terms and term frequencies.

3.2   The Indexing Process

Fig. 5 shows the indexing sequence diagram. The *Indexer* is the class that receives the method call `createIndex` from the user and is responsible for handling the process. The Indexer receives two parameters: *a list of documents to be indexed* and *the ontology that classifies them*. The first step is to parse each concept of the ontology and the related relations associated with each concept. During the concept parsing, the indexer parses each knowledge artifact to create the vocabulary and index terms. Once indexing is completed, the Indexer creates the vectors for the concepts and

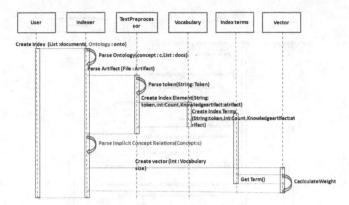

Fig. 5. Indexing sequence diagram

knowledge artifacts. During this step, the weight for each term in each document and concept is calculated and stored within the index.

3.3   The Searching Process

The searching sequence diagram is shown in Fig. 6. The major contribution is shown in the last part of the sequence diagram after returning the best matching concepts. Here, we use the implicit ontology relations saved during the indexing process and retrieve the corresponding concept vectors and their artifacts to include them in the calculation process. The similarity is calculated and the query vector and the list of the retrieved documents are ranked according to the similarity. The following is a description of the methods in Fig. 6.

- Query: The parameters for the query are the question itself, the vocabulary spoken by the questioning peer and the concept vector associated with the question.
- ParseQuery: is a method for pre-processing the question. It performs stemming and removes the stop words from the query.
- CreateQueryVector: compares the question with the knowledge artifacts. For this method to work, we must create a vector representation of the question itself.
- GetVector, StoreSimilarity and GetBestConcepts: the questioning concept vector is compared to each concept vector on the destination taxonomy. For that, we retrieve each vector and check its similarity with the questioning concept vector. At the end of the process, we are able to retrieve the best matching concepts with the questioning concept.
- NormalizeConceptVector: aligns the questioning concept vector with the targeted vectors and vocabulary.
- GetListOfRelatedConcepts: for each of the best matched concepts, we retrieve the related concepts to include in the next comparison.
- RecalculateSimilarity: in this step, we check the similarity between the newly related concepts and the best matched concepts.

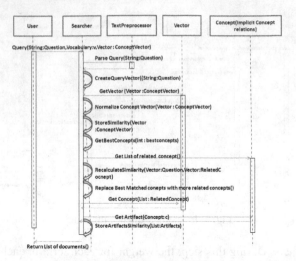

Fig. 6. Searching sequence diagram

- `ReplaceBestMatchedconcepts`: if the similarity calculation shows better concepts we reorder and replace the selected *three* concepts with better ones. The number three is arbitrary chosen and can be changed in the configuration files.
- `GetConcept` and `GetAtrifact`: retrieves the concept and its corresponding artifacts.
- `StoreArtifactSimilarity`: calculates the similarity among the documents from the related concepts and the question vector. The method returns the resulting documents with associated similarities.

The pseudo-code of the searching steps is shown in Fig. 7.

```
Process createAnswer (Vector: conceptVector, String: question)   {
// step 1:  Create the query vector
Get Question Terms
   Calculate Terms frequencies
   Set Terms frequencies in the query vector
   // step 2:  normalize the concept vector
Get Intersection between questioner and responder vectors
   Prepare projected responder document weight matrix
   // step 3:  select best matching concept
Loop on all responder concepts
   Compute similarity between responder vector and normalized questioner
          vector
   Sort the concepts list by similarity
   Return the best matched three concepts
// step 4:  Use indexed implicit ontology relations
Get the implicit indexed relations for the retrieved concepts
   Load the related concepts
   Compute similarity between responder vector and normalized related
          concept vector
   Add the concepts to the concepts list
   Resort the new list
   Return the three best matched  concepts
// step 5:  Load the artifacts and send results to user
   Load all the artifacts under the three best matched concepts
   Compute similarity between query vector and best matched concepts
          vectors
   Return the list to the user.
}
```

Fig. 7. Recommendation algorithm

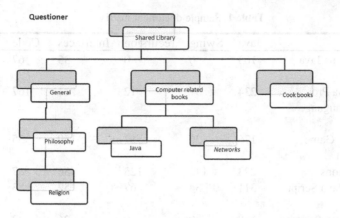

Fig. 8. Questioner library categorization

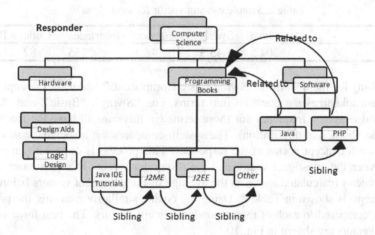

Fig. 9. Responder-specific implicit ontology relations

3.4   Example

The following is a simplified example of the recommendation process performed by our system. Fig. 8 shows the questioner categorization of its library while Fig. 9 illustrates that's of the responder after adding few implicit ontological relationships.

During the indexing process, concepts and document are parsed to build the document matrix. Assuming the following six books: "Design Patterns Java Workbook", "Effective Java Programming Language Guide", "Micro JAVA Game Development", "Java Collections", "Client-Side Java Script Reference", and "Java 2 Network Security" under the concept "Java", the resulting document matrix is illustrated in Table 1 and the concept vector for "Java" is illustrated in Table 2.

Table 1. Sample document matrix

	Java	Swing	recursion	Interfaces	Code	Pattern
Design Patterns Java Workbook	103	0	0	35	67	156
Effective Java Programming Language Guide	234	57	12	211	167	20
Micro JAVA Game Development	176	36	87	150	55	55
Java Collections	223	145	123	456	87	98
Client-Side Java Script Reference	311	34	67	52	83	45
Java 2 Network Security	180	0	0	26	33	22

Table 2. Sample concept vector for term "Java"

	Java	Swing	recursion	Interfaces	Code	Pattern
JAVA	204.5	45.3	48.1	155	82	66

Searching for the document "Swing Basic Components" under the concept "java", the system tokenizes the question into terms, i.e., "Swing", "Basic", and "Components" and assigns a frequency for those terms (in this case all have the frequency 1 due to the absence of duplication). The system compares the associated concept vector with each concept vector of the responder. This process is done with an intersection between the questioner vector and the responder vector. The questioner selected concept vector (calculated above as the average of the document vectors belonging to that concept) is shown in Table 2. Using the cosine similarity measure, the projected vector is compared to each of the responder concept vectors. The best three matched concept vectors are shown in Fig. 10.

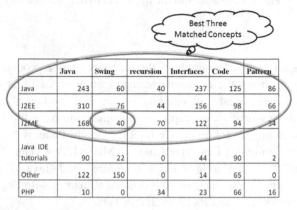

Fig. 10. Best three matched concepts

"Java", "J2EE", and "J2ME" are the three most related concepts considering the cosine similarity of the concept vectors. This list is updated taking into consideration the *related concept list* from the implicit ontological relations generated during the indexing phase. Assuming that "Programming books" is related to "Java", "other" to "J2EE", and "J2EE" to "J2ME", the concept called "other" represents one of the user classifications and indicates that matching concepts does not necessarily depend on the name of the classification and that all of them are included in the cosine similarity calculation. In this particular example, searching for a book about Swing, the system searches in all siblings of java, J2EE and J2ME concepts and replace related concepts. According to the cosine similarly, the system replaces "J2ME" with the concept "other" since it has a relative term frequency of 150 for the term "swing" as compared to 40 for the same term in the concept "J2ME" as illustrated in Fig. 11.

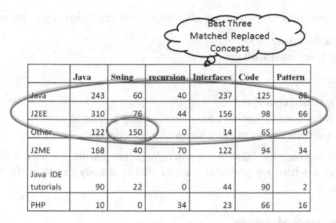

Best Three Matched Replaced Concepts

	Java	Swing	recursion	Interfaces	Code	Pattern
Java	243	60	40	237	125	86
J2EE	310	76	44	156	98	66
Other	122	150	0	14	65	0
J2ME	168	40	70	122	94	34
Java IDE tutorials	90	22	0	44	90	2
PHP	10	0	34	23	66	16

Fig. 11. Best three matched concepts after replacement

4   Assessment

4.1   Datasets

We use *two* different datasets. The first dataset is the same one used in evaluating the original KARe system. We insist on using the same dataset to quantify the increase in quality versus the increase in computation cost during indexing and searching. A summary of the dataset is found in Table 3 (Dataset I). Taxonomy *A* collects papers and classifies them according to user specific point of view. Taxonomy *B* is taken from the ACM Classification System. In our assessment, we simulate the questions and answers using the title and the body of the scientific papers. We test whether the algorithm is able to retrieve a paper giving its title or keywords from its abstract.

We also use a second dataset to ensure the generality of our solution. The second dataset is shown in Table 3 (Dataset II). The second dataset is a real library of programming books found at a medium sized software company. It is classified from two points of views and is used as questioner and responder exposed libraries. This dataset consists of 125 programming books as a questioner source and 206 as a responder target; thus slightly smaller that dataset I but has the advantage of having much higher

Table 3. Summary of datasets

Dataset	I		II	
Taxonomy	A	B	A	B
Number of documents	250	315	125	206
Number of concepts	28	15	32	24
Average doc/concept	9	21	4	9

terms frequencies, since every book contains hundreds of pages unlike the papers in dataset I with maximum of 20 pages per paper.

4.2   Input and Output Settings

To start the experiment execution, we give the system the following as input:

- the title of a document, and
- the concept associated with the paper.

The outputs are:

- the list of matched concepts,
- the list of documents classified under the resulting concepts, and
- the cosine similarity measure attached with each concept.

To distinguish between the false/true positive/negative alarms, we search for a specific paper or book where we previously know that it already exists in the target peer shared data pool.

4.3   Performance Measures

In our assessment, we use the following standard performance indices:

- Number of document *hits*: the average number of hits returned by the responder.
- *Recall*: is the fraction of the documents that are relevant to the query that are successfully retrieved.
- *Precision*: the fraction of retrieved documents relevant to the search.
- *F1-Measure*: is a measure of test accuracy. It considers both the precision and the recall The F1-Measure can be interpreted as a weighted average of the precision and recall, where an F1-Measure reaches its best value at 1 and worst score at 0.

In addition to scalability measures such as *Indexing time* vs. *the number of documents*, and *Searching time* vs. *the number of documents*.

4.4   Results of Dataset I

In this set of experiments, we perform 75 queries. A summary of the results is shown in Table 4. Under the original implementation, the number of documents found is 42. Using our ontology-based solution, the number increases to 59. The average recall is also increased from 0.573 to 0.786. Fig. 12 shows the detailed plotting of the recall

Table 4. Summary of the result of dataset I

	3 concepts (taxonomy)	3 concepts (ontology)
Document recall	0.573	0.786
Concept precision	0.153	0.238
F1-Measure	0.240	0.365

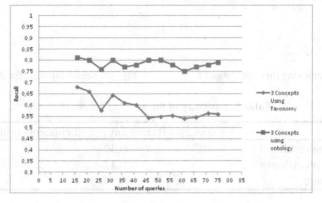

Fig. 12. Recall vs. number of queries for dataset I

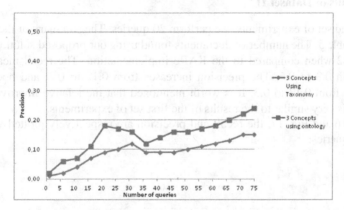

Fig. 13. Precision vs. number of queries for dataset I

versus the number of queries. The precision is also enhanced from 0.153 to 0.238. Fig. 13 shows the detailed plotting of the concept precision versus the number of queries. Consequently, the derived F1-Measure is enhanced from 0.24 to 0.365.

Considering the scalability measures, our proposed solution incurs a higher cost of computation - as expected - due to the increase in the result quality. The good news is that both indexing time, illustrated in Fig. 14, and searching time, illustrated in Fig. 15, increase with the same rate as the original KARe implementation. The relative increase in processing compared to KARe does not increase above 10% which is a fair price to pay for the improvement in quality especially that the absolute values for both indexing and searching times are very acceptable.

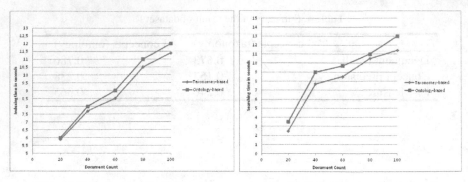

Fig. 14. Indexing time for dataset I   **Fig. 15.** Searching time for dataset I

Table 5. Summary of the result of dataset II

	3 concepts (taxonomy)	3 concepts (ontology)
Document recall	0.30	0.60
Concept precision	0.12	0.20
F1-Measure	0.17	0.30

4.5   Results of Dataset II

In this second set of experiments, we perform 20 queries. The summary of the results is shown in Table 5. The number of documents found using our proposed solution doubles from 6 to 12 when compared to the KARe implementation. The recall measure also doubles from 0.3 to 0.6. The precision increases from 0.12 to 0.2; and boosting the F1-Measure from 0.17 to 0.3. It is worth mentioned that the relative improvement for this dataset is very similar to the results of the first set of experiments.

In Fig. 16 and Fig. 17, the recall and precision are respectively plotted versus the number of queries.

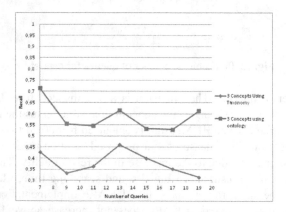

Fig. 16. Recall vs. number of queries for dataset II

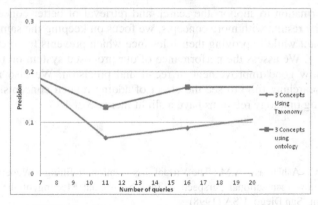

Fig. 17. Precision vs. number of queries for dataset II

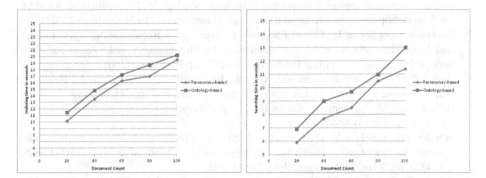

Fig. 18. Indexing time for dataset II   **Fig. 19.** Searching time for dataset II

The scalability measures reveal a slight increase in the computation time here too. Again, the relative increase in both the indexing time, illustrated in Fig. 18, and the searching time, illustrated in Fig. 19, are around 10% for all values of the document counts. An interesting observation, however, is made when comparing the absolute indexing times of experiment I, illustrated in Fig. 14, with that of experiment II, illustrated in Fig. 16. The large increase in indexing time is attributed to the large document size of dataset II (books) as compared to the document size of dataset I (papers). This difference is not present in the searching time due to the scalable nature of the B+-Trees of Lucene regarding retrieval.

5   Conclusions

Our main contribution in this work is integrating ontological concepts into the recommendation process. We extend the information retrieval part of multi-agent recommender system KARe by allowing the definition of simple ontological relations.

The simple and implicit ontological relations, such as sibling, parent/child and related-to relations, are presented as data properties in the OWL file. Saving those concepts during the indexing process and using them in the searching process gives additional information to support the search and retrieval of better concepts. Instead of increasing the results with more concepts, we focus on keeping the same number of concepts constant while improving their relevance which prevents the precision value from decreasing. We assess the performance of our proposed system on two datasets. The results show good improvement in recall and precision. We also measure the indexing and searching time to see the effect of adding related concepts. The results show that adding ontology relations have a slight increase of 10%.

References

1. Fuehrer, E.C., Ashkanasy, N.M.: The Virtual organization: defining a Weberian ideal type from the inter-organizational perspective. In: The Annual Meeting of the Academy of Management, San Diego, USA (1998)
2. Gomez Ludermir, P., Guizzardi-Silva Souza, R., Sona, D.: Finding the right answer: an information retrieval approach supporting knowledge sharing. In: Proceedings of AAMAS 2005 Workshop on Agent Mediated Knowledge Management, The Netherlands (2005)
3. Guizzardi-Silva Souza, R., Gomes Ludermir, P., Sona, D.: A Recommender Agent to Support Knowledge Sharing in Virtual Enterprises. In: Protogeros, N. (ed.) Agent and Web Service Technologies in Virtual Enterprises. Idea Group Publishing (2007)
4. Burke, R.: Hybrid web recommender systems. In: Brusilovsky, P., Kobsa, A., Nejdl, W. (eds.) Adaptive Web 2007. LNCS, vol. 4321, pp. 377–408. Springer, Heidelberg (2007)
5. Balabanovic, M., Shoham, Y.: Content-based, collaborative recommendation. Communication of ACM 40(3), 66–72 (1997)
6. Schafer, J.B., Frankowski, D., Herlocker, J., Sen, S.: Collaborative filtering recommender systems. In: Brusilovsky, P., Kobsa, A., Nejdl, W. (eds.) Adaptive Web 2007. LNCS, vol. 4321, pp. 291–324. Springer, Heidelberg (2007)
7. Bridge, D., Göker, M., McGinty, L., Smyth, B.: Case-based recommender systems. The Knowledge Engineering Review 20(3), 315–320 (2006)
8. Manning, C.: Introduction to Information Retrieval. Cambridge University Press, Cambridge (2008)
9. Hatcher, E., Gospodnetic, O.: Lucene in Action. Manning Publications (2004)
10. Deng, S., Peng, H.: Document Classification Based on Support Vector Machine Using A Concept Vector Model. In: The IEEE/WIC/ACM International Conference on Web Intelligence (2006)
11. Gomez Ludermir, P.: Supporting Knowledge Management using a Nomadic Service for Artifact Recommendation. Thesis for a Master of Science degree in Telematics, from the University of Twente Enschede, The Netherlands (2005)
12. OWL2 Web Ontology Language. Document Overview. In: W3C Recommendation (2009), http://www.w3.org/TR/owl2-overview/
13. The ACM computing classification system, http://www.acm.org/about/class/ccs98-html
14. Tudorache, T., Noy, N.F., Tu, S., Musen, M.A.: Supporting Collaborative Ontology Development in Protégé. In: Sheth, A.P., Staab, S., Dean, M., Paolucci, M., Maynard, D., Finin, T., Thirunarayan, K. (eds.) ISWC 2008. LNCS, vol. 5318, pp. 17–32. Springer, Heidelberg (2008)
15. Abdelazziz, L., Nagi, K.: Enhancing the Results of Recommender Systems using Implicit Ontology Relations. In: International Conference on Knowledge Engineering and Ontology Development (KEOD 2012), Barcelona, Spain (2012)

A Knowledge-Based Approach to the User-Centered Design Process

Stefan Negru and Sabin Buraga

Faculty of Computer Science Alexandru Ioan Cuza University of Iasi, Romania
{stefan.negru,busaco}@info.uaic.ro

Abstract. User-Centered Design is an approach for designing usable products and systems, which encompasses a collection of techniques, procedures and methods, placing the user at the center of all of them. Our current research is focused on two methods: Personas and Usability Testing. While one is used for communicating user requirements, the other one is used to evaluate a product's ease of use by observing the users behavior. Although both methods are used to collect data regarding user needs, preferences and behavior, little research has been conducted linking the collected data with other pieces of data concerning the product development and design process. This paper considers the use of semantic web technologies, such as microdata and ontologies, in order to provide a conceptual model as a basis for structuring, extracting and linking data collected via these two methods. Our approach consists of a set of HTML5 microdata schemas and an OWL specification, which include concepts and properties used to model personas and usability testing. In order to exemplify our model and extract desired data, we made use of semantically annotated templates for personas method and usability testing.

Keywords: Knowledge Modeling, User-centered Design, Ontology, Microdata, Personas, Usability Testing.

1 Introduction

Although numerous articles exist in literature regarding the User-Centered Design (UCD) and its corresponding techniques and methods, at its core it consists the following basic principles: "Early focus on users and their tasks; Evaluation and measurement of product usage; Iterating design" [1]. With these principles in mind we will be focusing on two widely used methods from UCD: *Personas* and *Usability Testing*.

A *persona* (term initially proposed by Cooper [2]) is regarded as a user archetype which can be used to "help guide decisions about product features, navigation, interactions, and even visual design" [3]. Ultimately, this archetype represents a group of users who share common behavioral or physical characteristics, goals, frustrations and preferences. Although it encompasses all this data regarding different users, it has one goal: to develop a profile of potential users for a certain product or system.

On the other hand "usability testing employs techniques to collect empirical data while observing representative end users using the product to perform realistic tasks" [1]. Users participating in a test are selected based on a persona, in order to identify direct users of that product/system.

A. Fred et al. (Eds.): IC3K 2012, CCIS 415, pp. 165–178, 2013.

As we can denote both methods are centered around the user, and although they are part of the same UCD process, it is difficult to establish connections and integrating the collected data (from both methods).

In order to tackle these issues, we consider the adoption of semantic web technologies [4]. RDFa[1] and microdata[2] enable the publication, extraction and reuse of data (in our case, (meta)data already stored into a HTML5[3] document). Our aim is to propose a conceptual model which consists of a set of HTML5 microdata schemas to be used in the personas and usability testing methods. By providing such an infrastructure of microdata schemas, we facilitate data extraction and – at the same time – linking that data extracted with data from other sources.

We also take into account ontologies, which provide semantics for humans and formalism for machine processing and reasoning. We propose *PersonasOnto* ontology as a way of modeling personas related concepts (such as usability testing). It provides an overview of the relationships between these methods and how they integrate in the UCD process.

This paper is an updated and revised extended version of the work presented in [5].

In the following, we first summarize related work in Section 2. Section 3 illustrates HTML5 microdata schemas and corresponding templates. Section 4 presents the *PersonasOnto* ontology and Section 5 illustrates a use case of the vocabulary. The paper ends with a conclusion and an outlook on future work.

2   Related Work

Several approaches on describing persons and user profiles have been presented in the last years, with vocabularies such as FOAF[4], SIOC[5] and Person[6]. At the same, the need for software systems [6] to automatically adapt to their users has increased, this resulted in an increased attention over the user's preferences and needs, disabilities and emotional status. In order to have a better understanding of the user, a number of domain specific ontologies concerning user profiles have been developed. Although most of these vocabularies focus on how the profile of a user can be described, most of them do not incorporate important aspects regarding the user requirements and user testing from a certain product.

Work has also been done in the direction of mapping between users, accessibility concepts and accessibility scenarios, in the AEGIS ontology[7].

A number of ontologies – as [7,8] – explore the implications of user models in the context of recommender systems, while others such as [9] present a generic ontology-based user modeling architecture applied in Knowledge Management Systems. Some of

---

[1] http://www.w3.org/TR/rdfa-core/
[2] http://www.w3.org/TR/microdata/
[3] http://www.w3.org/TR/html5/
[4] http://xmlns.com/foaf/spec/
[5] http://rdfs.org/sioc/spec/
[6] http://schema.org/Person
[7] http://www.aegis-project.eu/

the work in this direction is concerned with providing a standard ontology for modeling user profiles in order to facilitate communications between applications [10].

The review of the related work showed that most vocabularies emphasize the importance of context awareness, more details on this aspect and other relevant user modeling issues are reviewed in [11]. Furthermore, many of the proposed vocabularies focus on the user's profile, without exploring what are the implications of the user's preferences, goals and frustrations on the product or system.

3   Using Microdata to Structure Information

Even though a persona represents a fictional individual, it is created based on a group of persons. In most instances each person is interviewed, in order to establish the specific group it belongs to [12]. The data collected during these interviews is included into the persona document [13]. In this document, each archetype is developed in greater detail, along with scenarios which describe how it might interact with a product. The persona and their associated scenarios form the basis for specifying how users want to experience and interact with a certain product or application.

As with the *persona* method usability tests are based mainly on user interviews and observations, and are traditionally conducted with one test participant at a time ([14]), in the same manner as a controlled experiment. This data is compiled and interpreted using statistics or presented as is.

3.1   Personas and Usability Test Templates

As a first layer to structuring the information, we developed HTML5 templates for both the personas[8] and usability test[9], following existing guidelines [15,12,16] and the standards for usability test[10]. Both HTML documents contain information which can be semantically annotated by using microdata/RDFa, thus facilitating publication, consumption, and reuse of information.

The personas template is structured as follows:

1. *Type and Background Information* – contains details regarding the persona type and fictional background information such as birthday, name, gender, location and other information of interest;
2. *Main Details* – consists of a detailed description regarding this persona along with some characteristics (main points), goals and frustrations;
3. *Scenarios* – includes either a description of scenario tasks or images depicting scenarios;
4. *Other Details* – contains other relevant notes regarding the persona.

On the other hand, the usability test template is more complex focusing on of two main components:

---

[8] Publicly available at: http://blankdots.com/open/personas/
[9] Publicly available at: http://blankdots.com/open/usability/
[10] ISO/IEC 25062:2006 "Common Industry Format (CIF) for usability test reports".

1. *Test Plan* – contains usability testing procedures and objectives, tasks and scenario;
2. *Test Report* – includes findings, objectives achieved, answers to user questions, task completion time etc.

The *Test Plan* component is structured as follows:

1. *Main Details*: contains details regarding which component of a certain product or software application is being tested, the persona ([16,13]) type tested, context and duration of a test. Additional information regarding the tested person is gathered at this stage via the *User Background Information Form*;
2. *Overall Objectives*: consists of a list of objectives of a test, what we want to study or to evaluate;
3. *Scenario*: includes either a description of scenario tasks or images depicting a scenario;
4. *User Tasks and Post-Test Questions*: contains user tasks description and estimated completion time along with a series of post-test questions (which later will help to conclude the findings of the test);
5. *Other Notes*: denotes other details regarding the way this usability test is conducted.

The second component *Test Report* contains the following:

1. *Success Criteria*: the usability test is considered successful only if a certain percentage of users completed a pre-established criteria (for example: a certain task was completed in 2 minutes);
2. *Summary*: consists of a list of major findings/observations done by the test observer;
3. *Demographics*: includes personal details of the users that participated in the usability test;
4. *Interaction Notes*: a detailed descriptions of the participants' interactions. The notes can be separated by task, category, or whatever makes most sense for this test;
5. *Post-test Questions Answers*: answers to the post-test questions;
6. *Test Observations*: notes attached to the usability test (written, video or audio);
7. *Potential Design Improvements*: a list of potential design improvements as proposed by the tested user.

In the next section we will discuss how we can make use of this HTML template and annotate relevant data using existing vocabularies and our proposed vocabulary.

3.2 Microdata Schema

One advantage of the HTML5 microdata is that it is designed such that each piece of information in a document has assigned types from a single vocabulary, though each entity may have several types and properties from other vocabularies[11].

Because `schema.org` allows the extension of existing schemas, we made use of them in this section by proposing the Personas and Usability Testing schema[12] extension. As represented in Table 1, we proposed a set of new properties such as *personaType*, *tagLine*, *mainPoint*, *frustrationPoint*, *endGoal*, *scenario*, *context* and other.

Table 1. Persona Schema Properties and Description

Property	Expected Type	Description
personaType	Text	The type of Persona: Primary, Secondary, Negative, Supplemental, Served or Customer.
experienceGoal	Text	Experience goals are simple, universal, and personal.
endGoal	Text	End goals represent the user's motivation for performing the tasks associated with using a specific product.
lifeGoal	Text	Life goals represent personal aspirations of the user that typically go beyond the context of the product being designed.
businessGoal	URL or Text	Business goals represent the goals of the organization the persons works for.
technicalGoal	Text	Technical goals reflect technical aspects regarding an application/product for example: run in a variety of browser, data privacy etc.
disability	Disability or Text	A persona disability/disabilities relevant to accessibility aspects of the application/product.
myersBriggs	Personality or Text	Inspired by FOAF Myers Briggs personality classification which includes 16 4-letter textual codes [17].
topicInterest	Thing	A thing of interest for a certain persona, inspired by FOAF topic interest.
affectiveState	Emotion or Text	The affective state of the user at a certain moment, if a proper schema is not used, please provide a certain emotion using text format.
frustrationPoint	Text	Elements of a UI/UX or certain characteristics that frustrate the user or (s)he sees as pain points. These elements will help in usability evaluation.
userRole	Text	The role of the user in the application/product (if necessary – for example: admin, basic user etc.).
scenario	Scenario or URL	The scenario where the users represented by this persona will be used to test the usability.
context	Context	Description of the context that best fits a scenario recommended values: Physical/Virtual.

Some properties like *myersBriggs* and *topicInterest* were inspired from the FOAF vocabulary.

Properties like *minHeigh, maxWeight,minFeetSize* (along with others presented in Personas and Usability Testing schema) refer to a persona body measurement such as height, weight, bust size, waist size, and feet size. We used min and max values instead of a range, due to the fact that they are more precise than a range.

Along with some of the new properties, we proposed several additional schemas like *Disability, Personality, Emotion, Scenario* and *Context* (some of which are also used by the *UsabiliyTest* schema), in order to provide a more detailed vocabulary which fits our purposes. These schemas themselves come along with new properties and a range which specifies expected data types.

---

[11] http://www.w3.org/TR/html-data-guide/

[12] Publicly available at: http://blankdots.com/open/schema/

The *Scenario* schema has the following new properties and expected data types:

- *product*: A product (subject of a given usability test); for example, a software application or even a physical product such as mobile phone or car. Expected Type: Product or Application;
- *scenarioName*: Scenario name. Expected Type: Text;
- *userTask*: Tasks to be performed by the user. Expected Type: Text;
- *productTask*: Task performed by the product in response to the user. Expected Type: Text;
- *interactionMedium*: The medium of interaction; for example, Touch, Gestural, Mouse+keyboard etc. Expected Type: Text;
- *description*: Description of the scenario. Expected Type: URL or Text;
- *context*: Scenario context. Expected Type: Context;
- *participant*: Participants to this scenario. Expected Type: Person;
- *usabilityTest*: Usability test. Expected Type: UsabilityTest or Text;
- *evaluatedElements*: Focus/key elements in this scenario, tested later on, for example notifications element from the user interface. Expected Type: Text.

The *Disability* schema contains the following properties: *visualImpairment, hearingImpairment, gustatoryImpairment, somatosensoryImpairment, intellectualImpairment, mentalEmotionalDisorder, developmentalDisability*.

The *Context* schema has the following properties:

- *contextType*: Context type – recommended values: Physical/Virtual or Tangible/Intangible;
- *location*: Useful information about the location;
- *event*: An event has a location and a time;
- *geo*: The geographical coordinates of the context;
- *sensorData*: Sensors context data;
- *photos*: Photographs regarding the context;
- *reviews*: A set of reviews regarding the context.

The core of the *Usability Testing* described in Table 2. We introduced the *expectedAnswer* and *givenAnswer* properties, as some usability test scenarios require *control answers*, which can used as a reference point in order to compare them with the answers received during the usability test.

Alongside the core of the usability test schema, we identify the following properties for the *Task* schema (which in a bigger picture is part of the Workflow class):

- *taskType*: Could be an interactive task (performed by the user) or a non-interactive task (e.g. a feedback from the product/application).
- *taskPerformer*: A task could be performed by a Person, Product or Software Application.
- *estimatedTime*: The estimated length of time it takes to do the task, in ISO 8601 duration format.
- *completionTime*: The completion time of a task, in ISO 8601 duration format.
- *action*: A user or software/product performed action.
- *inputValue*: A user or software/product input value.
- *taskResult*: A user can complete the task, skip it or have trouble with it.

Table 2. Usability Testing Schema Properties and Description

Property	Expected Type	Description
productTested	Product or Application	A product is used in this scenario, it could be a software application or even a physical product (e.g. laptop or mobile phone).
componentTested	Text	A component of the product/sofware application that is being tested.
personaTested	Persona	The *persona* to be tested.
inspectionMethod	Text	Guidelines for examining the usability aspects of a UI design – inspired by [18].
testingMethod	Text	Method used for testing – inspired by [18].
participant	Person	Participant to the test (observer or user).
testDuration	Duration	The length of time it takes to do the test – ISO 8601 duration format.
context	Context	Description of the context that best fits a scenario.
scenario	Scenario	The scenario where the users represented by this persona will be used in the usability tests.
testObjective	Text	The objective/goal of the usability test.
userTask	Task or Text	Tasks to be performed by the user.
testQuestion	Text	A question presented to the user before/during/after the test.
successCriteria	Text	A successful design has been achieved when: 50% of users or 80% of users... etc.
testFinding	Text or URL	A finding extracted after performing the usability test. (e.g. "A number of users (specify the number) did/found..") – can be positive or negative finding.
recommendation	Text or URL	Recommendations to those findings can be issued, depending on the findings.
expectedAnswer	Text	An answer we expect to obtain to a certain question.
givenAnswer	Text	An answer that we obtained to a certain question.

With the inclusion of the *completionTime* property, we take into consideration that a user might complete a task in a shorter or greater amount of time than that estimated – *estimatedTime*. In case that the *completionTime* exceeds the *estimatedTime* then there is a problem with the task and/or the estimated completion time.

We must also take into account that a user might skip the current task, thus the *completionTime* would be **0**. For this reason, we added the *taskResult* property which provides a description the user thoughts on the task, how (s)he manages to complete it or why (s)he skipped or had trouble with the task. For example, a user might skip the task due to an unexpected error, inability to understand the task or due to a certain event that arises (e.g. the phone rings).

Although microdata is an easy way of annotating HTML document, it cannot express two aspects that RDFa supports: datatypes of literals and XML literals [19].

In the next section we will present the *PersonasOnto* ontology as an extension of the microdata schemas presented above.

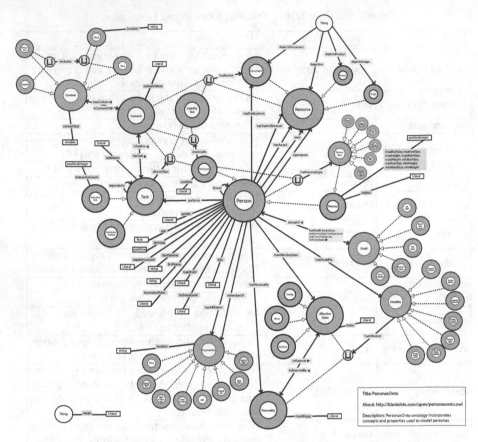

Fig. 1. Overview of the PersonasOnto ontology classes and properties using [20]

4   PersonasOnto Knowledge Engineering

As we previously mentioned, several widely used vocabularies like FOAF and schema.org/Person could be utilized, thus we applied such vocabularies to annotate information in our HTML template.

The *PersonasOnto*[13] ontology, provides a mean of annotating XHTML[14] and HTML documents with RDFa, but it can also be mapped in its RDF[15] representation to HTML5 microdata.

Table 3 presents an overview of the proposed ontology classes and subclasses and Figure 1[16] provides a graphical representation of the ontology classes and properties.

The *Person* class is the main one in the ontology, because it contains relevant characteristics such as age, gender, name, date of birth, concerning an individual but also

---

[13] Publicly available at: http://blankdots.com/open/personasonto.owl
[14] http://www.w3.org/TR/xhtml1/
[15] http://www.w3.org/RDF/
[16] http://blankdots.com/open/data/personasonto_vowl.svg

Table 3. PersonasOnto Ontology Classes

Class Name	Subclasses	Class Description
AffectiveState	Emotion, Feeling, Mood	The affective state of a person at a certain point.
Context	Intangible, Tangible, Place, Time	The Context in which a scenario takes place.
Disability	VisualImpairment, HearingImpairment, GustatoryImpairment, IntellectualImpairment, MentalEmotionalDisorder, Somoatosensory Impairment, Development-Disability	A person's disabilities mental and physical.
Goals	BusinessGoals, EndGoals, ExperienceGoals, LifeGoals, Technical Goals	The goals of a Persona.
Organization	Corporation, EducationalOrganization, GovermentOrganization, Group, MediumBusiness, NGO, SmallBusiness, SportsTeam	The type of Organization a Person belongs to.
Person	Participant, Persona	Basic information about a person and more precisely about a participant or a persona.
Personality	–	The MyersBriggs personality of a person.
Resource	Document, Image, Product	Type of resources available.
Scenario	–	A series of tasks the user performs in a certain context.
Task	InteractiveTask, NonInteractiveTask	Tasks performed by a Person in a Scenario. InteractiveTask could be certain action performed by a user, NonInteractiveTask could be a machine response to that action.
UsabilityTest	–	A usability test evaluates the usability of a certain Product in a Scenario.

is the super class of the *Persona* class. The*Persona* class could be regarded of being a separate identity of a person, although by definition has the same characteristics as the *Person* class. Also, as we previously mentioned, a persona represents a collection of preferences, goals, frustrations identifiable with a group of persons, and usually has a context associated.

Some of the classes like *AffectiveState*, *Personality*, *Disability*, *Organization* are used to describe several characteristics or a person like an emotion (e.g. anger), a disability (such as partial blindness), his/her personality (e.g. ESTJ – Extroversion, Sensing, Thinking, Judgment) and the organization (s)he works for, or (s)he is in.

Other classes like *Context*, *Scenario*, *Task*, *Goals*, *UsabilityTest* and subclasses *PersonaType* are connected with the *Persona* subclass. For example, a persona has a certain type (Primary, Secondary, etc.), has the characteristics of a (fictional) person, has certain life or experience *Goals*, (s)he performs certain tasks in a scenario. The scenario is placed in a certain context and can be also used in a usability test.

Many of the classes, subclasses and properties are the same as in the *Persona* schema described in the previous subsection. On the other hand, classes like *UsabilityTest*, *Person*, *Scenario* and *Task* have certain properties which better define the relationship between them. Such a property is expressed below in Turtle[17] format.

```
###  personasOnto.owl#isPartOfTest
:isPartOfTest rdf:type owl:ObjectProperty ;
     rdfs:range :UsabilityTest ;
     rdfs:domain [ rdf:type owl:Class ;
          owl:unionOf ( :Person
                        :Scenario
                        :Task )].
```

The *Context* Class includes a tangible (physical world) or intangible (virtual world) environment, but also could specify spatial and temporal concepts. This inclusion is expressed below.

```
### personasOnto.owl##includes
:includes rdf:type owl:ObjectProperty ;
   rdfs:range [
     rdf:type owl:Class ;
          owl:unionOf ( :Place
                        :Time )] ;
   rdfs:domain [
     rdf:type owl:Class ;
          owl:unionOf ( :Intangible
                        :Tangible )].
```

In the next section, we will focus on presenting a use case.

5   Use Cases

In order to illustrate how our model could be used, we provided both HTML templates (in a small scale experiment) to several teams working on different projects in order to integrate them in their UCD process and fill it up with data. The first challenge the team faced was to gather data for the personas document. The second challenge was extracting the user characteristics in order to select participants for the usability test.

The third and last challenge was to collect data for the usability test by interviewing and observing the test participants. At the end of each test, they compiled the data into the HTML template provided.

To exemplify how the conceptual model was used to structure data collected from the UCD process, we selected one of the projects from a single team. The purpose of the project we chose was to develop an application for monitoring hospital patients.

---

[17] http://www.w3.org/TR/turtle/

The data from the usability testing document for this project is presented below. We made used of Microdata to RDF Distiller[18] tool, in order to extract data from the template – expressed as RDF triples in the Turtle format.

```
[ a schema:UsabilityTest;
## general information regarding the test ##
    schema:productTested "Product"@en-us;
    schema:componentTested "Searching a patient
            and viewing his personal details"@en-us;
    schema:personaTested [ a schema:Persona;
        schema:personaType "Primary"@en-us ];
    schema:context [ a schema:Context;
        schema:location " University of Medicine
        "@en-us ];
    schema:testDuration "P5M"^^xsd:duration;
    schema:testObjective [ a rdf:Bag;
        rdf:_1 "Assess the overall effectiveness
            of the PaMI application for medical staff
            performing the search for a patient;"@en-us,
        rdf:_2 "Evaluate the time it takes users to
            find a patient through the search interface;
        "@en-us,
        rdf:_3 "Have an evidence of the users that
            needed the Online Help in order to do this
            task and ask them if it was helpful;"@en-us,
        rdf:_4 "Test if the terminology and the
            labels makes the application understandable;
        "@en-us ];

## details of some of participants in the test ##

    schema:participant [ a schema:Person;
        schema:gender "Male"@en-us;
        schema:name "User 1"@en-us ];
    schema:participant [ a schema:Person;
        schema:gender "Female"@en-us;
        schema:name "User 2"@en-us ];

## test report ##

    schema:successCriteria [ a rdf:Bag;
        rdf:_1 "50% of users did not used the Online
            Help in order to perform the task"@en-us,
        rdf:_2 "80% of users succeded in performing
            the task in a short time"@en-us ];
    schema:testFinding [ a rdf:Bag;
        rdf:_1 "Finding 1: 2/4 users found it very
            easy to perform the task."@en-us,
```

[18] http://www.w3.org/2012/pyMicrodata/

```
     rdf:_2 "Finding 2: 2/4 users used the Online
        Help in order to perform the task."@en-us,
     rdf:_3 "Finding 3: All the users found the
        application understandable."@en-us ];
  schema:testQuestion "Question 1: How easy was
        it for you to authenticate into the
        application?"@en-us,
  schema:givenAnswer "The interface of the
        application is simple and effective"@en-us;

## a list of given tasks and estimated
   completion time ##

   schema:userTask
     [ a schema:Task;
        schema:description "Searching a patient
              in application "@en-us;
        schema:estimatedTime "P2M"^^xsd:duration ],
     [ a schema:Task;
        schema:description "Tap on View patients
              button and then start typing the
              patient name "@en-us;
        schema:estimatedTime "P30S"^^xsd:duration ],
     [ a schema:Task;
        schema:description "Authenticate into the
              application "@en-us;
        schema:estimatedTime "P30S"^^xsd:duration ],

## results of the tasks performed by the users in the
   usability test ##

     [ a schema:Task;
        schema:completionTime "P3M"^^xsd:duration;
        schema:description "Searching a patient in
              application "@en-us;
        schema:taskPerformer [ a schema:Person;
           schema:name "User 1"@en-us ] ],
     [ a schema:Task;
        schema:completionTime "P2M"^^xsd:duration;
        schema:description "Searching a patient in
              application "@en-us;
        schema:taskPerformer [ a schema:Person;
           schema:name "User 2"@en-us ] ] ]
```

Although the conceptual model described in previous sections provides a guide for structuring and extract meaningful information from the templates, the challenge of linking information can only be solved by developing a software system.

Such a system would incorporates a knowledge base a TBox component [21] based on the *PersonasOnto* and the ABox populated with data extracted from the personas and usability testing documents. Furthermore, this system will aim to automatically to match persons to a personas document and recommend them for the usability test as a participant.

6 Conclusions and Future Work

In this paper we presented the *PersonasOnto* ontology along with the Persona and Usability Testing schemas. Both of them incorporate concepts and properties used to model the corresponding User-Centered Design methods. These vocabularies form a basis for structuring, extracting and linking information between the personas and usability testing methods.

We also explored the advantages of providing machine-readable HTML5 templates (personas and usability testing) which can be annotated information using microdata or RDFa. We illustrated the applicability of the conceptual model by integrated the machine-readable templates in the user-centered design process.

We plan to develop a system that that automatically validates and generates of personas and usability test based on existing social network user profiles, but also linking the collected information with other data for.

Acknowledgements. This work was partially supported by the European Social Fund in Romania, under the responsibility of the Managing Authority for the Sectorial Operational Program for Human Resources Development 2007-2013 [grant POSDRU/107/1.5/S/78342].

References

1. Rubin, J., Chisnell, D., Spool, J.: Handbook of Usability Testing: How to Plan, Design, and Conduct Effective Tests. John Wiley & Sons (2008)
2. Cooper, A.: The Inmates Are Running the Asylum. Macmillan Publishing (1999)
3. Goodwin, K.: Perfecting your personas (2005), http://bit.ly/2Syg7m
4. Allemang, D., Hendler, J.: Semantic Web for the Working Ontologist, 2nd edn. Effective Modeling in RDFS and OWL. Morgan Kaufmann Publishers Inc. (2011)
5. Negru, S., Buraga, S.: Persona modeling process: from microdata-based templates to specific web ontologies. In: International Conference on Knowledge Engineering and Ontology Development. KEOD 2012, pp. 34–42. SciTePress (2012)
6. Cheng, D.Y., Chao, K.M., Lo, C.C., Tsai, C.F.: A user centric service-oriented modeling approach. World Wide Web 14, 431–459 (2011)
7. Cena, F., Likavec, S., Osborne, F.: Propagating user interests in ontology-based user model. In: Pirrone, R., Sorbello, F. (eds.) AI*IA 2011. LNCS, vol. 6934, pp. 299–311. Springer, Heidelberg (2011)
8. Felden, C., Linden, M.: Ontology-based user profiling. In: Abramowicz, W. (ed.) BIS 2007. LNCS, vol. 4439, pp. 314–327. Springer, Heidelberg (2007)
9. Razmerita, L., Angehrn, A., Maedche, A.: Ontology-based user modeling for knowledge management systems. In: Brusilovsky, P., Corbett, A.T., de Rosis, F. (eds.) UM 2003. LNCS, vol. 2702, pp. 213–217. Springer, Heidelberg (2003)
10. Golemati, M., Katifori, A., Vassilakis, C., Lepouras, G., Halatsis, C.: Creating an ontology for the user profile: Method and applications. In: Proceedings of the First International Conference on Research Challenges in Information Science, RCIS 2007. IEEE (2007)

11. Fischer, G.: User Modeling in Human-Computer Interaction. User Modeling and User-Adapted Interaction 11, 65–86 (2001)
12. Pruitt, J., Adlin, T.: The Essential Persona Lifecycle: Your Guide to Building and Using Personas. Elsevier (2010)
13. Cooper, A., Reimann, R., Cronin, D.: About Face 3: The Essentials of Interaction Design. John Wiley & Sons (2007)
14. Lindgaard, G., Chattratichart, J.: Usability testing: What have we overlooked? In: Proceedings of the SIGCHI Conference on Human Factors in Computing Systems, CHI 2007, pp. 1415–1424. ACM (2007)
15. Long, L.: Real or Imaginary; the effectiveness of using personas in product design. In: Proceedings of the Irish Ergonomics Society Annual Conference, IES 2009 (2009)
16. Caddick, R., Cable, S.: Communicating the User Experience: A Practical Guide for Creating Useful UX Documentation. John Wiley & Sons (2011)
17. Myers, I.B., McCaulley, M.H., Quenk, N., Hammer, A.: MBTI Manual: A Guide to the Development and Use of the Myers-Briggs Type Indicator. Consulting Psychologists Press (1998)
18. Ivory, M.Y., Hearst, M.A.: The state of the art in automating usability evaluation of user interfaces. ACM Comput. Surv. 33, 470–516 (2001)
19. Vestlandsforsking: Semantic markup - report (2012), http://www.vestforsk.no/en/news/semantic-markup-report
20. Negru, S., Lohmann, S.: A visual notation for the integrated representation of owl ontologies. In: Proceedings of the 9th International Conference on Web Information Systems and Technologies, WEBIST 2013. SciTePress (to appear, 2013)
21. Baader, F., Calvanese, D., McGuinness, D., Nardi, D., Patel-Schneider, P.: The Description Logic Handbook: Theory, Implementation, and Applications. Cambridge University Press (2007)

Towards Automatic Ontology Alignment
for Enriching Sensor Data Analysis

Marjan Alirezaie and Amy Loutfi

Applied Autonomous Sensor Systems, Dept of Technology, Örebro University,
SE-701 82, Örebro, Sweden
{marjan.alirezaie,amy.loutfi}@oru.se

Abstract. In this work ontology alignment is used to align an ontology comprising high level knowledge to a structure representing the results of low-level sensor data classification. To resolve inherent uncertainties from the data driven classifier, an ontology about application domain is aligned to the classifier output and the result is recommendation system able to suggest a course of action that will resolve the uncertainty. This work is instantiated in a medical application domain where signals from an electronic nose are classified into different bacteria types. In case of misclassifications resulting from the data driven classifier, the alignment to an ontology representing traditional microbiology tests suggests a subset of tests most relevant to use. The result is a hybrid classification system (electronic nose and traditional testing) that automatically exploits domain knowledge in the identification process.

Keywords: Ontology Alignment, Sensor Data, Classification, Semantic Gap.

1 Introduction

We are surrounded by billions of sensors generating huge amount of data (1800 exabytes in 2011) in our daily lives [8]. There is a large potential to use this sensor data to provide a deeper and better understanding of the world around us. For this to be possible, data must be interpreted and represented in a manner that is compatible for humans. Typically, this interpretation is automated by using complex data driven analysis methods. The output of such methods can still contain inaccuracies due to the fact that low level sensor data is subject to shortcomings due to selectivity, uncertainties and errors. For specific domains e.g. medical domains, the high inaccuracy can hinder the uptake of using automatic data analysis, and thus new sensor technologies.

Infusing more knowledge into the domain generally helps to improve data interpretation. Nonetheless, data driven processes manipulating sensor readings are not able to automatically consider the wealth of high level domain-related knowledge for the integration. In other words, what is needed is a method that is able to automatically fuse the high level knowledge to low level data which are inherently unintuitive and difficult to interprete.

In this work, we extend the ontological alignment approach explained in [1] for improving signal level classificaion results so that the method is no longer dependent to the structure of the classifier. We show how to use ontology alignment methods to find

A. Fred et al. (Eds.): IC3K 2012, CCIS 415, pp. 179–193, 2013.

Table 1. Bacteria Species

Bacteria Name	Short Name	Bacteria Name	Short Name
Escherichia coli	EColi	Entercoccus faecalis	ENTFL
Pseudomonas aeruginosa	PSAER	Staphylococcus lugdunensis	STLUG
Staphylococcus aureus	STA	Pasteurella multocida	PASMU
Klebsiella oxytoca	KLOXY	Steptococcus pyogenes	STRPY
Proteus mirabilis	PRMIR	Hemophilus influenzae	HINFL

high level annotations for the results of signal data analysis in order to resolve ambiguities. Instantiating the methods using the electronic nose, we particularly examine the task of blood bacteria identification (10 blood bacteria species listed in Table 1). The alignment method presented here is applied over classification results of raw sensor data coming from the electronic nose and information pertaining to distinguishing bacteria laboratory tests which are modeled in an ontology. The ontology is aligned to the sensor data which is represented as a tree structure. Obtaining the tree structure depends on the classification method. A hierarchical classifier such as a decision tree, produces a tree structure that can be directly algined with the ontology. A non-hierarchical classifier, on the other hand, requires further processing to generate a tree structure that can be used in the proposed alignment method. Using this method, if the classifier does not have a hierarchical structure, we find cases causing uncertainties in the classification and make tree-shaped structures with them to feed the alignment process. This alignment technique also uses a heuristic to decrease the number of required labratory tests that should be applied on samples of bacteria. The result is an enriched classification method which is able to use information represented at a high-level (symbolic) to provide recommendations based on the automatised interpretation of the sensor data.

To familiarize the reader with the sensors used in this work we begin with a brief introduction of electronic noses and their applications to bacteria identification. The paper then proceeds to outline the related works in Sect.3. After that, the next section concentrates on details of the methodology. In Sect.5, our data set structure along with a short description about the sampling process is discussed. Then, Sect.6 represents results of each step of the methodology. The paper ends with the discussion and the conclusion.

2 Electronic Noses (E-Noses)

Comprising a set of chemical sensors, an electronic nose is a machine with the ability of detecting and discriminating odors or gases. It consists of an array of sensors each of with partial selectivity and a pattern recognition algorithm. Each odor based on its chemical characteristics makes a "fingerprint" from the sensor array which are converted to time series signals [24]. Figure 1 illustrates the response from one sensor in an e-nose when first exposed to the clean air (baseline phase: $t < T_0$), then to a target gas (sampling phase: $T_0 \leq t < T_R$) and finally again to the clean air (recovery phase: $t \geq T_R$).

Electronic noses have been applied to a number of different application domains, ranging from food process monitoring to environmental monitoring [24]. In this work

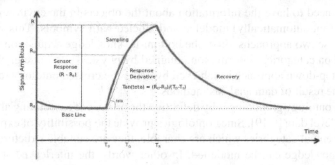

Fig. 1. A signal with three phases (Baseline, Sampling, Recovery)

an e-nose is used to provide a quick identification of bacteria existing in blood samples. This is an alternative method to traditional laboratory analysis where samples are first incubated and cultured requiring several days before an identification result is possible. With an electronic nose, it is possible to reduce this time significantly as the sampling of an odour requires anywhere from 3-6 minutes. More details on the use of electronic noses for the detection of bacteria can be found in [9,11].

The construction of an electronic nose instrument can vary depending on the number and types of sensors used. As described in Sect.5, two different types of e-noses are used in this work, resulting in two data sets, each of which is associated with a classififcation method.

3   Related Works

To clarify our alignment approach it is worth mentioning that it concerns fusion of information at different levels of abstraction in order to bridge the semantic gap which occurs between these levels [23,1].

Data fusion is known as one of the most popular solutions in data processing efforts. The common part of works related to integration of data is keeping synchronised the different types of data that come from various sensors measuring same environment [25]. Typically, in the area of sensor fusion, fusion methods consider homogeneous raw data sources i.e. numeric data, whereas the integration methodology in this work is applied on both numeric and symbolic data.

Works which consider fusing symbolic knowledge to numeric data have gained attention in AI fields related to robotics and physically embedded systems[17,18]. The symbol grounding problem [16] in general and the anchoring problem [10] in particular concentrate on the process of creating and maintaining the relation between a symbol chosen to label an object in the world and those data coming from sensors observing the same physical object in the environment. In most of works such as [15] the challenge is how to perform the anchoring in an artificial system and how to find relevant concepts related to symbols to improve the recognition process. In these efforts the association is done in two ways: grounding well-defined concepts in data (top-down process) and conceptualizing data that exemplifies the concept (bottom-up process). In these kinds of mapping we need to have the information about the objects in the environment which is manually (not automatically) modeled and labeled with symbols. This work leverages from these two approachs. First, the bottom-up knowledge acquisition is acquired solely based on categorical information obtained by physical sensors e.g. class labels. Secondly, a top-down approach is achieved by using the created ontology to find similarity with the result of data analysis methods.

Likewise, ontology grounding is by definition the process of associating abstract concepts to low level data [6,19]. Since ontologies provide the possibility of expressing entities in conceptual categories which are sharable, they are suitable structures whereby high level knowledge can be modelled. In other words, the interleaved structures of ontologies make it possible to enrich the symbolic representation of a concept by retrieving its different features reified in an ontology for provisioning a more accurate grounding [31]. Our reason to use ontologies, however, is not only for the purpose of grounding. In this work, the reusability of relevant concepts which are needed to make annotation and description for our final recommender system is basically considered.

For some situations where there are intelligible meanings for the measured features of data, it is certainly possible to extract feature-related concepts from ontologies and re-structure the data set according to them for the sake of improvement in data interpretation proceses [3,4]. In the case of electronic noses, because of specialty of sensors, there was no chance to find some extra meanings related to features measured by sensors to be intermixed with raw data.

From another point of view, it is worth to consider the automatic ontology development part of our work as well. According to the survey [7], the ontology generation tasks are categorized into 4 groups: conversion or translation, mining based, external knowledge based and frameworks; and our method belongs to the third one since we are utilizing a set of available ontologies [28] related to the domain. The first step of our ontology building uses the same approache applied by most of works such as [5] and [30] where they query the WordNet database to retrieve the initial synonyms or a simple definition for the concept. Instead of WordNet we use another repository that contains most updated biomedical ontologies being full of subsumption relations [28].

Finally, the alignment which is defined as the process of determining correspondences between concepts [26], is mostly used when two sides of the process are ontologies. However, in this work, we map an ontology with the result of a data driven method according to the names of bacteria.

4   Methodology

The methodology used in this work applies the following general steps which are discussed in next sections. First, we classify pre-processed sensor data. Second, a tree structure that contains the class labels (bacteria names), participating in misclassification situations, is built. Third, the resulted tree is aligned with the ontology containing relevant knowledge. Finally, the method replaces candidate parts of the ontology with their counterparts in the tree.

4.1 Classification of Sensor Data

As mentioned before, in this work we make the alignment process independent of the classificaton method. At this first step, to evaluate the resilience of the methdology both hierarchical and non-hierarchical classification techniques are used. For the hierarchical group the candidate method is the C4.5 decision tree and for non-hierarchichal the Dynamic Time Warping algorithm is evaluated.

The Decision Tree (D-Tree). A decision tree classification has the advantage that it provides transparency in the representation of the outputs [20] and has a suitable (hierarchical) structure for our alignment process [1]. The C4.5 algorithm is used and finds a feature of the training set providing the maximum degree of discrimination between different classes of bacteria. The algorithm iterates, each time splitting instances of the training set according to the most informative selected feature. Each feature value creates a decision node for the tree [20]. Using the confusion matrix[1] from the classification result, a second process finds misclassification positions among leaf nodes of the tree and assigns them all bacteria names sharing these nodes. During this process leaf nodes are divided into two groups A (containing all leaves without misclassification) and B (holding the rest).

Starting with group B nodes, a third process checks if the sibling node also belongs to the group B. In these cases where two sibling leaves belong to the group B, the process labels their parent by all bacteria names shared by them; otherwise, the process keeps the candidate B leaf node by its own bacteria names. Two procedures of our previous work, *RelabelDTree* and *CheckParent*, described in [1], show the details of the decision tree relabeling process. Once all nodes in group B or in the parents of group B are labeled, the tree is ready to be transfered to the alignment process.

Dynamic Time Warping (DTW). The accuracy of the d-tree is highly dependent on the selected features of the signal. Alternative approaches to classification such as Dynamic Time Warping are able to consider the entire signal data rather than extracting specific features from data [20]. The Dynamic Time Warping algorithm [29] is an extensively used technique to measure the similarity of two sequences of data over the time. Taking any two signals as input, the DTW calculates the (Euclidean) distance between them at each time point to see how dissimilar they are. As our sensor signals are the same size, they are suitable for the DTW method.

In order to apply our alignment method, a confusion matrix is needed. Following is a brief explanation on how we build the confusion matrix for the DTW signal processing results:

After preparing the training and test set, we build the $n \times m$ similarity matrix S, where n and m are the length of the test and the training set, respectively. The element $a_{i,j}$ of this matrix is the distance (dissimilarity) value between the ith test case and the jth training case. The confusion matrix C of dimension $l \times l$ is initialized to the zero matrix. Considering each row of the matrix S as r, we update the matrix C so that

---

[1] Confusion matrix is a table that visualizes the performance of a supervised learning algorithm so that rows and columns are labeled with actual and predicted classes, respectively [21].

its element $b_{i,j}$ receives 1-unit increase if i and j are r's test case label index and r's element label index that has the minimum distance value, respectively. To resolve the uncertainities in C, a binary tree is built from candidate labels[2].

To develop the binary tree, we start from the root node having been labeled with a set containing l bacteria names. These l names are those "actual" labels assigned to non-zero elements of a column of C that belongs to a "predicted" label. Creating two branch nodes of the root, the process has to label them so that each child node's label set is a subset of the parent's label set and the complementary set of its sibling as well. In order to keep the completeness[3] of the algorithm, in the process of creating children nodes we have to consider the power set[4] of the parent's label set. Consequently, we have $2^l/2$ different options to split the root node. However, for the sake of reducing the number of required microbiology tests in total (increasing the information gain), we prioritize those branches that are almost in balance in terms of the number of labels. In other words, we divide the number of labels l by 2 to find a laboratory test being able to differentiate between the two biggest subsets (among all elements of the power set) of bacteria names. The number of different branches holding $\lceil \frac{l}{2} \rceil$ labels is hence as follows:

$$\binom{l}{\lceil \frac{l}{2} \rceil} = \frac{l!}{(l - \lceil \frac{l}{2} \rceil)! \times \lceil \frac{l}{2} \rceil!} \tag{1}$$

The first subset of labels with length $\lceil \frac{l}{2} \rceil$ has a sibling set (a complementary label set) signed by the rest of input class labels. Once a branching-labeling step is finished, the alignment process (Sect.4.3) is called to check if there is a counterpart node for the root in the ontology. If the alignment returns a result, we proceed until a complete tree whose each leaf node is labeled by a single bacteria name is built; otherwise we switch to the next candidate for the parent node. In this way, since we divide the list by 2 at each step, the depth of the created tree is m if: $2^{m-1} < l \leq 2^m$.

Algorithm 1 represents the process of the building tree from the selected candidate class labels. Provided a confusion matrix, this method can indeed be applied on classification results of any classification method and make them ready for the alignment process with an ontology.

4.2   Microbiology Tests Ontology

The resulting tree from the classification process is aligned to an ontology that contains the high level knowledge related to the domain of bacteria laboratory tests. This ontology has been created in a semi-automatic way so that only the initial concepts such as bacteria represented in Table 1 and microbiology tests related to general categories of bacteria[5] with the positive/negative results have been manually modeled in the ontology.

---

[2] There is an uncertainity if at least one non-diagonal element of a confusion matrix is non-zero.

[3] An algorithm is complete when it finds the solution if there is one.

[4] The power set is a set of all subsets of a set.

[5] http://www.atsu.edu/faculty/chamberlain/Website/
lab/idlab/flowchp.htm

The automated phase of building the ontology regards knowledge existing in *BioPortal* [28]. *BioPortal* is the repository of biomedical ontologies among which *SNOMED CT* contains key terminologies in biomedicine (about 40,000 classes [12]).

Algorithm 1. Building tree from candidate class labels.

```
 1: function BUILDDTREE(labels)
 2:     root ← makeNode(labels)
 3:     if length(labels) = 1 then return root
 4:     end if
 5:     subSetList ← getSubSets(labels, length(labels)/2)
 6:     for all s in subSetList do
 7:         p ← getComplementarySet(s)
 8:         if ontologyAlign(root, s, p) ≠ null then
 9:             rightNode ← BuildTree(s)
10:             leftNode ← BuildTree(p)
11:             attachToTree(root, rightNode, leftNode)
12:         else return null
13:         end if
14:     end for return root
15: end function
```

This information[6], which is useful in finding the expected results of some microbiological tests for specific bacterial species, are certainly considered during traditional bacteria identification processes in laboratories [27].

Having the bacteria name as the input, the Java interface using the Jena API with the ARQ query engine runs a query in a loop to retrieve a hierachical list that contains the whole indirect super classes of this bacteria. This loop is run until it meets the class named *Bacteria* in the ontology. The returned list contains different (more general) categories of bacteria for which the microbiology tests with the positive/negative results are available in the ontology.

For example, the following SPARQL code returns the name of the first super class of the class which is labeled by the first bacteria name in the Table 1, *Escherichia Coli*.

```
SELECT DISTINCT ?ss0 ?nstep ?label
{
  GRAPH <SNOMEDCT URI>{
  snterm:112283007 rdfs:subClassOf ?ss0.
  ?ss0 skos:prefLabel ?label.
  BIND ('1' AS ?nstep).}
}
```

In an iterrative process, other classes are returned such as *GramNegative* or *Bacillus*, which state the gram stain and wall shape type of the bacteria, along with the laboratory tests such as the *Lactose* test which is positive. This extracted information is populated into our ontology after adding the object properties such as *hasTest* and *hasValue* for the *Escherichia Coli* class. Continuing this way of knowledge acquisition, we obtain an ontology that contains results of 7 different microbiology tests for all 10 types of blood bacteria.

---

[6] The chemical features of bacteria such as cell morphology and gram stain [22].

4.3 Ontology Alignment

The main task of the aligning step is comparing two types of entities, a tree node holding a set of class labels (bacteria name), and the ontology class(es) containing information related to the bacteria species. As the sensor data lacks semantics, we apply terminological and structural alignment methods [23].

The alignment method calculates the similarities between the two mentioned type of entities by using the Jaro-Winkler algorithm [13]. The algorithm is based on Jaro-Winkler distance (2) and counts the number of matching characters in two strings to measure the distance between them. The lower the JaroWinkler value, the more similar are the strings (bacteria names) [13].

$$distance = \frac{1}{3} \times (\frac{m}{|s_1|} + \frac{m}{|s_2|} + \frac{m-t}{m}) \tag{2}$$

Where:

m: number of matching characters.

t: half the number of transpositions.

s_i: length of ith string.

Because of having no isomorphism relation between the tree and the ontology [14], the graph inexact matching as the structural alignment method is used to determine if there is any similar structure between labeled nodes in the tree and the ontology.

Given the similarities between two structures, a replacement process copies the labels of the classes found in the ontology to relabel the counterpart nodes in the tree. In this way, the tree structure is an annotated structure that contains two types of information, sensor values and microbiology tests, that can act as a recommender in case of uncertainities caused by misclassifications.

5   DataSet

Ten types of bacteria species listed in Table 1, sub-cultured on blood and agar plates, are clinical samples in this scenario. In order to discriminate between them, we consider two sampling methods using two different electronic noses. The following section concisely describes the sampling configurations for these two e-noses:

5.1   Sampling Process (Bacteria in Blood and Agar)

The sampling system used for "sniffing" the bacteria in blood is a NST 3220 Emission Analyzer from Applied Sensors, Linköping, Sweden. This machine is composed of 22 chemical sensors[7]. In this experiment the baseline acquisition lasts for 10 seconds. Next, the headspace[8] gases are injected into the sensor chambers and the sensors are exposed for 30 seconds in the sampling phase which is followed by the recovery phase lasting

---

[7] 10 MOS and 12 MOSFET sensors [24].

[8] The headspace is the space just above the liquid sample in a bottle [24].

for 260 seconds (Fig.1). Each of the 10 bacteria has been sampled 60 times with the 600 "sniffs". Further details of the sampling process and preparation are given in [11].

To make a suitable structured training set for the classification, we pass the continuous time series data generated by 22 sensors through a pre-processing phase that includes two steps: Baseline manipulation normalizing the sensor data and compression extracting informative descriptors of signals to make feature vectors [20]. In this way, we replace each sensor signal with its two descriptors indicated in Fig.1. The static response calculating the difference between the end point of the sampling phase and the baseline, gives one single parameter; and the response derivative which is equal to the slope of the line contiguous to the third second of the the sampling phase. A total of 44 feature values are produced for the dataset of 600 samples accompanied by a label list containing bacteria species names listed in the Table 1.

The second data set was collected using a Cyranose 320, which contains a sensor array of 32 conducting polymer sensors. This electronic nose samples bacteria which has been cultured on agar plates. The sampling and recovery phases are 20 and 80 seconds long, respectively. Each of the 10 bacteria has been sampled 40 times resulting in 800 "sniffs". Additional details of this sampling method are given in [9]. Due to the similarities of these signals, we keep the whole signals without any feature extraction phase.

6   Results

Based on the aforementioned assumption about dividing the classification methods into hierarchical and non-hierarchical groups, we first discuss on results of the two methods separately and then concentrate on the alignment technique results for both.

6.1   Hierarchical Classifier Results

The C4.5 decision tree algorithm was applied for the first data set read by the NST 3220 Emission Analyzer electronic nose. This data set as described in Sect.5.1 contains 600 instances with 44 features. After applying a 10-fold cross validation on this data set in order to generalize the error estimation of the classification [21], two thirds (400 cases) of the samples were chosen to form the training set and the rest were considered as test cases. Figure 2 shows the result of the classification fed by the training set. Decision nodes of the tree have been labeled by feature names and criteria values. Leaf nodes of the tree have also been marked by bacteria species names. The confusion matrix of this classification is depicted in Fig.3(a). According to this matrix and (3), among the 200 test cases, there are 39 misclassifications corresponding to an accuracy of 80%.

$$accuracy = \frac{tp + tn}{tp + tn + fp + fn} = \frac{\sum_{i=1}^{l} a_{ii}}{sum_{i=1}^{l} sum_{j=1}^{l} a_{ij}} \tag{3}$$

Table 2 shows the details of group B nodes which hold misclassified cases. For example, as we can see in the table, the node number 10 is shared with two types of bacteria, the predicted type 2 (according to the training set) and the actual type 7 (according to the test set). In the same way, the details about node number 43 which is shared by bacteria type 10, 2, 3 and 7 are listed shown by 11th, 12th and 13th items in Table 2.

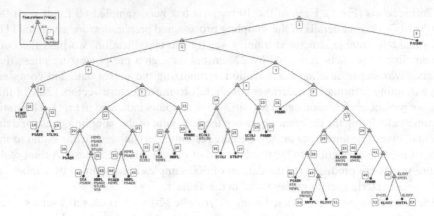

Fig. 2. The Decision Tree Showing Misclassification Cases with Red Labels

```
16   0   4   0   0   0   0   0   0   0          20   0   0   1   0   2   0   3   0   2
 0  15   2   0   0   0   0   0   0   3           2  17   0   0   0   0   1   0   0   1
 0   1  15   0   3   0   0   0   0   1           0   3  19   0   3   0   0   2   3   0
 0   0   0  17   0   3   0   0   0   0           0   0   1  23   1   2   0   0   0   0
 0   0   0   1  18   1   0   0   0   0           0   0   4   1  23   3   4   1   0   1
 2   0   0   3   2  13   0   0   0   0           2   0   0   2   1  17   0   1   0   0
 2   3   0   0   0   0  12   0   0   3           0   2   0   0   0   0  14   0   3   2
 0   0   0   0   0   0   0  20   0   0           0   1   7   0   0   0   0  18   0   0
 0   0   0   0   0   0   0   0  20   0           0   0   1   2   0   2   2   1  19   0
 0   2   3   0   0   0   0   0   0  15           2   1   0   0   0   0   0   0   0  16
```

(a) DTree Confusion Matrix.   (b) DTW Confusion Matrix.

Fig. 3. Resulted Confusion Matrices

The details of this table are visually depicted in Fig.2. For example, the subtree containing node 49 as the parent and nodes 52 and 53 as children belonging to group B and are sharing bacteria number 4 (*Klebsiella Oxytoca* or *KLOXY*) and 6 (*Entercoccus faecalis* or *ENTFL*).

The misclassification caused by the inconsistencies between the predicted and actual classes is resolved by utilizing the ontology information which is about the discriminating laboratory tests. The alignment results is explained in Sect.6.3.

6.2 Non-hierarchical Classifier Results

We chose 260 test cases and 540 samples for the training set. The confusion matrix from the DTW method is shown in Fig.3(b). It shows that there are 74 misclassifications among 260 test cases that lead to an accuracy of 72%.

Table 3 indicates how incorrect predictions correspond to the actual classifications. For example, for some situations where the test case actual class is bacteria number 1 (*EColi*), 3 (*STA*), 5 (*PRMIR*), 6 (*ENTFL*) or 9 (*STRPY*) the classifier has predicted this test case as the bacteria number 8 (*PASMU*) (column 8 in Figure 3(b)). Therefore, the label set *EColi, STA, PRMIR, ENTFL, PASMU, STRPY* will be the input for the mentioned building tree process. One of the built tree hierarchical structure returned for this example is represented in Figure 5(a). The size of the label set is $l = 6$ and the depth of the tree is $m = 3$ ($2^2 < 6 \leq 2^3$). Having this tree, the alignment process can follow the similarity checking process explained in the following section.

Table 2. Decision Tree, B-leaf Nodes (Prctd = Predicted, Actl = Actual)

Item#	Node	Prctd	Actl	Number	Item#	Node	Prctd	Actl	Number
1	10	2	7	3	11	43	10	2	2
2	23	5	3	3	12	43	10	3	1
3	24	1	7	2	13	43	10	7	3
4	26	1	6	2	14	44	2	10	1
5	33	3	1	4	15	45	10	2	1
6	34	3	10	3	16	46	2	3	1
7	39	4	5	1	17	50	6	4	2
8	39	4	6	1	18	50	6	5	1
9	40	5	6	2	19	52	4	6	2
10	42	3	2	2	20	53	6	4	1

Table 3. DTW Misclassification Cases (Prctd = Predicted, Actl = Actual)

Prctd	Actl	Actl	Actl	Actl	Actl	Prctd	Actl	Actl	Actl	Actl	Actl
1	2	5	6	10	-	6	1	4	5	9	-
2	3	7	8	10	-	7	2	5	9	-	-
3	4	5	8	9	-	8	1	3	5	6	9
4	1	5	6	9	-	9	3	7	-	-	-
5	3	4	6	-	-	10	1	2	5	7	-

6.3   Alignment Results

By the string matching method, the alignment process finds all bacteria names in the ontology that are similar to the candidates. Table 4 demonstrates some parts of Jaro-Winkler distances between bacteria names in the classifier and in the ontology. As we can see, the minimum value of each column is located in the diagonal position which proves the correctness of the mapping of the bacteria names. Considering these values, the graph matching method then extracts the most similar parts of the ontology to the tree in terms of the bacteria names labeling the nodes.

The alignment result related to the decision tree classifier is depicted in Fig.4(a). The laboratory test candidate for the parent of nodes 52 and 53 (Table 2) sharing *KLOXY* and *ENTFL* is the *Catalas* test which discriminates between *ENTFL* and *KLOXY* with its negative and positive response. Therefore, the ontology suggestion can annotate the subtree holding information about these leaf nodes. By applying the alignment process on the whole nodes in group B, we will finally have an annotated decision tree demonstrated in Fig.6.

Similarly, the mapping of laboratory tests on nodes of the binary trees built based on the Algorithm 1 is represented in Fig.4(b). In this figure, we can see the *MannitolFermentation* is the best microbiology test candidate to divide the 6 bacteria labeling the leaf nodes of the tree (Fig.5(a)). *EColi*, *PASMU* and *PRMIR* have the negative reaction to the *Mannitol* test in contrast to *STA*, *STRPY* and *ENTFL* which have positive reaction to it. It means that the *Mannitol* test is chosen to annotate the root of the tree which is

Table 4. Jaro-Winkler distances of names between the classifier and the ontology

D-Tree / Ontology	EColi	PSAER	STA	KLOXY	PRMIR	ENTFL	STLUG	PASMU	STRPY	HINFL
Escherichia coli	**0.364**	0.515	0.535	1	0.515	0.521	0.579	0.579	0.569	0.492
Pseudomonas ae...	0.503	**0.259**	0.414	0.585	0.379	0.503	0.503	0.352	0.414	0.585
Staphylococcus a...	0.641	0.530	**0.200**	0.530	0.584	0.502	0.279	0.503	0.397	0.502
Klebsiella oxytoca	0.522	0.522	0.407	**0.397**	0.581	0.663	0.663	0.663	0.407	0.663
Proteus mira...	0.519	0.439	0.572	0.580	**0.242**	0.661	0.519	0.364	0.536	1
Entercoccus fa...	0.477	0.502	0.540	0.584	0.502	**0.293**	0.502	0.668	1	0.584
Staphylococcus lu...	0.650	0.539	0.206	0.539	0.587	0.508	**0.269**	0.515	0.406	0.508
Pasteurella mul...	0.502	0.289	0.397	0.584	0.377	0.530	0.420	**0.224**	0.579	0.584
Steptococcus py...	0.532	0.506	0.331	0.585	0.532	0.670	0.337	0.532	**0.238**	1
Hemophilus infl...	0.423	0.423	0.581	0.670	0.532	0.503	0.532	0.391	0.359	**0.379**

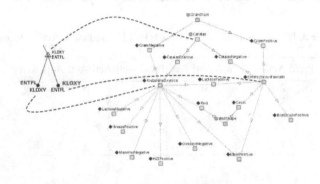

(a) Decision Tree- Ontology Alignment

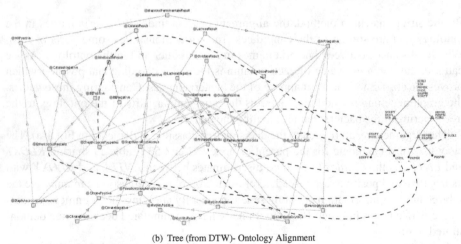

(b) Tree (from DTW)- Ontology Alignment

Fig. 4. Alignment Process (Between Candidate SubTree and Matched Sub-Ontology)

pointing to the 8th column of the confusion matrix. In other words, whenever the result of the classifier is bacteria number 8 (*PASMU*) the *MannitolFermentation* test is recommended to be applied first. In the same way, we find the second best laboratory

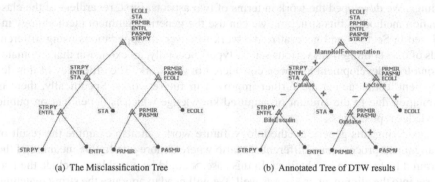

(a) The Misclassification Tree   (b) Annotated Tree of DTW results

Fig. 5. The Tree Built from Misclassification of DTW (Predicted Bacteria PASMU)

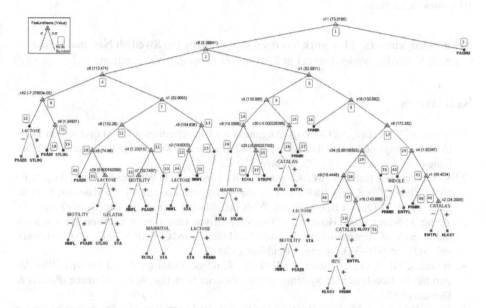

Fig. 6. Annotated Decision Tree by Laboratory Test Information

test for each branch of the tree. *Lactose* and *Catalase* tests are suggested for annotating the right and left child of the root, in order. *EColi* has the positive reaction to the Lactose test whereas the *PASMU* and *PRMIR* result to the negative one. By follwoing all the object properties named *hasReaction* (green arrows in the Fig.4(b)) we obtain the annotated tree demonstrated in Fig.5(b).

7   Conclusions

This work is as the extension of our previous work [1] where we implemented an ontology alignment method to improve classification results of electronic nose sensors readings. We developed the work in terms of two aspects. First, regardless of the classification method and its structure, we can use the generic alignment methodology introduced in Sect.4.3 and generalize the work to cover all applications using different kinds of data coming from various sensor types. Secondly, an extension that automates the ontology development has been considered in this work. The efficiency of this development technique can be further improved in future works. Specifically, there is a limitation due to the amount of acquired knowledge which is dependent on public knowledge repositories.

Considering this generic methodology, future work can also examine the result of the alignment process for a different scenario where a more intelligible meaning of the measured features can be obtained. In this case, it could be possible to include the feature set into the alignment process as well. We will need to appraise the string matching methods if we confront more labels and names than the current label set containing only 10 names of bacteria.

Acknowledgements. This work has been supported by the Swedish National Research Council, Vetenskapsrådet, project nr 2010-4769, on cognitive olfaction.

References

1. Alirezaie, M., Loutfi, A.: Ontology Alignment for Classification of Low Level Sensor Data. In: Proceedings of 4th KEOD International Conference on Knowledge Engineering and Ontology Development, pp. 89–97. Springer (2012)
2. Salvadores, M., Horridge, M., Alexander, P.R., Fergerson, R.W., Musen, M.A., Noy, N.F.: Using SPARQL to Query BioPortal Ontologies and Metadata. In: Cudré-Mauroux, P., et al. (eds.) ISWC 2012, Part II. LNCS, vol. 7650, pp. 180–195. Springer, Heidelberg (2012)
3. Zhang, J., Silvescu, A., Honavar, V.: Ontology-Driven Induction of Decision Trees at Multiple Levels of Abstraction. In: Koenig, S., Holte, R. (eds.) SARA 2002. LNCS (LNAI), vol. 2371, pp. 316–323. Springer, Heidelberg (2002)
4. Bouza, A., Reif, G., Bernstein, A., Gall, H.: SemTree: Ontology-Based Decision Tree Algorithm for Recommender Systems. In: International Semantic Web Conference (Posters & Demos) (2008)
5. Kong, H., Hwang, M., Kim, P.: Design of the automatic ontology building system about the specific domain knowledge. 8th ICACT International Conference on Advanced Communication Technology, International Symposium on High Performance Distributed Computing (2006)
6. Jakulin, A., Mladenić, D.: Ontology Grounding. In: Proceedings of 8th International Multi-Conference Information Society, pp. 170–173 (2005)
7. Bedini, I., Nguyen, B.: Automatic Ontology Generation: State of the Art. Technical report, University of Versailles (2007)
8. Gantz, J.F., Chute, C., Manfrediz, A., Minton, S., Reinsel, D., Schlichting, W., Toncheva, A.: The diverse and exploding digital universe: An updated forecast of worldwide information growth through 2011. Technical report, emc, IDC (2008)
9. Längkvist, M., Loutfi, A.: Unsupervised feature learning for electronic nose data applied to bacteria identification in blood. In: NIPS Workshop on Deep Learning and Unsupervised Feature Learning (2011)

10. Loutfi, A., Coradeschi, S., Saffiotti, A.: Maintaining Coherent Perceptual Information using Anchoring. In: The 19th International Joint Conference on Artificial Intelligence (IJCAI), pp. 1477–1482 (2005)
11. Trincavelli, M., Coradeschi, S., Loutfi, A., Söderquist, B., Thunberg, P.: Direct identification of bacteria in blood culture samples using an electronic nose. IEEE Trans. Biomedical Engineering 57 (2010)
12. Price, C., Spackman, K.: SNOMED clinical terms. British Journal of Healthcare Computing & Information Management 17(3), 27–31 (2000)
13. Jaro, M.: Advances in Record-Linkage Methodology as Applied to Matching the 1985 Census of Tampa, Florida. Journal of the American Statistical Society 84, 414–420 (1989)
14. Hlaoui, A.: A new algorithm for inexact graph matching. Object recognition supported by user interaction for service robots, vol. 4, pp. 180–183 (2002)
15. Melchert, J., Coradeschi, S., Loutfi, A.: Knowledge Representation and Reasoning for Perceptual Anchoring. Tools with Artificial Intelligence 1, 129–136 (2007)
16. Harnad, S.: The Symbol Grounding Problem. Physica D: Nonlinear Phenomena 42, 335–346 (1990)
17. Sossai, C., Bison, P., Chemello, G.: Fusion of symbolic knowledge and uncertain information in robotics. Int. J. Intell. Syst. 16, 1299–1320 (2001)
18. Chella, A., Frixione, M., Gaglio, S.: Anchoring symbols to conceptual spaces: the case of dynamic scenarios. Robotics and Autonomous Systems 43, 175–188 (2003)
19. Fiorini, S.R., Abel, M., Scherer, C.M.S.: An approach for grounding ontologies in raw data using foundational ontology. In: Information Systems. Elsevier (2012)
20. Quinlan, R.: C4.5: Programs for Machine Learning (Morgan Kaufmann Series in Machine Learning), 1st edn. Morgan Kaufmann (1992)
21. Bishop, C.M.: Pattern Recognition and Machine Learning (Information Science and Statistics). Springer (2006)
22. Seltmann, G., Holst, O.: The Bacterial Cell Wall. Springer (2002)
23. Ehrig, M.: Ontology Alignment: Bridging the Semantic Gap. Springer (2007)
24. Pearce, T.C., Schiffman, S.S., Nagle, H.T., Gardner, J.W.: Handbook of machine olfaction: electronic nose technology. Wiley-VCH (2003)
25. Joshi, R., Sanderson, A.C.: Multisensor Fusion: A Minimal Representation Framework. Series in Intelligent Control and Intelligent Automation. World Scientific (1999)
26. Euzenat, J., Shvaiko, P.: Ontology matching. Springer (2007)
27. A national clinical and anatomic pathology reference laboratory (2006), http://www.aruplab.com
28. The BioPortal Metadata Ontology (2012), http://www.aruplab.com
29. Ratanamahatana, C.A., Lin, J., Gunopulos, D., Keogh, E.J., Vlachos, M., Das, G.: Mining Time Series Data. In: Data Mining and Knowledge Discovery Handbook, pp. 1049–1077 (2010)
30. Moldovan, D., Girju, R.: Domain-Specific Knowledge Acquisition and Classification using WordNet (2000)
31. Jakulin, A., Mladenić, D.: Ontology Grounding. In: SIKDD at Multiconference IS (2005)

Performing Ontology Alignment via a Fuzzy-Logic Multi-layer Architecture

Susel Fernández, Ivan Marsa-Maestre, and Juan R. Velasco

Department of Computer Engineering, University of Alcalá, Alcalá de Henares, Madrid, Spain
{susel.fernandez,ivan.marsa,juanramon.velasco}@uah.es

Abstract. Data integration is becoming increasingly critical due to the vast amounts of information available in the Web and to the need for services that use information from different sources. Within the semantic Web, ontologies are crucial to provide data sharing and operability. However, when applications and services produced by different developers interact, we need to allow data to be shared and reused across distinct ontological frameworks. The process of establish "agreements" between different ontologies is called alignment, and is usually achieved by finding correspondences between their entities. In this paper we present an improvement of a fuzzy multi-layer architecture to perform ontology alignment. We use fuzzy logic techniques to combine different similarity measures among ontology entities, taking into account criteria such as the terminology, and the internal and relational structure of the concepts. This work was validated using the tests of the *Ontology Alignment Evaluation Initiative (OAEI)*. The results show that the proposed techniques outperform previous approaches in terms of precision and recall.

Keywords: Semantic Web, Ontology Mapping, Fuzzy Logic.

1 Introduction

Ontologies are one of the crucial components of the Semantic Web, which enable the design of exhaustive conceptual schema to facilitate communication and exchange of information between different systems and institutions. An ontology is a formal and explicit description of the elements belonging to a domain, such as concepts, properties, relations, functions, and axioms. The classes or concepts in the ontology represent any structured meaningful entity (i.e. one that provides information and contain properties). Relations represent interactions between classes. The most common relation we can find is inheritance, which is usually called taxonomic. Axioms are used to define the meaning of ontological components, and individuals are concrete instances of particular classes.

Heterogeneity in the representation of knowledge in different ontologies hampers interoperability between applications that use this knowledge. Semantic interoperability can be achieved by finding relationships between entities belonging to different ontologies and using these relations. The process of finding correspondences between entities of different ontologies is called ontology alignment, and can be applied in intelligent agent technology in order to enable agents to collect and

A. Fred et al. (Eds.): IC3K 2012, CCIS 415, pp. 194–210, 2013.
© Springer-Verlag Berlin Heidelberg 2013

integrate information from different sources, to find and use web services without user intervention; in electronic commerce, unifying descriptions of products, processes and transactions, etc. In addition, alignment allows possibilities such as facilitating several business applications to work automatically with different data sources, or to compare between different ontologies in order to exchange information with each other without having to previously agree on the meaning of the data.

Despite the many contributions that have been developed in the context of ontology alignment, there is no integrated solution that has proven to be robust and effective enough to be taken as a basis for future development and can be used by non-experts. One of the current challenges in ontology alignment is managing uncertainty. This problem has been addressed in several ways, but in general all approaches can be improved because none has yet fully automatically solved the problem of uncertainty in ontology alignment. Another front in this field is the improvement of similarity measures in order to obtain more precise values minimizing human intervention in the process.

In this paper, we present a multilayer fuzzy rule-based system for ontology alignment, combining several similarity measures for establishing correspondences between ontologies entities. The first measure is the terminological similarity, which takes into account the linguistic and semantic information of the context of the concept's names, and the second is the structural similarity, which uses both the taxonomic hierarchy of concepts and the internal structure of them, represented by their properties, types and cardinality restrictions. Due to the fuzzy combination of linguistic methods with semantic and evolutive learning on a significant number of test ontologies we have obtained very accurate alignments in several ontologies, outperforming most of the existing methods.

The rest of the paper is organized as follows: Section 2 describes the related work; then in section 3 and 4 we discuss the similarity measures. Section 5 presents the architecture of the fuzzy rule-based system; Section 6 is dedicated to the evaluation and experimental results. Finally the last section summarizes our conclusions and enumerates some future lines of research.

2   Related Work

There are some previous works aimed at ontology alignment, which have made interesting contributions, but so far none offers a complete matching due to the structural complexity of the ontologies.

SMART [9], PROMPT [10] and PROMPTDIFF [11] are tools that have been developed using linguistic similarity matches between concepts and a set of heuristics to identify further matches. These tools do not take into account the properties or relations of concepts and assume that all ontologies belong to a specific knowledge model, which makes the systems not applicable to other knowledge models.

Other developments use probabilistic methods, such as CODI [8], which produces mappings between concepts, properties, and individuals. CODI is based on the syntax and semantics of Markov logic. GLUE [3] employs machine learning techniques to find mappings. In [12] a probabilistic framework for automatic ontology mapping based on Bayesian Networks is proposed. This approach only takes into account the probability of occurrence of concepts in the web, which makes it fail if two very similar concepts have not the same level of popularity.

There are more recent works that combine lexical similarity with other techniques; one of them is *ASMOV* [7], which iteratively calculates the similarity by analyzing lexical elements, relational structure, and internal structure; *AgreementMaker* [2] comprises several matching algorithms that can be concept-based or structural. The concept-based matchers support the comparison of strings and the structural matchers include the descendants' similarity inheritance. In AgreementMaker the structural similarity depends absolutely on linguistic relationships, which means that correct results are not obtained in ontology concepts which are not linguistically similar, although the ontologies are very similar in structure.

In *Eff2Match* [18] the alignment process consists of four stages: Anchor Generation, where entities are identified using an exact string matching technique; Candidates Generation, where they find for entities using a vector space model approach; Anchor Expansion, to identifies more equivalent pairs of entities using terminological methods and Iterative Score Boosting to identify more pairs of equivalent concepts using the expanded anchor set. This system yields poor results in ontologies which differ linguistically and structurally, and thus it is difficult to get correct results in heterogeneous ontologies.

GeRMeSMB [14] is the integration of two tools; *GeRoMeSuite* offers a variety of matchers which can match ontologies and schemas in other modelling languages such as *XML* or *SQL*; and *SMB* mainly works on the similarity matrices produced by *GeRoMeSuite*. It improves the clarity of the similarity values by reinforcing "good" values and penalizing "bad" values for increase the precision of the match result. This is a very generic system and does not work very well on real ontologies in specific domains.

SOBOM [19] deals with ontology from two different perspectives: ontology with hierarchical structure and ontology with other relationships, combining the results of every step in a sequential way. If the ontologies have regular literals and hierarchical structures, the system can achieve satisfactory alignments and avoid missing alignment in many partitioning matching methods. If the literals of concept missed, the system will get bad results.

3   Terminological Similarity Measures

The terminology of the names of concepts and properties in ontologies provide valuable information about the entities that they model, and therefore it is the first item to consider when attempting matching them. Most ontology alignment methods process names as separate elements using string-based techniques, such as distance functions. Authors have also employed other techniques combined with the use of external linguistic resources such as thesauri and specialized directories in different domains. But still there are very few studies that use semantic context information to calculate the terminological similarity among ontology entities. A contextual approach could improve the values of similarity between the entity's names taking into account not only their construction from the lexical point of view but also the context in which the terms are used within the ontologies. In the following, we briefly describe the measures we use assess the terminological similarity between concepts.

3.1   Semantic Similarity

The semantic similarity is computed using the *Jaccard coefficient* [15] that is one of the most used binary similarity indexes. Given two sets of data this coefficient is defined as the size of the intersection of the sets divided by the size of their union. For two binary observations *i* and *j*, the *Jaccard coefficient* is calculated by:

$$S_{Jaccard}(i,j) = \frac{a}{a+b+c} \qquad (1)$$

where *a* is the number of times that both observations have the value 1, *b* is the number of times observation *i* has value 1 and observation *j* has value 0, and *c* is the number of times observation *i* has value 0 and observation *j* has value 1.

For the semantic similarity calculation we make successive searches of documents from the Web, specifically in *Wikipedia*[1]. In a similar way to [12], to ensure that the search only returns relevant documents to the entities, the search query is formed by combining all the terms on the path from the root to the current node in the taxonomy. Let us assume that the set A^+ contains the elements that support entity *A*, and the set A^- contains the elements that support the negation of *A*. Elements in A^+ are obtained by searching for pages that contain *A* and all *A*'s ancestors in the taxonomy, while elements of A^- would be those where *A*'s ancestors are present but not *A*. For each pair of entities *A* and *B*, three different counts are made: (*a*) the size of $A^+ \cap B^+$, (*b*) the size of $A^+ \cap B^-$, and (*c*) the size of $B^+ \cap A^-$. Once these values are obtained for each pair of origin and destination ontology entities their similarity is calculated using Equation 1. Figure 1 shows fragments of two ontologies and the search queries that would be formed to calculate the semantic similarity between the concepts *Book* and *Proceedings*.

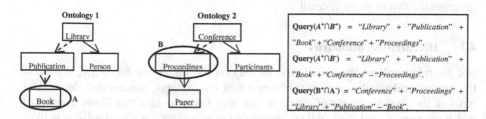

Fig. 1. Parts of two ontologies and search queries for semantic similarity

3.2   Lexical Similarity

The lexical similarity is the strongest indicator of correspondence between entities, because usually the ontology developers within the same domain use linguistically related terms to express equivalent entities [6]. In this work two types of lexical similarity are considered: the first one is based on synonyms, and the second one is based on the derivationally related forms of the words. Lexical similarity index is computed as follows:

---

[1] *Wikipedia*: www.wikipedia.org

1. Given the concepts A and B, the first step is to remove the meaningless words (stop words).
2. Then we obtain lists of synonyms and words derived from each one using *WordNet*[2]. In the specific case of medical ontologies we use the specialized directory *UMLS*[3] as lexical tool.
3. Next, we apply the *Porter Stemming Algorithm* [13] to remove the morphological ends of the words from the lists of synonyms and derivative words.
4. Let A and B be the sets of roots of synonyms or words derived from each concept obtained in this step, we can calculate the linguistic similarity of synonymy and derivationally related words between the concepts according to the following equation:

$$S(a,b) = \min\left[\frac{|A \cap B|}{|A|}, \frac{|A \cap B|}{|B|}\right] \tag{2}$$

The first ratio indicates the fraction of overlap of the set A with respect to set B, and the second ratio indicates the overlap fraction of the set B relative to the set A. As both sets are not necessarily having the same number of elements, we choose the minimum value among their degrees of overlap as an indicator of their similarity.

4   Structural Similarity Measures

The structural similarity among the entities in ontologies is bases on two key issues: the relational structure, which considers the taxonomic hierarchy of concepts; and the internal structure, comprising property restrictions of concepts. In the following we describe the two of them in detail.

4.1   Hierarchical Similarity

For the relational structure similarity, we rely on the taxonomic hierarchy. We define the hierarchical similarity as the influence that the siblings, parents and descendants have on the final similarity of concepts. We start from the idea that if two concepts A and B are similar, and their siblings, descendants, or parents are also similar, it is likely that A and B are equivalents. Figure 2 shows an example of how to calculate the hierarchical similarity of siblings. Let m be the number of siblings of the concept A, n the number of siblings of the concept B and let A_i and B_j be the ith and j^{th} siblings of concepts A and B, respectively. The hierarchical similarity would be the average of the maximum of the similarities between all the siblings of A and all of B (Equation 3).

$$Sim(A, B) = \frac{1}{n}\sum_{i=1}^{n}\max\left\{Sim(A_i, B_j)\right\}_{j=1}^{m} \tag{3}$$

---

[2] *WordNet*: http://wordnet.princeton.edu/
[3] *UMLS* (*Unified Medical Language System*):
   http://www.nlm.nih.gov/research/umls/

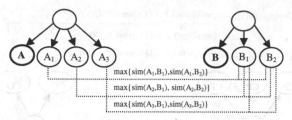

Fig. 2. Hierarchical similarity of the concepts A and B based on the similarity of their siblings

4.2   Internal Similarity

This similarity is based on the internal structure of concepts, which is influenced by other factors besides their names and relationships such as the similarity of their properties, and cardinality. Similar to the calculation of the relational structure similarity, here we start from the idea that if two concepts are similar, they have the same number of properties, and these properties have a high degree of similarity, then it is very likely that they are equivalent concepts and we reinforce the value of their similarity. In contrast, if two classes have some resemblance, but they have not the same number of properties or these properties are not similar, we decrease their similarity value. For each pair of classes A and B we call *property-wise similarity* to the value that the properties bring to the final class similarity. It is calculated as the same way as in the taxonomic hierarchy, averaging the maxima of the similarities between the properties of the concept A and the properties of the concept B (Equation 3).

Property Similarity. The similarity between two properties is influenced by three factors: the lexical similarity of their names, the similarity of the classes to which they belong (*Domain*), and the similarity of their types (*Range*).

Domain Similarity is the similarity between the classes to which the properties belong. Since in ontology the same property can belong to several classes, this is a similar case to the calculation of the hierarchical similarity of parents. Figure 3 shows an example of the calculation of the domain similarity of the properties P and Q.

Range Similarity is the similarity between the types of the properties. In the case of object properties, we directly find the similarities between the classes those objects belong to. If we are dealing with data properties, we must take into account the compatibility between the corresponding XML data types. If both are primitive data types, the correspondence between them has to be exact. If at least one type is a derived XML type, we need to look their primitive root type and the problem is reduced to comparing two primitive types. In this paper we consider that if two data types are derived from the same primitive type, then they are compatible.

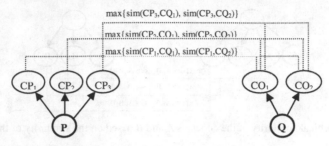

Fig. 3. Domain *Similarity* for the properties *P* and *Q*

5   A Multi-layer Fuzzy Rule-Based System

Fuzzy Rule-based Systems constitute an extension to classical rule-based systems. They deal with "IF-THEN" rules whose antecedents and consequents are composed of fuzzy logic statements instead of classical logic ones. They have been successfully applied to a wide range of problems in different domains with uncertainty and incomplete knowledge [1]. A Fuzzy Rule-based System consists of 4 parts: the knowledge base; the inference engine that is responsible for drawing the conclusions from the symbolic data that have arrived using the rules governing the system in which it works; and the fuzzification and defuzzification interfaces which have the function of converting real input values into fuzzy values and the other way around.

We defined *FuzzyAlign*, a multi-layer fuzzy rule-based system. The system is composed by four layers. The output values of each one serves as input to the upper layer and each layer provides an improvement in the calculation of the similarity: the first one is the lexical similarity layer, the second one is the terminological similarity layer, the third is the structural layer and the latter is the alignment layer. Figure 4 shows the architecture of the system.

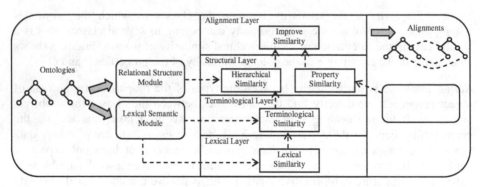

Fig. 4. Architecture of *FuzzyAlign*

5.1   Lexical Layer

In the first layer of the fuzzy system we calculate the lexical similarity. The two input variables represent the similarities of synonyms and derivations, respectively, and the output variable represents the overall linguistic similarity. Computation is performed using the Lexical-Semantic module, where lexical similarities are calculated in the manner explained in Section 3 using *WordNet* and *UMLS* as lexical directories. Because of the distribution of lexical similarity values, equally spaced fuzzy sets were defined for the three variables with triangular membership functions (Figure 5). Figure 6 shows the architecture of the lexical layer.

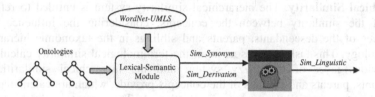

Fig. 5. Fuzzy triangular-shaped membership functions for linguistic similarity variables

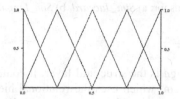

Fig. 6. Lexical layer of the fuzzy rule-based system

5.2   Terminological Layer

The terminological layer of the fuzzy system is responsible for carrying out the computation of the similarity between the names of the entities, combining lexical similarity with the semantic elements of the context. This layer receives as input the similarity value obtained in the lexical layer and the semantic similarity calculated in the Lexical-Semantic module using *Jaccard coefficient*. The output is the terminological similarity. Figure 7 shows the architecture of the terminological layer. To define the membership functions, we used equally spaced fuzzy sets, except in the case of the variable representing the semantic similarity due to the distribution of the values. In this case we use the quartiles of the data to narrow the membership triangles. Figure 8 shows the triangular membership functions defined for the three variables.

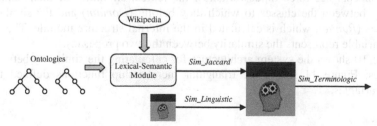

Fig. 7. Architecture of the Terminological layer

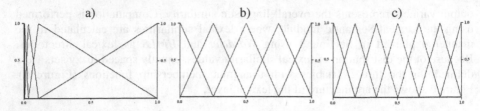

Fig. 8. Membership functions for: a) *Sim_Jaccard*, b) *Sim_Linguistic*, c) *Sim_Terminologic*

5.3 Structural Layer

The third layer of *FuzzyAlign* is the structural layer. It contains two subsystems: The hierarchical similarity system, which uses the relational hierarchy of the ontologies; and the property similarity system, which uses the internal structure of the concepts.

Hierarchical Similarity. The hierarchical similarity system is entitled to refine the value of the similarity between the concepts, considering the influence of the similarities of the descendants, parents and siblings in the taxonomic hierarchy of each ontology. This task receives as input the terminological similarity calculated in the previous layer of the fuzzy system and the hierarchical similarities that descendants, parents and siblings of the concepts provide, which are calculated in the relational structure module as seen in Section 4. The output is the hierarchical similarity of the concepts. The architecture of the system for calculating the hierarchical similarity is shown in Figure 9. For the 5 variables was defined the same fuzzy sets and membership functions shown in Figure 8 c).

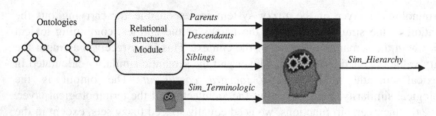

Fig. 9. Architecture of the Hierarchical Similarity System

Property Similarity. The system for calculating the similarity between the properties of the concepts has three input variables: the lexical similarity of their names, the similarity between the classes to which they belong (*Domain*) and the similarity of their types (*Range*), which is calculated in the internal structure module. The system output variable represents the similarity between the two properties.

Figure 10 shows the system architecture for calculating the similarity between the properties. Figure 11 shows the triangular membership functions defined for the system variables.

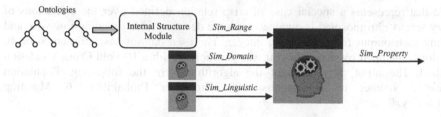

Fig. 10. Architecture of the Property Similarity System

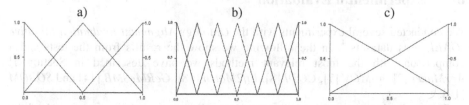

Fig. 11. Membership functions for: a) *Sim_linguistic*, b) *Sim_Domain* and c) *Sim_Range*

5.4   Alignment Layer

The alignment layer is the final layer of the fuzzy rule-based system and aims to provide the final similarity index between the concepts taking into account the influence of the number of properties and the value of similarity that properties bring to the similarities between them. The *property-wise similarity* is calculated on the internal structure module. Depending on these values, the hierarchical similarity is reinforced or weakened, resulting in a more accurate indicator of the similarity between the concepts. Figure 12 shows the architecture of the alignment layer of the fuzzy rule based system. After obtaining the similarities between all concepts and properties of the two ontologies, the alignment module is responsible for selecting the equivalent entities and constructs the final set of alignments. To carry out this process we have chosen a confidence threshold of 0.8, which means that two entities are considered equivalent, if their similarity value is equal to or greater than 80%. The matching rules are constructed via the *Java API Alignment Format* [4], allowing the generation of outputs in different formats such as SWRL, OWL and RDF.

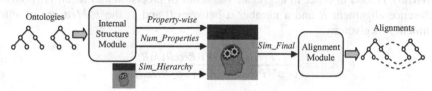

Fig. 12. Architecture of the Alignment Layer

5.5   Evolutive Learning of the Fuzzy Rule Bases

The rule bases of the fuzzy system were deduced using the genetic algorithm *THRIFT* [16] for the learning of rule bases. This method works by using a complete decision table that represents a special case of crisp relation defined over the collections of fuzzy sets. A chromosome is obtained from the decision table by going row-wise and coding each output fuzzy set as an integer. The used dataset has information of 40 ontologies mapped by experts and it was partitioned with a 10-Fold Cross Validation method. The input parameters of the algorithm were the following: Population Size=61, Number of Evaluations=1000, Crossover Probability=0.6, Mutation Probability=0.1.

6   Experimental Evaluation

We conducted several experiments with the *Ontology Alignment Evaluation Initiative (OAEI)* tests datasets [4]. In the following we show the results from the tests and a comparison with the most relevant methods we have described in Section 1: *AgrMaker* [2], *ASMOV* [7], *CODI* [8], *Eff2Match* [18], *GeRMeSMB* [14] and *SOBOM* [19].

The evaluation measures used in ontology alignment systems are typical of the field of information retrieval. They are *Precision, Recall* and *F-Measure*. Given a reference alignment *R* and the resultant alignment *A*, these similarity measures are defined as follows:

Precision is the fraction of correct instances among those that the algorithm believes to belong to the relevant subset [15]. Given a reference alignment *R*, the precision of some alignment *A* is given by:

$$Precision(A, R) = \frac{|R \cap A|}{|A|} \tag{4}$$

Recall is computed as the fraction of correct instances among all instances that actually belong to the relevant subset [15]. Given a reference alignment *R*, the recall of some alignment *A* is given by:

$$Recall(A, R) = \frac{|R \cap A|}{|R|} \tag{5}$$

F-Measure is used in order to aggregate the result of precision and recall [15]. Given a reference alignment *R* and a number α between 0 and 1, the *F-Measure* of some alignment *A* is given by:

$$F_\alpha(A, R) = \frac{Precision(A, R) \cdot Recall(A, R)}{(1 - \alpha) \cdot Precision(A, R) + \alpha \cdot Recall(A, R)} \tag{6}$$

The higher α, the more importance is given to precision with regard to recall. Often, the value $\alpha = 0.5$ is used. This is the harmonic mean of precision and recall.

---

[4] *OAEI. Ontology Alignment Evaluation Initiative*:
   http://www.oaei.ontologymatching.org

6.1   Benchmark Test

The domain of this first test [5] is Bibliographic references. It is based on a subjective view of what must be a bibliographic ontology. The systematic benchmark test set is built around one reference ontology and many variations of it. The ontologies are described in *OWL-DL* and serialized in the *RDF/XML* format. The reference ontology contains 33 named classes, 24 object properties, 40 data properties, 56 named individuals and 20 anonymous individuals. The tests are organized in three groups: *Simple tests (1xx)* such as comparing the reference ontology with itself, and with another irrelevant ontology; *Systematic tests (2xx)* obtained by discarding features from some reference ontology. It aims at evaluating how an algorithm behaves when a particular type of information is lacking; *four real-life ontologies (3xx)* of bibliographic references found on the web and left mostly untouched. Table 1 shows the results of the alignment methods that performed the benchmark test by group of test.

Table 1. Benchmark test results for the alignment methods in terms of Precision, Recall and F-Measure

Test	1xx			2xx			3xx			H-Mean		
System	P	R	F	P	R	F	P	R	F	P	R	F
AgrMaker	0.98	1.00	0.99	0.95	0.84	0.89	0.88	0.53	0.66	0.93	0.74	0.82
ASMOV	1.00	1.00	1.00	0.99	0.89	0.94	0.88	0.84	0.86	0.95	0.91	0.93
CODI	1.00	0.99	0.99	0.70	0.42	0.53	0.92	0.43	0.59	0.85	0.52	0.65
Eff2Match	1.00	1.00	1.00	0.98	0.63	0.77	0.89	0.71	0.79	0.95	0.75	0.84
GeRMeSMB	1.00	1.00	1.00	0.96	0.66	0.78	0.79	0.38	0.51	0.91	0.58	0.71
SOBOM	1.00	1.00	1.00	0.97	0.94	0.95	0.77	0.70	0.73	0.90	0.86	0.88
FuzzyAlign	**1.00**	**1.00**	**1.00**	**0.98**	**0.95**	**0.97**	**0.88**	**0.84**	**0.86**	**0.95**	**0.93**	**0.94**

It can be seen in the simple's test (*1xx*) that the performance of all the systems was optimal. For the systematic tests (*2xx*) the *FuzzyAlign* system had a high precision, surpassed only by *ASMOV*, however we have obtained the best value of recall and f-measure. For real cases (*3xx*) we have obtained the same precision as *AgrMaker* and *ASMOV*, being surpassed by *Eff2Match* and *CODI*, however we have obtained the best recall and f-measure like *ASMOV*. Finally looking at the harmonic means (H-Mean) of precision, recall and f-measure of the three phases, can be observed that our system achieved the highest precision, recall and f-measure average results, outperformed all the other systems. The confidence threshold used for the selection of the valid alignment was 0.8.

6.2   Conference Test

Conference test [5] contains quite real-case ontologies suitable because of their heterogeneous character of origin. The goal of this experiment is to find all correct matches within a collection of ontologies describing the domain of organizing

Table 2. Conference test results of *FuzzyAlign* in terms of Precision, Recall and F-Measure

Ontology 1	Ontology 2	P	R	F
cmt	*Conference*	0.85	0.85	0.85
cmt	*Confof*	0.83	0.29	0.43
cmt	*Edas*	0.88	0.88	0.88
cmt	*Ekaw*	0.83	0.83	0.83
cmt	*Iasted*	1.00	1.00	1.00
cmt	*Sigkdd*	1.00	0.47	0.64
Conference	*Confof*	1.00	1.00	1.00
Conference	*Edas*	0.75	0.75	0.75
Conference	*Ekaw*	0.80	0.70	0.85
Conference	*Iasted*	0.97	1.00	0.99
Conference	*Sigkdd*	1.00	1.00	1.00
Confof	*Edas*	1.00	0.42	0.59
Confof	*Ekaw*	1.00	0.35	0.52
Confof	*Iasted*	1.00	1.00	1.00
Confof	*Sigkdd*	0.66	1.00	0.80
Edas	*Ekaw*	1.00	0.28	0.44
Edas	*Iasted*	1.00	0.44	0.61
Edas	*Sigkdd*	0.88	0.44	0.59
Ekaw	*Iasted*	1.00	1.00	1.00
Ekaw	*Sigkdd*	1.00	1.00	1.00
Iasted	*Sigkdd*	1.00	0.86	0.92

Table 3. Conference test results for the alignment methods in terms of confidence threshold, Precision, Recall and F-Measure

System	Threshold	P	R	F
AgrMaker	0.66	0.53	0.62	0.58
ASMOV	0.22	0.57	0.63	0.60
CODI	-	0.86	0.48	0.62
Ef2Match	0.84	0.61	0.58	0.60
GeRMeSMB	0.87	0.37	0.51	0.43
SOBOM	0.35	0.56	0.56	0.56
FuzzyAlign	**0.80**	**0.93**	**0.74**	**0.83**

conferences. In table 2 we show the results of applying our system with 21 reference alignments, corresponding to the complete alignment space between 7 ontologies from the confer ence data set. Table 3 shows the values of precision, recall and f-measure obtained by the 7 systems that we compared, and the confidence threshold set by each of them to provide the highest average of F-Measure. In the case of *CODI* they not provided a confidence threshold because their results were the same regardless of the threshold. We can observe that with a confidence threshold of 0.8 our system scored precision, recall and f-measure much higher than others. This means that we are considering as valid alignment only those mappings whose similarity value is greater than 80%, which shows that the system has shown better results with a greater level of rigor in the selection of alignments.

6.3   Anatomy Test

This track consists of two real world ontologies to be matched [5]. The source ontology describes the Adult Mouse Anatomy (with 2744 classes) while the target ontology is the *NCI Thesaurus* describing the Human Anatomy (with 3304 classes). The anatomy test consists of four tasks: *Task #1*, which emphasizes f-measure, *Task #2*, which emphasizes precision, *Task #3*, which emphasizes recall, and *Task #4*, which tests the capability of extending a partial reference alignment. We performed only the Tasks#1 through #3 and use the following configuration parameters:

Task #1. The optimal solution alignment is obtained by using the default parameter settings. Confidence threshold value was 0.8.

Task#2. The alignment with optimal precision is obtained by changing the threshold for valid mappings to 0.9.

Task#3. The alignment with optimal recall is generated by changing the threshold to 0.6.

Considering that the ontologies used in this test are very specific to the medical field, we performed the experiment in two parts: first using *WordNet* as lexicon to calculate the linguistic similarity, and then using the *UMLS* medical databases instead of *WordNet*. Now we explain the results of both cases.

Anatomy Test Using Wordnet. The results of this test for *FuzzyAlign* using *WordNet* were not the best. This is mainly due to the fact that our system gives much weight to the lexicon and these specific medical terms are not all in *WordNet*. In this case the use of general purpose lexical tool as *WordNet* instead of a medical board causes that the system did not achieve optimal lexical similarities and the lack of this information affected the overall result. In Table 4 we can observe the results of the 7 systems in the anatomy test per task. In the case of *SOBOM* authors have only performed Task#1.

Table 4. Anatomy test results for the alignment methods in terms of precision, recall and F-measure

System	Task #1			Task #2			Task #3			H-Mean		
	P	R	F	P	R	F	P	R	F	P	R	F
AgrMaker	0.90	0.85	0.87	0.96	0.75	0.84	0.77	0.87	0.82	0.87	0.82	0.84
ASMOV	0.79	0.77	0.78	0.86	0.75	0.81	0.71	0.79	0.75	0.78	0.77	0.79
CODI	0.96	0.65	0.77	0.96	0.66	0.78	0.78	0.69	0.73	0.89	0.66	0.76
Ef2Match	**0.95**	**0.78**	**0.85**	**0.96**	**0.74**	**0.84**	**0.95**	**0.78**	**0.85**	**0.95**	**0.77**	**0.85**
GeRMeSMB	0.88	0.31	0.46	0.88	0.31	0.46	0.08	0.89	0.15	0.20	0.39	0.27
SOBOM	0.95	0.78	0.86	-	-	-	-	-	-	-	-	-
FuzzyAlign	0.72	0.74	0.73	0.75	0.45	0.56	0.44	0.76	0.56	0.61	0.62	0.64

Anatomy Test Using UMLS. In this test we use the *UMLS metathesaurus* instead of *WordNet* to perform the lexical similarity. The *UMLS* or *Unified Medical Language System* is a set of files and software that brings together many health and biomedical vocabularies and standards to enable interoperability between computer systems. *UMLS* provides a list with unique identifiers (*CUIs*) which can be associated to entities belonging to different sources, including *SNOMED CT, NCI, MeSH*, and *RxNorm* vocabularies. The pairs of entities from different sources with the same *CUI* are considered synonyms. As we can see in Table 5 the results of *FuzzyAlign* in Anatomy test improve considerably with the use of *UMLS*. This is due to the fact that the linguistic similarity values obtained are more appropriate and therefore also improve the final similarity values between the ontology's entities.

Table 5. Anatomy test results for the alignment methods in terms of precision, recall and F-measure

System	Task #1			Task #2			Task #3			H-Mean		
	P	R	F	P	R	F	P	R	F	P	R	F
AgrMaker	0.90	0.85	0.87	0.96	0.75	0.84	0.77	0.87	0.82	0.87	0.82	0.84
ASMOV	0.79	0.77	0.78	0.86	0.75	0.81	0.71	0.79	0.75	0.78	0.77	0.79
CODI	0.96	0.65	0.77	0.96	0.66	0.78	0.78	0.69	0.73	0.89	0.66	0.76
Ef2Match	0.95	0.78	0.85	0.96	0.74	0.84	0.95	0.78	0.85	0.95	0.77	0.85
GeRMeSMB	0.88	0.31	0.46	0.88	0.31	0.46	0.08	0.89	0.15	0.20	0.39	0.27
SOBOM	0.95	0.78	0.86	-	-	-	-	-	-	-	-	-
FuzzyAlign	**0.97**	**0.91**	**0.94**	**1.0**	**0.85**	**0.92**	**0.85**	**0.92**	**0.88**	**0.94**	**0.89**	**0.91**

7   Conclusions and Future Work

In this article we have described our work aimed to provide a method for the ontology alignment process using fuzzy logic techniques. We have presented *FuzzyAlign*, a Multi-Layer Fuzzy System which computes the similarities between entities from different ontologies, taking into account semantic and lexical elements and also the relational and the internal structures of the ontologies. The system has been tested in the basic tests proposed for *OAEI* to evaluate the performance of ontology alignment methods, showing better results than other systems in general purpose ontologies and ontologies from real life with correct lexical constructions. Through our experiments yield satisfactory results in general proposed ontologies, there were some limitations in our approach. The most important of them was that due to the importance of lexical similarities in the process of matching, the use of *WordNet* as lexicon caused the system not to provide optimal results in very specific domain ontologies such as medical ones. To address this problem we have incorporated the use of the specialized medical metathesaurus *UMLS*. Using this tool we managed to improve the results of the alignment system in this type of ontologies. Another of the drawbacks of using *WordNet* is that is it only has English language dictionaries, making it impossible to obtain valid alignments between ontologies in other languages. To solve this problem we intend to explore the use of *EuroWordNet* tool, which emerged as a project to interconnect several dictionaries of European languages with the same structure as the original Princeton *WordNet*.

In addition, the execution time of the system increases when processing too large ontologies due to the high amount of information, so we need to improve the scalability of the application. Finally, we are also interested in extending the technique to propose an integration model that allows matching taking into account the use of other relations between entities in real domains instead of just equivalence.

References

1. Cordón, O., Herrera, F., Hoffman, F., Magdalena, L.: Genetic Fuzzy Systems. In: Evolutionary Tuning and Learning of Fuzzy Knowledge Bases. World Scientific, Singapore (2001)
2. Cruz, I.F., Palandri, A.F., Stroe, C.: AgreementMaker Efficient Matching for Large Real-World Schemas and Ontologies. In: International Conference on Very Large Databases, Lyon, France, pp. 1586–1589 (September 2009)
3. Doan, A., Madhavan, J., Domingos, P., Halevy, A.: Ontology Matching: A Machine Learning Approach. In: Staab, S., Studer, R. (eds.) Handbook on Ontologies in Information Systems, pp. 397–416. Springer (2004)
4. Euzenat, J., Shvaiko, P.: Ontology Matching. Springer, Heidelberg (2007)
5. Euzenat, J., Shvaiko, P., Giunchiglia, F., Stuckenschmidt, H., Mao, M., Cruz, I.: Results of the Ontology Alignment Evaluation Initiative 2010. In: Proceedings of the 5th International Workshop on Ontology Matching, OM-2010 (2010)
6. Fernández, S., Velasco, J.R., López-Carmona, M.A.: A Fuzzy Rule-Based System for Ontology Mapping. In: Yang, J.-J., Yokoo, M., Ito, T., Jin, Z., Scerri, P. (eds.) PRIMA 2009. LNCS, vol. 5925, pp. 500–507. Springer, Heidelberg (2009)
7. Jean-Mary, Y., Shironoshita, E.P., Kabuka, M.: Ontology Matching with Semantic Verification. Journal of Web Semantics. Sci. Serv. Agents World Wide Web (2009), doi:10.1016/j.websem.2009.04.001
8. Noessner, J., Niepert, M., Meilicke, C., Stuckenschmidt, H.: Leveraging Terminological Structure for Object Reconciliation. In: Aroyo, L., Antoniou, G., Hyvönen, E., ten Teije, A., Stuckenschmidt, H., Cabral, L., Tudorache, T. (eds.) ESWC 2010, Part II. LNCS, vol. 6089, pp. 334–348. Springer, Heidelberg (2010)
9. Noy, N.F., Musen, M.A.: SMART: Automated Support for Ontology Merging and Alignment. In: 12th Workshop on Knowledge Acquisition, Modelling and Management (KAW 1999), Banff, Canada (October 1999)
10. Noy, N.F., Musen, M.A.: The PROMPT suite: Interactive tools for ontology merging and mapping. International Journal of Human-Computer Studies 59(6), 983–1024 (2003)
11. Noy, N.F., Musen, M.A.: PROMPTDIFF: A Fixed-Point Algorithm for Comparing Ontology Versions. In: 18th National Conference on Artificial Intelligence (AAAI 2002), Edmonton, Alberta, Canada (August 2002)
12. Pan, R., Ding, Z., Yu, Y., Peng, Y.: A Bayesian Network Approach to Ontology Mapping. In: Gil, Y., Motta, E., Benjamins, V.R., Musen, M.A. (eds.) ISWC 2005. LNCS, vol. 3729, pp. 563–577. Springer, Heidelberg (2005)
13. Porter, M.F.: An Algorithm for Suffix Stripping. Program 14(3), 130–137 (1980)
14. Quix, C., Gal, A., Sagi, T., Kensche, D.: An integrated matching system: GeRoMeSuite and SMB– Results for OAEI 2010. In: Proceedings of the 5th International Workshop on Ontology Matching, OM 2010 (2010)
15. Van Rijsbergen, C.J.: Information Retrieval, 2nd edn. Butterworths, London (1979)

16. Thrift, P.: Fuzzy Logic Synthesis with genetic algorithms. In: Proceedings of the 4th International Conference on Genetic Algorithms, pp. 509–513. Morgan Kaufmann (1991)
17. Wang, S., Wang, G., Liu, X.: Results of the Ontology Alignment Evaluation Initiative. In: Proceedings of the 5th International Workshop on Ontology Matching, OM 2010 (2010)
18. Watson Wey, K., Jun Jae, K.: Eff2Match results for OAEI 2010. In: Proceedings of the 5th International Workshop on Ontology Matching, OM 2010 (2010)
19. Xu, P., Wang, Y., Cheng, L., Zang, T.: Alignment Results of SOBOM for OAEI. In: Proceedings of the 5th International Workshop on Ontology Matching, OM 2010 (2010)

An Answer Set Programming Solution
for Supply Chain Traceability

Monica L. Nogueira and Noel P. Greis

Center for Logistics and Digital Strategy, The University of North Carolina at Chapel Hill,
Kenan Center CB#3440, Chapel Hill, 27713 U.S.A.
{monica_nogueira,noel_greis}@unc.edu

Abstract. Developing measures to improve the traceability of contaminated
food products across the supply chain is one of the key provisions of the 2011
FDA Food Safety Modernization Act (FSMA). In the event of a recall, FSMA
requires companies to provide information about their immediate suppliers and
customers—what is referred to as "one step forward" and "one step backward"
traceability. In this paper we implement the logic-based approach called answer
set programming that uses inference rules to trace the flows of contaminated
products—both upstream to the source of the contamination and downstream to
consumer locations. The approach does not require common standards or
unique product identifiers for tracking individual products. This elaboration-
tolerant method can accommodate changes in the supply chain such as: 1) the
addition of new multiple product pathways; 2) consideration of multiple
ingredients in a single product; and 3) multiple products with multiple
pathways. We demonstrate this highly flexible methodology for pork and
peanut products.

Keywords: Answer Set Programming, Traceability, Supply Chain, Food Recall
Process.

1   Introduction

Food safety is a challenging problem that has been growing worldwide due to the
globalization of the food supply chain, internationalization of trade, and new eating
habits, among other factors. The lack of a consistent, unified, and standardized
tracking and tracing system for food manufactured, produced, processed, packed,
held, distributed, and sold in the United States is a major pain point of the American
food safety system, but this is a problem that affects most countries, if not all. The
food supply chain consists of many entities from producer/grower and processor to
distributor and retailer. Each of these entities is linked to one another through the
food chain. Contamination can enter the food chain at any point due to a range of
causes from improper processing or handling to intentional contamination.

In the U.S. once public health officials have determined that a foodborne disease
event has occurred and identified the offending product and its manufacturer, a
product recall may be issued by the U.S. Food and Drug Administration (FDA)

A. Fred et al. (Eds.): IC3K 2012, CCIS 415, pp. 211–227, 2013.

agency. This recall signals the launch of a series of actions by state food safety departments to remove any contaminated products from retail shelves within their states. State agencies must quickly determine whether any recalled products are being sold by retail enterprises or whether the contaminated products have been used as ingredients in any products being sold. At the point of recall, state agencies are required to piece together information from enterprises across the food chain in an environment where there is not a uniform system for linking this information, nor accepted standards for identifying products, nor any central place where this information is stored and accessible.

The difficulty of the task is complicated by (1) the complexity of the food chain where a single food product can be made of hundreds of ingredients which each may be supplied by multiple suppliers; (2) the fact that uniform standards for data collection in the food industry do not exist, making it difficult to re-create the food chain for contaminated products; and (3) the fact that companies are often reluctant to make public proprietary information about their supply chain suppliers and customers. Further, traceability across enterprise boundaries requires agreements and coordination among suppliers and customers that can be difficult to achieve.

This paper is organized as follows. Section 2 discusses the motivation behind our work. Section 3 presents background information about the problem being solved. Section 4 describes existing traceability schemes. An ASP program encoding the traceability problem for a simple and a more complex supply chain is discussed in Sections 5 and 6, respectively. Conclusions appear in Section 7.

2   Motivation

The lack of track-and-trace capability has received considerable attention recently due to several high-profile and costly incidents of foodborne disease in the United States (c.f. peanut butter, spinach, jalapenos peppers) and abroad (c.f. milk, pork, sprouts). New studies from the U.S. Centers for Disease Control and Prevention (CDC) estimate the total effect of contaminated food consumed in the United Sates as follows: 47.8 million illnesses, 127,839 hospitalizations and 3,037 deaths per year [1, 2]. The total cost of food contamination in the U.S. was recently estimated to be $152 billion a year including health and human welfare costs, as well as economic damage to companies and entire industries [3]. In 2009, the Peanut Corporation of America (PCA) peanut butter contamination alone sickened more than 700 people in 44 states and was associated with nine deaths—and also resulted in the largest dollar-valued food recall in U.S. history. More than 3,000 products were recalled. Early estimates of the costs to the peanut butter industry due to lost peanut butter and peanut sales were more than $1 billion.

The PCA peanut contamination also illustrates the problems of determining both the source and the location of contaminated foods in the food chain. Difficulties are complicated when the contamination is ingredient-driven, that is when the contaminated product is an ingredient in a large number of different products that are sold in many different channels.

Traceability refers broadly to the ability, for any product at any stage within the food chain, to identify the initial source (backward tracing) and, eventually, its final destination (forward tracing) [4]. Tracking refers to the ability to identify, for any product, its actual location at any given time. Together these two capabilities provide the functionality of a "track-and-trace" system for the food supply chain.

A 2009 traceability exercise conducted by the U.S. Department of Health and Human Services (DHHS) illustrated the gaps in the current system. Investigators purchased 40 different products and attempted to trace each through the supply chain back to the farm or the border, in the event of an imported food. Of the 40 products, only five could be traced back completely to the point of origin; 31 could be traced back only partially; and four of the products could not be traced back at all [5].

3   Background

The FDA Food Safety Modernization Act (FSMA) that was signed into law by President Obama in January 2011 was the first major overhaul of food safety law in the U.S. in decades. It set the stage for a new era in food safety regulation that moves FDA towards new risk based approaches. FSMA includes several key provisions that position the FDA to improve its ability to respond to a food recall. First, the FDA now has the authority to issue a mandatory recall when it has been determined that there is a reasonable probability that a food poses a threat to human health. Previously, FDA could only request a voluntary recall.

FSMA also requires that the FDA establish, as appropriate, within the FDA "a product tracing system to receive information that improves the capacity of the Secretary to effectively and rapidly track and trace food that is in the United States or offered for import into the United States" [6]. FSMA does not specify the details of such a traceability system or the technology to be used, but directs the FDA to conduct at least two pilot projects to evaluate methods for improving traceability. On September 2011, the FDA announced that the Institute of Food Technologists (IFT) will "carry out two new pilot projects at the direction of FDA to explore and demonstrate methods for rapid and effective tracking and tracing of food, including types of data that are useful for tracing, ways to connect the various points in the supply chain and how quickly data can be made available to FDA" [7].

In addition, FSMA expands the registration requirements established by the U.S. Congress in the 2002 Bioterrorism Act that required all facilities that manufacture, process or pack food to register with the FDA, but exempted farms and retail food establishments, by limiting that exemption only to family and smaller growers.

Finally, in support of traceability, FSMA requires companies to provide for all food products "one step forward" and "one step backward" traceability. Food facilities are not required to provide full traceability for their products "from farm to fork" but only from/to their immediate suppliers and immediate customers. If every food facility maintains such records it should be possible to trace the entire food chain.

4   Existing Traceability Schemes

The ability to reduce the costs, both human and financial, in the event of a food recall event depends directly on the ability to locate, or trace, contaminated food products across the food chain. Any traceability solution should address the need by food safety personnel, in the event of a food product recall, to quickly identify companies within their jurisdiction that have a high likelihood of possessing contaminated products. The efficiency and effectiveness of a traceability system depends on the ability to collect, transmit, and analyze information about the handling of food products across all stages of the food chain.

A wide range of traceability schemes are currently in use by food system stakeholders [8, 9, 10, 11, 12, 13]. These systems range from paper-based records to bioactive labeling technology to an array of IT-based solutions from bar codes and radio-frequency identification (RFID) technologies supported by software systems to database management systems. Across the food chain, companies use a variety of these systems which may not be interoperable. An efficient traceability system should be able to link all these different monitoring techniques into an integrated, unified and consistent system.

A necessary requirement to accomplish this integration is the availability of a common standard identification system that is recognized across all stakeholders, or a system to create these translations. Thus, when a contaminated product is confirmed, it would be possible to trace the unique identifier (RFID) or product code (bar code) for that product with all of the companies that were involved in the creation of that food product. In the case of RFID, the tag on the contaminated product would contain the entire history/pedigree for that product. The Global Traceability Standard, a full supply chain traceability solution proposed by the universal standard committee GS1 (General Standard One), recommends the use of Global Location Numbers (GLN), a universal trade unit identification scheme based on the Global Trade Item Number (GTIN), and Electronic Product Codes (EPC) to enable the use of RFID tags to trace products [14, 15]. A methodology for modeling traceability information using the Electronic Product Code Information Service (EPCIS) framework and statecharts in the Unified Modeling Language (UML) to define states and transitions in food product has recently been proposed [16]. While progress has been made in achieving this integration, mostly within large vertically integrated multi-nationals, the difficulties of achieving such a system based on standard codes have been noted above.

In this paper, which extends the work of [17], we explore a different logic-based approach that uses inference rules to determine the set of all companies that may be linked to a contaminated product. Our approach does not depend on the availability of a common standard or unique identifier. Rather, the proposed approach utilizes information about the primary suppliers and customers for all food companies, along with their products—consistent with the "one step forward" and "one step backward" required under FMSA as noted above. In the event of a recall for Product A manufactured by Company X, we use logic programming to compute the set of all companies that are linked to the dyadic unit food-company across the entire supply chain. Using rules, we can trace backward to the set of likely companies that are the possible source of the contamination and can trace forward to identify the destination and location of similarly contaminated products.

We use a form of declarative programming – Answer Set Programming (ASP) [18], to represent complex pathways of the food supply chain and to track-and-trace recalled products and other information of interest to public health officials. ASP has been applied to industrial problems, but to the best of our knowledge it has not been used in food supply chain applications before.

5   ASP Program Encoding

The ASP paradigm is based on the stable models/ answer sets semantics of logic programs [19, 20] and has been shown to be a powerful methodology for knowledge representation, including the representation of defaults, inheritance reasoning, and multiple interesting aspects of reasoning about actions and their effects, as well as being particularly useful to solve difficult search problems. In the ASP methodology, search problems are reduced to the computation of the stable models of the problem. Several ASP solvers—programs that generate the stable models of a given problem encoded in the ASP formalism—have been implemented, e.g. ASSAT, clasp, Cmodels, DLV, GnT, nomore++, Pbmodels, Smodels, etc. In what follows we provide the basic syntactic constructs and the intuitive semantics of the ASP language used in this work. A complete formal specification of the syntax and semantics of the language can be found in [20, 21].

A signature Σ of the language contains constants, predicates, and function symbols. Terms and atoms are formed as is customary in first-order logic. A literal is either an atom (also called a positive literal) or an atom preceded by $\neg$ (classical or strong negation), a negative literal. Literals l and $\neg l$ are called contrary. Ground literals and terms are those not containing variables. A consistent set of literals does not contain contrary literals. The set of all ground literals is denoted by $lit(\Sigma)$. A rule is a statement of the form:

$$h_1 \vee ... \vee h_k \leftarrow l_1, ..., l_m, \text{not } l_{m+1}, ..., \text{not } l_n. \tag{1}$$

where h_i's and l_i's are ground literals, *not* is a logical connective called negation as failure or default negation, and symbol $\vee$ corresponds to the disjunction operator. The head of the rule is the part of the statement to the left of symbol $\leftarrow$, while the body of the rule is the part on its right side. Intuitively, the rule meaning is that if a reasoner believes $\{l_1, ... , l_m\}$ and has no reason to believe $\{l_{m+1}, ..., l_n\}$, then it must believe one of the h_i's. If the head of the rule is substituted by the falsity symbol $\perp$ then the rule is called a constraint. The intuitive meaning of a constraint is that its body must not be satisfied. Rules with variables are used as shorthand for the sets of their ground instantiations. Variables are denoted by capital letters. An ASP program is a pair of $\langle \Sigma, \Pi \rangle$, where Σ is a signature and Π is a set of rules over Σ, but usually the signature is defined implicitly and programs are only denoted by Π. A stable model (or answer set) of a program Π is one of the possible sets of literals of its logical consequences under the stable model/answer set semantics.

Our encoding—the set of rules of program Π—contains roughly 25 rules, while event records (in ASP, rules with an empty body, also called "facts") and the ontologies describing facts, utilized for experiments, are in the thousands. We use the DLV system [22] as our ASP solver.

Advantages of applying the ASP formalism to the food supply chain traceability problem include: (1) ASP can easily encode many forms of domain knowledge, including hierarchical ontologies and heuristics. As shown by some previous works [23, 24], ASP allows generating ontologies for different types of information relevant to this domain, e.g. food, geographical, disease, etc. Encoding of heuristics makes it possible to prune the search space and increase the efficiency of tracking and tracing a contaminated product in the supply chain; (2) ASP is well-suited to represent action and change. A food supply chain is an intrinsically dynamic enviroment where food products move from one node, or food operator, to the next node in the chain, and the track-and-trace of contaminated products posing risk to human lives should be highly efficient to curb a contamination event that may spread very rapidly; and (3) ASP is well-suited to deal with incomplete information—an inherent problem of this domain as food enterprises are averse to sharing information about their supplier and customer bases since it constitutes competitive advantage to their business.

5.1   Domain Representation

Given the proprietary nature of supplier/customer base information and the difficulty to obtain this data directly from private sector companies, we turned to data publicly available on the World Wide Web and using web scrapping techniques downloaded and assembled a database of suppliers of food and agricultural products. This database contains more than 6,000 American companies located in all 50 states, the District of Columbia and Puerto Rico, with firms encompassing the whole food supply chain, including: grower, manufacturer, processor, packer, distributor, wholesaler, retailer, etc. Each firm is classified as at least one of these types, but a firm may have more than one role in the supply chain, e.g. it may be a processor and also a wholesaler of its products. Besides the standard information about a firm, i.e. name, address, the database contains a list of the product categories that the firm commercializes, e.g. salad dressing, juice mixed, peanut butter.

We demonstrate the power of using ASP to solve the traceability problem by showing an example involving pork products. For simplicity sake, in this example we assume that the supply chain for pork sausages, shown on Figure 1, encompasses: (a) farmers supplying fresh pork meat to (b) processors supplying chilled or frozen pork to (c) manufacturers of pork sausages supplying (d) wholesalers of pork sausages supplying (e) retailers who sell pork sausages to consumers. A small number of companies that populate this supply chain, as identified in our assembled supplier database, are also shown in Figure 1 in the form of a directed graph. In this graph, each node corresponds to a company identified by an id code, and an edge originating from a company/node A and connecting it to a company/node B expresses a supplier-customer relationship where A supplies certain food product to B. In addition, each type of company/ node aligns vertically with its role or category in the pork supply chain represented at the top of Figure 1. For example, company "cp3092" corresponds to a farmer who supplies fresh pork meat to three processors identified by codes "cp123", "cp393", and "cp684"; processor "cp123" supplies chilled or frozen pork meat to four manufacturers, e.g. "cp273"; and so forth. In the ASP knowledge base, each company is modeled by three types of "facts," rules (2)-(4).

```
company(Idcode,Name,State).
```
(2)

```
type_company(Idcode,Type).
```
(3)

```
prod_supplied(Idcode,Product).
```
(4)

In our model, for the purpose of this example, each company is represented by a single rule (2), which identifies it by an id code, its name, and the state where the company is located. For simplification, we assume that each company has a single facility and this is the state where the supplied product originates and is shipped to others. Rules of type (3) indicate the role each company exerts in the supply chain. As mentioned before, it is not uncommon that a given company may have more than one role, e.g. a wholesaler may also be a retailer who sells directly to consumers. Thus, such company will have at least two rules of type (3), one to indicate that the company is of type "wholesaler" and the other that the company is a "retailer". It is very common for a given company to supply several products, and thus, our knowledge base contains a rule of type (4) for each of these products. Once a recall of a product sold by a certain company is issued, this information is added to the knowledge base in the form of rule (5), with the company being identified by the id code.

```
recall(Product,Idcode).
```
(5)

The expected course of action at this point is that the contaminated product, and its derivative products, are taken out of the market and destroyed. Since only limited information is made available to food safety officials about which companies may be affected—those who received the tainted product or supplied a related contaminated product—delays in the recall process put in risk human lives. Our approach works to reduce these latencies by generating all possible paths this product may have travelled through the supply chain graph. We generate each complete path, from farmer to retailer, for the product in question, as described in the next section.

First, assume that wholesaler company "cp1050" recalls their "porksausage" product. Our knowledge base contains a simple ontology which models the main stages of a food product as it evolves from raw, unprocessed food at the farmer/ grower level of the supply chain, to a processed food ready for consumption at the retail point-of-sale. At each stage of the supply chain the product supplied from a company A becomes an ingredient to the company B to which it has been supplied. In the case of pork sausages, the ontology contains facts (6)-(11) which express the production process sequence for pork products illustrated on Figure 1.

```
is_ingr(porkfresh,porkchilled).
```
(6)

```
is_ingr(porkchilled,porkfrozen).
```
(7)

```
is_ingr(porkchilled,porksausage).
```
(8)

```
is_ingr(porkfresh,porkfrozen).
```
(9)

```
is_ingr(porkfrozen,porksausage).
```
(10)

```
is_ingr(porksausage,porksausage).
```
(11)

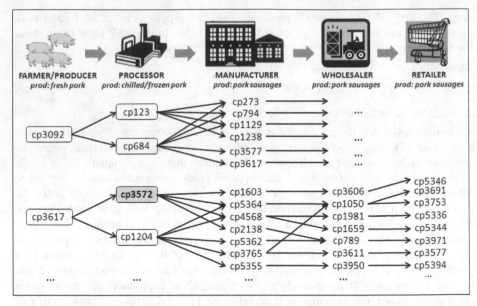

Fig. 1. Illustrative Supply Chain for Pork Sausages

5.2   Generating Supply Chain Paths

We use a two-step approach to solve the problem of identifying companies affected by a food recall when incomplete information may hinder the process and create delays. In the first step, we generate all supply chain paths for pork products with rules of type (12)-(17), where the supplied pork product at each level of the supply chain is used to prune the search among all other possible combinations of food products represented in our knowledge base. Intuitively, (12) means that a five-tuple *supply_chain(G,P,M,W,R)* represents the complete path of production of a given final product, e.g. pork sausages, from grower/producer to processor to manufacturer to wholesaler to retailer. Rules (13)-(17) compute the individual supplier-customer relations, or edges of the supply chain graph.

$$
\begin{aligned}
&\texttt{supply\_chain(G,P,M,W,R) :-}\\
&\quad\texttt{produces(G,porkfresh),processes(P,porkchilled),}\\
&\quad\texttt{manufactures(M,porksausage),wholesells(W,porksausage),}\\
&\quad\texttt{sells(R,porksausage).}
\end{aligned} \tag{12}
$$

$$
\begin{aligned}
&\texttt{produces(C,F) :-}\\
&\quad\texttt{company(C,\_,\_),type\_company(C,grower),}\\
&\quad\texttt{prod\_supplied(C,F), F==porkfresh.}
\end{aligned} \tag{13}
$$

$$
\begin{aligned}
&\texttt{processes(C,F) :-}\\
&\quad\texttt{company(C,\_,\_),type\_company(C,processor),}\\
&\quad\texttt{prod\_supplied(C,F), F==porkchilled.}
\end{aligned} \tag{14}
$$

```
manufactures(C,F) :-
    company(C,_,_),type_company(C manufacturer),          (15)
    prod_supplied(C,F), F==porksausage.

wholesells(C,F) :-
    company(C,_,_),type_company(C,wholesaler),            (16)
    prod_supplied(C,F), F==porksausage.

sells(C,F) :-
    company(C,_,_),type_company(C,retailer),              (17)
    prod_supplied(C,F), F==porksausage.
```

In the second step, each such supply chain path is broken down and ex-pressed as individual supplier-customer relations by rules (18)-(21). The reason for converting the supply chain back to these relations is to improve the efficiency of the computation during the tracing stage. Rule (18), and similarly rules (19)-(21), intuitively expresses that a grower / producer company G supplies fresh pork to a processor company P which utilizes this product as the main ingredient to produce and supply chilled pork to its customers. Rule (18), as well as (19)-(21), also enforces that companies G and P are not the same to ensure that the supply chain graph is cycle free.

```
supplies(G,porkfresh,P) :-
    supply_chain(G,P,M,W,R),company(G,_,_),
    type_company(G,grower),prod_supplied(G,porkfresh),    (18)
    company(P,_,_),type_company(P,processor),
    prod_supplied(P,porkchilled), G!=P.

supplies(P,porkchilled,M) :-
    supply_chain(G,P,M,W,R),company(P,_,_),
    type_company(P,processor),prod_supplied(P,porkchilled), (19)
    company(M,_,_),type_company(M,manufacturer),
    prod_supplied(M,porksausage), P!=M.

supplies(M,porksausage,W) :-
    supply_chain(G,P,M,W,R),company(M,_,_),
    type_company(M,manufacturer),
    prod_supplied(M,porksausage),                         (20)
    company(W,_,_),type_company(W,wholesaler),
    prod_supplied(W,porksausage), M!=W.

supplies(W,porksausage,R) :-
    supply_chain(G,P,M,W,R),company(W,_,_),
    type_company(W,wholesaler),
    prod_supplied(W,porksausage),                         (21)
    company(R,_,_),type_company(R,retailer),
    prod_supplied(R,porksausage), W!=R.
```

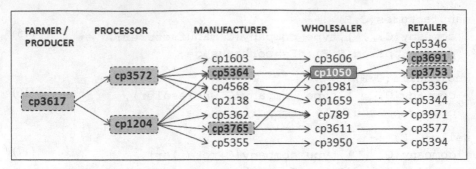

Fig. 2. Tracing Contaminated Pork Sausages in the Supply Chain

5.3   Tracing Contaminated Products

The goal of tracing the contamination forward in the supply chain from the point of recall, e.g. wholesaler "cp1050," is achieved by rules (22) and (23). Rule (22) says that if recalling company C, located in state LC, supplies its recalled food product F to company A, located in state LA, then LA may be affected by the recall and is part of the contamination. Thus, company A must be inspected by food safety officials to verify that its entire contaminated product is taken out of the market. Rule (23) propagates this trace to the next forward stage of the supply chain. Figure 2 shows an example of firms affected by a recall after tracing back and forward in the supply chain such product. Similarly, rules (24) and (25) trace back the contaminated product through the supply chain.

```
forward_trace(C,LC,F,A,LA)  :-
    recall(F,C),supplies(C,F,A),                              (22)
    company(C,_,LC),company(A,_,LA),C!=A.
```

```
forward_trace(B,LB,F1,A,LA)  :-
    company(B,_,LB),company(A,_,LA),company(C,_,LC),
    supplies(B,F1,A),is_ingr(F,F1),                           (23)
    forward_trace(C,LC,F,B,LB),  B!=C,  B!=A,  A!=C.
```

```
backward_trace(A,LA,F,C,LC)  :-
    recall(F,C),supplies(A,F,C),                              (24)
    company(C,_,LC),company(A,_,LA),C!=A.
```

```
backward_trace(B,LB,F1,C,LC)  :-
    company(B,_,LB),company(A,_,LA),company(C,_,LC),
    supplies(B,F1,C),is_ingr(F1,F),                           (25)
    backward_trace(C,LC,F,A,LA),  B!=C,  B!=A,  A!=C.
```

Finally, when these rules are submitted to the answer set solver DLV, we obtained the following list of atoms which corresponds to the solution of the traceability problem illustrated in Figure 2. In addition to the rules listed above a couple of other rules are used to retrieve the name of the recalling company, companies supplied the contaminated product directly, and their downstream customers. These companies are named in atoms of the type *affected_comp(Idcode, Name, State)*. Company names and codes appearing in this example are for illustrative purposes only and do not correspond to real company names in the knowledge base. Note that using ASP we can further focus the search and obtain a list of affected companies on a given state.

```
{recalling_comp(cp1050,atrading,ca),
  forward_trace(cp1050,ca,porksausage,cp3691,il),
  forward_trace(cp1050,ca,porksausage,cp3753,il),
  affected_comp(cp3691,gustopack,il),
  affected_comp(cp3753,apacking,il),
  backward_trace(cp3617,il,porkfresh,cp3572,il),
  backward_trace(cp3617,il,porkfresh,cp1204,ca),
  backward_trace(cp3572,il,porkchilled,cp5364,il),
  backward_trace(cp1204,ca,porkchilled,cp3765,il),
  backward_trace(cp5364,il,porksausage,cp1050,ca),
  backward_trace(cp3765,il,porksausage,cp1050,ca).}
```

Assume now that processor firm "cp3572" is recalling its chilled pork product. To find a solution to this new contamination problem one needs only to add new fact (26). The list of atoms below shows a portion of the results computed by DLV.

```
recall(porkchilled,cp3572).
{recalling_comp(cp3572,ainc,il),
  forward_trace(cp3572,il, porkchilled,cp5364,il),
  forward_trace(cp3572,il, porkchilled,cp4568,la),
  forward_trace(cp3572,il, porkchilled,cp2138,wi),
  forward_trace(cp3572,il, porkchilled,cp1603,ok),
  forward_trace(cp5364,il,porksausage,cp1050,ca),
  forward_trace(cp4568,la,porksausage,cp1981,wi),
  forward_trace(cp4568,la,porksausage,cp1659,co),
  forward_trace(cp2138,wi,porksausage,cp789,fl),            (26)
  forward_trace(cp1603,ok,porksausage,cp3606,il),
  forward_trace(cp1050,ca,porksausage,cp3691,il),
  forward_trace(cp1050,ca,porksausage,cp3753,il),
  forward_trace(cp1981,wi,porksausage,cp5336,ca),
  forward_trace(cp789,fl,porksausage,cp3971,il),
  forward_trace(cp3606,il,porksausage,cp5346,ga),
  forward_trace(cp1659,co,porksausage,cp5344,ny),
  backward_trace(cp3617,il,pork-fresh,cp3572,il),
  affected_comp(cp1603,afoods,ok),
  affected_comp(cp1050,atrading,ca), ...}
```

6   Complex Supply Chains

Simple and linear supply chains such as the pork example depicted in Figure 1 are not common. Increased market forces, as trade and globalization of products, have added high complexity to supply chains. Thus, the solution implementation discussed on the previous sections may not be as useful or adequate.

A more realistic and complex supply chain for peanut products is presented in Figure 3. Similar to the pork example, this supply chain includes a linear trajectory of transformation of a commodity, i.e. raw peanuts, into a final product, i.e. peanut butter. This (national) supply chain moving in a straight line from farmer/producer to retailer is shown at the top portion of Figure 3. In addition, several other sup-ply chain pathways are presented at the bottom of Figure 3 including export outside the U.S. It includes other stakeholders such as (a) brokers who market or sell raw peanuts supplied by wholesalers, blanched peanuts supplied from processors, and/or peanut butter supplied from manufacturers to (b) distributors supplying these products, and others commercialized directly from manufacturers, to national wholesalers and to (c) exporters supplying products to international distributors selling both commodities and final products through the supply chain of another country.

Such more complex supply chains allow a better demonstration of the expressive power of ASP. Even though now the supply chain has a larger number of ramifications and new stakeholders, modeling the additional elements is straightforward. Each firm belonging to the peanut supply chain in our supplier database is still modeled in the ASP knowledge base by rules (2)-(4). Three new types of companies are included: "distributor", "broker," and "exporter". As before, some companies may have more than one role in the supply chain, e.g. a broker company may also be a distributor of products, and thus, those will be modeled by as many rules of type (3) as the roles it performs. Similarly, to the pork example, companies may commercialize several products which will be modeled by as many rules of type (4).

The production stages of the supply chain, where raw peanuts are supplied to a processor to be transformed into prepared and preserved peanuts used to produce crude peanut oil, or transformed into blanched peanuts, then supplied to a manufacturer of peanut oil or peanut butter, are captured by (27)- (31) in food our ontology.

```
is_ingr(peanutraw,pnprepared).
```
(27)

```
is_ingr(pnprepared,pncrudeoil).
```
(28)

```
is_ingr(peanutraw,pnblanched).
```
(29)

```
is_ingr(pnblanched,pnbutter).
```
(30)

```
is_ingr(pncrudeoil,peanutoil).
```
(31)

6.1   Modeling Complex Supply Chains

Generating all possible production pathways for the peanut supply chain using the previous implementation is possible, but cumbersome. It is clear that this approach would be highly inefficient when encoding larger and more complex supply chains.

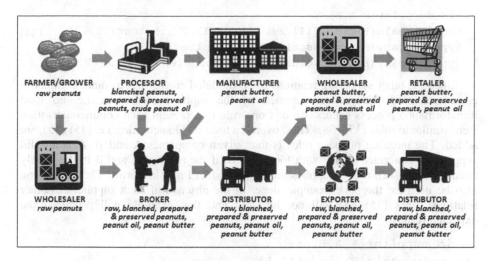

Fig. 3. More Complex Example of Supply Chain for Peanut Products

Rules (13)-(17) are also insufficient to express all supplying relationships among stakeholders of this example and additional rules are required. As shown in Section 5.2, encoding each supplier-customer relation as a rule is more efficient and corresponds to modeling each arrow appearing in the supply chain illustration of Figure 3. Thus, we substitute rules (18)-(21) by new rules (32)-(45).

Rule (32) is the "de facto" generator of all supplier-customer relations. Intuitively it is read as: company A supplies ingredient and/or product I1 to company B if, and only if, A is a valid supplier of subproducts of food F to B, where I1 (supplied by A) and I2 (supplied by B) are both subproducts of F.

```
supplies(A,I1,B) :-
    is_of(I1,F),is_of(I2,F),valid_supplier(F,A,B),          (32)
    prod_supplied(A,I1),prod_supplied(B,I2).
```

By using a single, generic rule (32) to generate all supplying relations, we allow for a more general modeling of all valid supplier-customer connections, further encoded by rules with head predicate *valid_supplier(F,A,B)*, or (33)-(45). Rules (33)-(36) encode the linear supply chain path from grower to retailer, similarly to (18)-(21) in the pork example, while (37)-(45) represent supply relations to the new types of stakeholders appearing in the peanuts example.

Rules (33)-(34) model the transformative stages of the supply chain and express that a company A is a valid supplier, of a product made with food F, to a company B if the type of roles A and B have in the supply chain of F are such that product I1 supplied by A to B is the ingredient for product I2 supplied by B to other firms.

```
valid_supplier(F,A,B) :-
  supply_chain(F),is_of(I1,F),is_of(I2,F),is_ingr(I1,I2),
  type_company(A,grower),prod_supplied(A,I1),                    (33)
  type_company(B,processor),  prod_supplied(B,I2),A!=B.

valid_supplier(F,A,B) :-
  supply_chain(F),is_of(I1,F),  is_of(I2,F),is_ingr(I1,I2),      (34)
  type_company(A,processor),  prod_supplied(A,I1),
  type_company(B,manufacturer),prod_supplied(B,I2),A!=B.
```

Since the other supplying relations to be modeled consist of customer companies commercializing the product supplied by the supplier companies, i.e. no food transformation process occurs, we add one rule to represent each commercialization step, similar to rules (35)-(38). Thus, overall a total of 11 such rules, i.e. (35)-(45), are added. The meaning of these rules is that, given companies A and B, A is a valid supplier of a product I made with food F to B, if the roles of A and B in the supply chain of F are such that B supplies the same product I to other firms. These rules can also be used for the pork example since, in the absence of such supplier-customer relations, rules (35)-(45) will not be fired and that example will produced the expected results.

```
valid_supplier(F,A,B) :-
  supply_chain(F),is_of(I,F),
  type_company(A,manufacturer),prod_supplied(A,I),              (35)
  type_company(B,wholesaler),prod_supplied(B,I),A!=B.

valid_supplier(F,A,B) :-
  supply_chain(F),is_of(I,F),
  type_company(A,wholesaler),prod_supplied(A,I),               (36)
  type_company(B,retailer),prod_supplied(B,I),A!=B.

valid_supplier(F,A,B) :-
  supply_chain(F),is_of(I,F),
  type_company(A,grower),prod_supplied(A,I),                   (37)
  type_company(B,wholesaler),prod_supplied(B,I),A!=B.
```

During a food recall, a supply chain of a particular food product can now be targeted, among a number of existing food chains, simply be adding a fact of type *supply_chain(S)* to the ASP knowledge base. Fact (46) means that the only supplying relationships to be proven by the ASP solver are those belonging to the peanut supply chain. For the pork example, we would add a similar fact where *S=pork*.

```
supply_chain(peanut).                                          (46)
```

In addition to (27)-(31), raw ingredients and other subproducts of peanuts—the food product of this supply chain—are modeled in the ASP knowledge base by facts (47)-(52). They express that certain raw ingredients or subproducts, i.e. raw peanuts, crude peanut oil, are part of the peanut supply chain, since they are made from peanuts. New facts of type *is_of(Prod,S)* would also be added for the pork example.

```
is_of(peanutraw,peanut).                                          (47)

is_of(pnprepared,peanut).                                         (48)

is_of(pncrudeoil,peanut).                                         (49)

is_of(pnblanched,peanut).                                         (50)

is_of(peanutoil,peanut).                                          (51)

is_of(pnbutter,peanut).                                           (52)
```

Alternatively, facts describing that a product I1 is an ingredient of a product I2, i.e. facts of type *is_ingr(I1,I2)*, could be combined with the facts associating each ingredient/product I to a specific food supply chain S, i.e. *is_of(I,S)*, to form a new predicate that would substitute those: *is_ingr_of(I1,I2,S)*. This would provide a more concise encoding, especially for foods with a large number of subproducts, but still be less general than a more developed ontology for describing this supply chain's food.

The use of the complete peanut ontology would have "peanut" at its root node and thus, the ASP program would entail that products I1 and I2 are in fact subproducts of peanut. Thus, the root node of each food hierarchy would characterize each corresponding supply chain. As we model only a very small portion of the peanut hierarchy in this example, we have opted to present a simpler, but less elegant, representation in this program. To include a complete ontology for peanut would require additional rules and the partial modification of the ASP program rules.

6.2   Tracing Contamination in Complex Supply Chains

Rules (22)-(23) for tracing forward from the point of recall require no modification and are included in the new program as rules (53)-(54). As in the pork example, (54) ensures that the ASP program will recursively trace forward companies affected by the recall which are (a) more than one step removed from the company recalling the contaminated product and (b) use product F received from a supplier as an ingredient to produce their product F1. However, since it is valid for some firms to re-sell products, as expressed by (35)-(45), we need to add recursive rule (55) to capture the situation where both a supplier and its customer supply the same product F. Similarly, contaminated products are traced backward in the supply chain by rules (56)-(58). Rules (24)-(25) used in the pork example are renamed as (56)-(57), and (58) is a new rule added to encode cases of supplier and customer supplying a same product.

```
forward_trace(B,LB,F,A,LA):-
    supplies(B,F,A),forward_trace(C,LC,F,B,LB),
    company(B,_,LB),company(A,_,LA),company(C,_,LC),          (55)
    B!=C, B!=A, A!=C.

backward_trace(B,LB,F,C,LC)  :-
    supplies(B,F,C), backward_trace(C,LC,F,A,LA),
    company(B,_,LB),company(C,_,LC),company(A,_,LA),          (58)
    B!=C, B!=A, A!=C.
```

While the above encoding has been developed for a complex peanut supply chain it can easily accommodate the previous pork example simply by including facts of type *is_of(I,S)* for pork. ASP programs that require only small changes to accommodate new or changing circumstances, as this one does, are called "elaboration tolerant". Programs exhibiting such property are highly preferable.

7   Conclusions

This paper demonstrates the utility of answer set programming in identifying not only the source of a food contamination but also the location of contaminated products across complex food chains for pork and peanut products. We represent all possible paths of a contaminated product across the supply chain as a sequence of stages by which a food product evolves from raw, unprocessed food at the farmer/grower level of the supply chain, to a processed food ready for consumption at the retail point-of-sale. Using rules of inference, we then reduce the set of all possible pathways of contamination based on information contained in the recall. We are also able to capture the process by which contaminated products become ingredients in other products during sequential stages of production. The logic-based approach developed herein is well-suited to be used by state agencies charged with inspecting food production, distribution and retail facilities in the event of a national recall. The approach is particularly useful for ingredient-driven contaminations in which the contaminated product is used as an ingredient in a broad set of secondary products.

References

1. Scallan, E., Hoekstra, R.M., Angulo, F.J., Tauxe, R.V., Widdowson, M.-A., Roy, S.L., Jones, J.L., Griffin, P.M.: Foodborne Illness Acquired in the United States—Major Pathogens. Emerging Infectious Diseases 17(1), 7–15 (2011)
2. Scallan, E., Griffin, P.M., Angulo, F.J., Tauxe, R.V., Hoekstra, R.M.: Foodborne Illness Acquired in the United States—Unspecified Agents. Emerging Infectious Diseases 17(1), 16–22 (2011)
3. Scharff, R.L.: Health-Related Costs From Foodborne Illness in the United States. The Produce Food Safety Project at Georgetown University (2010)
4. Fritz, M., Schiefer, G.: Tracking, tracing and business process interests in food commodities: A multi-level decision complexity. Int'l Journal Production Economics 117(2), 317–329 (2009)
5. U.S. Department of Health and Human Services: Office of Inspector General. Traceability in the Food Supply Chain, Report OEI-02-06-00210 (March 2009)
6. FDA Food Safety Modernization Act of 2010, S. 510, 111th Congress, 2nd Session. Signed into law by President Barack Obama (January 4, 2011)
7. U.S. Food and Drug Administration. Pilot Projects for Improving Product Tracing along the Food Supply System (September 2011)
8. Buhr, B.L.: Traceability and information technology in the meat supply chain: implications for firm organization and market structure. Journal of Food Distribution Research 34(3), 13–26 (2003)

9. Raschke, A., Strich, S., Huppke, S., Neugebauer, M., Geuther, E., Bertling, W., Walders, B., Reiser, C., Hess, J.: Induction and detection of long-lasting peptide-specific antibody responses in pigs and beef cattle: A powerful technology for tracing meat processing chains from stock farmers to sales counters. Food Control 17(1), 65–74 (2006)
10. Regattieri, A., Gamberi, M., Manzini, R.: Traceability of food products: general framework and experimental evidence. Journal of Food Engineering 81, 347–356 (2007)
11. Bulut, H., Lawrence, J.D.: Meat Slaughter and Processing Plants' Trace-ability Levels: Evidence From Iowa. Working Paper #08015, Working Paper Series, Department of Economics, Iowa State University (April 2008)
12. Shanahan, C., Kernan, B., Ayalew, G., McDonnell, K., Butler, F., Ward, S.: A framework for beef traceability from farm to slaughter using global standards: an Irish perspective. Computers and Electronics in Agriculture 66(1), 62–69 (2009)
13. Souza-Monteiro, D.M., Caswell, J.A.: Traceability adoption at the farm level: An empirical analysis of the Portuguese pear industry. Food Policy 34(1), 94–101 (2009)
14. Fritz, M., Schiefer, G.: Tracking and tracing in food networks. In: World Conference on Agricultural Information and IT (IAALD AFITA WCCA 2008), August 24-27, pp. 967–972. Tokyo University of Agriculture, Tokyo (2008)
15. GS1. Business Process and System Requirements for Full Supply Chain Traceability. GS1 Global Trace-ability Standard, Issue 1.2.2 (March 2010)
16. Thakur, M., Sørensen, C.-F., Bjørnson, F.O., Forås, E., Hurburgh, C.R.: Managing food traceability information using EPCIS framework. Journal of Food Engineering 103(4), 417–433 (2011)
17. Nogueira, M.L., Greis, N.P.: Recall-driven Product Tracing and Supply Chain Tracking using Answer Set Programming. In: Filipe, J., Dietz, J.L.G. (eds.) Proc. 4th Int'l Conf. Knowledge Eng. and Ontology Development (KEOD 2012), pp. 125–133. SciTePress (2012)
18. Marek, V.W., Truszczynski, M.: Stable models and an alternative logic programming paradigm. In: The Logic Programming Paradigm: A 25-Year Perspective, pp. 375–398. Springer, Berlin (1999)
19. Gelfond, M., Lifschitz, V.: The stable model semantics for logic programming. In: Kowalski, R., Bowen, K. (eds.) International Logic Programming Conference and Symposium, pp. 1070–1080. MIT Press (1988)
20. Gelfond, M., Lifschitz, V.: Classical negation in logic programs and dis-junctive databases. New Generation Computing 9, 365–385 (1991)
21. Niemela, I., Simons, P.: Extending the Smodels System with Cardinality and Weight Constraints. In: Logic-Based Artificial Intelligence, pp. 491–521. Kluwer Academic Publishers (2000)
22. Leone, N., et al.: The DLV system. In: Flesca, S., Greco, S., Leone, N., Ianni, G. (eds.) JELIA 2002. LNCS (LNAI), vol. 2424, pp. 537–540. Springer, Heidelberg (2002)
23. Nogueira, M.L., Greis, N.P.: Rule-Based Complex Event Processing for Food Safety and Public Health. In: Bassiliades, N., Governatori, G., Paschke, A. (eds.) RuleML 2011 - Europe. LNCS, vol. 6826, pp. 376–383. Springer, Heidelberg (2011)
24. Nogueira, M.L., Greis, N.P.: Application of Answer Set Programming for Public Health Data Integration and Analysis. In: Tjoa, A.M., Quirchmayr, G., You, I., Xu, L. (eds.) ARES 2011. LNCS, vol. 6908, pp. 118–134. Springer, Heidelberg (2011)

Modelling Services with DEMO

Carlos Mendes[1], Mário Almeida[1], Nuno Salvador[2], and Miguel Mira da Silva[1]

[1] Instituto Superior Técnico, Technical University of Lisbon,
Avenida Rovisco Pais, Lisboa, Portugal
[2] Câmara Municipal de Pombal, Pombal, Portugal
{carlos.mendes,mario.almeida,mms}@ist.utl.pt,
nuno.salvador@cm-pombal.pt

Abstract. According to data from the Portuguese Association of City Councils (DGAL – Portuguese abbreviation) the 308 city councils in Portugal employ about 135 000 people and spend about 3.8 billion euros a year [1]. The Portuguese government is trying to reduce those numbers due to the Troika memorandum demands. In this paper we demonstrate a proposal to identify services that do not create new original results, therefore services with great potential to be automated. This demonstration was carried out in a Portuguese city council named Pombal (CMP). We evaluated the proposal by collecting feedback from CMP employees and customers. We identified some improvements that could save millions of euros to the Portuguese state and, therefore, contribute to the Troika memorandum demands.

Keywords: Enterprise Ontology, DEMO, Service Catalog, Service Level Agreements.

1   Introduction

In previous research we focused on closing the gap between customers' expectations and the perceived service [2] by formally specifying the customers' expectations into Service Level Agreements (SLAs) [3], [4], [5], using as a foundation the DEMO methodology and respective Enterprise Ontology theory [6].

DEMO (Design & Engineering Methodology for Organizations) is a methodology for modelling, (re)designing and (re)engineering organizations and networks of organizations. The theory that underlies this methodology is called Enterprise Ontology (EO) that by itself is based on the speech act theory. We have chosen EO because this theory can help us expand the expressiveness of the service descriptions and, consequently, allow a better alignment between expectations and perceptions [5].

In this paper we applied DEMO-based Service Level Agreement (SLA) [5] in the Portuguese city council of Pombal (CMP - Portuguese abbreviation), i.e. we specified the service catalogue of CMP including the services CMP provides to Pombal residents and the SLAs they use to comply with the residents' expectations.

To evaluate the SLA attributes we collected feedback from 23 employees of CMP and 7 customers. The majority of the proposal attributes (14 in 16) was classified as important since they received a minimum score of 7.6 in 10 while the remaining two

A. Fred et al. (Eds.): IC3K 2012, CCIS 415, pp. 228–242, 2013.

received an score of 5.9. Besides these findings, as our proposal identifies services that do not create new original results, we also identified services with great potential to be automated. We present an example of a possible improvement that could save millions of euros to the Portuguese state.

Our study was conducted using the Design Science Research Methodology (DSRM) that aims at creating and evaluating IT artefacts intended to solve identified organizational problems [7]. These artefacts include constructs (vocabulary and symbols), models (abstractions and representations), methods (algorithms and practices) and instantiations (implemented and prototype systems). This research method comprises the following phases [8]: problem identification, objectives definition, design and development, demonstration, evaluation and communication.

The paper is structured as follows. We will start by providing a brief overview of the literature on the problem area (Section 2). In Section 3, we introduce the theoretical background of this research, the Enterprise Ontology theory. Afterwards, we introduce the DEMO-based solution to specify the services quality (Section 4). In Section 5, we describe an experiment at CMP. In Section 6, we explain the evaluation process, which uses data from the experiment, and specify the lessons learned. Finally, we present our conclusions (Section 7).

This section corresponds to the problem identification and motivation phase of DSRM. It also corresponds to the objectives definition phase.

2   Related Work

This section describes the current solutions for specifying services quality and explains why these solutions do not solve the gaps problem [2].

We analysed several solutions to specify the Service Quality: Service Level Management best practices, web services based solutions and the Generic Service Specification Framework (GSSF). In spite of the different backgrounds, all contributed to the service quality specification. The first solution is proposed by many best practices frameworks, such as ITIL [9] or CMMI [10], the second represents the solutions focused on web services and the third is an Enterprise Ontology-based approach (even though the main goal of the GSSF was to specify the services and not the service quality itself, this framework also contributed to the problem area).

Service Level Management is one of the key processes by which organizations manage their services, because it acts as the interface between the customer and the provider. At its most basic level, Service Level Management is involved in the following activities: define, agree, record and manage levels of service. There are a number of key elements required to ensure that services are fit for purpose and use, and remain so throughout their lifetime: service level requirements, targets and agreements [9].

Basically, to understand the Service Level Requirements (SLR) means that the customers' needs and wants are understood, i.e. an SLR is a customer requirement for an aspect of a service. SLRs are based on business objectives and are used to negotiate Service Level Targets (SLT) which are commitments documented in Service Level Agreements (SLAs). SLTs are based on SLRs and are needed to ensure that the service is fit for purpose. SLTs should be SMART: specific, measurable, attainable, realistic and timely. Finally, SLA is an agreement between a provider and a customer that describes the service; it documents the SLTs and specifies the responsibilities of the provider and customer. Over the years it has also been the chosen concept to specify services quality [9].

Regarding Service Level Management solutions, current approaches have two main flaws. First, they lack a strong conceptual foundation because they were derived from best practices of several years of implementations - not from a well-founded theory. Consequently, the inexistence of a theory may cause incoherencies among those solutions (second flaw). Service Level Management solutions are process-driven and not service-driven. These solutions are designed to work individually as processes but the interactions between these processes (such as Request Fulfilment, Service Level Management and Incident Management) are usually unclear. For instance, the connection between an incident and an SLA is neither clearly explained in ITIL nor in CMMI.

There are some solutions to specify the services quality that originated in the web services community. In [11] the authors show how to use Web Service Description Language (WSDL) and Web Service Flow Language (WSFL) to specify SLAs. However, this work suffers from the web vision tunnel as it is focused on the web services and does not try to specify business services. For instance, the specifications do not include penalties or prices. The researches in [12], [13] and [14] have the same bottleneck. Despite this trend in the web service community, there are some recent researches that try to overcome the mentioned web service tunnel vision. In [15] a novel framework for specifying and monitoring SLAs for Web Services is introduced: the Web Service Level Agreement (WSLA) framework. This framework is applicable to any inter-domain management scenario such as business process and service management or the management of networks, systems and applications in general. In [16] and [17] business criteria is also included in SLAs. These three solutions represent a new movement in the web service community; however, none is based on a strong conceptual foundation.

Another contribution to the gaps problem is the Generic Service Specification Framework (GSFF) [18], which is based on the following generic service definition [19]: *a service is a universal pattern of coordination and production acts, performed by the executor of a transaction for the benefit of its initiator, in the order stated in the standard pattern of a transaction.*

We adopted this service definition in our research since this definition is the only one that, as our research, uses DEMO as a conceptual foundation.

The GSSF defines four main areas of concern for each service: the service executor, the service production, the service coordination and the service contract option. The first one defines who the provider of the service is. The second focuses on the production act to be performed by the executor. The third gives the consumer all the information required for conducting a successful communication with the provider. And finally, the service contract option specifies one or several contract options from which service consumers can choose.

Even though the quality aspects are very basic, the Generic Service Specification Framework represents a large contribution to the service specification research area. However, the level of service quality specification is not always sufficient, because sometimes customers and providers have different expectations due to a lack of specification [18].

3   Theoretical Foundation

This section briefly describes the Enterprise Ontology theory (the theory that supports our proposal).

Enterprise Ontology [6] is based on four axioms – operation, transaction, composition and distinction – and the organization theorem. The operation axiom states that the operation of an enterprise is constituted by the activities of actor roles that are elementary chunks of authority and responsibility, fulfilled by subjects. In doing so, these subjects perform two kinds of acts: **production acts** and **coordination acts**. These acts have definite results: production facts and coordination facts, respectively. By performing production acts (P-acts) the subjects contribute to bringing about the goods and/or services that are delivered to the environment of the enterprise. By performing coordination acts (C-acts) subjects enter into, and comply with, commitments towards each other regarding the performance of production acts.

The transaction axiom states that coordination acts are performed as steps in universal patterns. These patterns, also called **transactions**, always involve two actor roles (initiator and executer) and are aimed at achieving a particular result. A transaction develops in three phases: the order phase (O-phase), the execution phase (E-phase), and the result phase (R-phase). In the O-phase the two actors agree on the expected result of the transaction; in the E-phase the executer executes the production act needed to create the expected result; and in the R-phase the two actors discuss if the transaction result is equal to the expected result.

The composition axiom establishes the relationships between transactions. This axiom states that every transaction is either a) enclosed in another transaction, b) is a customer transaction of another transaction, or c) is a self-activation transaction. The latter case refers to transactions that give rise to further transactions of the same type.

The distinction axiom states that there are three distinct human abilities playing a role in the operation of actors, called **performa**, **informa**, and **forma**. An ontological act (performa) is an act in which new original things are brought about. Deciding and judging are typical ontological production acts. Regarding the coordination between people, typical ontological acts are requesting and promising. An infological production act is an act in which one is not concerned about the form but, instead, about the content of the information. Typical infological acts are inquiring, calculating, and reasoning. Regarding the coordination between people, formulating thoughts (in written or spoken sentences) and interpreting perceived (through listening or reading) sentences are typical infological coordination acts. Acts like copying, storing, and transmitting data are typical datalogical acts, while speaking, listening, writing, and reading are typical datalogical coordination acts.

4   Proposal

This section corresponds to the design and development step of DSRM. Our proposal is composed by the following steps:

1. Identify the services;
2. Specify the executor, production and coordination of the services;
3. Specify the SLAs for each identified transaction/service.

The first step is to identify the services of the provider, using for that purpose a process based on the methodology proposed in [6]. This process is composed by six steps: Enterprise Description, Performa-Informa-Forma Analysis, Coordination-Actors-Production Analysis, Transaction Pattern Synthesis, Result Structure Analysis and Actor Transaction Diagram/Service Identification.

The second step of the proposal is to apply part of the GSSF [18]: service executor, service production and service coordination. The service executor area defines who the provider of the service is; the service production area focuses on the production act to be performed by the executor; and the service coordination area gives the consumer all the information required for conducting a successful communication with the provider.

The fourth area, the contract options, is replaced by our definition of SLA presented in Fig. 1. That means for each identified service, one should specify the list of associated SLAs using our SLA definition (step three).

Our SLA proposal considers four main areas of concern with their respective attributes that we will now explain. The first section is called **Generic SLA Information** and it defines the name of the SLA (*SLA Name*) and the SLA purpose (*SLA Description*). Additionally, the Generic SLA Information describes who owns this SLA (*SLA Owner*), it also provides a contact of this person (*SLA Owner Contact Information*) and, finally, this section defines the name of the service that the SLA applies to (*SLA Service*).

The second considered section contains information concerning the dates of the SLA and is called **Temporal SLA Information**. In this section the date on which the SLA was established (*SLA Creation Date*) is defined as well as the time interval on which the SLA is valid (*SLA Validity Period*), the information related to the SLA modification dates by the customer (*SLA Version Control Information*) and the information concerning the SLA review dates performed by an entity related to the service provider (*SLA Review Period Information*).

Next, we define a section called **Responsibility SLA Information** that regards the information about the responsibilities of each actor in the execution of this SLA. In this section two attributes are specified concerning the obligations and duties of the customer (*SLA Customer Responsibilities*) and the service provider (*SLA Provider Responsibilities*).

Finally, the last section is called **Specific SLA Information** and for each type of SLA (*SLA Type*) it specifies six different types of targets (*SLA Targets*), which can give rise to actions if they are not fulfilled (*SLA Penalties*), but if they are fulfilled, this should be rewarded (*SLA Bonuses*). Each type of SLA is also associated to a price (*SLA Price*).

Several of these attributes can be gained from DEMO diagrams, such as, for example, the SLA Owner that can be identified by the Actor Transaction Diagram (ATD) or the SLA Penalties and the SLA Bonuses that can be gained from the Action Model (AM).

Thus, this research intends to reduce the gaps by formally specifying the SLAs, using EO theory as a foundation.

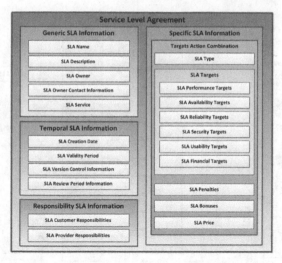

Fig. 1. Structure and attributes of the DEMO-based SLA

5   Demonstration

This section corresponds to the demonstration phase of DSRM. We evaluated the proposal using an experiment in order to validate its applicability. The demonstration occurred in a Portuguese city council named Pombal (CMP). Pombal is located in Leiria District and is composed of 17 parishes. It has a total area of 626.1 km² and a total population of 58,617 inhabitants. The population of the city of Pombal is about 18,000 inhabitants. CMP employs a total of 389 people with 203 men and 189 women and has five major departments divided into divisions, units and sections. In 2010 CMP spent a total of 20 553 200 € from which 7 542 250 € in human resources [20].

In order to identify the services (step one of the proposal) we interviewed individually 17 employees from CMP. With the purpose of having an overall perspective of the entire CMP we have selected employees from all the departments. During the interviews participants were asked to describe the activities performed by CMP. The interviews were recorded and transcribed as well as checked and discussed by two interviewers each ensuring unbiased findings and avoiding misinterpretation as specified in [21].

The interviews allowed us to develop an enterprise description of CMP that was used as input for the service identification step (proposal first step). We do not fully describe the six sub steps of the service identification step due to space limitation, nevertheless these sub steps are based on DEMO [6] and are described in previous publications [4] [3]. The result of this first step is called the Actor Transaction Diagram (ATD) and we decided to show only the ATD with information regarding some departments due to space limitations (Fig. 2).

In the ATD, a transaction/service is represented using a diamond in a disk that contains the respective combination of C-acts and a P-act. Each transaction is connected to two boxes, representing the initiator and executor actor roles. The initiator is

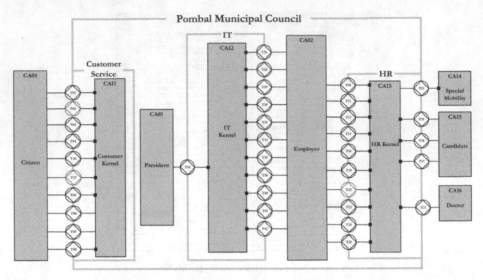

Fig. 2. ATD with HR and IT

connected to the transaction symbol using a solid line, while the executor is connected to the transaction using a solid line ending in a black square. The grey boxes refer to composite actor roles, i.e. elements whose exact structure is not known. All the environmental elements, i.e. elements outside the organization that we are studying, are represented with grey boxes for that reason. This also means that we can represent the studied organization with a grey box when referring to the kernel of the organization, which can be further specified by using elementary actor roles represented by white boxes.

We identified 173 services of which 145 are ontological, 17 are infological and 11 are datalogical (see Section 3). These services correspond to all services provided by the five major departments that constitute the City Council. Figure 2 illustrates seven major Composite Actor Roles (Citizen, Customer Kernel, President, IT Kernel, Employee, HR Kernel, Special Mobility, Candidate and Doctor) on a total of nearly 50 across the entire City Council.

The IT division operates and maintains the computer equipment, develops new tools, supports their applications, and conducts courses to enhance learning of the new features. We identified 12 services provided by this division: *Network Configuration* (T26), *Hardware Installation* (T28), *Hardware Uninstallation* (T29), *Application Development* (T30), *Incident Resolution* (T31), *Database Management* (T32), *Software Installation* (T35), *Software Uninstallation* (T36), *Backup Realization* (T40), *Handbook Definition* (T41), *Training* (T42) and *Business Intelligence Study Realization* (T43).

In order to proceed to the second step of the proposal (specify the executor, production and coordination of the services), first we had to model the Process Model, the Action Model, and the State Model of Pombal City Council, since some aspects of the GSSF (used in the second step of the proposal) depend on these models. We do not present all of these models due to space limitations.

Table 1. Service *"Application Development"* Specification

Service Specification - *Application Development (T30)*	
Service Executor	
Actor Role	*Developer (A02)*
Contact Information	General Email: suporte@cm-pombal.pt Email: xxxxx Phone: xxxxx
Service Production	
Production Act	Application Development is the act of designing and developing new applications or features required by other units of the Municipal Council or by the IT Division.
Production Information Used	The object classes used in this services are: Application, Handbook, Employee, Course
Production Fact	Application App has been developed
Production Kind	The production kind of this transaction is: Ontological
Production World Semantics	Application, Handbook and Employee (See **Fig. 3**)
Preconditions	Pre and post-conditions are gained from the Action Model. There are no preconditions for this service.

Postconditions	Regarding the post condition, we have the following action rule:
	when development of [Application] is accepted
	    then training of [Application] must be requested
	    and definition of [Handbook] must be requested

Service Coordination	
Coordination Acts	See
Coordination Kind	This service is a Human Service
Protocol	1. Contact the IT Division; 2. Specifies the Application requirements; 3. Wait for the end of development; 4. Test Application; 5. Application available to the target audience.
Location	Email, Phone

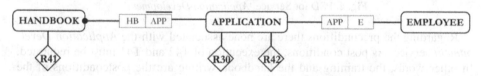

Fig. 3. State Model for Service *'Application Development'*

We applied the Generic Service Specification Framework (GSSF) to specify the services of Fig. 2 (second step of the proposal). An example of this specification is illustrated in Table 1. This table describes the attributes of the service *Application Development* that implements the T30 transaction and is provided by the IT Division.

The service specified above is called *Application Development* and is carried out by a Developer whose contact is available in the Service Executor section. In the Service Production area, it is specified that a new application is produced, which makes this an ontological service and, based on the State Model (Fig. 3), we found the information classes used in this service: Application, Handbook and Employee.

The ontological coexistence rules between these classes are the following: an Application may have a Handbook and Handbook may describe several applications. A Handbook always requires the existence of an Application. An Application may be used by several Employees and an employee may have access to several applications.

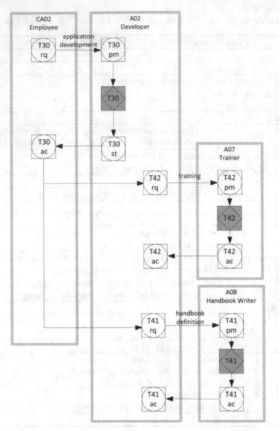

Fig. 4. PSD for Service '*Application Development*'

Regarding the preconditions there are none associated with the *Application Development* service. As post conditions, the execution of T42 and T41 must be requested. In other words, the training and the handbook writing are the postconditions of the *Application Development* service.

Concerning the Coordination area, the coordination acts involved in this service are illustrated by the Process Structure Diagram (PSD) (Fig. 4). In this diagram we see that the employee (CA02) makes a request for a new application (T30 rq) and this request is handled by the developer (A02). Which means that A02 may promise to develop the new application (T30 pm), develop the application (T30 ex) and state it (T30 st). Then, the employee (CA02) may accept the result of the development (T30 ac). When this happens, the developer (A02) has to start two new transactions: write the user manual (T41 rq) and schedule a training on this new application (T42 rq). At that moment, the standard pattern may be repeated for these two transactions (T41 and T42). In other words, T41 and T42 may be promised, executed, stated and accepted.

In addition, in the Coordination area the procedure or protocol to successfully contact the service provider is specified as well as the location of the service that, in this case, is by email or phone.

In the third step of the proposal we specified the SLAs associated to the CMP services using our SLA proposal (Figure 1). Table 2 illustrates an example of this specification for the service *Application Development*.

This SLA concerns the development of an application named WebDoc2.0 and the SLA Owner is the IT Division Chief. Multiple contacts of the IT Division Chief are specified in order to be contacted by the service customer at any time. This SLA was made on the first day of January of 2011, was valid until 31 of November of 2011 and was changed in July 15. To fulfil this SLA, the service provider needs to complete all the points specified in the SLA Provider Responsibilities (completion of ERP integration, improvement of communication between citizens and putting into production all features developed).

Table 2. SLA *"WebDoc 2.0 Development"* Specification

Service Level Agreement Specification	
Generic SLA Information	
SLA Name	**WebDoc 2.0 Development**
SLA Description	SLA concerning the development of a tool for document management.
SLA Owner	Nome: Nuno Salvador Category: IT Division Chief Organic Unit: IT Division
SLA Owner Contact Information	Email: xxx Phone: xxx
SLA Service	*Application Development* (T30)
Temporal SLA Information	
SLA Creation Date	January 1, 2011
SLA Validity Period	Until November 31, 2011
SLA Version Control Information	July 15, 2011
SLA Review Period Information	NA
Responsibility SLA Information	
SLA Customer Responsibilities	NA
SLA Provider Responsibilities	1 - Finish ERP Integration; 2 - Improve communication between the citizens; 3 - Place in operation all the features developed.
Specific SLA Information	
SLA Type	**Overcome Goal**
SLA Targets	
Performance	Until November 15, 2011
SLA Penalties	NA
SLA Bonuses	Evaluation Score: 5 Career: Allows career evolution
SLA Price	0 €

Table 2. (*continued*)

SLA Type	Fulfillment Goal
SLA Targets	
Performance	Until November 31, 2011
SLA Penalties	NA
SLA Bonuses	Evaluation Score: 3
SLA Price	0 €
SLA Type	**Non Fulfillment Goal**
SLA Targets	
Performance	After November 31, 2011
SLA Penalties	Evaluation Score: 1 Career: Can be fired with probable cause
SLA Bonuses	NA
SLA Price	0 €

This SLA has three types that depend on the date of completion of the development. Penalties and bonuses are translated into career points that influence the career development.

On the one hand, in case of SLA Type "Overcome goal", the SLA Owner wins five career points and can evolve in his career.

On the other hand, if the service provider does not meet the deadline the SLA Type "Non Fulfilment Goal" applies and the SLA Owner only wins one career point and can be fired with probable cause.

Note that this SLA has no price defined because this is an internal service to the CMP and no chargeback is made among the CMP departments.

6   Evaluation

This section corresponds to the evaluation phase of DSRM and in order to explain the evaluation we use the framework proposed in [22]. This framework identifies what is actually evaluated, how it is evaluated and when the evaluation takes place.

Table 3 illustrates the answers to the three main questions that this framework proposes to answer:

- **What is Actually Evaluated?** The artifact evaluated is the proposed set of steps of Section 4 (a design process) and the results of applying these steps to the CMP (Services and SLAs; a design product);
- **How is it Evaluated?** We used CMP employees' feedback to evaluate the DEMO-based SLA structure and the CMP services and SLAs. This represents a naturalistic evaluation since it was conducted using a real artifact in a real organization facing real problems;
- **When was it Evaluated?** It was evaluated ex post (after the design artifact was developed).

Table 3. Evaluation strategy

	Ex Ante	Ex Post
Naturalistic	Design Process / Design Product	P: CMP employees' feedback & customers' feedback / C: SLA Attributes Quality
Artificial	Design Process / Design Product	Design Process / Design Product

P summarizes the essential characteristics of the evaluation Process, while C indicates the evaluation Criteria.

The evaluation was naturalistic since we applied our proposal in a real organization with real data. The evaluation was ex post since it occurred after the demonstration in the CMP. We evaluated both the proposed set of steps of Section 4 (including the DEMO-based SLA structure) and the results of applying these steps to the CMP (Services and SLAs specification). In order to evaluate the DEMO-based SLA structure (see Fig. 1) and the Services and SLAs specification from CMP we collected feedback from 23 CMP employees and 7 CMP customers. They were arbitrarily chosen and were asked to classify the attributes of the SLA proposal from 1 to 10 according to the importance (being 1 irrelevant and 10 essential). Figure 5 illustrates the average and standard deviation per attribute. As can be seen there was little variation in the answers of the interviewees in most attributes.

The first 14 attributes had high average scores (from 7.60 to 8.40) and the remaining two (Bonuses and Price) had lower classifications (5.90 and 6.85).

These results indicate that the majority of the proposed attributes (14 in 16) were classified as important since they scored a minimum of 7.60 in 10 possible points. The remaining two attributes (Bonuses and Price) scored 5.90 and 6.85 revealing that they were classified as less important when comparing to the first 14. These results can be explained by the fact that there are no chargeback among the departments of the 30 inquired persons. Therefore, they value more the attributes that describe the service quality than the ones that capture the costs.

Besides this validation, we also validated the service catalogue from the IT Division with the IT chief that agreed with all 12 identified services. The same validation method is being used for other units of the City Council hoping to have the same acceptance that in the IT Division.

The definition of a city council service catalogue has great potential, because the services that this type of organizations provides to the citizens are similar. For instance, in Portugal there are 308 city councils and in theory they all have the same purpose. Having identified the service catalogue of one city council we can validate if it is applicable to other city councils and eventually find some services that could be provided in cooperation.

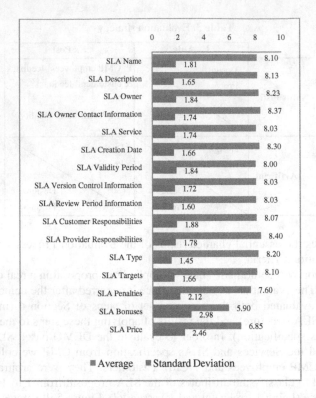

Fig. 5. Rating of proposal attributes

Knowing the services and the type of services provided allows one to understand how the service provider can improve his performance. For instance, the service *Collect Water Usage* – T155 is a datalogical transaction since it neither involves the creation of new original facts (ontological) nor information processing (infological). Hence, it has large potential to be optimized, since technology can be used to reduce the effort needed to execute datalogical acts [6].

CMP has four employees dedicated to collect water usage. Assuming that CMP spends 77 555 € a year with these four employees ((7 542 250 € / 389 employees) * 4 employees) and the other 307 city councils in Portugal do not have this service automated and use similar resources on it, then this would represent an expense of 23 809 468 € a year in a service that has great potential to be automated. This value is estimated using only the costs of the employees' wages so if we add the supporting costs (IT support, HR support, etc.) then the estimated value would certainly be higher.

This kind of analysis has special value because of the current situation Portugal is in since it points some solutions to a number of current challenges imposed by the Troika memorandum.

7   Conclusions

There are several solutions that contributed to closing the gaps, but none solved the problem completely. Some lacked detail in specifying the services' quality (like the Generic Service Specification Framework), others were not based on a strong conceptual foundation (such as ITIL, CMMI or WSLA) and the majority of the web services based solutions suffer from the web service tunnel vision.

In this paper we apply our most recent proposal to specific services in a Portuguese city council. This proposal is based on the Enterprise Ontology theory and identifies services that do not create new original results, therefore services with great potential to be automated. In this particular city council we found services that if automated could save the Portuguese state millions of euros.

Our new SLA proposal revealed to be more flexible and mature when comparing to the previous one. The employees' and customers' feedback revealed that in the CMP context 14 in 16 of the proposal attributes were considered important. This indicates that this version of DEMO-based SLAs has more potential to capture the customers' expectations than the older version of the proposal that had only 5 attributes. By specifying these attributes, customers can structurally define their expectations which may help the alignment between customers and service providers. The specification of the customers' expectations into SLAs helps the service providers to understand those expectations and consequently reduce the gaps among the two.

The last step of DSRM, communication, is being achieved through scientific publications aimed at the practitioners and researchers within the service science area.

As future work, we intend to evaluate the proposal using practitioners' feedback. Presently, we are interviewing recognized practitioners involved in SLAs specification. Furthermore, we will apply our proposal in an organization with a more complex service exchange.

References

1. Ministry of Finance and Public Administration, 2011 State Budget (October 2010), http://static.publico.clix.pt/docs/economia/PropOERel2011.pdf (accessed April 09, 2012)
2. Parasuraman, A., Zeithaml, V., Berry, L.: A Conceptual Model of Service Quality and its Implication for Future Research. Journal of Marketing (1985)
3. Mendes, C., Ferreira, J., Mira da Silva, M.: Identifying Services from a Service Provider and Customer Perspectives. In: Fred, A., Dietz, J.L.G., Liu, K., Filipe, J. (eds.) IC3K 2011. CCIS, vol. 348, pp. 307–322. Springer, Heidelberg (2013)
4. Mendes, C., Ferreira, J., Mira da Silva, M.: Comparing Service Using DEMO. In: 3rd International Joint Conference on Knowledge Discovery, Knowledge Engineering and Knowledge Management - Special Session on Enterprise Ontology, Paris (2011)
5. Mendes, C., Mira da Silva, M.: DEMO-based Service Level Agreements. In: Snene, M. (ed.) IESS 2012. LNBIP, vol. 103, pp. 227–242. Springer, Heidelberg (2012)
6. Dietz, J.: Enterprise ontology - theory and methodology. Springer (2006)
7. Hevner, A., March, S.T., Park, J., Ram, S.: Design Science in Information Systems Research. MIS Quarterly (2004)

8. Peffers, K., et al.: A Design Science Research Methodology for Information Systems Research. Journal of Management Information Systems (2008)
9. Office of Government Commerce, ITIL v3 – Service Design, The Stationery Office (2007)
10. CMMI for Services, Version 1.3, Software Engineering Institute - Carnegie Mellon University (2010)
11. Sahai, A., Durante, A., Machiraju, V.: Towards Automated SLA Management for Web Services. Technical Report, Hewlett-Packard Company (2002)
12. Tosic, V., Patel, K., Pagurek, B.: WSOL – Web Service Offerings Language. In: Bussler, C.J., McIlraith, S.A., Orlowska, M.E., Pernici, B., Yang, J. (eds.) CAiSE 2002 and WES 2002. LNCS, vol. 2512, pp. 57–67. Springer, Heidelberg (2002)
13. Dobson, G.: Quality of Service in Service-Oriented Architectures (2004), http://digs.sourceforge.net/papers/qos.html (accessed June 2, 2011)
14. Frolund, S., Koistinen, J.: QML: A Language for Quality of Service Specification, HP Software Technology Laboratory (1998)
15. Keller, A., Ludwig, H.: The WSLA Framework: Specifying and Monitoring Service Level Agreements for Web Services. Journal of Network and Systems Management 11(1), 57–81 (2003)
16. Andrieux, A., Czajkowski, K., Dan, A., Keahey, K., Ludwig, H., Nakata, T., Pruyne, J., Rofrano, J., Tuecke, S., Xu, M.: Web Services Agreement Specification (WS-Agreement). Open Grid Forum (2007)
17. Liu, Y., Ngu, A., Zeng, L.: QoS Computation and Policing in Dynamic Web Service Selection. In: 13th International World Wide Web Conference on Alternate Track Papers & Posters, pp. 66–73. ACM, New York (2004)
18. Terlouw, L., Albani, A.: An Enterprise Ontology-Based Approach to Service Specification. IEEE Transactions on Services Computing (2011)
19. Albani, A., Terlouw, L., Hardjosumarto, G., Dietz, J.: Enterprise Ontology Based Service Definition. In: 4th International Workshop on Value Modeling and Business Ontologies, Amsterdam, The Netherlands (2009)
20. Pombal City Council, 2011 Pombal City Council Budget, Pombal (2011)
21. Kvale, S.: Doing interviews. Sage Publications, London (2007)
22. Pries-Heje, J., Baskerville, R., Venable, J.: Strategies for Design Science Research Evaluation. In: 16th European Conference on Information Systems (ECIS), pp. 255–266 (2004)

A Method for Reengineering Healthcare Using Enterprise Ontology and Lean

David Galego Dias, Carlos Mendes, and Miguel Mira da Silva

Instituto Superior Técnico, Technical University of Lisbon
Av. Rovisco Pais, Lisboa, Portugal
{david.n.dias,carlos.mendes,mms}@ist.utl.pt

Abstract. Global healthcare spending has increased in the last decades, and there is data showing inefficiency in resource consumption that is not reflected in healthcare improvement. To overcome these problems, this paper proposes a method based on Enterprise Ontology and Lean to find non value-added transactions, and redesign them to improve healthcare. Demonstrations of the method were accomplished using field studies within the national health system, making it possible to find transactions that can be refined or improved. Results and evaluation prove that the proposed method yields an adequate and clear process view, and is reliable when it comes to reengineer healthcare.

Keywords: Enterprise Reengineering, Enterprise Ontology (EO), Design Methodology for Organizations (DEMO), Healthcare, Emergency Department.

1 Introduction

In a world of growing business dynamics, high rates of technological advances and organizational changes, organizations need to be effectively and continuously redesigned and reengineered in order to achieve strategic and operational success. The inefficiency of processes and the lack of innovation are the main reasons for strategic failures, entailing serious consequences for business and its competitiveness [1] [2].

These strong external forces and the need for innovation also challenge healthcare. Its organizations need to improve treatments, eliminate non value-added activities, reduce waiting time and expenses, treat more patients, and implement new technological services. Besides these challenges, the healthcare system suffers from problems of operational management, and its processes are considered inefficient [3] [4].

A frightening factor is that not only its expenditure accounts for 10% of the Gross Domestic Product (GDP) in developed countries, but there is also an increasing trend, as depicted in Figure 1. Other than that, there is data indicating that cost and quality are not correlated, and showing inefficiency in resource consumption, which is not reflected in improved quality of care. Consequently, the quality of life may be affected because of a knock-on effect on the economy, increase in taxes and insurance contributions, disinvestment in other public services, and increased difficulty to afford healthcare [4], [5]. Hence, this paper stems from the assumption that many healthcare processes have become inefficient and unsustainable, which affects the whole system.

A. Fred et al. (Eds.): IC3K 2012, CCIS 415, pp. 243–259, 2013.

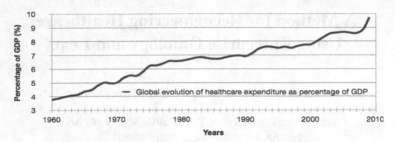

Fig. 1. Global evolution of healthcare expenditure – adapted from [6]

Although the problem is identified as a need for redesign and reengineering, some authors argue that there is no strong and reliable method to solve this problem [7]. It is estimated that over 70% of strategic initiatives such as Total Quality Management, Business Process Reengineering, and Six Sigma, among others, tend to fail [8] [9]. In this context, we addressed three main reasons for this: 1) The lack of integration among the various enterprise elements at the design level; 2) The inability to deal with the enterprise dynamics at the operational level due to weak enterprise construction models; and 3) The need to change management that advocates the development of self-awareness within the organization [2], [10], [11].

Following this, we propose an approach that combines Enterprise Ontology (EO) with Lean to reengineer healthcare. We chose this approach as a foundation for our proposal, because, on one hand, EO is deemed able to provide a better understanding of organizations' dynamics, has a strong and well-formed theory, allows a good alignment between the enterprise design and operation, and enables a unified reengineering strategy [12], [13]. On the other hand, with the addition of Lean we take advantage from its proven benefits to manage healthcare, for the quality management and the continuous improvement [14]. Therefore, we are addressing the following research question: **can EO and Lean contribute to improve the healthcare system?**

The research was conducted by using the Design Science Research Methodology (DSRM) that aims at creating and evaluating artifacts to solve relevant organizational problems [15]. The obtained artifact is a method that provides guidance on how to find improvements through a set of steps, and it was demonstrated using field studies in an Emergency Department (ED), a Primary Healthcare Center, and a Pharmacy. Besides the possible improvements in each organization, we are also interested in analyzing the interactions between them, so that we can conclude how they can improve their cooperation. To evaluate the proposed artifact we used: 1) The framework proposed in [16], 2) Demonstrations of the method; 3) Interviews with practitioners; 3) The Four Principles from [17]; and 4) The Moody and Shanks Quality Management Framework [18] to evaluate the produced models.

The steps from the DSRM are reflected upon in the sections of this paper, which is structured as follows. In Section 1 we just introduced our problem and motivation. Then, a brief overview of the literature is provided (Section 2). Afterwards, we identify the objectives of the solution and describe the proposal to redesign the healthcare processes (Section 3). Next we present field studies where the proposal was applied as demonstration (Section 4). In Section 5, we describe the evaluation strategy and discuss the results of applying the proposal. Finally, we draw conclusions in Section 6.

2   Related Work

This section presents a brief overview of the Quality Management (QM) and EO.

2.1   Quality Management

Edwards Deming, one of the main and originator sources in QM, defended that organizations could increase quality and reduce costs by adopting appropriate principles of management. He identified seven constructs as main drivers: visionary leadership, internal and external cooperation, learning, process management, continuous improvement, employee fulfillment, and customer satisfaction [19]. Hence, authors defend that these topics are considered crucial not only to compete and prosper, but also to merely survive against external forces [1]. In response to the need of QM and continuous improvement, different methodologies and strategies appeared, such as Organization Design and Engineering methodologies, Lean, Six Sigma, Total Quality Management, among others [8], [9].

Lean is considered one of the most used in the management of the healthcare system [14]. It is typically grounded in the PDCA Operating Framework, and focus on the waste removal to deliver an improved flow time. The PDCA cycle suggests that all work should be measured and performed to standards, and it is composed by the following steps: 1) Plan: recognize an opportunity and plan a change, its needed steps, and results' prediction; 2) Do: test the change using small-scale studies as trials under controlled conditions; 3) Check or study: changes are tested in small-scale studies to examine its results, and if process improvements were verified, it should be considered the implementation on a broader scale; 4) Act: implement the changes in a broader scale and then repeat the cycle again with a differ plan [20].

Some of the benefits of the QM and particularly Lean in the healthcare management are the reduction of processing and waiting time, decline in the mortality rate, increase in quality through a reduction of errors, decrease in the service costs and resource expenditure, better warehouse management, and increased employee motivation and customer satisfaction [21], [22], [23]. On the other hand, some authors point out some drawbacks, such as the high rate of failed implementations, the mischaracterization or degradation of services, and the loss of organization's essence [14].

The Improvement Quantification is considered another topic related with the QM, which helps to make decisions and prioritize improvements based on the expected return and feasibility. There are different approaches that may help a manager to make informed and just-in-time decisions about improvements. For example, costing models that may help to identify the cost from each activity, allowing for a greater knowledge about its indirect and variable costs [4].

2.2   Enterprise Ontology

Enterprise Ontology (EO) is a theory that has its roots in the PSI-Theory (Performance in Social Interaction), and is perceived as a model for describing and understanding the enterprise construction and operation at the level of human interactions, allowing a better understanding of the operation. Dietz brings a complementary view of the EO, in which ontology is viewed as the "highest level" conceptual model, fully independent of how the enterprise is implemented. It is an enterprise context based concept that is considered the highest conceptual model and helps ensure integrated enterprise. It also guides the transition from ontological models to construction models, which means that it assists in engineering activities [12], [2].

Unlike other methodologies, EO is considered to provide a deep understanding of the dynamics of an organization with a strong and well-formed theory that allows a good alignment between the enterprise design and the enterprise operation [2].

Its particular methodology, DEMO, provides a structured working approach for modeling, (re)designing and (re)engineering of organizations by layering it into three parts, and focusing only on the one that refers directly to the complete knowledge of the enterprise – the *Ontological or Essential Layer*, which is considered to affect the other two layers (*Informational* and *Documental*), as illustrated in Figure 2. Focusing only on the essence conducts to a reduction in the complexity of the obtained diagrams, considered in over 90% [13], [7].

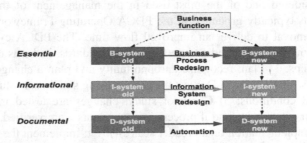

Fig. 2. The layered integration of an enterprise and its transformation activities [13]

Regarding DEMO methodology, it consists of four interrelated aspect models, represented by particular diagrams, lists and tables, as illustrated in Figure 3. The Construction Model (CM) details the identified transactions types and associated actor roles, as well as the information links between the actor roles and the information banks. The Process Model (PM) specifies the state and transaction spaces, and it is partially based on the information defined on the CM concerning which actor roles perform the *coordination acts*. In addition, PM also contains the causal and conditional relationships between transactions, which determine the possible trajectories between transactions. The State Model (SM) specifies the information banks and the state space of the *production world*: the object classes, the fact types, and the result types, as well as the existential laws that hold. The Action Model (AM) specifies the action rules that serve as guidelines for the actors in

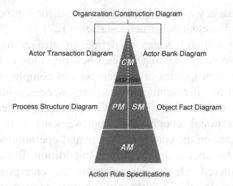

Fig. 3. The ontological triangle with aspect models and diagrams of DEMO [12]

dealing with every coordination step, which are grouped according to the distinguished actor roles. The bottom layers from the ontological triangle integrate concepts defined in the upper aspect models, as depicted in Figure 3. For further reading about the EO, DEMO methodology, and the four axioms significant to understand the methodology we refer [12].

There are some examples in the healthcare system in which EO was applied to study its internal transactions and simplify their analysis. These contributions validated that EO avoids the lack of integration among the various enterprise elements at the design level and produces strong enterprise construction models [24], [25], [26]. In addition, we can find examples of using EO to improve operational processes [13] due to its differentiated and structured working approach focused on the essential design of the organization.

3   Proposal

This section corresponds to the *definition of the objectives for the solution* and the *design and development* steps of DSRM.

3.1   Objectives of the Solution

In order to overcome the problem statement about the inefficiency and unsustainability of the healthcare system, different approaches are identified. Nevertheless, some authors still argue that there is not a reliable method to solve these problems. It is estimated that over 70% of strategic initiatives such as Total Quality Management, business process reengineering (BPR), and Six Sigma, among others, tend to fail [8], [7], [9]. Furthermore, Dietz also adds that the current literature on enterprise engineering consists merely of best practices, without an integrating theory and a clear definition of the field [12]. Inline with that conclusion, *Caetano et al.* demonstrated that when comparing BPMN and DEMO models, there was a set of implicit and missing actions in BPMN, proving that it does not provide means to assess the actual consistency and completeness of a business process, due to the lack of formal semantics and unclear construct description [27].

Following this, our research seeks to define an artifact method based on the theories of EO because of the strengths described previously, namely the benefits previously described, the properties of correctness and completeness it assures in its models, and the properties of essentialness and conciseness, which help to construct and analyze (more) models, making it possible to design the healthcare system and seek for inter-organizational cooperation improvements between its units. The improved alignment between the enterprise design and operation leads to an improved self-awareness within healthcare organizations. In addition, EO clearly defines three notions that we considered relevant in governing the enterprise dynamics and to identify improvements in the healthcare system: competence, authority, and responsibility, as explained in the Operation Axiom [12]. Most of these notions are absent or not clear defined in other enterprise modeling techniques [12], [2].

To take advantage from some already proven benefits from Lean for the QM and Continuous Improvement, particularly in the healthcare system, we intend to combine the analysis from EO with the improvement identification from Lean. This way, the EO may be considered as input for the Plan step of the PDCA Operating Framework, to help with the identification of opportunities. In other words, from DEMO models one may identify improvements (as suggested in the Plan step), and in the end produce and Organization Redesign model that reflects the change plan. To identify improvements one should consider the existing standards on Healthcare Management, BPR, and improvement quantification. The following steps from PDCA cycle are out of the thesis' scope, as they need the creation of prototypes and implementation in a broader scale.

In short, our main objective is to **propose a method based on EO and Lean to find non value-added transactions, and redesign them to improve the healthcare**. Other goals are to demonstrate, evaluate and communicate the artifact, to show its efficiency and efficacy. To do that, we are applying the proposal to different units of the healthcare system. Besides the possible improvements in each healthcare unit, we are also interested in analyzing the interactions between them so that we can conclude how they can improve cooperation, as part of the demonstration.

3.2   Proposed Artifact Method

This section belongs to the *design and development* step of DSRM, in which we present a *different* artifact [17] to identify innovations to improve the healthcare. It considers the contributions from EO [12] and from Lean [20], [22].

The proposal starts with the **Modeling Phase,** which uses EO to study the organization and its processes. To construct its diagrams, it consists of a defined sequence of steps (illustrated in Figure 4) that begins with a textual or process representation of an organization, and ends with an aspect model. As result, this phase provides a structured working approach by layering the organization into three parts, and focusing only on the one that directly refers to the complete knowledge of the

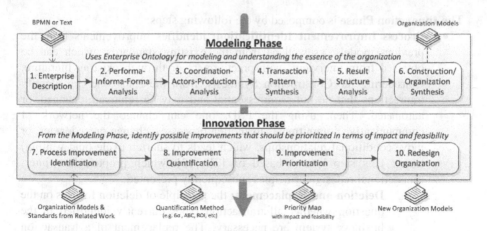

Fig. 4. Graphical representation of the proposed method

organization and independent of the implementation – the *Ontological Layer*. In this research we focus on the *Construction* and *Process Models*.

The proposal then continues with the **Innovation Phase**, which is based on four additional steps from Lean to assist in the Continuous Improvement and the QM process. These steps consist on the Plan step of the Lean PDCA Operating Framework that recognizes an opportunity and plans the change. Therefore, this phase identifies possible improvements from the previous models, prioritizes them in terms of impact and feasibility, and then proposes redesigned models for the organization. As result, this phase gives the appropriate tackle to handle the transformation process, and helps to choose the most profitable improvements first. Figure 4 illustrates the proposal including its inputs and outputs, and subsequently we describe its steps.

The Modeling Phase follows the steps described in [12], [28]. It starts by analyzing the Enterprise Description to look for *ontological actions* (or *performa* abilities). This leads to a reduction of the complexity relatively to other methodologies since the *infological* and *datalogical* actions are excluded. Afterwards, the Coordination-Actor-Production Analysis identifies the coordination and production acts. Then, in the Transaction Patter Synthesis the coordination and production acts are clustered into transaction types together with the corresponding results, and dependencies between the transaction types are identified during the Result Structure Analysis. Finally, in the Construction Synthesis, the initiator and executor roles of each transaction type are identified, and graphical representations are developed. These first six steps result in different diagrams, such as the Actor Transaction Diagram (ATD) and Process Structure Diagram (PSD), obtained by applying the four axioms described in [12].

Finishing the Modeling Phase, one can proceed with the Innovation Phase, or deepen some analysis, including further details on the Enterprise Description or other aspect models.

The Innovation Phase is composed by the following steps:

- **Process Improvement Identification:** identifies improvements from the previous models, as well as from the existing standards, which can be obtained from interviews with practitioners or through existing literature. Considering the Construction Synthesis, from the ATD one can identify transactions that do not seem essential and may be removed, changed, or automated. Then, using the PSD, one can change the network of communicative commitments to shorten processes, change precedencies, or move conditional relationships, which leads to shorten cycle (and waiting) times. This step is also based on [7], [13], which propose four inter-transactional redesign principles subsequently presented:

    a. **Deletion and Replacement:** the principle of deletion focuses on the question whether all transaction types currently identified in the business system are necessary. The replacement of a transaction type implies a fundamental change to the preposition of the transaction type;

    b. **Change of Optimal Relationships:** this principle assumes that all the identified transaction types are necessary, and focuses on finding the optimal structure of them, changing optional and causal relationships. The application of this principle may result in causal relationships becoming optional. This leads to shorten cycle (or waiting) times;

    c. **Advancing Initiating Points:** inline with the previous technique, the advancement of the point of initiation is concerned with the assumption it can lead to a reduction of the total lead-time;

    d. **Parallelization of Transaction Types:** this principle examines the conditional relationships between the start of transaction types and explores the possibility to start others in parallel. This is possible removing conditional relationships that allow one transaction to start before the previous one is finished. The reduction of conditional restrictions improves the parallelization between transactions and prevents deadlocks.

- **Improvement (or *kaizen*) Quantification:** after identifying possible improvements, some metrics must be established to quantify them in terms of feasibility and impact. Some *kaizen* examples can be: time invested in each transaction compared to the total time spent on the whole service, people involved, management frameworks, associated defect, or other analytical methods (e.g. costing models, financial analysis, Six Sigma, etc.);

- **Improvement Prioritization:** in this step improvements are prioritized in terms of impact and feasibility, which helps to choose the most important improvements. This is then represented in a map divided into four quadrants, being the X-axis the feasibility to accomplish it, and the Y-axis the quantified impact. Each improvement is placed in a particular quadrant, being the ones that fit into the superior diagonal also the ones that are more important to implement (the ones with higher impact and feasibility);

- **Redesign Organization:** after choosing the most profitable improvements the organization is redesigned to include the decisions. Alternatively, one can deepen some analysis by including more information in the Enterprise Description, or producing other aspect models from DEMO.

Having the redesigned organization models with the results from the Innovation Phase, it should be prepared a proposal with specific implementation strategies (i.e. the plan with the needed steps). Afterwards, one should continue with the next steps from the PDCA cycle (Do, Check and Act) to implement the planned changes. The implementation itself is beyond of the scope of this paper.

To sum up, this method replaces the analysis from Lean by a Modeling Phase based on EO, incorporating its contributions to achieve models considered formally correct, easier to analyze, and enabling a unified reengineering strategy [12].

4   Demonstration

This section corresponds to the *demonstration* step of DSRM. To demonstrate the proposal we applied it to different healthcare units, including a hospital ED, a Pharmacy, and a Primary Healthcare Center. In this paper, we present the demonstration at the ED and a summary with other demonstrations to redesign the healthcare system. To conduct these demonstrations we interviewed different practitioners, namely to obtain input to apply the proposal, feedback about the obtained outputs (models and improvements obtained), and validation on the proposal. This information was collected through field visits on location at each participating organization, as well as follow-up phone calls and e-mails. In this paper we do not present the six steps of the modeling phase due to space limitations, but these steps were demonstrated in previous publications [28], [29].

4.1   Emergency Department (ED)

To demonstrate the method, we applied it to the internal operation of an ED in a hospital near Lisbon with more than 100,000 admissions per year, expecting that by eliminating wasteful transactions, it would be possible to improve processes without compromising the organization. To conduct the demonstration, we interviewed 5 patients and 10 practitioners (the ED director, physicians and nurses, and health services researchers), namely to obtain the enterprise description.

From the enterprise description and after the first two analyses from the proposed method, we defined the transactions by clustering the identified *acts* and *facts* in what is denominated by *Transaction Pattern Synthesis*. The results are presented below in the ATD (Figure 5), according to the Transaction Axiom from EO [12].

In the ATD, a transaction is represented using a diamond in a disk. Each transaction is connected to two boxes, representing the initiator and executor actor roles. The initiator is connected to the transaction symbol using a solid line, while the executor is connected to the transaction using a solid line ending in a black square. The grey boxes refer to composite actor roles, i.e. elements whose exact structure is not known. All the environmental elements, i.e. elements outside the organization that we are studying, are represented with grey boxes for that reason. This also means that we can represent the studied organization with a grey box when referring to the kernel of the organization, which can be further specified by using elementary actor roles represented by white boxes.

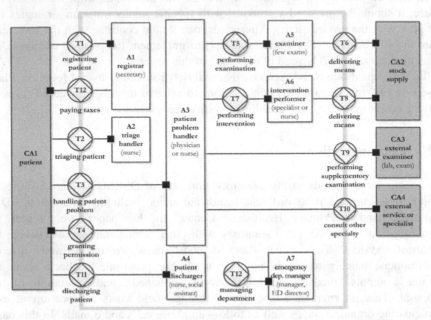

Fig. 5. Actor Transaction Diagram (ATD) of the Emergency Department (ED)

As depicted in this diagram, new patients are registered to the hospital (T1); then they go through a triage process (T2); after that, patients' problems are handled (T3); and finally, they are discharged (T11). These four transactions are initiated by an external actor, the *patient*. They are respectively requested to the *registrar*, *triage handler*, *patient problem handler*, and *patient discharger* that execute them.

The handling of the patients' problems may lead to the following actions: performing some urgent internal examinations (T5); performing medical interventions (T7); performing supplementary examinations (T9); and consulting another external specialty (T10). Since these tasks have different responsibilities, four different actors are discerned: *examiner, intervention performer, external examiner,* and *external service or specialist.* The first two are internal actors, used for urgent examinations and interventions (i.e. specific interventions may need specialists, such as a surgery or a psychiatry episode). The last two are used for non-urgent situations, such as some extended interventions or supplementary examinations. In addition, there are two transactions concerning the delivery of means (T6 and T8), a transaction concerning the patients' permission (T4), and finally the payment transaction (T12).

In the **Innovation Phase**, one must identify process improvements from the obtained diagrams. First, after some analysis from the ATD, one may conclude that transaction T1 can be improved or even removed. In the former, the registration can be implemented by the triage actor role (improvement A from Table 1). Alternatively, registration can be removed if patients' records were shared, or the *production* can be automated through a standardized electronic form with the necessary reasoning to become an *infological* action (improvement B) [13] [12]. Practitioners agreed that this transaction can be considered as a non-value added task, and consequently its actors could be allocated to other activities.

With the PSD (obtained from the ATD), one can conclude that it is <u>not efficient having to go through several iterations and actors to be forwarded to another external service</u> (specialist or examiner in T10). For example, instead of being forwarded immediately after triage, patients need to be admitted (T1), triaged (T2), and seen by a physician (T3) to be finally forwarded to another specialty outside the ED. This leads to unnecessary consumption of resources, higher waste of time, and the patient leaves without being treated in the ED. There is some literature suggesting strategies of *Fast-Tracking* (improvement C) and *Provided Directed Queuing* (improvement D) to anticipate the resolution of some patients' problems. These strategies are claimed to improve waiting time, customer satisfaction, length of stay, and efficiency [30].

In Table 1 we quantify the improvements in which we want to work at. To infer the level of impact, we consider that the elimination of a transaction has a higher impact than a precedence change. Avoiding a transaction conducts to the same classification as its elimination or automation. Avoiding an actor has even higher impact, because it eliminates the transaction and reduces the costs with human and physical resources. Finally, to assess the feasibility we considered that more changes to the service leads to lower feasibility (i.e. hardware, software or people involved). The presented values were obtained with the help of the interviewed practitioners for this demonstration.

Table 1. Improvements identification with its impact and feasibility (from 1 to 5)

#	Improvement	Impact	Feasibility	Impact description	Feasibility description
A	Patient registers in triage	4	2	Avoid T1 and transfer responsibility to T2	Triage should be fast to respect the service level
B	Remove or automate registration	5	4	Avoid transaction T1 and actor A1	Computer terminal with new hard and software
C	*Provided Directed Queuing*	5	5	May eliminate transaction to reduce flow	Reallocate only one physician
D	*Fast-Track System*	4	4	May eliminate transaction to reduce flow	Reallocate physician and a new space

The priority map (Figure 6) addresses the impact and feasibility levels from the last step: D shows large impact and feasibility, followed by B and C.

We could apply a more formal method for the improvement quantification, but this would not change the method itself. For example, in [4] the authors use TDABC (Time-Driven Activity Based Cost) to quantify costs from each transaction. Using the formulas from [4] and the time we measured for each transaction, we concluded that

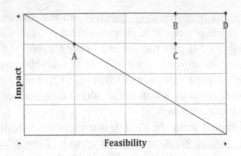

Fig. 6. Priority map of the emergency department

the registry itself (T1 implied in improvement A) accounts for 180,000€ (EUR) per year, and that a low-acute episode accounts for 27.6€ (EUR) for the whole visit (implied in improvements C and D). This numbers could be even worse if we consider the overheads' costs, the wasted periods and the inventory consumption. These calculations were presented in [31].

4.2   Other Demonstrations in the Healthcare System

By analyzing the improvements from the three demonstrations altogether, we found that it is possible to improve their cooperation since some overlapped transactions were found, as demonstrated in [31], [32] and illustrated in Table 2.

Table 2. Overlapped transactions in ED, Pharmacy and Primary Healthcare Center

ED Transactions	Pharmacy Transactions [31]	PHC Transactions [32]
T1 – Registering patient	T1 – Creating profile	T1 – Register patient
T3 – Handling patient	T9 – Medical consultation T2 – Filing prescription	T5 – Emergency consultation T10 – Filing prescription
T5 – Performing examination	T3 – Performing examination	T6 – Performing examination
T7 – Performing intervention		T8 – Performing intervention
T10 – Consult other specialty		T11 – Consult other specialty

From the previous table, one may identify, for example, that a patient has to register and create healthcare records in every healthcare organizations, consuming much time, replicating resources and repeating information. Other than that, we may find an overlap of responsibilities between healthcare units when it comes to handle a patient, as different professionals and organizations handle the same healthcare issues.

These results, supported by practitioners, suggest that: A) The relationship between these entities need to be redesigned; B) There is a lack of definition regarding the responsibilities, authorities and competencies; and C) There are inefficiencies in the patient referral between different units. These conclusions are also aligned with[5].

5   Evaluation

This section corresponds to the *evaluation* step of DSRM. It explains how we proceeded to evaluate the proposal (Section 5.1), and then demonstrates that the proposal addresses the research problem (Section 5.2).

5.1   Evaluation Strategy

To evaluate the proposal, we used the framework proposed in [16], which aims to help researchers to build strategies for evaluating the outcome of a DSRM. This framework identifies **what is actually evaluated, when the evaluation takes place**, and **how it is evaluated**. To answer the third question, we based on different authors to propose a strategy with steps outlined to evaluate a DSRM artifact method. The evaluation strategy entails the following steps:

1. Constructing scenarios to **demonstrate the artifact** and how to use it to solve the research question, which is considered an approach to validate a design science research artifact of type method [15];
2. Gathering **feedback through interviews with practitioners** about the artifact, ability to follow its steps, and potential to obtain relevant results;
3. The **Moody and Shanks Quality Management Framework** to assess the quality of the models produced in the demonstration phase [18];
4. The **Four Principles proposed by** [17] to evaluate a DSRM artifact.

This evaluation method follows the **design evaluation guideline within DSRM** [15]. In this research we have mainly used the *descriptive* evaluation method to assess the artifact, which uses relevant research to build a convincing argument for the artifact's utility, and constructs detailed scenarios around the artifact to demonstrate the utility. Nevertheless, the observational and analytical methods could also be used, but this would involve introducing observable metrics, conducting socio-technical experiments, and selecting modeling tasks that would allow such measurement [15]. Such evaluation is beyond the scope of this research.

5.2   Evaluation Results

The framework proposed in [16] is formulated as follows:

- **What was actually evaluated?** The evaluated artifact was the method described in Section 3, which is a DSRM artifact method. This evaluation represents an *artifact design process*, since it is defined as a set of activities, methods and practices that can be used to guide a procedure workflow to improve the healthcare;
- **When was it evaluated?** It was evaluated after the artifact construction, and after the demonstration. Therefore, the evaluation strategy is *ex post*, since it was performed after the design artifact development;

- **How is it evaluated?** To evaluate the artifact and its results we used the described strategy, which are applied below. This represents a *naturalistic* evaluation since it is conducted using a real artifact in a real organization facing real problems as a case of study.

The **demonstration** reveled that: A) The proposal is generic to be applied in different healthcare organizations; B) It is a formal method, with a list of specific steps to follow; C) From a given enterprise description one can achieve similar enterprise models, as Dietz suggests [12]; D) From the obtained models it is possible to find non value-added transactions and from them suggest and prioritize improvements; and F) It is possible to obtain a redesigned organization. In other words, it was possible to demonstrate the artifact's utility, and how to use it to solve the research problem.

The **feedback from the interviews** (with the same practitioners from demonstration) was rather positive because: 1) They validated the importance of the research problem and the motivations behind the proposal; 2) They understood and agreed with the obtained models (after explaining them), which were considered to properly depict the studied organizations; 3) Improvements were discussed and the interviewees agreed that the ones we identified were sometimes similar to those suggested by them; 4) Practitioners concluded that the proposal could be applied effectively and efficiently to solve the research problem, regardless of whom applies it. Overall, practitioners showed a good acceptance and enthusiasm for this innovative approach.

From the **Moody and Shanks Quality Framework**, almost all quality factors were accomplished. Only *understandability* was partially, and *implementability* was not. The first factor as practitioners find models difficult to interpret needing an adaptation period. The second one as models are implementation independent (describing only the essence of organizations).

The **Four Principles proposed by** [17] were also accomplished. 1) Abstraction: the artifact can be applied to any healthcare service from a given enterprise description; 2) Originality: the proposed artifact is not present in the body of knowledge of the domain since it was designed by relating different subjects, such as healthcare management, BPR, EO and Lean; 3) Justification: the artifact is supported by the related work, described by textual and graphical representations, and it was justified and validated in different ways; 4) Benefit: the artifact provides a structured working approach for reengineering, it leads to differentiated and well-grounded improvements, and provides a better understanding of the dynamics of an organization, among other benefits when compared to existing methodologies.

6   Conclusions

This research addresses healthcare management problems, in which its processes have become inefficient and unsustainable. To overcome these problems, this research proposes a method based on EO and Lean to find non value-added transactions, and redesign them to improve the healthcare.

We chose the EO as foundation for our proposal as it is deemed able to provide a better understanding of the dynamics of an organization, allows a good alignment between the enterprise design and operation, and enables a structured reengineering strategy. On the other hand, with the addition of Lean, we intended to take advantage from its proven benefits to manage the healthcare industry, for the Quality Management and the Continuous Improvement, therefore considering the combination of EO with the PDCA Operating Framework to identify, quantify, and plan improvements.

Considering this research's objectives and evaluation, we may conclude that the expectations were achieved since it was possible to: A) Formulate the method; B) Demonstrate its use in real field studies; C) Find non value-added transactions when applying it; D) Suggest redesign improvements; and E) Get validation and positive feedback from practitioners about the method and its results. Thereby, we may answer to the raised research question that: it is possible to associate EO with Lean to propose improvements and redesign healthcare.

Some of the main contributions of this research are the proposal that is different from the current state-of-the-art approaches, its practical demonstrations in real healthcare units, and the improved inter-organizational cooperation and self-awareness obtained from models. Therefore, it is expected that healthcare organizations may use some of the described advantages of the proposal to solve the problems of inefficiency and unsustainability in the healthcare system. In addition, it can also be a contribution towards helping healthcare professionals to validate processes and improve their way of working.

As future work, further research is being performed to better quantify the impact and feasibility of the proposed improvements during the demonstration, namely by including costing models to the obtained DEMO diagrams. Furthermore, the proposal may be expanded to consider the remaining application of Lean PDCA cycle and other EO models, such as *Action* and *Interstiction* Models, which can be useful in the redesign of information systems, inline with previous researches [13].

References

1. Kotter, J.: Leading Change: Why Transformation Efforts Fail. Harvard Business School Press, Boston (1996)
2. Henriques, M., Tribolet, J., Hoogervorst, J.: Enterprise Governance and DEMO: a reference method to guide enterprise (re)design and operation with DEMO. In: Conferência da Associação Portuguesa de Sistemas de Informação, Viana do Castelo (2010)
3. Christensen, C., Grossman, J., Hwang, J.: The Innovator's Prescription - A disruptive solution for Health Care. McGraw Hill, New York (2009)
4. Kaplan, R., Porter, M.: How to solve the cost crisis in health care. Harvard Business Review (September 2011)
5. Walshe, K., Smith, J.: Healthcare Management. Open University Press, Berkshire (2010)
6. OECD. In: Organization for Economic Co-operation and Development statistics, http://stats.oecd.org/, http://stats.oecd.org/ (accessed 2012)

7. Dietz, J., Hoogervorst, J.: Enterprise Ontology in Enterprise Engineering. In: Proceedings of the 2008 ACM Symposium on Applied Computing, New York, pp. 572–579 (2008)
8. Mintzberg, H.: The rise and fall of strategic planning, New York (1994)
9. Lifvergren, S., et al.: Lessons from Sweden's first large-scale implementation of Six Sigma in healthcare, pp. 117–128. Springer (2010)
10. Dias, D.G., Lapão, L.V., Silva, M.M.D.: Using Enterprise Ontology for Improving Emergency Management in Hospitals. In: 24th European Medical Informatics Conference (MIE 2012), Pisa (2012)
11. Dias, D., Xie, S., Silva, M., Helfert, M.: Using Enterprise Ontology Methodology to Assess the Quality of Information Exchange. In: 18th Americas Conference on Information Systems (AMCIS), Seattle (2012)
12. Dietz, J.: Enterprise Ontology - Theory and Methodology. Springer, Delft (2006)
13. Reijswoud, V., Mulder, H., Dietz, J.: Communicative Action Based Business Process and Information Systems Modelling with DEMO. International Journal of Information Systems (1999)
14. Burgess, N., Radnor, Z.J.: How is lean being applied to health? Classifying approaches to lean implementation in the NHS. In: The 14th Annual International Research Society for Public Management (IRSPM) Conference - The Crisis: Challenges for Public Management, Berne (2010)
15. Henver, A., et al.: Design Science in Information Systems Research. MIS Quartely 28(1), 75–105 (2004)
16. Pries-Heje, J., Baskerville, R., Venable, J.: Strategies for Design Science Research Evaluation. In: 16th European Conference on Information Systems (ECIS), pp. 255–266 (2004)
17. Österle, H., Becker, J., Frank, U., Hess, T., Karagiannis, D., Krcmar, H., Loos, P., Mertens, P., Oberweis, A., Sinz, E.: Memorandum on Design-Oriented Information Systems Research. European Journal on Information Systems 20, 7–10 (2011)
18. Moody, D.L., Shanks, G.G.: Improving the Quality of Data Models: Empirical Validation of a Quality Management Framework. Information Systems 28, 619–650 (2003)
19. Rungtusanatham, M., Forza, C., Filippini, R., Anderson, J.: A Replication Study of a Theory of Quality Management Underlying the Deming Management Method. Journal of Operations Management 17(1), 77–95 (1998)
20. Womack, J., Jones, D.: Lean thinking. Simon & Schuster, London (2003)
21. Fillingham, D.: Can lean save lives? In: Leadership in Health Services, UK, vol. 20, pp. 231–241 (2007)
22. García-Porres, J., et al.: Lean six sigma applied to a process innovation in a mexican health institute's imaging department. In: 30th Annual International IEEE EMBS Conference, pp. 5125–5128 (2008)
23. Radnor, Z.: Transferring Lean in to Government. Journal of Manufacturing and Technology Management 21, 411–428 (2010)
24. Maij, E., et al.: A process view of medical practice by modeling communicative acts. Methods of Information in Medicine, 56-62 (2000)
25. Habing, N., et al.: Activity patterns in healthcare - indentifyinf building blocks for the CPR. In: ACM SIGGROUP Bulletin, vol. 22 (August 2001)
26. Maij, E., et al.: Use cases and DEMO: aligning functional feautures of ICT-infrastructure to business processes. International Journal of Medical Informatics, 179-191 (2002)
27. Caetano, A., Assis, A., Tribolet, J.: Using Business Transactions to Analyse the Consistency of Business Process Models. In: 5th Conference on Research and Practical Issues of Enterprise Information Systems, Denmark (2011)

28. Mendes, C., Ferreira, J., Silva, M.: Identifying Services from a Service Provider and Customer Perspectives (2012)
29. Mendes, C., Ferreira, J., Silva, M.: Comparing Service Using DEMO, Paris (2011)
30. Medeiros, D., et al.: Improving patient flow in a hospital emergency department. In: Winter Simulation Conference, Austin, pp. 1526–1531 (2008)
31. Dias, D.: Method for Improving Healthcare Management Using Enterprise Ontology. Master Dissertation from Instituto Superior Técnico (2012)
32. Dias, D., Mendes, C., Mira da Silva, M.: Using Enterprise Ontology for Improving the National Health System. In: 4th International Conference on Knowledge Engineering and Ontology Development, Barcelona (2012)

Interactive Exploration of Structural Concepts in Code

Paul Heckmann and Daniel Speicher

Institute of Computer Science III, University of Bonn, Germany
heckmann@uni-bonn.de, dsp@cs.uni-bonn.de

Abstract. Understanding a software system is the first task in any reengineering activity. For this very challenging task one effective approach is to identify interesting and reoccuring structures in the software and to study these structures individually. In object-oriented software such structures typically consist of a few classes. The well known among them are called design pattern. Yet, which structures to look at in particular? Can we identify interesting structures that are not that well known? Which structures to be a clue to start with?

In this paper we extend a previously suggested approach of pattern mining using Formal Concept Analysis. We propose a way to eliminate redundant information in the overall analysis result. Besides that, we introduce two new features: The first feature is a *filtering element* that allows us to interactively and dynamically narrow the analysis space. The second is the *prominence* of a class - a measurement of the importance of the class to the overall system.

In an experimental evaluation we applied our approach on two software projects. In the first, JUnit, our tool guided the experimenter to central structures that can be found in the online documentation yet was unknown to the experimenter. In the second the tool led us to core structures of our own software.

Keywords: Design Pattern, Pattern Mining, Formal Concept Analysis, Interactive Software Exploration.

1 Introduction

An object oriented software system essentially can be seen as a composition of structural concepts. Concepts in which classes and interfaces are connected with each other using building mechanisms like abstraction, inheritance, and composition to realize a certain functionality important to the respective part of the system. Some of these concepts reoccur over the entire project and constitute to a programs unique character, others reoccur yet only in certain parts suggesting core concepts of the program. Hence revealing any of these structural concepts can be a first important step to understanding the software itself, its character and its core functionality. But not all of these concepts arise by design, e.g. by using design patterns as introduced by [1]. They may arise implicitly and strongly depend on the developers style of solving a certain design problem.

In this work we propose an approach to mine structural concepts using a bi-clustering technique called Formal Concept Analysis (FCA) [2], building on a previous approach proposed by [3]. This technique allows us to group structures in source code into meaningful groups without requiring any knowledge on the to-analyze program nor the existence of a reference library of structures. We then improve this approach by the use of

A. Fred et al. (Eds.): IC3K 2012, CCIS 415, pp. 260–270, 2013.

a more efficient mining algorithm and extend it by adding filtering features that, on one hand, allows us to interactively explore the structures in a program, and on the other hand supports us in finding those structural concepts constituting to its core functionality.

In section 2 we give a brief introduction to the very basic idea of FCA. In section 3 we reproduce the approach firstly introduced by [3] to apply FCA on our problem of mining structures in source code. In section 4 we present our extensions to this approach. Finally, in section 5 we validate the performance improvement on three software projects of different size and conduct two experiments to examine the practicability of our extensions.

2   Formal Concept Analysis

Formal Concept Analysis (FCA) [2][4] is a branch of lattice theory that allows us to identify meaningful groupings of *objects* G, i.e. quantities in a data set, that have common *attributes* M. In all the extent of this work, we are going to explain FCA on a very simple yet illustrative example in which we pick a set of birds as G and a set of bird characteristics as M. The triple (G, M, I) is called a *formal context*, where $I : G \times M \longrightarrow \{0, 1\}$ is an incidence function which returns 1 for a pair $(g, m) \in G \times M$ if g has attribute m, 0 otherwise. This formal context can be organized in an incidence matrix $\mathcal{M}$, as depicted in Table 1[1]. Here, $\mathcal{M}_{(i,j)}$ has an entry if object $I(i, j) = 1$.

Table 1. FCA bird example context

	can fly	can swim	sings	migratory	monogamous
Ara	×				×
Bluejay	×		×		×
Kiwi					×
Mallard	×	×			
Pelican	×	×		×	

Using this context, FCA groups the objects and their attributes into *formal concepts*, listed in Table 2. Such a formal concept consists of two sets, an *extent* and an *intent*. The intent contains all *common* attributes that apply to the objects in the extent. In the same way all objects contained in the *extent* share all properties contained in the *intent*. Therefore a concept is a maximal collection of elements sharing common properties. Adding an attribute to a concept's intent there would be at least one object in the extent that does not have this attribute. Adding an object to the extent there would be at least one attribute in the intent this object does not have. As a consequence the formal concepts build a *complete partial order* that can be written as a lattice. Table 2 in some way suggests this order by the increasing number objects and the decreasing number attributes from top to bottom.[2]

---

[1] It needs to be noted that ornithology usually is not part of our research. The data shown in Table 1 may not be entirely correct.

[2] In the context of this particular work, we are not making any use of this partial order yet.

Table 2. Formal concepts for the context in Table 1

i	Extent E_{c_i} ($\downarrow$)	Intent I_{c_i} ($\uparrow$)
c_1	{Pelican}	{can fly, can swim, migratory}
c_2	{Bluejay}	{can fly, sings, monogamous}
c_3	{Ara, Bluejay}	{can fly, monogamous}
c_4	{Mallard, Pelican}	{can fly, can swim}
c_5	{Kiwi, Ara, Bluejay}	{monogamous}
c_6	{Ara, Bluejay, Mallard, Pelican}	{can fly}

3   FCA Application

3.1   Setup of the Formal Context

We apply FCA on an object oriented software system by considering structures between classes and interfaces as the set G of FCA objects and class relationships that constitute to a structure as the set M of FCA attributes. A first logical step hence will be to define a structure model, i.e. the set of types of relationships we are looking for in a system. For this, we adapt relationships from LePUS3 [5], a modeling language for design patterns, and classify them into three structural and four behavioral relationship types, as listed in Table 3. All relationships are orthogonal to each other. For instance, the *calls* relationship between two classes A and B only applies if there is no forwards relationship between A and B.

Table 3. Our set of relationships used to describe structures in source code

Relationship r	A class A is related to a class/interface B by r if
	structural
has	A has a one-to-one object association to B.
aggregates	A has a one-to-many object association to B.
specializes	A extends or implements B.
	behavioral
calls	a method in A calls a method in B.
forwards	a method in A calls a method in B that shares the same signature.
creates	a method in A calls the constructor of a class and binds the new instance to a variable of type B.
produces	a non-private method in A creates a new instance of B and returns it.

Using this model to describe structures, we then adopt the FCA approach introduced by [3] and later improved by [6]. Here, having as input the set P of all classes in a software and a fixed order $n \in \mathbb{N}$ we gather those class substructures that contain n classes. Such a substructure can be described by two components, namely its n-tuple of classes in P^n (our FCA object) and a set of relationships between these classes (our FCA attribute). We illustrate this approach by the example structure in Figure 1 with $P = \{A, B, C, D, E\}$ and $n = 3$. The corresponding formal context to Figure 1 is shown in Table 4 while Table 5 lists the resulting formal concepts.

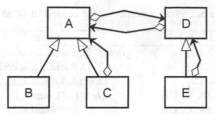

Fig. 1. A simple class structure that could be an input to our FCA approach. It contains five classes related to each other by two of our relationship types, *has* (simple association, diamond at source) and *specialization* (inheritance, triangle at target).

Table 4. Formal context for the structure in Figure 1 and $n = 3$. Four connected substructures can be found. The attributes represent relationships and refer to the classes in the tuples by their indexes. For instance, the attribute *spec*(2, 1) can be read as "the element at index 2 specializes the element at index 1".

	spec(2,1)	spec(3,1)	has(3,1)	has(2,1)	spec(3,2)	has(3,2)	has(1,3)	has(1,2)
(A, B, C)	×	×	×					
(A, B, D)	×		×				×	
(A, C, D)	×		×	×			×	
(A, D, E)					×	×	×	×

3.2   Iterative Concept Analysis

The problem of generating all concepts given a set its complexity is $\#P$-complete. More specifically, for a given context (G, M, I) the algorithm we used for our approach [7][8] has time complexity $O(|G|^2 \cdot |M| \cdot |C|)$, where C is the set of all concepts for the context. The space complexity is $O(|G| \cdot |M| \cdot |C|)$.

In order to reduce the set of concepts computed by FCA as well as its expected runtime, we apply FCA in two iterations. In the first iteration, only the three structural relationships in Table 3 are considered, creating concepts revealing *the definition of a system part*. In case we want to further examine *how such a definition works* with regard to its calling behavior, we apply FCA on it in a second iteration. This time we consider both, structural and behavioral relationships as attributes, however, vastly reduce the set of FCA objects as only the current concept's extent is used as input.[3]

3.3   Postprocessing

Removing Disconnected Structures. Due to the way we construct the formal context basically two post-processing steps have to be taken. The first one is to remove concepts whose intent describe a disconnected graph. In Table 5 concept c_5 represents such a graph, as the nodes with the indexes 1 and 2 are connected with each other, the node with index 3 is not connected to any of the other two.

---

[3] The idea was already proposed by [3], with the difference that they used as attribute augmentation the methods calling each other and their names.

Table 5. Formal concepts for the context in Table 4

i	Extent E_{c_i} ($\downarrow$)	Intent I_{c_i} ($\uparrow$)
c_1	$\{(A, D, E)\}$	$\{has(1, 2), has(3, 2), spec(3, 2), has(2, 1)\}$
c_2	$\{(A, C, D)\}$	$\{has(3, 1), has(2, 1), spec(2, 1), has(1, 3)\}$
c_3	$\{(A, B, C)\}$	$\{spec(3, 1), has(3, 1), spec(2, 1)\}$
c_4	$\{(A, B, D), (A, C, D)\}$	$\{has(1, 3), has(3, 1), spec(2, 1)\}$
c_5	$\{(A, D, E), (A, C, D)\}$	$\{has(1, 2)\}$
c_6	$\{(A, B, D), (A, C, D), (A, B, C)\}$	$\{spec(2, 1), has(3, 1)\}$

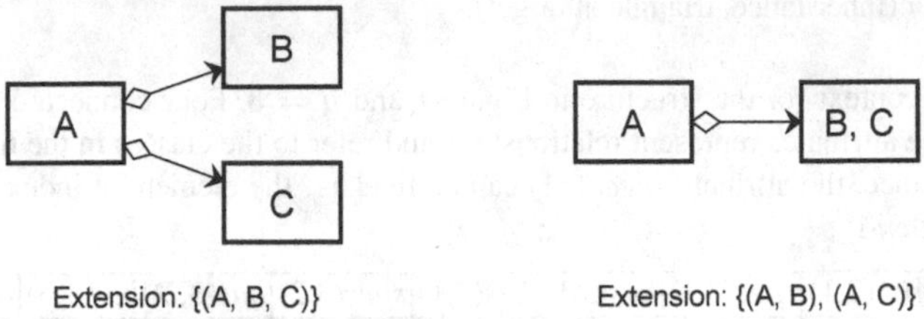

Extension: {(A, B, C)}   Extension: {(A, B), (A, C)}

Fig. 2. A "star pattern" containing redundant information (left) and its reduction (right)

Merging Equivalent Structures. For a context like ours FCA may produce concepts that are basically equivalent and can be merged. That is the case if there exists a permutation of the indexes of two concepts such that the intent of one concept can be mapped into the other. For instance, in Table 5 concept c_1 is equivalent to c_2 by the mapping $1 \rightarrow 3, 2 \rightarrow 1$ and $3 \rightarrow 2$. To find such a mapping is a graph matching problem and hence not trivial. We used the VF2 algorithm proposed by [9] to accomplish this task, however, since the graphs we are trying to match are relatively small a naive depth-first search would work just as well.

Removing Redundant Information. As an additional step we remove those concepts that contain redundant information. This is the case if the structure the concept describes contains a symmetric subgraph. A prominent example is the "star pattern" as depicted in Figure 2 (a). Here we can reduce the pattern to the one in Figure 2 (b) without losing any information. Since the reduced concepts represent structures of a lower order, we can ignore them.

4 Filtering Features

4.1 Filter Elements

In order to dynamically change the space FCA is applied on we make use of *Filter Elements*. When gathering all substructures of given order in the setup of the formal context we proceed inductively, i.e. first compute all structures of order $n = 2$, then augment them to structures of order $n = 3$, etc. Before starting an analysis run we can declare classes as filter elements and aggregate them in a list. It is then guaranteed that in the first inductive step each structure of order $n = 2$ consists of at least one element from the filter element list. As a consequence, the structures serving as FCA objects then are the union of all structures that evolve around the classes in our list of filter elements.

4.2 Class Prominence

Taking a look at the extent of a formal concept we gain interesting information on single classes. One of such is the *prominence* of a class. For the extent $E_c \subset P^n$ of a formal concept c we can consider an index $r \in \{1, \ldots, n\}$ of the n-tuples in E_c as a *role* of concept c. A class $p \in P$ in any of the n-tuples in G_c at the index r then can be seen as a role player of role r. The set of all role players for a role r in a concept c we further refer to as $P_{c,r}$. In Table 5, concept c_6, we have three roles[4] according to their indexes in the 3-tuples, played by the following classes: $P_{c_6,1} = \{A\}$, $P_{c_6,2} = \{B, C\}$, $P_{c_6,3} = \{D, C\}$.

Given a class $p \in P$ and a role r in a concept c, we define by (3) its prominence $u(p, r, c)$ in r simply as the scaled frequency $\eta(p, r, c)$ (1) of the class playing this role multiplied by the size of E_c.

$$\eta(p, r, c) := \frac{|\{e \in E_c \,|\, e_r = p\}|}{|E_c|} \tag{1}$$

$$\varphi(x) := x/(1 - x) \tag{2}$$

$$u(p, r, c) := \varphi(\eta(p, r, c)) \cdot |E_c| \tag{3}$$

$$u(p) := \sum_{c,r} u(p, r, c) \Big/ \sum_{c,r,p} u(p, r, c) \tag{4}$$

In a final step we compute the absolute prominence $u(p)$ (4) of a class p simply by summing up the prominence values for a class over all roles over all concepts and normalize the outcome over the absolute prominence values of all classes.

It is worth noticing that $u(p)$ as defined in (3) has two factors. Previously proposed measures for interestingness are often composed of two values. One value measure the relevance of a finding the other the unexpectedness [10]. In our case the first factor emphasizes classes that are prevalent players of a role (of which only a few exist and therefore are unexpected). The second factor simply measures the size of the extent of the concept and thus its relevance.

Throughout the concepts listed in Table 5, apparently the class A is more prominent than any of the other four classes. Pretending that we replace class A with another class X in concepts c_1-c_3, class A would still gain the highest prominence value as it is the only player of role $r = 1$ in concepts c_4-c_6, which all have a larger extent.

The prominence of a class can serve us in two ways: First, the more prominent a class the higher the probability that it plays a role in a core concept of a software project.

---

[4] The number roles is determined by the order n.

One can see the prominence as a gravitation of a class. The higher, the greater is the part of a software project that is 'attached' to this class. Second, it could help us determining appropriate filter elements. The lower the prominence of a class, the smaller the expected number of computed concepts.

5 Case Studies

5.1 Our Tool

We implemented our approach as part of the *Cultivate*[5] plugin for the *Eclipse*[6] IDE. Cultivate is a code analysis tool for Java programs. It in turn bases on the *JTransformer*[7] plugin which provides a Prolog fact base that represents the full abstract syntax tree of the to-analyze Java program. Cultivate implements several program analyses written in Prolog that are applied on this fact base.

To compute the formal concepts we use a relatively young algorithm proposed by [7]. This algorithm is particularly suitable to our approach compared to the algorithm [2] used by previous works because it saves time not carrying the hierarchical order of the concepts required to build the concept lattice. Due to our post-processing reorganizing the computed concepts this additional information is useless to us anyway. Secondly, this algorithm allows us to parallel the computation, distributing the computational load over several CPUs.

5.2 Data Set

We apply our tool on three different Java projects:

- *JUnit*[8] 4.7, which is a testing framework for Java code of smaller size.
- *Cultivate*, the framework our tool is based on. It consists of a platform providing utility classes and the engine on one hand and several addons on the other hand that build upon this platform but not upon each other. As a consequence the overall cohesion in this project is very low. Also we are familiar with its domain, which makes it easier to evaluate our findings.
- *JHotDraw*[9] 7, a free Java-based framework for creating graphical editors. In contrast to Cultivate we are not familiar with this project, yet enjoy it to be well documented. Also the overall cohesion of the project compared to Cultivate is fairly high.

For the sake of simplicity we consider only the set of core classes (ignoring external library classes) that fulfill the following requirements: The class neither is of a basic type (*Integer*, *Double*, ...), an enumeration type nor an anonymous class.

---

[5] http://sewiki.iai.uni-bonn.de/research/cultivate/start
[6] http://www.eclipse.org/
[7] http://sewiki.iai.uni-bonn.de/research/jtransformer/start
[8] http://www.junit.org
[9] http://sourceforge.net/projects/jhotdraw

Table 6. Observed runtime behavior of our tool in seconds and number concepts computed in the first iteration

	JUnit 4.7		Cultivate		JHotDraw 7	
#classes	143		607		625	
	runtime	concepts	runtime	concepts	runtime	concepts
$n = 2$	<1s	5	~2s	7	~3s	9
$n = 3$	<1s	16	~3s	45	~14s	38
$n = 4$	~1s	24	187s	154	130s	141

5.3 Performance

We ran our analyses on an Intel Quad Core @2.83 GHz with 4 GB RAM under normal load. Table 6 shows the runtime behavior we observed and number concepts computed for each of the three sample projects and with regard to the order n.

For $n \geq 5$ on Cultivate our tool failed due to lack of memory. This may or may not be caused by Prolog and the fact it loads the entire fact base into the main memory as well as it caches its query results. The reason why the analysis on JHotDraw is faster than on Cultivate for $n > 3$ we ascribe to Cultivate being way less cohesive than JHotDraw, which eventually leads to a faster growth of concepts in n. Yet the actual number concepts in both projects are insignificantly different what suggests a fairly large impact of our additional post-processing.

Recalling previous observations made by [6], using the Ganter algorithm [2] the analysis of a sample project written in Smalltalk with 167 classes and $n = 4$ took approximately two days.[10] Compared to this, our own results by far excel our expectations and prove this technique to be a time-efficient way to analyze software projects even of larger scale.

6 Example Applications

6.1 Experiment 1: JUnit

In a first experiment we pick the smaller of our sample projects, JUnit, and let one of the authors apply our tool on it with the goal to yield most relevant structures of the project in at most five analysis steps. The project is well-documented and makes extended use of design patterns, however, the experimenter neither is familiar with the project nor with its documentation at the time of execution.

As a first step, the experimenter runs an analysis on structures of order $n = 2$. Despite our expectations we found $n = 2$ particularly instructive, as its corresponding concepts are small in number, easy to understand and most often already reveal those atomic relationships between two classes larger patterns are only based on. The analysis computes five concepts of which the one depicted in Figure 3 (a) catches the experimenters attention as it is one of two with more than one relationship. This pattern

---

[10] A less advanced hardware may have an impact on these stats, too, considering the previous observations date back eight years.

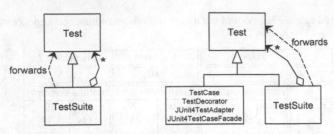

Fig. 3. Composite pattern candidate (left) and verified instance (right) in JUnit 4.7

suggests the implementation of a tree structure using the *Composite* pattern [1]. In order to check this presumption the experimenter runs an analysis on order $n = 3$ using the class TestSuite as a filter element. Since this class has a significantly smaller prominence than Test, it is more suitable as a filter. Two concepts are computed, one of them actually representing the Composite pattern as depicted in Figure 3 (b).

As a specialization of Test we find a class called TestDecorator in Figure 3 (b) which suggests the implementation of the *Decorator* pattern [1]. Following the same procedure as before (using TestDecorator as filter element) we can verify our presumption.

Rechecking our findings so far with the JUnit documentation we can verify the composite pattern instance as one most relevant to the base framework, while the decorator pattern instance is particularly important to the extensions part of JUnit which can be used by developers to implement and plug-in custom test definitions.

6.2   Experiment 2: Cultivate

In a second experiment we wanted to examine the precision of the prominence calculation and how it can be exploited to find core concepts in a software project. For this, we use our own project, Cultivate. Firstly, because we are familiar with it and can assess the validity of the computed prominence values. Secondly, Cultivate basically is a platform with a few core classes that, however, are extensively used by the addons that build on the platform.

We run an analysis on structures of order $n = 3$ and retrieve the list of all occurring classes ranked by their prominence. The one with the highest prominence value is CultivateViewPart ($\sim 12\%$), the next-prominent class is BaseQuery ($\sim 3\%$). In both cases we agree with the tool: CultivateViewPart is a class used as an abstract view part that follows selections in the workbench and manages the subscription of analyses on the corresponding software projects. It is basically inherited from all add-on projects that provide a workbench view and in fact is one of the central classes in Cultivate. BaseQuery is *the* abstract class to query analyses on Prolog side, specialized by 63 different classes inside the add-on parts.

We declare CultivateViewPart as filter element and run the analysis again. The result is a set of 14 concepts of which two describe exactly the main responsibilities of this class, depicted in Figure 4.

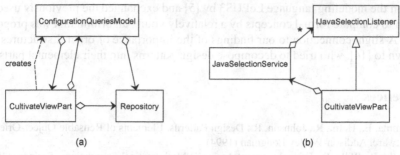

(a)   (b)

Fig. 4. Two core concepts of the Cultivate project. In Figure (a) the `CultivateViewPart` creates and attaches to a `ConfigurationQueriesModel` object which then retrieves a `Repository` for the currently selected project and handles query subscriptions on that repository. (b) describes an *Observer* pattern [1], in which `CultivateViewPart` is an observer, `JavaSelectionService` handles workbench selections.

7 Discussion

Judging from the observed runtime performance, we can see that FCA is a practical technique for mining structures in software projects even of larger size. We find a promising approach to assess the relevance and importance of certain classes of the software project which may lead us to those parts of the software that constitute its most important functionality. Yet we conclude that this assessment still requires some fine-tuning. Having such an assessment we can exploit it either as a clue to search and identify core concepts of the corresponding project or as an assistance in choosing appropriate filter elements to narrow down the space the analysis is applied on.

In this work, we did not yet make use of the actual strength of FCA: That is, to create a *reverse partial order* between intent and extent of all concepts[11]. This feature may allow us to introduce two further measurements: Abstractness, resp. specificity of a software structure. Apart from that, it seems reasonable to apply our idea of class and structure measurement using clustering techniques with less high complexity [11][12][13] and may be addressed in future work.

8 Related Work

Formal Concept Analysis (FCA) was firstly proposed by [2] as a branch of lattice theory. The first effort towards structure mining in source code using FCA was achieved by [3]. Their approach then was later refined by [6] who reduced the number of FCA objects and hence improved the overall runtime. Further structure mining efforts for object-oriented systems have been achieved by [11][12][13] who used subgraph matching to group same structures formed by classes. We adopted the approach by [3] refined by [6], enhanced the set of FCA attributes, i.e. class relationships, using a set of relationships

---

[11] As you can see in Table 2, the number attributes of all concepts stand in an inverse relation to the number objects.

based on the modeling language LePUS3 by [5] and exchanged the previously used algorithm to compute formal concepts by a relatively young algorithm that was proposed by [7]. A slight connection to our findings of the importance of order 2 structures can be drawn to [14], who tried to decompose design patterns into their elemental parts.

References

1. Gamma, E., Helm, R., Johnson, R.: Design Patterns. Elements of Reusable Object-Oriented Software. Addison-Wesley Longman (1994)
2. Ganter, B., Wille, R.: Formal Concept Analysis: Mathematical Foundations. Springer (1998)
3. Tonella, P., Antoniol, G.: Object Oriented Design Pattern Inference. In: Proceedings of ICSM, p. 230. IEEE Computer Society Press (1999)
4. Carpineto, C., Romano, G.: Concept Data Analysis, Theory and Applications. Wiley & Sons (2004)
5. Eden, A.H., Hirshfeld, Y., Yehudai, A.: LePUS - A Declarative Pattern Specification Language. Technical report (1998)
6. Arévalo, G.: High Level Views in Object-Oriented Systems using Formal Concept Analysis. PhD thesis, University of Bern (2004)
7. Krajca, P., Outrata, J., Vychodil, V.: Parallel Recursive Algorithm for FCA. In: Concept Lattices and Their Applications (CLA), pp. 71–82 (2008)
8. Kuznetsov, S.O.: Learning of Simple Conceptual Graphs from Positive and Negative Examples. In: Żytkow, J.M., Rauch, J. (eds.) PKDD 1999. LNCS (LNAI), vol. 1704, pp. 384–391. Springer, Heidelberg (1999)
9. Cordella, L.P., Foggia, P., Sansone, C., Vento, M.: An improved algorithm for matching large graphs. In: 3rd IAPR-TC15 Workshop on Graph-based Representations in Pattern Recognition, Cuen, pp. 149–159 (2001)
10. Exman, I., Amar, G., Shaltiel, R.: The Interestingness Tool for Search in the Web. In: Proceedings of the Third International Workshop on Software Knowledge (SKY 2012), Barcelona, Spain (2012)
11. Zhang, Z. X., Li, Q. H., Ben, K.-R.: A New Method for Design Pattern Mining. In: International Conference on Machine Learning and Cybernetics (2004)
12. Gupta, M., Pande, A.: Design patterns mining using subgraph isomorphism: Relational view. International Journal of Software Engineering and Its Applications (IJSEIA) (2011)
13. Belderrar, A., Kpodjedo, S., Guéhéneuc, Y.G., Antoniol, G., Galinier, P.: Sub-graph Mining: Identifying Micro-architextures in Evolving Object-oriented Software. In: Europ. Conference on Software Maintenance and Reengineering (2011)
14. Smith, J.M., Stotts, D.: Elemental Design Patterns: A Formal Semantics for Composition of OO Software Architecture. In: IEEE/NASA Software Engineering Workshop, pp. 183–190 (2002)

ROM: An Approach to Self-consistency Verification of a Runnable Ontology Model

Iaakov Exman and Reuven Yagel

Software Engineering Department, The Jerusalem College of Engineering,
POB 3566, Jerusalem, 91035, Israel
{iaakov,robi}@jce.ac.il

Abstract. In the quest for the highest possible abstraction of software systems, Runnable Knowledge has been proposed for MDA. But in order to be useful in practice as a system design, it must be amenable to verification. This work precisely describes the necessary steps by which ROM – a Runnable Ontology Model tool – concurrently generates a running model and its respective test script from the designed Runnable Knowledge, allowing verification that the model is self-consistent. The novel implementation idea is to use ready-made mock object libraries to efficiently obtain the code for a running model. Detailed examples are provided to illustrate each of the ROM generation steps.

Keywords: Runnable Knowledge, Ontology, Ontology States, Model Testing, Mock Objects, Self-consistency, Verification.

1 Introduction

It is common wisdom that software development costs are reduced when one discovers errors as early as possible along the development. Within an MDA – Model Driven Architecture – approach, early means in a higher level of abstraction. Within software engineering lifecycles, early means already at the requirement phase.

Elsewhere Exman et al. [1] have proposed Runnable Knowledge as the highest possible software abstraction level, viz. one starts from the bare system concepts – an ontology – and the respective ontology states.

In order to have a useful system at the Runnable Ontology abstraction level, this model must be self-consistent. This means to have the necessary and sufficient set of concepts. The choice of concepts for the system ontology is analogous to the choice of software classes at the (e.g., UML) design level.

This paper proposes an agile approach to Model Testing for self-consistency by running the model. A special purpose tool generates from the ontology and its states the classes of the system under development (SUD). The tool also generates in parallel the tests to be applied to the SUD.

ROM – a Runnable Ontology Model tool – is described. It is expected to concurrently generate a running model and its tests, based on higher level knowledge extracted from specifications and input ontologies. The tool feasibility is demonstrated

A. Fred et al. (Eds.): IC3K 2012, CCIS 415, pp. 271–283, 2013.

by substituting its modules by the concatenation of existing tools, each performing the role of one of its modules.

Next we compare executable approaches to specification and design testing.

1.1   Executable Approaches to Specification and Design Testing

One could roughly divide executable approaches to specification and design testing for software systems into two categories.

In the first approach one gradually builds an executable specification which can exercise the developing system. The final product of this approach is a full system specification that while it has been built it has also validated that the running system behaves as expected. A typical example is using tools like Cucumber, see e.g. the book by Wynne and Hellesoy [2].

In the second approach one essentially stays at the software model level, and the final product of this approach is an executable model, which must later on be converted into the actual software system. A typical example is Executable UML in the book by Mellor and Balcer [3].

Our novel proposed approach is at a higher abstraction level than any of the previous approaches. It is conceptually above the usual UML model, while being runnable – thus verifiable – without getting down to production code.

1.2   Related Work

Here we present a concise review of the relevant literature.

In recent years the Agile software community emphasizes various early testing methods, e.g. Freeman and Pryce [4]. One of its main purposes is to obtain short feedback loops in order to properly guide the development of software systems in a world of fast changing environments and technologies.

Those testing methods can be considered an evolution of the unit-testing practices of Test Driven Development (TDD) by Beck [5]. These methods develop scripts demonstrating the various system behaviors, instead of code specification of just the interface and required behavior of certain modules. These are also known as automated functional testing, as the referred scripts' execution can be automated.

One such method is known as Acceptance Test Driven Development (ATDD) or alternatively just as Agile Acceptance Testing, e.g. Adzic [6]. Another group of methods extending TDD is Behavior Driven Development (BDD) e.g. North [7]. It stresses stakeholder readability and shared understanding. Recent representatives of the above approaches are e.g. Story Testing, Specification with examples Adzic [8] or just Living/Executable Documentation, e.g. Brown [9].

Some related tools for implementing these practices are FitNesse and Cucumber. Fitnesse (see Adzic [10] and its predecessor FIT) are wiki-based web tools allowing non-developers to write acceptance tests, in a formatted manner, e.g. tabular example/test data.

Cucumber, see e.g. the book by Wynne and Hellesoy [2] and related tools, e.g. SpecFlow [11], directly support BDD. Their main feature is the ability to run stories written in plain natural language (originally English but by now a few dozen others). This tool is also highly connected to unit testing tools and user/web automation tools. In a previous work by Yagel [12] we have discussed more extensively these practices and tools.

A representative reference for Executable UML is the book by Mellor and Balcer, [3]. An introductory overview of ontologies in the software context is found in Calero et al. [13].

A classification of "Test Doubles" (differing mock objects) is given at [14].

In the remaining of the paper we introduce the Runnable Knowledge abstraction level (section 2), describe agile design and testing for Runnable Knowledge (section 3), propose an approach for code generation from Runnable Knowledge (section 4), describe a case study (section 5) and conclude with a discussion (section 6).

2   The Runnable Knowledge Abstraction Level

Runnable Knowledge (Exman et al. [1]) is an abstraction level above standard UML models. The latter usually separates modeling of structure and behavior, typically into class diagrams and statecharts.

UML classes usually contain detailed attributes, functions with their arguments. UML statecharts may contain detailed conditions and parameters.

In contrast, Runnable Knowledge is based upon concepts alone, their relationships and possible states, abstracting away detailed attributes, functions, conditions and parameters.

In the next sub-sections we describe concepts and relationships within an ontology and their respective possible states.

2.1   Ontology Concepts and States

To gradually illustrate our ideas, step by step, we have recourse to a widely used example – cash withdrawal from an Automatic Teller Machine (ATM).

Two ontologies, ATM and bank account, are used in the ATM example in Fig. 1.

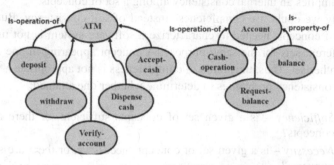

Fig. 1. ATM and bank Account Ontologies – The ATM ontology concepts (left) are machine operations. The Account ontology concepts (right) are one property (*balance*) and operations.

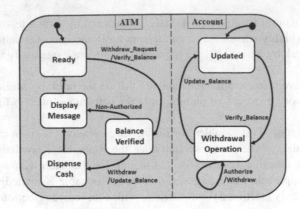

Fig. 2. ATM and bank Account Ontologies' States –These are the parallel states for the ATM and the account. This diagram only displays the states for a cash withdrawal operation. Other operations, such as a deposit or retrieval of account balance have similar diagrams.

In Fig. 2 one can see the corresponding ontologies' states.

All the ontologies in this paper – e.g. in Fig. 1 – are nano-ontologies [15], a term referring to their relatively very small size. Their main purpose is a compact defini- tion of the concepts relevant to a given software system. Their compactness is also important in terms of computation efficiency.

Domain ontologies found in the technical literature are much larger. Their purpose is usually stated as to define a domain vocabulary. Nano-ontologies may be derived from the larger domain ontologies, but this issue is out of the scope of this work.

2.2   The Self-consistency Problem

A Runnable Ontology Model – a system in the highest software abstraction level – consists of an ontology and the respective ontology states.

A Runnable Ontology model is said to be self-consistent if it contains a necessary and sufficient set of concepts to characterize the desired software system. Self- consistency implies an internal consistency among a set of concepts.

Apparently we could use completeness instead of self-consistency. But we claim that the set of concepts needed to characterize a software system is not unique. One may use different sets for the same purpose. A concept appearing in one set may be absent in another set. Thus, the notion of completeness is not appropriate.

The self-consistency problem is to determine whether one attained:

- *Sufficiency* – is a given set of concepts sufficient, or there are lacking concepts?
- *Necessity* – is a given set of concepts necessary, or there are superfluous concepts?

The self-consistency problem is a higher-level abstraction of the problem of choice of UML classes.

In this work we propose a runnable iterative solution to the problem.

3   Testing Runnable Knowledge

In this section we handle the same ATM example from an agile testing angle. In Fig. 3 one sees two specific scenario examples for authorized and non-authorized cash withdrawal from an ATM system. The motivation for using two scenarios is to enable testing of the two different paths in the l.h.s of the Ontologies' States in Fig. 2. These two paths differ by the branching out from the "BalanceVerified" sub-state of ATM. One branch is the authorized Withdraw transition and the other is the Non-Authorized transition.

The idea is that a potentially runnable ontology and its respective states turns into an actually runnable model by means of a specific scenario. In other words, a path is chosen among the possible states and conditions are provided for transitions between states.

```
Feature:  Account Withdrawal

Scenario: Successful withdrawal from an account
    Given an account has a balance of $100
    When $20 are withdrawn from an ATM
    Then the account balance should be $80

Scenario: Unsuccessful withdrawal from an account
    Given an account has a balance of $100
    When $120 are withdrawn from an ATM
    Then A non-authorized message is displayed
    And the account balance should be $100
```

Fig. 3. Initial Requirements for ATM – These are specific scenarios for successful/unsuccessful cash withdrawal from an ATM. They are expressed in the Gherkin style used by the Cucumber tool.

The example in Fig. 3 uses a simple syntax called Gherkin (see Chelimsky et al. in ref. [16]). Its main feature is the ability to describe specification examples in close to plain natural language but in a way that will be easy to transform it into a runnable test script.

The keywords shown here in blue are: a) Feature: for giving a title to this specification part; b) Scenario: a title for a specific walk through; c) Given: the pre-conditions before some action is taken; d) When: an action that triggers the scenario; e) Then: the expected outcome; f) Optional: And, But: additional steps happening in one of the previous stages.

Some other keywords are available which are not covered here.

3.1   Agile Design and Testing

The requirements written in a specification file – such as in Fig. 3 – are usually developed together with the system's stakeholders or business experts.

At first, running this specification will fail since there is no code supporting it. This will lead the developers to design a domain model to satisfy this specification.

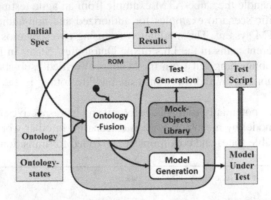

Fig. 4. ROM Software Architecture – modules are represented as round (green and white) rectangles, while input and output are plain (yellow) rectangles. The wide arrow pointing from the Model Under Test (MUT) to the Test Script, means that the latter is used to test the MUT.

A tool like cucumber can suggest needed steps for satisfying the suggested scenarios. Mock objects could stand for the missing concepts and allow to re-run the feature.

Usually an initial version of a specification is developed and tested. This specification is iterated and refined until a necessary and sufficient specification is achieved. This specification is realized by detailed runnable test scripts which demonstrate the self-consistency of the system. They can even catch software regressions while adding new features to a system.

4   Generation of Running Model from Runnable Knowledge

In this section we carefully describe ROM, the tool for generation of a testable running model from the Runnable Knowledge level of abstraction.

4.1   Rom Software Architecture

Software system development begins with the elicitation of system requirements into the initial specification.

The initial specification, still needing verification, together with the ontology and its states are the input to the ontology fusion module of ROM (in Fig. 4). The specifications are elaborated to validate the correct interactions between the various model states and transitions.

The fusion module uses the concepts in the ontology to generate a model with interfaces. It uses the ontology states to generate the test script.

From the latter one generates, using a mock object library, the mock objects which constitute the runnable Model under test (MUT). This is tested by unit tests generated in parallel.

If the tests results are negative, one modifies the specifications and/or the ontology and repeats the loop. Otherwise the system model is approved.

The Runnable Knowledge model – consisting of the ontologies and their states – is the utmost abstract level in the software layers hierarchy. It is potentially runnable in the sense that, by use of a suitable tool, one can make transitions between states.

The essential role of mock object libraries, proposed in this work, is to obtain a fast and efficient translation of Runnable Knowledge into an actually running model. This is illustrated in the following case study.

5   Case Study: Internet Purchase

Here we describe in detail a case-study – an internet purchase case – having a clearly different nature from the above ATM cash withdrawal example, in order to show the range of applicability of our approach.

At the first stage we identified that the Internet Purchase case study requires modeling of two classes: the shopping cart and products that can be put in the cart for later purchase. Testing of these classes is schematically represented by a transaction in which two products are purchased.

We also demonstrate the iterative nature of the development process of working with ROM until a satisfactory model is developed. In this case study three iterations are shown:

I.   Iteration I – the initial ontology, ontology states and example scenarios as inputs; the intermediate Mock expectation setups; the Extracted Runnable model and Runnable test script as outputs; one learns that a TotalQuantity concept is lacking;

II.   Iteration II - the lacking concept is added to the ontology and to the Runnable model and test script;

III.   Iteration III – additional correct values are inserted in the Mock setups and the corrected test script is run.

Iteration I

Internet Purchase Runnable Knowledge

The internet purchase Runnable Knowledge consists of the ontology (seen in Fig. 5) and its states (seen in Fig. 6) of two classes: 1) Shopping cart – that may contain products; 2) Product – with properties, say its Price.

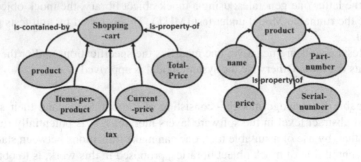

Fig. 5. Shopping cart and Product Ontologies – The Shopping cart ontology (left) shows objects contained by the cart (product and items-per-product) and purchase properties (total-price, current-price and tax). The Product ontology (right) concepts are just its properties.

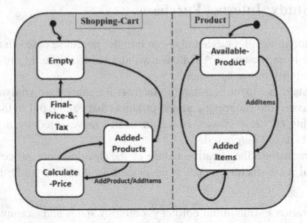

Fig. 6. Shopping-cart Ontology States – it shows two parallel states: *shopping-cart* and *product*. The product is either available (default state) or added to the cart. The cart default is empty. A product can be added, its price or final price-&-tax calculated, ending the transaction.

Internet Purchase Model Testing

Example scenarios in a feature file are seen in Figure 7.

```
Feature: Adding to a shopping cart

Scenario Outline: Add items to shopping cart
        Given An empty shopping cart
        When I add <x> items of Product A at ($10)
        And I add <y> items of Product B at ($20)
        And the tax is 8%
        Then the shopping cart contains <z> items
        And the total price is <total>$

Examples:
        | x | y | z | total|
        | 0 | 0 | 0 | 0.0  |
        | 1 | 0 | 1 | 10.8 |
        | 0 | 2 | 2 | 43.2 |
        | 1 | 2 | 3 | 54.0 |
```

Fig. 7. Shopping-cart: Example Scenarios, using a table notation – Adding items to a shopping cart

Internet Purchase Model and Running Implementation

An example of ROM's output uses C# code with the following tools/libraries:

- SpecFlow (a .NET Cucumber variant) for running the feature [11];
- NUnit (a .Net unit testing library) used here for assertions [17] (with the test runner of Resharper [18]);
- NSubstitute (a .Net mock library) for setting up domain mocked objects [19].

Running ROM on the above specification and ontology will output:

- model with interfaces containing properties and method signatures (Fig. 8),
- mock expectation setups (Fig. 9),
- a runnable test script (Fig. 10).

```
interface IProduct
{
    string Name { get; set; }
    double Price { get; set; }
    int PartNumber { get; set; }
    string SerialNumber { get; set; }
}

public interface IShoppingCart
{
    double TotalPrice { get; set; }
    double CurrentPrice { get; set; }
    double TaxRate { get; set; }

    int ItemsPerProduct { get; set; }
    IProduct Product { get; set; }

    void AddProducts(IProduct product, int quantity);
}
```

Fig. 8. Shopping-cart: Extracted model – two interfaces correspond to the ontology concepts: IShoppingCart and IProduct. For each property, are generated getter and setter methods.

```
[Given(@"An empty shopping cart")]
public void GivenAnEmptyShoppingCart()
{
    shoppingCart = Substitute.For<IShoppingCart>();
}

[When(@"I add (.*) items of Product A at \(\$(.*)\)")]
public void WhenIAddItemsOfProductAAt(int quantity, int price)
{
    product = Substitute.For<IProduct>();
    product.Name = "A"; product.Price = price;
    shoppingCart.AddProducts(product, quantity);
}

[When(@"I add (.*) items of Product B at \(\$(.*)\)")]
public void WhenIAddItemsOfProductBAt(int quantity, int price)
{
    product = Substitute.For<IProduct>();
    product.Name = "B"; product.Price = price;
    shoppingCart.AddProducts(product, quantity);
}
```

Fig. 9. Shopping-cart: Mock expectation setups – for each of the mock objects expectations are setup: the shopping cart's quantity and total price, the product A price and the product B price

Figure 9 shows the expectation setup script for the properties of the mock objects, viz. the shopping cart and the products A and B.

Note that the test script could be further refactored by using regular expression and joining methods, e.g. adding items with different prices (we left that out for simplicity).

Figure 10 displays the test script generated for the Shopping cart case study. ROM extracts a concept called "ContainsItems" from the test script. This concept is missing from the ontology of Figure 5 and subsequenly was not generated for the running model of Figure8. Thus it is currently marked with red italic font in Figure 10.

```
[Then(@"the shopping cart contains (.*) items")]
public void ThenTheShoppingCartContainsItems(int quantity)
{
    Assert.That(shoppingCart.ContainsItems, Is.EqualTo(quantity));
}

[Then(@"the total price is (.*)\$")]
public void ThenTheTotalPriceIs(double price)
{
    Assert.That(shoppingCart.TotalPrice, Is.EqualTo(price)
      .Within(0.0001));
}
```

Fig. 10. Shopping-cart: Runnable test script – the script reveals that there is a missing concept. It his hinted by the *"ContainsItems"* marked in red italic font.

Iteration II

The first iteration revealed that the ontology model (Fig. 5) and the generated running model (Fig. 8) miss some concept for *"ContainsItems"* (as marked in red italic font in Fig. 10) in the shopping cart. The lacking concept should be added.

In the second iteration the ontology model was enlarged with a new concept of "Total Quantity" (Figure 11).Thus this property is added to the model and finally the script is runnable.

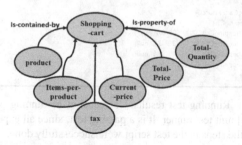

Fig. 11. Shopping Cart Ontology – enlarged with new concept "TotalQuantity"

```
[Then(@"the shopping cart contains (.*) items")]
public void ThenTheShoppingCartContainsItems(int quantity)
{
    Assert.That(shoppingCart.TotalQuantity,
        Is.EqualTo(quantity));
}

[Then(@"the total price is (.*)\$")]
public void ThenTheTotalPriceIs(string price)
{
    Assert.That(shoppingCart.TotalPrice, Is.EqualTo(price));
}
```

Fig. 12. Test script completed with the shopping cart missing concept, TotalQuantity (in green font)

Iteration III

One can further check the generated model consistency by setting up the mocks to return correct values (Fig. 13) as expected according to the specification of the fully developed implementation.

```
[BeforeStep("then")]
public void MockSetup(int quantity, double price)
{
    shoppingCart.TotalQuantity.Returns(quantity);
    shoppingCart.TotalPrice.Returns(price);
}
```

Fig. 13. Added TotalQuantity and TotalPrice to the Mock Setup

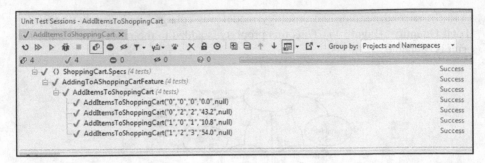

Fig. 14. Shopping-cart: Running test results – screenshot of running of the above test script within Resharper's [18] unit test runner. It is a passing test, since all expectations where met by the models, and all of the steps in the test script were successfully done.

Once the mock expectations were set and the test script is ready, there only remains to actually run it in a test runner tool, as seen in the screenshot in Fig. 14. This test script can later be reused and re-issued to check correctness of the actual developing implementation.

6   Discussion

The agile process embodied in ROM contributes significantly to the understanding of all aspects of a system under development. As an example, we have learned by working on the shopping cart system's specification that one also needs a "TotalQuantity" property that was previously lacking. This was therefore added to the initial ontology.

In the common agile practice – test driven development – there is a cycle red-green-refactor: red (i.e. failure) after developing the specification; green (i.e. success) when the implementation code fulfills the specification; and a refactor stage to improve the design of the code, prior to moving to the next requirement. Here the red phase is happening while the model is being developed. A green phase is achieved when the runnable model is complete, even though it is realized with mock objects.

6.1   Future Work

Among future issues to be investigated is to what extent ROM can be made fully automatic, or it will remain a useful, but quasi-automatic tool. A more extensive response should deal with each of the ROM modules in separate.

Concerning implementation, ROM will be configured to produce code using other languages\libraries, e.g. Ruby which is usually more concise than, e.g., C#/Java. The tool can also use a specific language to improve the produced scripts, e.g., using partial classes in C# to separate expectations from the test script.

To further integrate and streamline the suggested process with the whole development cycle, ROM could be also implemented as an IDE plug-in.

6.2   Main Contribution

The main contribution of this work is the usage of mock objects as a fast implementation means to verify the system design while in the Runnable Knowledge level, the highest abstraction level for software systems.

References

1. Exman, I., Llorens, J., Fraga, A.: Software Knowledge. In: Exman, I., Llorens, J., Fraga, A. (eds.) Proc. SKY 2010 Int. Workshop on Software Engineering, pp. 9–12 (2010)
2. Wynne, M., Hellesoy, A.: The Cucumber Book: Behaviour Driven Development for Testers and Developers, Pragmatic Programmer, New York, USA (2012)
3. Mellor, S.J., Balcer, M.J.: Executable UML – A Foundation for Model-Driven Architecture. Addison-Wesley, Boston (2002)
4. Freeman, S., Pryce, N.: Growing Object-Oriented Software. Addison-Wesley, Boston (2009)
5. Beck, K.: Test Driven Development: By Example. Addison-Wesley, Boston (2002)
6. Adzic, G.: Bridging the Communication Gap: Specification by Example and Agile Acceptance Testing. Neuri, London (2009)
7. North, D.: Introducing Behaviour Driven Development. Better Software Magazine (2006), http://dannorth.net/introducing-bdd/
8. Adzic, G.: Specification by Example – How Successful Teams Deliver the Right Software, Manning, New York, USA (2011)
9. Brown, K.: Taking executable specs to the next level: Executable Documentation, Blog post (2011), http://keithps.wordpress.com/2011/06/26/taking-executable-specs-to-the-next-level-executable-documentation/
10. Adzic, G.: Test Driven.NET Development with FitNesse. Neuri, London (2008)
11. SpecFlow – Pragmatic BDD for .NET (2010), http://specflow.org
12. Yagel, R.: Can Executable Specifications Close the Gap between Software Requirements and Implementation? In: Exman, I., Llorens, J., Fraga, A. (eds.) Proc. SKY 2011 Int. Workshop on Software Engineering, pp. 87–91. SciTePress, Portugal (2011)
13. Calero, C., Ruiz, F., Piattini, M. (eds.): Ontologies in Software Engineering and Software Technology. Springer, Heidelberg (2006)
14. Meszaros, G.: xUnit Test Patterns: Refactoring Test Code. Addison-Wesley, Boston (2007)
15. Exman, I.: Web Search of New Linearized Medical Drug Leads. In: Exman, I., Llorens, J., Fraga, A. (eds.) Proc. SKY 2011 Int. Workshop on Software Engineering, pp. 108–115. SciTePress, Portugal (2011)
16. Chelimsky, D., Astels, D., Dennis, Z., Hellesoy, A., Helmkamp, B., North, D.: The RSpec Book: Behaviour Driven Development with RSpec, Cucumber, and Friends, Pragmatic Programmer, New York, USA (2010)
17. NUnit (2012), http://www.nunit.org
18. Jetbrains. Resharper – productivity add-on for Microsoft Visual Studio (2013), http://www.jetbrains.com/resharper/
19. NSubstitute. A friendly substitute for .NET mocking frameworks (2013), http://nsubstitute.github.com/

Topology Labeling: An Indexing Structure
to Find Complex Relationships within Ontologies

Karina Robles[1], Alejandro Ruiz[2], Anabel Fraga[1], and Juan Llorens[1]

[1] Carlos III University of Madrid, Spain
[2] University of Piura, Peru
{krobles,afraga,llorens}@inf.uc3m.es, alejandro.ruiz@udep.pe

Abstract. Semantic Web technologies have contributed mainly to organize the knowledge and to search about this organized knowledge. One of the most important and complex kinds of search is to know if two entities are related within an ontology. These are called Semantic Associations, which have been classified using ρ operators: ρ-path, ρ-join and ρ-iso. Then, a ρ-query will solve any of them.

Studies about this area offer low performance execution times, while others increase the performance with pre-processing using complex structures in memory.

We focus on design of a simplified representation of the ontology that facilitates the graph traversal and reduces the algorithms complexity to solve these operators, starting from the first of them: ρ-path. We propose a topology labeling: we create a tree structure and identify each node with an interval index besides the level of root dependence. We will leave some space within the interval to manage future ontology modifications.

To validate this technique we will create ontologies of the order of thousands to 10000 nodes and a framework test with the implementation of the technique proposed in this paper. We will implement other techniques in order to compare execution times and performance.

Keywords: Semantic Search, Semantic Associations, Rho-Operator, Ontology, Ontology Navigation.

1 Introduction

Semantic Web technologies have contributed mainly to organize the knowledge and to search about this organized knowledge. Due to the increased importance of the Semantic Web, there is a growing interest in search not only about the concepts but also about the relationships between them, i.e. finding how two entities are related. This kind of search is important when the meaningful information is the direct or indirect relationship between entities, instead of the entities themselves.

Although, ontologies lead the researches to find new ways of searching, such as new visual interfaces to help the user in semantic queries [1–7] or semantic query languages (e.g. RQL [8], SquishQL [9], TRIPLE [10] and others [11–13]), these kinds of search do not solve the problem of how two resources are connected.

A. Fred et al. (Eds.): IC3K 2012, CCIS 415, pp. 284–294, 2013.

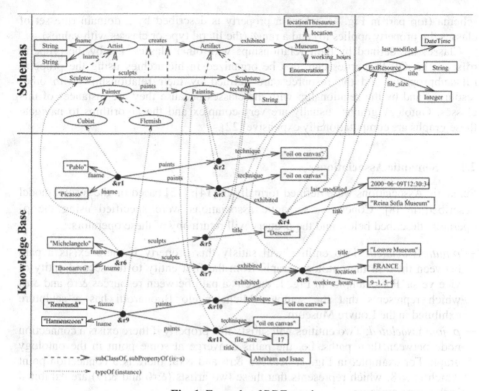

Fig. 1. Example of RDF graph

Two entities can be connected by relationships, through other intermediate entities, especially when these two entities are too far apart within the ontology. These complex relationships are called *Semantic Associations* and they have been studied and classified by using the ρ operators: ρ-path, ρ-join and ρ-iso [14].

To solve them, ontology relationships in the path between entities X and Y should be evaluated, [14–17]. Therefore, the problem is about to finding paths between nodes in a graph[9], [18–20]. A new approach has been developed in this paper based on a transformation of the ontology graph, which we call topology labeling. A new representation of the ontology graph is proposed to facilitate paths localization simplifying the algorithms to solve ρ operators.

2   The Ontology Graph

The RDF[1] ontology can be modeled by a tagged directed graph, where triplets (Subject, Property, Object) are represented by an arc tagged with the name Property connecting the nodes Subject and Object [8], [9], [20], [21]. Ontology graphs are built following the RDF definitions: Property types and classes are defined in the RDF

---

[1] Resource Description Framework is a family of World Wide Web Consortium (W3C) specifications, originally designed as a metadata model.

schema (top part in Fig. 1), where a property is described by a domain (the set of classes the property applies to) and a range (the literal type or classes with values).

Classes are defined by their relationships with other classes, i.e. by the property rdfs:subclassOf. Properties can also be organized in hierarchies using the property rdfs:subproperty, and the resources are defined by their relationships with other resources and by the relationship with the classes because they are instances of this classes. Ontology graphs usually are very complex and the algorithms to navigate these graphs are computationally expensive [22].

2.1   Semantic Associations

Semantic Associations were defined formally in [14], [15] based on the formal model described in [8]. Concisely, semantic associations were specified using the ρ-operators described below and the ρ-query will return any of these operators:

— *ρ-pathAssociated*: Two entities will satisfy this property if there exists a path between them in the ontology graph, from the first entity to the second entity or vice versa. For example, in Fig. 1, there is a path between resources &r6 and &r8 which represents that the sculptor Michelangelo Buonarotti has a sculpture exhibited in the Louvre Museum.

— *ρ–joinAssociated*: Two entities will satisfy this property if there exists a connection node between their paths, i.e. the paths converge at some point in the ontology graph. For example, in Fig. 1, between &r6 and &r9, there is a connection point which is &r8, which represents that these two artists (&r6 and &r9) are exhibited in the same museum.

— *ρ-cpAssociated*: Two entities will satisfy this property if they are descendants of the same class and they are at the same level in hierarchy. For example, in Fig. 1, &r2 and &r3 satisfy this property because they are both paintings.

— *ρ-isoAssociated*: Two entities will satisfy this property if they have similar characteristics, i.e. their properties and classes are similar. For example, in Fig. 1, &r1 and &r6 are ρ-isomorphic, because they both represent an artist, which creates works of art exhibited in a museum.

We will focus on solving these operators. We start by offering a solution for the first operator (ρ-pathAssociated) which is the most important because other operators are specializations obtained by adding some restrictions. For example, to get the ρ-joinAssociated between &r6 and &r9 in Fig. 1, we have to evaluate if the path from &r6 and the path from &r9 converge in some point.

3   Topology Labeling

The main goal of this work is to return the path that connects the terminal nodes. Therefore, we present a different transformation based on intervals to search easily a path between two nodes. Those intervals are assigned to each node after a transformation of the graph into a DAG is made in order to get a tree structure.

3.1   Transforming into a DAG

In this process, we will obtain a tree structure or a forest of trees, in which there is a virtual root node and all the real nodes are located as dependents of it. First, we make copies of the important nodes inside the strongly connected components as we show in Fig. 3, a) part, and locate these copies under the node on which they depend. Second, we also make copies of nodes that have multiple classifications as we show in Fig. 3, b) part.

In that way, we can easily check if one node is connected with another, because the connected node will be in one branch of the first node within the tree. This avoids a recursive process.

For instance, a node that belongs to two or more classifications, i.e. two or more classes, will be duplicated in the new graph and each copy will be located depending on the corresponding classification. As it is shown in Fig. 2, the nodes Sculpture and Painting are duplicated due to their multiple classifications, and also the node String that depends on Painting.

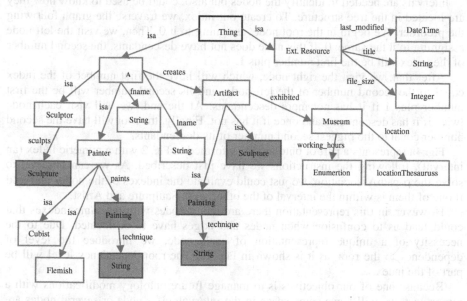

Fig. 2. Example of Transformation into DAG – this is a transformation of the schema part of the ontology in Fig. 1

Fig. 2 is an example of a transformation of the schema part of the ontology in Fig. 1. The corresponding data layer can be transformed by the same method. The root node will always be the node Thing. Therefore, at the end we obtain one model which represents all the ontology.

Graph cycles which may exist in the ontology [16], should be recognized. In this approach, a node will be copied as many times as necessary to represent its dependence with respect to other nodes, as it is shown in Fig. 3.

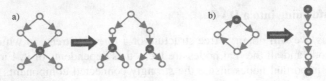

Fig. 3. Transformation of graph cycles – The a) part shows the transformation of a cycle with appended nodes. The b) part shows the transformation of a pure cycle.

3.2 Assignment of Intervals

This section will provide us a unique representation of each node as well as how we can locate paths between evaluated nodes. The work in [23], based on solving the transitive of hierarchical relationships (is-a) in KOS (Knowledge Organization Systems), suggested us the idea of its application to semantic associations, especially to solve the ρ-pathAssociation.

Intervals are needed to identify the nodes but also, could be used to know how they are located in the tree structure. To create our index, we traverse the graph following the preorder. We begin in the root node, assigning to it 0. Then, we visit the left node assigning to it the same: 0. If this node does not have descendants, the second number of the index will be the first number plus 1.

After that, we visit the right node, which will have the first number of the index equal to the second number of the left node, and its second number will be the first number plus 1 if it has not more descendants. At the end, we will visit each node twice if it has descendants and once if it has not. Finally, the root will have the second number equal to the biggest second number of its descendants.

Fig. 4a represents a part of ontology transformed in Fig. 2 with a numeric index (an interval), following the instructions we have just described. As we can observe, to solve the ρ-pathAssociation, we just could evaluate the indexes of the nodes and find if one of them is within the interval of the other, e.g., Sculpture and Artist.

However, in this representation there are many nodes with the same indexes that could lead us to confusion when nodes and edges have to be located. Due to the necessity of a unique representation of each node, we introduce the level of dependency to the root, as it is shown in Fig. 4b. The root dependency level will be part of the index.

Because one of our objectives is to manage future ontology modifications with a minor cost, we will leave some space in the intervals. If a node or several nodes are added as descendants of a given node, they could take as intervals the free numbers left in the space. Fig. 5 shows the new representation.

4   Validation Technique

In this section we describe the validation technique in detail. We have to use ontologies with high numbers of nodes in order to test out technique. Therefore, a graph with say 5000 nodes will be suitable to start with the test. Having this graph, different queries have to be executed. These queries will return different paths with different lengths. We begin with 10 queries, and eventually, we will increase this number as needed.

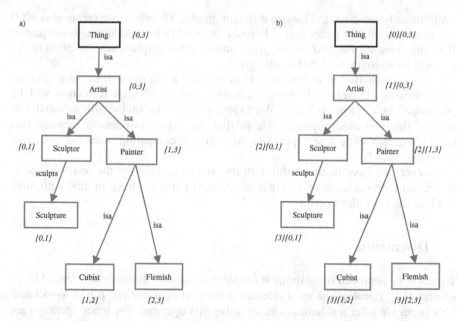

Fig. 4. Interval representation - a) Interval representation of an extract of the ontology. b) Root dependency level added to the interval representation.

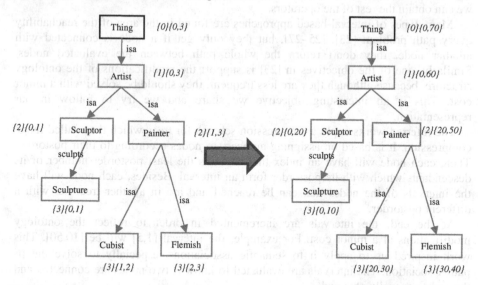

Fig. 5. Topology labeling: the interval representation will have some space within the boundaries to manage future ontology modifications

Additionally, we have to evaluate different graphs. Therefore, we create also 8000 nodes graph and 10000 nodes graph. In order to simplify the validation, those queries will be the same for the 5000 nodes graph and n-nodes graph. Thus, the 5000 nodes graph will be a subgraph of the n-nodes graph.

We choose the Barton technique [24] to perform a comparison because it is the most promising. Therefore, both our technique and the Barton technique will be implemented in a framework test. We expect that our technique will demand less execution time answering queries, i.e. to find the ρ-pathAssociation between two entities. We also expect less pre-computing times comparing with the recursive transformation of Barton.

However, we have to be careful with the graphs created for the test, because if they are not representative of all kinds of problem paths we have to deal with, they could bias the algorithms performance.

5   Discussion

We present an approach to solve the semantic associations, formally introduced by ρ-operators [14]. Because, the most relevant is the first operator viz. ρ-pathAssociated, in this paper we offer a solution to easily solve this operator. The other operators are only specializations: if some restrictions are added to the path between two entities, such as common join node (ρ-joinAssociated), or common ancestor (ρ-cpAssociated), we can obtain the rest of the operators.

Many types of interval-based approaches are found in the area of the reachability query path problem [23], [25–27], but they only get if a node is connected with another node, they don't return the whole path between the evaluated nodes. Similarly, one of the objectives in [23] is support the modifications of the ontology structure, because although they are less frequent, they should be solved with a minor cost. This is an interesting objective we share and we try to follow in our representation.

The referred works apply a compression schema for trees which they called range compression. It is based on assigning numbers to nodes according to their postorder. Then, each node will have an index that contains the least postorder number of its descendants which with its postorder form an interval. Besides, each node will have the intervals of the nodes that can be reached and are in another tree, i.e. with a different postorder.

At the end, the intervals are incremented in order to reflect the ontology modifications at a minor cost. For example, the interval [1,5] will be [10,50]. This work inspired us to apply it to semantic associations, especially to solve the ρ-pathAssociation. The intervals are evaluated to know if two nodes are connected and the paths are easily obtained.

Another approach [27], assigns intervals according to the preorder and also the postorder numbers. Nodes with more than one predecessor nodes, will have more than one interval in this representation. This idea is very close to our representation.

We have put together different ideas used previously within different approaches in order to obtain faster the paths between two entities, i.e. to find the semantic associations. The expected results of the validation experiments will probe whether the execution time in our technique will indeed be less than others.

Future work will use the same test framework for the development of the rest of the ρ operators.

6   Related Work

In this section we describe some approaches that try to solve the ρ operators. The studies in [14], [15] use a Schema Path Index (SPI) which provides fast access to all possible paths between two classes in a schema. The paths that involve resources that belong to the data layer are in the InterClass Index (ISI). This index stores the information about the schemas that are linked due to multiple classifications. They export the nodes at the data layer to the schema layer using artificial nodes that collapse the two class nodes.

Their idea is finding all alternative paths that reach the terminus node from the origin node, and evaluating each one to get the final result according the ρ operator searched. They do a pre-process to store all paths between classes at the schema level in matrices. The computational complexity is $O(|V|^{|V|})$ [28].

A biomedicine research [29], [30] implements a retrieval system for biomedical patents, by identifying the most meaningful relationships between them, i.e. by using semantic associations. To solve the ρ-operators they transform the ontology into two graphs: the isaGraph and the propertyGraph. They use ranking algorithms based on PageRank [31] and measures the nodes by subjectivity and objectivity measures related to other nodes connected to them.

Another research [16], [17] proposes an index of a condensed graph. They focus on the design of a new index structure and specially on solving ρ-pathAssociated and ρ-joinAssociated under the idea that searching a tree is easier than searching a graph. The complete graph is transformed recursively into a tree or a forest of trees, by collapsing the strongly connected components into a single node. They use an incidence matrix to represent the transitive closure in each transformation.

The information of the transformation is stored in two inverted files: one with the multiple nodes and their corresponding signatures, and the second with the signature with the list of multiple nodes. The transforming time of the ontology into the forest of tress is about $O(2n)$ and the creation time of the signature for each node is $O(n)$. These times are smaller than those in [15]. However, the computational time depends of the size of the matrices and inverted files in each transformation. This approach is called ρ-index [32].

In this paper, we consider these approaches [16], [17], but we follow a different idea. Instead of collapsing the strongly connected components, we make copies of the important nodes within them, as seen in Section 3.1. This action will make easy to find the paths without a recursive process.

This method was improved in [24], [28] by using a clustering technique to obtain a condensed graph. The matrices in each step are called Path Type Matrix because instead of an adjacency matrix which contains Boolean values, they contain the edge that links two nodes.

Study [33] presents an improvement of the Barton technique [24] by using less memory. They do not need to store all the matrices in each step. They proved that the first matrix and the last one join to the inverted files have the necessary information to solve the semantic associations.

Semantic associations are also used to locate interesting and meaningful relations in an e-learning environment [34], by using one or more known intermediate entities introduced by the user. They use a modified bidirectional Breadth-First Search algorithm to find the paths and rank them according to the users' domain of interest.

7   Conclusions

To solve Semantic Associations is a very complex task that needs high computational capabilities, because of the nature of the ontology graph. A transformation of this graph is presented in this paper which simplifies the navigation. This transformation is called topology labeling and it is obtained by two phases: first, we transform the graph into a DAG, and second, we assign an interval representation to each node.

The created index is based also on the level of root dependence. It facilitates the construction of algorithms regarding to the resolution of ρ operators. An example of the procedure of one algorithm for ρ-pathAssociation is mentioned: a path between two entities is finding when one of them is within the interval of the other and with different level of dependence.

References

1. Athanasis, N., Christophides, V., Kotzinos, D.: Generating On the Fly Queries for the Semantic Web: The ICS-FORTH Graphical RQL Interface (GRQL) 1. In: McIlraith, S.A., Plexousakis, D., van Harmelen, F. (eds.) ISWC 2004. LNCS, vol. 3298, pp. 486–501. Springer, Heidelberg (2004)
2. Catarci, T., Dongilli, P., Di Mascio, T., Franconi, E., Santucci, G., Tessaris, S.: An ontology based visual tool for query formulation support. In: Proceedings of the 16th Eureopean Conference on Artificial Intelligence, pp. 308–312 (2005)
3. Zhang, L., Yu, Y., Zhou, J., Lin, C., Yin, Y.: An Enhanced Model for Searching in Semantic Portals. In: WWW 2005: Proceedings of the 14th International Conference on World Wide Web, pp. 453–462 (2005)
4. Koutsomitropoulos, D.A., Domenech, R.B., Solomou, G.D.: A Structured Semantic Query Interface for Reasoning-Based Search and Retrieval. In: Proceedings of the 8th Extended Semantic Web Conference on The Semantic Web: Research and Applications, pp. 17–31 (2011)
5. Möller, K., Dragan, L., Handschuh, S.: A Visual Interface for Building SPARQL Queries in Konduit. In: 7th International Semantic Web Conference (2008)
6. Smart, P.R., Russell, A., Braines, D., Kalfoglou, Y., Bao, J., Shadbolt, N.R.: A Visual Approach to Semantic Query Design Using a Web-Based Graphical Query Designer. In: Gangemi, A., Euzenat, J. (eds.) EKAW 2008. LNCS (LNAI), vol. 5268, pp. 275–291. Springer, Heidelberg (2008)

7. Zenz, G., Zhou, X., Minack, E., Siberski, W., Nejdl, W.: From Keywords to Semantic Queries — Incremental Query Construction on the Semantic Web. Web Semantics: Science, Services and Agents on the World Wide Web 7(3) (2009)
8. Karvounarakis, G., Alexaki, S., Christophides, V., Plexousakis, D., Scholl, M.: RQL: A Declarative Query Language for RDF. In: Proceedings of the 11th International World Wide Web Conference, WWW 2002 (2002)
9. Miller, L., Seaborne, A., Reggiori, A.: Three Implementations of SquishQL, a Simple RDF Query Language. In: Horrocks, I., Hendler, J. (eds.) ISWC 2002. LNCS, vol. 2342, pp. 423–435. Springer, Heidelberg (2002)
10. Sintek, M., Decker, S.: TRIPLE - a query, inference, and trans- formation language for the semantic web. In: Proceedings of the First International Semantic Web Conference on The Semantic Web, pp. 364–378 (2002)
11. Fikes, R., Hayes, P., Horrocks, I.: OWL-QL – A Language for Deductive Query Answering on the Semantic Web. Web Semantics: Science, Services and Agents on the World Wide Web 2(1), 19–29 (2004)
12. Haarslev, V., Ralf, M., Wessel, M.: Querying the Semantic Web with Racer + nRQL. In: Proceedings of the KI-2004 International Workshop on Applications of Description Logics (2004)
13. Guha, R.V.: rdfDB: An RDF Database, http://www.guha.com/rdfdb/ (accessed: June 30, 2012)
14. Anyanwu, K., Sheth, A.: The ρ Operator: Discovering and Ranking Associations on the Semantic Web. ACM SIGMOD Record 31(4), 42–47 (2002)
15. Anyanwu, K., Sheth, A.: The ρ-Operator: Enabling Querying for Semantic Associations on the Semantic Web. In: Proceedings of the 12th International Conference on World Wide Web (2003)
16. Barton, S.: Designing Indexing Structure for Discovering Relationships in RDF Graphs. In: Database, Texts, Specifications and Objects Workshop (DATESO), pp. 7–17 (2004)
17. Barton, S.: Indexing Structure for Discovering Relationships in RDF Graph Recursively Applying Tree Transformation. In: Semantic Web Workshop at 27th Annual International ACM SIGIR Conference, pp. 58–68 (2004)
18. Wu, G., Li, J., Feng, L., Wang, K.: Identifying Potentially Important Concepts and Relations in an Ontology. In: Sheth, A.P., Staab, S., Dean, M., Paolucci, M., Maynard, D., Finin, T., Thirunarayan, K. (eds.) ISWC 2008. LNCS, vol. 5318, pp. 33–49. Springer, Heidelberg (2008)
19. Rada, R., Mili, H., Bicknell, E., Blettner, M.: Development and Application of a Metric on Semantic Nets. IEEE Transactions on Systems, Man and Cybernetics 19(1) (1989)
20. Hayes, P.: RDF Model Theory, http://www.w3.org/TR/rdf-mt (accessed: May 30, 2012)
21. Lassila, O., Swick, R.: Resource Description Framework: Model and Syntax Specification (1999)
22. Hildebrand, M., Van Ossenbruggen, J., Hardman, L.: An Analysis of Search-based User Interaction on the Semantic Web. Information Systems, no. INS-E0706, pp. 1386–3681 (2007)
23. Agrawal, R., Borgida, A., Jagadish, H.V.: Eficient Managemente of Transitive Relationships in Large Data and Knowledge Bases. In: Proceedings of the 1989 ACM SIGMOD International Conference on Management of Data, pp. 253–262 (1989)
24. Barton, S., Zezula, P.: Indexing Structure for Graph-Structured Data. In: Zighed, D.A., Tsumoto, S., Ras, Z.W., Hacid, H. (eds.) Mining Complex Data. SCI, pp. 167–188. Springer, Heidelberg (2009)

25. Wang, H., He, H., Yang, J., Yu, P.S., Yu, J.X.: Dual Labeling: Answering Graph Reachability Queries in Constant Time. In: 22nd International Conference on Data Engineering (ICDE 2006), p. 75 (2006)
26. Chen, L., Gupta, A., Kurul, M.E.: Stack-based Algorithms for Pattern Matching on DAGs. In: Proceedings of VLDB (2005)
27. Trißl, S., Leser, U.: Fast and Practical Indexing and Querying of Very Large Graphs. In: SIGMOD 2007 (2007)
28. Barton, S.: Indexing Graph Structured Data. Masaryk University (2007)
29. Mukherjea, S., Bamba, B., Kankar, P.: Information Retrieval and Knowledge Discovery Utilizing a BioMedical Patent Semantic Web. IEEE Transactions on Knowledge and Data Engineering 17(8), 1099–1110 (2005)
30. Mukherjea, S., Bamba, B.: BioPatentMiner: An Information Retrieval System for BioMedical Patents. In: Proceedings of the Thirtieth International Conference on Very Large Data Base, VLDB 2004, vol. 30, pp. 1066–1077 (2004)
31. Page, L., Brin, S., Motwani, R., Winograd, T.: The pagerank citation ranking: Bringing order to the web (1999)
32. Barton, S., Zezula, P.: ρ-index – An Index for Graph Structured Data. In: 8th International Workshop of the DELOS Network of Excellence on Digital Libraries, pp. 57–64 (2005)
33. Hassan, Z., Qadir, M.A.: MinG: An Efficient Algorithm to Mine Graphs for Semantic Associations. In: International Conference on Computer Networks and Information Technology (ICCNIT), pp. 59–64 (2011)
34. Viswanathan, V., Krishnamurthi, I.: Finding relevant semantic association paths through user-specific intermediate entities. Human-centric Computing and Information Sciences 2(1), 9 (2012)

Part III

Knowledge Management
and Information Sharing

Reference Operation Generation Method on Project Manager Skill-up Simulator

Keiichi Hamada[1], Masanori Akiyoshi[2], and Masaki Samejima[1]

[1] Osaka University, 2-1 Yamadaoka, Suita-shi, Osaka, 565-0871, Japan
[2] Hiroshima Institute of Technology,
2-1-1 Miyake, Saeki-ku, Hiroshima-shi, Hiroshima, 731-5193, Japan
{hamada.keiichi,samejima}@ist.osaka-u.ac.jp,
m.akiyoshi.we@cc.it-hiroshima.ac.jp

Abstract. In order to enable project managers to develop practical skills for solving problems in project management without costly training such as "on the job training", we have developed the project manager skill-up simulator. By inputting actions for improving the project to the simulator, the result of the project is simulated based on a project model and displayed to the project manager. For the purpose of giving feedback for learning actions to improve the project result, we propose a generation method of the reference operation that is a sequence of the actions to derive good results of the project. The proposed method generates the project principle from optimal operations derived by reinforcement learning on automatically generated operations. And, the reference operation is generated by applying the project principle to a certain project model. Experimental results show that the proposed method can automatically generate the reference operation as well as manual generation.

Keywords: Project Manager Skill-up Simulator, Reinforcement Learning, Decision Tree, Reference Operation Generation.

1 Introduction

Along with increase of difficult projects where a complicated IT system is developed in a short time, it becomes important to educate project managers that have enough skills to manage the projects. Many project managers learn project management skills through PMBOK(Project Management Body Of Knowledge)[11] and literatures which indicate success and failure factors of projects[8]. In addition, in order to make project managers acquire skills of planning and carrying out the project, it is important for them to face actual projects and solve issues of projects. Thus, many educational methods have been proposed such as OJT(On the Job Training), PBL(Project Based Learning)[3,13], and RP(Role Playing)[7]. But, effects of their methods depend on trainers and they take a lot of time to educate learners.

Recently, the efficient educational methods using simulators have been proposed such as SESAM[5] and PMT[4]. Thus, we have developed a project manager skill-up simulator which can interact with a learner in order to take an action against issues such as explosive occurrence of bugs[9,1]. In this simulator, the learner can input actions

A. Fred et al. (Eds.): IC3K 2012, CCIS 415, pp. 297–307, 2013.

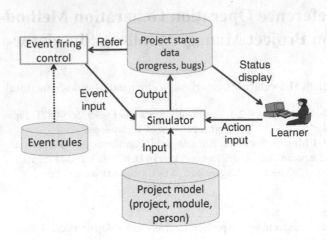

Fig. 1. Outline of project manager skill-up simulator

such as "overtime directive" and "collaborative work with expert members" based on the simulated situations in a project model. After finishing the simulation of the project model, this simulator outputs QCD(Quality, Cost, Delivery) as a result of an "operation" defined as a sequence of the actions. But, this simulator can not give feedback to the learner about whether each action in the operation was good or not. Trainers have to take much time to investigate the reference operation in order to evaluate each learner's operation from practical viewpoints.

This paper addresses a generation method of the reference operation to support the evaluation. First, the proposed method generates the project principle based on operations that are automatically generated on various projects. Applying the generated project principle to a certain project model, the proposed method generates the reference operation.

2 Learning Process with Project Manager Skill-up Simulator

2.1 Project Manager Skill-up Simulator

In order to realize the efficient education of the project mangers, we have developed the project manager skill-up simulator shown in Figure 1.

The project manager skill-up simulator loads a project model where modules to be developed and human resources are defined. After loading the project model, the simulator displays the status of the project to the learner. Finding some problems in the status, the learner inputs the actions to the simulator. Receiving the actions from the learners, the simulator decides the next status of the project and displays again to the learner. The trainer can set events, such as occurrence of bugs, to adjust the difficulty of the project. The simulator can change the status randomly based on the events.

The details of the main components of the project manager skill-up simulator are described in the followings:

Project Model. When this simulator starts, this simulator shows the learner a project model that the trainer defines. The project model consists of modules and persons as they exist in the real project:

– Module
The module is a part of the information system and developed by a person. When all the modules are developed, the simulation is finished. In the module, the initial development time and the initial quality are given by the trainer. The development time and quality changes by "necessary technical domain" and the "difficulty" of the module. So, we define the module as a set of the domain and the difficulty ranked with {A(difficult), B(normal), C(easy)}.

For example, a project has module 1 of "database: A, 10 days for development with no bugs" and module 2 of "interface: C, 20 days for development with no bugs".

– Person
Who develops the module is an important factor to decide the progress of the project. The person that is an expert for the domain of the module can develop more rapidly. So, we define the person as sets of the domain and the skill level with {A(expert), B(normal), C(novice)}. For example, Person 1 is defined as "database: A, interface: A, server: A, security: C".

Project Status Data. In simulating the project, this simulator calculates parameters of the project status such as the progress and the quality. The person develops the module at a constant rate everyday, but causes some bugs in the module. The progress is defined as the accumulated rate of the development and the quality is defined as the number of bugs. Both of the progress and the quality are invisible without any actions from the learner.

Events. In order to adjust difficulty to a learner's performance, this simulator causes "Events" which harm the project status by sudden bug explosion, sudden man-hour increasing and so on. Before the simulation, the trainer defines pairs of the event and the occurrence condition. When the occurrence condition is satisfied in the project status, the event happens at the rate that depends on the project status; if the project goes smoothly, the event often happens.

2.2   Learning Process

Figure 2 shows the user interface of the simulator. Checking the project status, the learner can stop the simulation any time for inputting actions through the button menu. The information of the project model is also displayed to the leraner. The module status is displayed with "Gantt chart" including two bars that represent initial planned duration and changed duration from starting date to completion date. When the module is too difficult for a person to develop, the schedule delays and the quality is lost.

The most important task of a project manager is to detect and predict such schedule delays and quality losses. The learner can take "check progress" to check the number

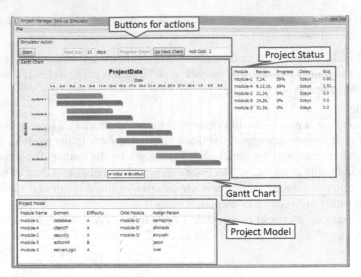

Fig. 2. An interface of simulator

of existing bugs and whether the schedule is delayed or not. If project managers judge to improve the delay, project managers can input the actions of "overtime directive" and "collaborative work with expert members" after taking "check progress". To take "overtime directive" increases working hours, and to take "collaborative work with expert members" debugs detected bugs and does not generate new bugs.

After completing all modules in the project model, this simulator outputs QCD as a result of the simulation. Q(Quality) is influenced by the quality losses including latent bugs. C(Cost) is an additional cost calculated from days spent for "overtime directive" and "collaborative work with expert members". D(Delivery) is influenced by the schedule delay.

2.3 Problem on Generating Reference Operation for Feedbacks of Evaluation

A learner gets the result of the operation by QCD, but can not know whether each action in the operation is good or not. So, it is required that the simulator shows feedbacks of the evaluations of the actions. Here, excellent project managers do not fail to conduct a operation based on the project principle that is a group of rules to improve the status of projects. The status is shown by QCD, and the project principle sets which targets of QCD is improved. So, the rule in the project principle is a set of a status to be improved and a necessary action to improve the target of QCD called "learning target" for projects. A learning target is defined as which of QCD to be acquired by the learner. The decision making by project managers consists of two stages of operations: project managers check the progress and take actions based on the progress. From this viewpoint, there are two kinds of project principles: "check progress" and "take action".

This simulator can give feedback to the learner whether the learner's operation follows the project principle or not. But the learner's operation can not be compared directly with a project principle because situations and actions for them vary with projects.

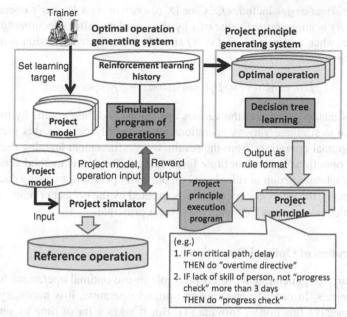

Fig. 3. Outline of proposed method of the reference operation

In order to compare the learner's operation, it is necessary to generate the reference operation which is an operation following the project principle.

So, the project principle is necessary to generate the reference operation. But, to generate the project principle takes a lot of time because trainers simulate and analyze many operations to get a good result for each project in order to generate the project principle which must be useful for various projects. This paper addresses how to generate such project principle and reference operation.

3   Reference Operation Generation Method Using Reinforcement Learning

3.1   Outline of Automatic Generation Method

We propose a generation method of the reference operation using reinforcement learning[14,10] and decision tree[2,12]. Figure 3 shows the outline of the generation method.

In order to generate the reference operation based on the project principle, this method uses optimal operations which lead to the best results corresponding to a learning target set in advance. The optimal operations can be considered to include correct judgments and actions to improve the status of the project. Because QCD should be kept smaller by the project manager, an optimal operation is defined as the operation to minimize the objective function $f(operation)$ defined as follows:

$$f(operation) = \sum_{i \in LearningTarget} f_i(operation) \qquad (1)$$

where $LearningTarget$ includes Q, C or D, $operation$ is a learner's operation, and $f_i(operation)$ is an outputted value of i by the simulator after executing $operation$. For example, when a learning target is "Q and D", the objective function is decided as the following:

$$f(operation) = f_Q(operation) + f_D(operation)$$

First, this method generates the various operations and their results by the simulation program to simulate various operations automatically. Second, this method generates the optimal operations from the results by reinforcement learning that can find the optimal oprtations faster than brute force search. Third, this method generates the project principle as a group of rules based on optimal operations by using decision tree learning. Finally, the reference operation is generated by the project principle execution program which takes action following the project principle.

3.2 Generation of Optimal Operation

It is necessary for generating a project principle to use optimal operations for various types of projects. In order to generate an optimal operation, it is necessary to minimize the objective function as formula (1). But it takes a lot of time to generate the optimal operation using brute force search. So, the proposed method searches an operation to improve the status of the project. Figure 4 shows the generation of the optimal operations.

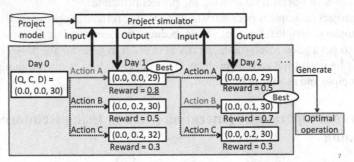

Fig. 4. Generation of the optimal operations

As shown in Figure 4, the proposed method tries all the actions for the project model and gets the results of the actions by the simulator. Then the proposed method decides the reward that depends on the result. By reinforcement learning which determines future actions so as to maximize reward based on past actions, the proposed method searches the optimal operations. How to decide the reward by the project status is important for searching the optimal operations. Let s_t (t as time) denote the project status as a set of QCD given by simulator and action a denote one of selectable actions in simulator. The reward $r(s_t, a)$ is defined as formula (2) and Table 1.

$$r(s_t, a) = - \sum_{i \in LearningTarget} h_i(s_t, a) \qquad (2)$$

Table 1. Description $h_i(s_t, a)$

$h_i(s_t, a)$	Description
$h_Q(s_t, a)$	Variation of bugs based on s_{t-1} divided by bug rate per day in taking no action
$h_C(s_t, a)$	Variation of costs based on s_{t-1} divided by costs per day in taking "overtime directive" and "collaborative work with expert members"
$h_D(s_t, a)$	Expected finishing dates on s_t divided by expected finishing dates in taking no action

In the reward, $h_Q(s_t, a)$ or $h_D(s_t, a)$ is given in comparison with Q or D in taking no action, and $h_C(s_t, a)$ is given in comparison with costs per day in taking "overtime directive" and "collaborative work with expert members". This formula means that the reward $r(s_t, a)$ gets better when the value of bugs or costs in s_t is smaller than one in s_{t-1}, or the expected finishing dates on s_t is early. For example, when learning target is "Q and D", reward $r(s_t, a) = -(h_Q(s_t, a) + h_D(s_t, a))$.

3.3 Generation of Project Principle

Figure 5 shows the outline of the project principle generating method. In order to generate the project principle corresponding to all aspects, the proposed method uses decision tree learning based on optimal operations. In decision tree learning, it is possible to find the characteristic attributes as nodes of the decision tree to classify data to the classes. So, the proposed method constructs the decision tree by using the project status as the attributes and the action as the classes. The constructed decision tree indicates the most frequent action on a certain project status in the optimal operations. Therefore, the proposed method generates the project principle from two decision trees, "check progress" and "take action", which are stages of the learner's decision-making in project management.

From the decision tree, a learner gets the information of class labels (leaf nodes shown in Figure 5) as actions and attributes(nodes in Figure 5) as shown Table 2. Based on these decision trees, the project principle execution program written is automatically generated. The generated programs apply the actions based on the decision tree. And the results of the actions by the programs are decided as the reference operations.

4 Evaluation Experiment

4.1 Results of Experiment

In order to generate a project principle, we used 10 project models that have 4.5 modules in average, manpower of 4 people, 10 days for each module, and 30 days for development. We set the learning target to "Q and D" because we assumed that a learner is experienced the project contained high standards for quality and delivery. And we experimented to compare the reference operation by manually generated project principle to one by the proposed method. For decision tree learning, we used WEKA which is one of the open source data mining tools[6].

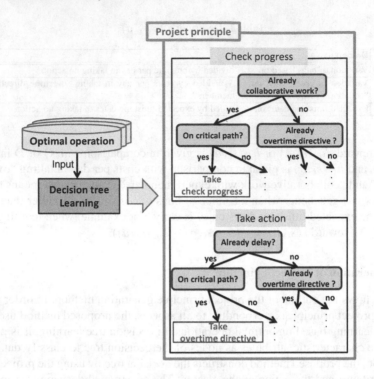

Fig. 5. Outline of project principle generating method

Table 2. The information displayed to the learner

The getting information before "check progress"	
1.	The number of days since the last "Check progress"
2.	The difference between a module's difficulty and a programmer's skill
3.	Whether the module is on critical path
4.	Has "collaborative work" been taken?
5.	Has "overwork" been taken?

The getting information after "check progress" with one before "check progress"	
6.	Is the project delayed?
7.	Whether additional man-hour exist
8.	The number of bugs

Figure 6 shows the project model includes difficulty and skill of person assigned for each module, and Figure 7 shows the result of experiment. Figure 7 shows the result with the project principle generated by the proposed method is better than one with manually generated project principle. To generate the project principle manually took 360 minutes, while to generate one by the proposed method took 240 minutes. So, this experimental result showed the proposed method generated the project principle and reference operation in reasonable time.

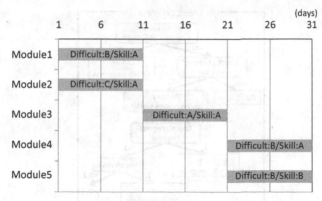

Fig. 6. Detail of project model

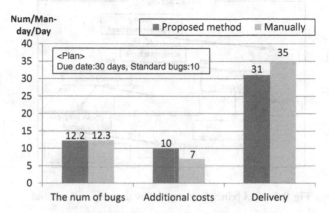

Fig. 7. Result of experiment

4.2 Discussion

In this experiment, the project principle generated manually has 13 rules, while the project principle generated by the proposed method has 24 rules. Figure 8 shows a part of the project principle by the proposed method. The manually generated project principle shows that "check progress" is necessary every three days. However the project principle by the proposed method reflects project status in detailed manner. Expert project managers confirmed that it is possible to apply the project principle by the proposed method to various types of projects. But, they said that an operation for a real project does not contain many "check progress". On the other hand, the reference operation by the proposed method indicates that "check progress" should be taken many times, which may be improper actions in the operations. Here, we confirmed the result for the modified project principle that the number of "check progress" is decreased. The result in decreasing "check progress" is shown in Figure 9. Figure 9 shows that the result in applying the project principle generated by the proposed method is better than one manually in terms of Q and C. Thus, it is confirmed that the project principle arranged manually is more useful.

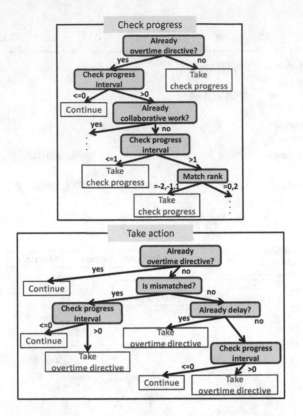

Fig. 8. Project principle generated by the proposed method

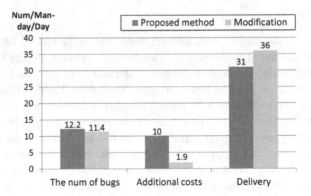

Fig. 9. The result in decreasing "check progress"

5   Conclusions

This paper addressed a generation method of reference operation to improve project results using reinforcement learning. The proposed method generates the project principle which is a group of rules to indicate what to do in situations of the project automatically. And the reference operation is generated by applying the project principle to project models. The project principle is generated by optimal operations using reinforcement learning and decision tree learning. The experimental result with the project principle generated by the proposed method is better than one with manually generated project principle. To generate the project principle manually took 360 minutes, while to generate one by the proposed method took 240 minutes. So, this result showed that the proposed method generated the project principle and reference operation efficiently. Our future work is how to generate more useful project principle.

Acknowledgements. This work was partially supported by KAKENHI; Grant-in-Aid for challenging Exploratory Research (23650537).

References

1. Akiyoshi, M., Samejima, M., Komoda, N.: A Project Manager Skill-up Simulator Towards Problem Solving-based Learning. In: Proceedings of 4th International Conference on Knowledge Management & Information Sharing (KMIS 2012), pp. 190–195 (2012)
2. Agrawal, R., Imielinski, T., Swami, A.: Database mining: A performance perspective. IEEE Transactions on Knowledge and Data Engineering 5(6), 914–925 (1993)
3. Boud, D., Feletti, G.: The Challenge of Problem-based Learning. Routledge (1998)
4. Davidovitch, L., Shtub, A., Parush, A.: Project Management Simulation-Based Learning For Systems Engineering Students. In: Proceedings of International Conference on Systems Engineering and Modeling 2007 (ICSEM 2007), pp. 17–23 (2007)
5. Drappa, A., Ludewig, J.: Simulation in software engineering training. In: Proceedings of the 22nd International Conference on Software Engineering (ICSE 2000), pp. 199–208 (2000)
6. Hall, M., Frank, E., Holmes, G., Pfahringer, B., Reutemann, P., Witten, I.H.: The WEKA Data Mining Software: An Update. ACM SIGKDD Explorations Newsletter 1(11), 10–18 (2009)
7. Henry, T.R., LaFrance, J.: Integrating role-play into software engineering courses. Consortium for Computing Sciences in Colleges 22(2), 32–38 (2006)
8. Horine, G.: Absolute Beginner's Guide to Project Management. Que Publishing (2005)
9. Iwai, K., Akiyoshi, M., Samejima, M., Morihisa, H.: A Situation-dependent Scenario Generation Framework for Project Management Skill-up Simulator. In: Proceedings of the 6th International Conference on Software and Data Technologies, vol. 2, pp. 408–412 (2011)
10. Kaelbling, L.P., Littman, M.L., Moore, A.W.: Reinforcement learning: A survey. Journal of Artificial Intelligence Research 4, 237–285 (1996)
11. Project Management Institute: The PMBOK Guide (2009)
12. Quinlan, J.R.: Induction of decision trees. Machine Learning 1(1), 81–106 (1986)
13. Savery, J.R., Duffy, T.M.: Problem Based Learning: An Instructional Model and Its Constructivist Framework. Educational Technology 35(5), 31–38 (1995)
14. Watkins, C.J.C.H., Dayan, P.: Q-learning. Machine Learning 8(3-4), 279–292 (1995)

Exploiting and Reusing Collaborative Traces to Facilitate Sharing Experiences in Groups

Qiang Li, Marie-Hélène Abel, and Jean-Paul A. Barthès

UMR CNRS 7253 Heudiasyc, Université de Technologie de Compiègne,
Centre de Recherches de Royallieu 60205, Compiègne, France
{liqiang,marie-helene.abel,barthes}@utc.fr
http://www.hds.utc.fr

Abstract. In the context of a web-based collaborative working environment, any interactive activity produces a set of Collaborative Traces (CT). Such traces reflects the group's working experience from past actions and can be used to capitalize knowledge. This paper proposes a method and fundamental principles to enhance the exploitation of collaborative traces. Grounded on our previous work that defined a collaborative trace and proposed a corresponding model, we define a model of complex filter and discuss its possible functionalities. We demonstrate how the model can be used in the E-MEMORAe2.0 collaborative platform.

Keywords: Collaborative Working Environment, Trace-based System, Collaborative Trace, Collaborative Engineering, Experience Management.

1 Introduction

Due to the rapid changes in information technology, people can work together using new and faster web-based collaborative working environment (CWE) with less restrictions due to time or geographic position, and even to language or culture. Such environments can strongly promote and enhance different aspects of computer-supported cooperative/collaborative work, e.g. the process of organizational knowledge management, group communication or decision making. In a typical collaborative workspace, users can send email, edit wikis, share documents or have a video conference. Such interactions with the system or with other members of the group leave traces that contain information about the collaborative activities as indicated by Li et al. [1,2]. Research issues concerning traces are at the intersection of three fields of study: Knowledge Management(KM), Information Sharing (IS) and Experience Management (EM).

In this article, we do not intend to discuss the three concepts, but accept the following common definitions: (i) information is "process data" [3]; (ii) knowledge is "authenticated information" [4,5]; and (iii) experience is "a special case or a refined form of knowledge at a higher level" (see Sun and Finnie [6] or Schneider [7]). According to Clauzel and his colleagues [8], traces can be viewed as a kind of "knowledge sources" representing users' experiences in synchronous collaborative learning activities. Moreover, Mille and his team [9] claim that the knowledge of both individual and group can be captured from the modeled traces. Later, they explain that traces of interaction reflect

A. Fred et al. (Eds.): IC3K 2012, CCIS 415, pp. 308–322, 2013.

experience more than simple knowledge for supporting complex tasks in a computer-mediated environment. They propose a framework to assist the creation of Trace-Based Systems [10]. In addition, Laflaquière et al. [11] state that traces from past interactions can be used to measure personal experience.

In the domain of personal experience management, the above research works greatly enrich the trace theory and provide new directions for practical applications. However, not enough attention has been given to the issue of experience sharing and reuse for *group* in *collaborative working environments*. In the context of CWE, this issue concerns three aspects of group experience management: (i) defining different kinds of traces in a group; (ii) modeling such traces with a view to support collaborative work; (iii) exploiting the defined traces according to the group or personal needs. Although the interactive activities in CWE are numerous and intricate, the main one is collaboration. We define a Collaborative Trace (CT) as a set of collaborative activities. This article is based on our previous research results (the CT definition and the CT model [1,2]), and concentrates on some possible methods to exploit and reuse collaborative traces.

In the paper we first briefly report some remarkable definitions and projects of Trace from the literature, and revisit the definition of collaborative trace. We then explain the collaborative activities in shared workspaces and recall the main features as well as some basic notations of our proposed model of Collaborative Trace. Then, we analyze various possibilities for exploiting collaborative traces in group spaces giving some examples. Then we present an evaluation of our model and show how it is exploited in the E-MEMORAe 2.0 collaborative platform. Finally, we conclude with a summary and discuss future work.

2   Definition of a Trace

In nature, a trace is a mark, an indication or an object denoting the existence or passing of activities, e.g. a series of animal footprints in the wood. It results from effective "actions" in the surrounding "environment". In the domain of IT the concept of trace has been defined by several authors to report interaction activities between user and system. Almost a decade ago, Mille and his colleagues proposed an approach named MUSETTE (Modeling USEs and Tasks for Tracing Experience [9,12]). The objective of MUSETTE was to "capture a user trace according to a general use model describing the objects and relations handled by the user of the computer system". In this case, primitive traces are collected and analyzed as a "task-neutral knowledge base" for reusing experience. A generic framework for experience modeling and experience management are mentioned and discussed in details by Champin et al. [12,9].

On the basis of their results, Laflaquière and his colleagues [10] found that the trace can be used to solve some important issues in experience management, e.g. "activity reflexivity and experience reuse." They defined a trace as "temporal sequences of observed items." They introduced a framework to support the Trace Based Systems (TBS) focused on the processing of trace exploitation. With minor variance, Clauzel and his colleagues [13] defined an Interaction Trace as: "histories of users' actions collected in real time from their interactions with the software." Zarka et al. [14] defined a trace of interaction as "a record of the actions performed by a user on a system, in other words, a trace is a story of the user's actions, step by step." In the TRAIS project (Personalized and Collaborative Trails of Digital and Non-Digital Learning Objects)[1], the researchers analyze a trace that can be regarded as a sequence of actions in a hypermedia environment to identify the users' overall objectives. From another perspective, Settouti et al. [15] defined a numerical trace as a "trace of the activity of a user who uses a tool to carry out this activity saved on a numerical medium." They applied the framework of trace-based system in the Technology-Enhanced Learning (TEL) Systems that can provide personal services.

From the definition of interaction trace, we introduced the new concept of Collaborative Trace (CT) and recall its definition: "A Collaborative Trace is a set of traces produced by a user belonging to a group and aimed at that group." In the following section, a brief summary of our collaborative trace model [1,2] is provided with some basic notations.

3   Collaborative Actions in Group Shared Spaces

3.1   Collaborative Activities

A collaborative working environment (CWE) represents a kind of computer-supported working environment that, according to Angelaccio and d'Ambroggio [16], "consists of a network of spatially dispersed actors (either humans or not) that play different roles and cooperate to achieve a common goal." It stems from the concept of "virtual workspaces" (see Schaffers et al. [17]) and can be used to assist both individual work and cooperative work, e.g., e-work and e-professional as defined by Prinz et al. [18]. With various information and communication technologies and tools, group users could conduct their collaborative work through the CWE (Ballesteros and Prinz [19]). Actually, very basic factors found in CWE facilitate knowledge and information sharing in group as shown by Patel and Wilson [20].

In software engineering principally, collaborative activities can be divided into four types: "Mandatory, Called, Ad hoc, and Individual" as mentioned by Robillard and Robillard [21], e.g. scheduled video conferences, sending e-mails, or managing documents. In a typical CWE, most of these activities take place in the collaborative workspace (shared workspace) as remarked by Martinez et al. [22]. With the development of Internet and of wireless technology, time and space are no longer a strong constraint, therefore, CWE inherits and extends the concept of groupware. In the early research stages, a shared workspace is designed as "a form of electronic white-board" that helps collaborators draw or write as mentioned by Whittaker et al. [23]. No doubt that communication (e.g. video and audio conference) and information sharing (e.g. exchanges of messages or files) are elementary parts of shared workspace functions. In addition, during the past decade, knowledge management (e.g. document management, group wikis and task management) and application sharing (work in the same application in real-time) have extended functionalities of CWE, adding new features to the shared workspace.

---

[1] http://www.noe-kaleidoscope.org/telearc/

All interactions or actions that concern diverse functionalities of CWE in the shared workspace can be recorded as traces. Thus, a trace model is necessary and strongly required in the process of experience management. It not only constitutes the historical list showing the user's past actions, but also reports the previous "experiences" helping to perceive and interpret his interactions with the system. The trace model proposed by Clauzel and his colleagues [13] for the project ITHACA represents and visualizes traces in the context of synchronous collaborative learning platforms. To address similar issues, Lafifi [24] and his colleagues introduced a trace model for the project SYCATA, concentrating on the global architecture of the collaborative learning system. In a different approach, the trace model proposed by Sehaba [25] dealt with the transformation process for the adaptation of the shared trace in accordance with the user's profile. For CWE, a collaborative trace model could facilitate the analysis and reuse of knowledge and experience in groups. It focuses on the activities that involve or engage the collaborators in group shared workspace.

3.2 Collaborative Trace Model

Before explaining our model, a simple example is introduced. Assume that in an established CWE, some engineers collaborate within a project. John finds a crucial problem that may be helpful for all the group members. So, first of all, he sends a mail to the group, then creates a new entry on this issue in group's wiki, and finally shares his solution (a pdf document) in the group workspace. In the meantime, Tom and Peter, whose views are similar but different from John's on this problem, both ask for a video conference with John in a reply email. John receives the emails and agrees on a video conference with Tom and Peter. Finally, they obtain a satisfactory answer to the problem in the subgroup meeting.

Thinking back about the meaning of an interaction trace, apparently, there are three basic factors concerning the trace: (i) "Emitter" who produces the trace; (ii) "Receiver" who obtains the trace (the target of the trace); (iii) "Information as a set of properties and corresponding values," that are the elements of the active environment in which the trace is generated and utilized. In a practical web-based CWE, the definition of "Emitter" and "Receiver" depends on the structure of the collaborative group. A collaborative group is generally defined as *a set of users with the same collaborative objective* and can be expressed as:

$$g_j = \{u_i, u_k, ..., u_m\}$$

It may contain several subgroups and independent users. Moreover, a single user can be considered as a special type of collaborative group (a group of one person): $g_i^0 = \{u_i\}$.

A trace is formally defined as:

$$t_{i,j}^k = < E_i, D_j, Q_k >$$

where $t_{i,j}^k$ is the kth trace sent by the ith Emitters E_i (a set of users), and received by the jth Receivers D_j (a set of users), and Q_k is a subset of pairs of the set Q, each element including a property and some value. Different situations of Emitters and Receivers lead to identify three types of traces (see Li et al. [1]).

Collaborative Trace. A collaborative trace can be regarded as a type of trace that satisfies the conditions:

$$E_i = g_i^0 = \{u_i\}$$

and

$$D_j \neq g_i^0$$

This means the trace is the result or the effect of an operation that has been done by an "Emitter" and then flows to another user or to a group. In particular, we can classify different types of collaborative traces based on the relations between "Emitter" and "Receiver":

i) The trace is produced and transferred within a group:

$$u_i \in g_k, D_j \subset g_k$$

That is to say, the emitter belongs to the receivers group. However, the relations between D_j and g_k indicate that two types of sub-situations exist:

(a) The trace is between the subgroups:

$$u_i \in g_k, D_j \subseteq g_k$$

For instance: a member sends an email to several group members that constitute a subgroup. This type of trace records the collaborative activities in subgroups.

(b) The trace is inside the whole group:

$$u_i \in g_k, D_j = g_k$$

For example: a member announces the details of the project schedule in group (that a message is sent to all the group members). All the group's activities are recorded by this type of trace.

ii) The trace is between two groups:

$$\exists g_k, u_i \notin g_k$$

and

$$D_j \subseteq g_k$$

For instance: some groups work together in a project, and documents or messages are shared between groups. In Figure 1 shows the differences between such traces.

In order to analyze and reuse collaborative traces, a *filter* is applied as a tool or a pattern in the CT model. The basic component of a filter is an extractor (operators to access some part of the trace), then elementary filters, and last, a complex filter (a combination of elementary filters). In practice, the most important part is the design of elementary filters. An elementary filter can be considered as a predicate testing the value associated with a specific property. Any given property may have many elementary filters. Formally, an elementary filter is defined as:

$$\xi : V \times V \to B, \; where \; V = \{true, false\}$$

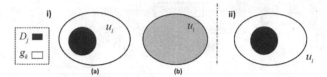

Fig. 1. Example of two types of collaborative trace

With elementary filters, we can analyze a set of particular traces according to our needs: $\{t \,|\xi_j^k(\alpha(t, p_j), v_m)\}$, where α is the value extractor, and ξ_j^k is one of the operators associated with property p_j. v_m is a reference value. For example: we'd like to find the traces that mention female members in the group. We apply

$$\xi_{sex}^{member} \equiv femaleEqual(\alpha(t, sex), female)$$

Concisely, a collaborative trace model is a triple structure: $(G, Q, \varXi)$, where G is the set of users: $G = \{g_j\}$, that for $\forall E_i \subset G, \forall D_j \subset G$, they meet the conditions: $E_i = g_i^0 = \{u_i\}$ and $D_j \neq g_i^0$. Q is a set in which each element includes a property and a value: $Q = P \times V = \{< p_l, v_m >\}$. P is a set of properties (attributes in the environment) and V is a set of values : $p_l \in P$ and $v_m \in V$. Z is a set of elementary filters: $\varXi = \{\xi\}$. Indeed, programming can be greatly simplified using this model of collaborative trace.

4   Exploitation of Collaborative Traces

Continuing the example above: (i) Naturally, the email sent to the group by John is stored in the group shared workspace, but has it been read by all the group or just by a single person? Same question for the shared pdf document: did they open and view it or not? (ii) If Tom or Peter were absent, it would affect the results of the video conference with John? In other words: do Tom and Peter have the same competence on this problem and any one of them could be substituted for the other? (iii) Actually, John, Tom and Peter collaborate together and can be regarded as a subgroup. Were the others in the group satisfied by their answers to the problem? Is the new added entry in the group wikis really helpful for their project? In CWE, such questions are common but difficult to answer. They are directly relevant to the issue of CT exploitation.

As we explained in Section 3, collaborative traces record past interactive activities in a group shared workspace and can be used to enhance an application, to generate adaptive scenarios and to assist members. In general, collaborative activities produce more information and knowledge than personal states. Therefore they may create a large number of CTs in the group space. Elementary filters are limited, when screening and analyzing a large amount of CTs against actual demands. A complex filter is thus proposed and designed to help addressing this issue. It is defined as a logical combination of elements of $\varXi$ ($\varXi$ is the set of elementary filters, $\varXi = \{\xi\}$).

Thus,

$$\zeta : T \times \varXi \times P \times V \to B$$

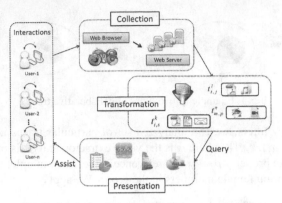

Fig. 2. A primary framework of trace-based system

An example of group collaborative trace would be

$$\{t \mid t \in CT_{i,l} \wedge \xi_j^k(\alpha(t, p_j), v_l) \wedge \ldots \wedge \xi_m^n(\alpha(t, p_m), v_s)\}$$

This allows selecting for example traces emitted by a user, mentioning the concept of "culture", or traces sent to a particular group during a specific week, or traces of messages sent by a specific user to a specific group, etc.

Three foundational parts constitute a primary framework of trace-based systems ([10] and [11]): (i) Collection; (ii) Transformation; and (iii) Presentation. One can clearly see this architecture Figure 2:

– Collection: this process uses diverse sensors and collectors, in a web-based CWE, the main data consists of text documents, hypertext documents, linked structures, server logs, browser logs and so on. The level of capture determines the diversity of the values. Collecting can be done on-line or off-line;
– Transformation: this part includes three functions: filtration, calculation and analysis. The output (CT) from the first process can be classified, analyzed, merged and edited automatically or manually. The programming language and practical system environment (e.g. the number of users, the hardware support, etc.) directly affect the efficiency and accuracy;
– Presentation: the last process utilizes the outcomes from the transformation procedure. The object is to explain the users' finished "interactive activities" and assist them in their future work. To make the modeled traces easier to understand and reuse, the interface design and the mode of presentation (for instance: visualization, audio or video) require serious consideration.

In CWE, the exploitation of collaborative traces is principally focused on the transformation and the presentation process. It is an important issue concerning experience reuse in EM theory (for the general experience reuse, refer to Bergmann [26] or

Tautz [27]). CWE, like other application scenarios, e.g. electronic commerce, diagnosis of complex technical equipment or electronics design, has the following characteristics:

- Knowledge intensive: Collaborative knowledge, e.g., about group project (e.g., project description and budgeting, task management, human resources, re-set target), group member (e.g., background, competence and character etc.) and group management (e.g., leadership and hierarchical relationships) directly influences every phrase and is enriched with group needs;
- Vague collaboration description: the goal of group collaboration are often vague, incompletely specified. To clarify the problem and the objective, regular meetings are recommended for all the group members.
- Large collaboration/solution space: the more possible collaborations and solutions there are, the larger the space would be in CWE and a single collaboration or solution is not enough for a complex project. Normally, these solutions depend on the number of tasks and involved people;
- Group size: different kinds of people (e.g., engineers, experts or manager) are needed in every process of problem solving and act in a collaborative task. However, for the this issue, most of the research works examine small group sizes as done by Steiner [28] or Ellis et al. [29]. A great challenge for CWE is the large size of collaborative groups.
- Highly dynamic: the rapid change and development of technology has a great effect on the renewal of knowledge, the people involved, the potential collaboration, the working style and so on.

Like the sketched situations above, a complex project is heavily based on collaborative experience. Collaborative traces sharing and reuse enable helping individuals and groups to avoid making same mistakes over again. To understand the process of exploiting collaborative traces, four basic scenarios are introduced as follows:

- Record and analyze the finished collaborative activities: this scenario can be characterized as "a dictionary of group collaborative activities in accordance with the chronological order". All the members in a group could distinctly see their interactions and the corresponding results in the group shared space, e.g. the usage status of shared documents or the sent email may be opened by the others;
- Assist group future work: in this case, the filtered CTs can be distinguished as "a guide" for the future collaborations in groups. For example, if a task that failed in several ways, we can avoid doing the same mistake in the future. Besides, some potential collaborations may be found by their similar CTs, e.g. the comparable preference of shared documents or entries in group wikis;
- Support group decision making: in this situation, collaborators can review all their past decisions with their consequences and the project progress in the group. They can make a better decision with such classified CTs. For instance: during Tendering, the analysis of customer RFP (Request For Proposal) within collaborators;
- Enrich group knowledge: in this circumstance, CTs reflect the needs and preferences of groups, with recommendation strategies, new knowledge is gained and shared in the collaborative workspace. For example: from the preferred books, links or videos, it is easy to recommend more items with similar topics.

In CWE, the benefits of CT exploitation can be summarized as follows:

- Shorter Project Completion Time/Cycle: e.g. the cost of time or group work efficiency;
- Improved Project Quality: e.g. from the reuse of CTs, we would make less mistakes but have more potent collaborations.
- Reduced Project Expenditure: e.g. from the analysis of CTs, it is not difficult to identify collaborators' contributions and attitudes.

5   Application

In this section, we evaluate our CT model and several complex filters in a web-based collaborative platform E-MEMORAe2.0 (Leblanc and Abel [30]). Within the MEMO-RAe approach, E-MEMORAe2.0 [2] (Figure 3) is combined with: (i) Models stem from knowledge engineering to support Knowledge Management; (ii) Semantic Web technologies to facilitate sharing and interoperability; (iii) Web 2.0 technologies to promote the social processes. Via this platform, both the fields of organizational collaboration and expertise can be enhanced by means of ontologies that define knowledge in organizations (Abel and Leblanc [32]). In a shared workspace, the users can exchange messages, edit and annotate shared documents, write wikis, share calendar and so on. For the personal use, the user can navigate through shared concepts (or ontologies); moreover he can also organize and capitalize resources (e.g. documents, links, etc.). Until now, within the range of this platform, only two kinds of personal interactions are recorded: (i) access to concepts; (ii) access to resources; then presented in the "Navigation history." In order to facilitate collaborative activities, apparently, the personal traces are limited and weak. The application of the CT model of complex filters directly is easy to do and constitutes a significant improvement.

In our application, firstly, the CTs are stored in accordance with the CT model conditions; then the queries are done through the designed complex filters; lastly, the results are presented in a chart or graph. Summarizing our case: the collaborative group has four members: Qiang, Étienne, Marie-Hélène and Jean-Paul, formally, $g_1 = \{u_1, u_2, u_3, u_4\}$. They cooperate in a project called "Trace". The group has two subgroups. Recall the general framework of trace-based systems, the proposed model and filters are mainly in the transformation and presentation parts. Three factors of CT are collected and stored in a MySQL database: the list "per_id" from the table "mem_personne" is used to identify the members (e.g. the E_i and D_j); the values and properties are decided by the needs of practical scenarios (refer to different methods of CT exploitation), but the "time" and "date" of past interactions is determined as the "Index" of CT (geographical position could be another choice). We discuss respectively the two cases ("Concepts" and "Resources") in the following part:

- For the case: "Concepts," the three components of collaborative trace are written as: (i) E_i ="The administrator (one of the members who is in charge of building the ontology, e.g. creation, connection of concepts etc.)"; (ii) D_j ="All the

---

[2] The address of platform E-MEMORAe2.0 : http://www.hds.utc.fr/memorae/
[3] This is a new interface of E-MEMORAe2.0, see also:
http://www.hds.utc.fr/memorae/.

The focus box   The core concept   The shared concepts map   The User's box

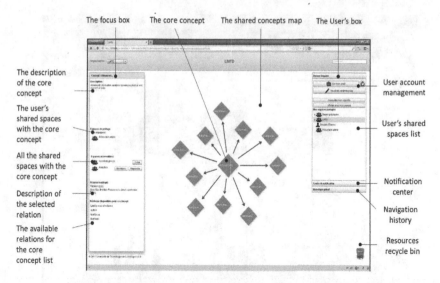

The description
of the core
concept

The user's
shared spaces
with the core
concept

All the shared
spaces with the
core concept

Description of
the selected
relation

The available
relations for
the core
concept list

User account
management

User's shared
spaces list

Notification
center

Navigation
history

Resources
recycle bin

Fig. 3. The collaborative platform E-MEMORAe2.0 (in French)[3]

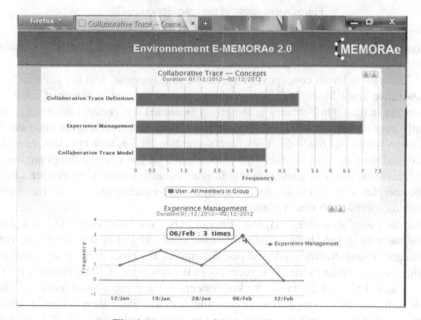

Fig. 4. An example of the case "Concepts"

group members" (g_1), e.g. every member can view and check the shared ontology
in the group workspace; (iii) Apart from "time" and "date" (formally, which can
be written as v_1), the frequency/times (that can be considered their preferences) is
intended as another value (v_2) for the property (p_1) "the concerned concepts in the
group shared ontology"; the upper part of Figure 4 shows the three most consulted

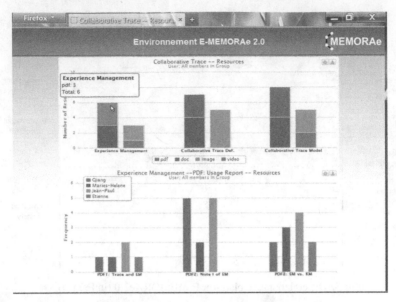

Fig. 5. An example of the case "Resources"

concepts: "Collaborative Trace Definition", "Experience Management", "Collaborative Trace Model" during one month (from 01/12/2012 to 02/12/2012).

The most relevant concept is "Experience Management" and the least is "Collaborative Trace Model". The lower chart presents the trail of the most consulted concept in time. On February 6, the group examined this concept three times, but only once on January 12. Therefore, we can clearly obtain their preference and attention within this ontology. The group's attention for this concept changes and is reduced after a period of time. This phenomenon may imply that the group has an extensive understanding and shares common conclusions about this concept. Here, the complex filter (ζ_1) is designed to analyze the frequency and the change of consulted concepts with time variation;

– For the case: "Resources", (i) the "Emitter" (E_i) and "Receiver" ($D_j = g_1$) are the same as in the above situation, but the traced object is the shared file (two categories: one involving pdf and doc documents, the other concluding video and images) that concerns the three concepts in the ontology. (ii) The property (p_2) is "the shared files for the most checked concept in the group shared ontology". Besides, for the values, one (v_3) is the situation of shared files (file types and quantity) and another (v_4) is the frequency/times of the service situation for each type of the file. As shown in Figure 5, the upper chart demonstrates the quantity of each type file that has been shared in group workspace during one month (same as the case "Concepts": from 01/12/2012 to 02/12/2012).

We can see that the most frequently consulted concept (from the first case), Experience Management (EM), is connected with several files: three pdf, three doc, one video and two images in this period. And the least interesting concept, "Collaborative Trace Model," has the most connected files: four pdf, four doc, two videos and three images. The lower figure presents the state of service for the three shared pdf documents that is associated with the concept EM. The "frequency" signifies the number of times the file has been opened ("open this file"). For the "PDF2" ("Note I of EM"), it is obvious to see that Étienne (u_2) had a lack of interest and has never opened this file. However, it was certainly read several times by Qiang (u_1) and Jean-Paul (u_4). Moreover, Marie-Hélène (u_3) is more interested in "PDF3" ("EM vs. KM") than in other documents ("Note I of EM" or "Trace and EM"). In this case, the complex filter (ζ_2) is used to help observe, compare and analyze the group's preference and members' contributions in collaborative workspace.

As a consequence of the filtered CTs, some potential collaborative relations that tightly rely on their "preferences" and "contributions" will be recommended within group members, for example: Qiang (u_1) and Jean-Paul (u_4) collaborate with the subject of "PDF2". Furthermore, the competence or knowledge background within group members can be identified with more complex filters, e.g. from the similarity of the shared files. It is helpful to allocate the tasks or replace a member in some particular situation. For instance if we are missing an expert in a group, we could propose another expert for this task. Without a doubt, the group's knowledge is enriched by these shared files and the ontology in the group workspace. Using the filtered CTs, we could understand the service state of the shared knowledge, e.g. the level of knowledge usage and the type of knowledge requested in the group.

In the E-MEMORAe2.0 platform, the group collaborative working experiences are modeled and reused by the application of collaborative traces. CTs model and the complex filters can also be applied to other ends, like in supporting the Tendering process (in railway applications) (Penciuc [33]) and the organizational Content Management (Étienne [34]). Moreover, our model can be expanded to different collaborative platforms, e.g. in an agent-based CWE, or collaborative learning systems.

6 Conclusions and Future Work

In this paper, we proposed using a complex filter approach to facilitate group experience management and support collaborative works in the context of web-based CWE (Section 4). This approach has been developed from the results of our previous work: Collaborative Trace Definition and Model. A literature review of traces and some basic notations of our CT model were introduced and discussed in Section 2 and Section 3. Furthermore, to validate this method and some principles of exploitation of CTs, two typically use cases based on the collaborative platform E-MEMORAe2.0 were compared and explained in Section 5. Exploiting collaborative traces concerns several critical issues in the different fields of EM study, which can be summarized into three principal points: (i) Experience reuse in collaborative groups, not only for personal usage; (ii) Analysis and modeling of the finished/past collaborative activities and interactions; (iii) Enriching organizational knowledge and supporting the Knowledge Management process in CWE. Using the CT model and the complex filters are also interesting in other fields, such as: expert systems, multi-agent systems, or social systems. More specifically, the exploited CTs could serve the group structured planning and decisioning process since all of the group collaborative interactions (e.g. group decision and the corresponding result) are completely recorded by CTs. As a very important strategic tool, SWOT (Strengths, Opportunities, Weaknesses and Threats) Analysis would be an ideal object to reuse these filtered CTs.

Actually, SWOT Analysis is fundamentally composed of a series of questions (e.g. for Strengths: Do you have immensely experts in your group for a new project?) that once the answers are found, we will have a equivalent SWOT Matrix. We are now attempting to build a type of complex filters with the purpose of retrieving a particular set of collaborative traces that could automatically be applied to the process of answering the SWOT questions. To reconstruct the questions, we can refer to some approaches in QA (Question Answering) systems (see Lehnert[35] and Mollá [36]) and then combine with our CTs-based SWOT filters. Moreover, there exists many other collaborative interactions in the platform E-MEMORAe 2.0, for example: in a group shared work space that adds concepts or resources, or creates comments in wikis. We are interested in testing our model and filtering more varieties of collaborative interactions. Besides, in the coming semester, we will apply our application to the collaborative learning scenarios via the platform E-MEMORAe 2.0 at the University of Technology of Compiègne. It will be used to support the students' collaborative activities that can be traced and analyzed by our application.

References

1. Li, Q., Abel, M.H., Barthès, J.-P.: Sharing working experience: Using a model of Collaborative Traces. In: 2012 IEEE 16th International Conference on Computer Supported Cooperative Work in Design (CSCWD), pp. 221–227. IEEE (May 2012)
2. Li, Q., Abel, M.H., Barths, J.-P.: A Model of Collaborative Trace to Enrich Group Experience. In: The 5th International Conference of the World Summit on the Knowledge Society, Rome, Italy (July 2012)
3. Zins, C.: Conceptual approaches for defining data, information, and knowledge. Journal of the American Society for Information Science and Technology 58(4), 479–493 (2007)
4. Dretske, F.I.: Knowledge and the Flow of Information. Australasian Journal of Philosophy 61, p. xiv, 273 (1981),
   http://mitpress.mit.edu/item/default.asp?tid=7275&ttype=2
   (retrieved)
5. Machlup, F.: Knowledge: Its creation, distribution, and economic significance, vol. 3. Princeton University Press (1984)
6. Sun, Z., Finnie, G.: Experience management in knowledge management. In: Khosla, R., Howlett, R.J., Jain, L.C. (eds.) KES 2005. LNCS (LNAI), vol. 3681, pp. 979–986. Springer, Heidelberg (2005)
7. Schneider, K.: Experience and knowledge management in software engineering. Springer (2009)
8. Clauzel, D., Sehaba, K., Prié, Y.: Enhancing synchronous collaboration by using interactive visualisation of modelled traces. Simulation Modelling Practice and Theory 19(1), 84–97 (2011)
9. Champin, P.A., Prié, Y., Mille, A.: Musette: a framework for Knowledge Capture from Experience. Extraction et Gestion des Connaissances, 2167 (2004)

10. Laflaquière, J., Settouti, L.S., Prié, Y., Mille, A.: Trace-based framework for experience management and engineering. In: Gabrys, B., Howlett, R.J., Jain, L.C. (eds.) KES 2006. LNCS (LNAI), vol. 4251, pp. 1171–1178. Springer, Heidelberg (2006)
11. Laflaquiére, J., Mille, A., Ollagnier-Beldame, M., Prié, Y.: Modeled traces for systems allowing reflection on personal experience. International Journal of Human-Computer Studies, 12 p. (2010)
12. Champin, P.A., Prié, Y., Mille, A.: Musette: Modelling uses and tasks for tracing experience. In: ICCBR, vol. 3, pp. 279–286 (June 2003)
13. Clauzel, D., Sehaba, K., Prié, Y.: Modelling and visualising traces for reflexivity in synchronous collaborative systems. In: International Conference on Intelligent Networking and Collaborative Systems, INCOS 2009, pp. 16–23. IEEE (November 2009)
14. Zarka, R., Cordier, A., Egyed-Zsigmond, E., Mille, A.: Trace replay with change propagation impact in client/server applications. In: Conference AFIA (2011)
15. Settouti, L.S., Prie, Y., Marty, J.C., Mille, A.: A trace-based system for technology-enhanced learning systems personalisation. In: Ninth IEEE International Conference on Advanced Learning Technologies, ICALT 2009, pp. 93–97. IEEE (July 2009)
16. Angelaccio, M., D'Ambrogio, A.: A model transformation framework to boost productivity and creativity in collaborative working environments. In: International Conference on Collaborative Computing: Networking, Applications and Worksharing, CollaborateCom 2007, pp. 464–472. IEEE (November 2007)
17. Schaffers, H., Brodt, T., Pallot, M., Prinz, W.: The Future Workplace-Perspectives on Mobile and Collaborative Working. Telematica Instituut, The Netherlands (2006)
18. Prinz, W., Loh, H., Pallot, M., Schaffers, H., Skarmeta, A., Decker, S.: ECOSPACE–Towards an Integrated Collaboration Space for eProfessionals. In: International Conference on Collaborative Computing: Networking, Applications and Worksharing, CollaborateCom 2006, pp. 1–7. IEEE (November 2006)
19. Ballesteros, I.L., Prinz, W.: New collaborative working environments 2020. Report on industry-led FP7 consultations and 3rd Report of the Experts Group on Collaboration@ Work, European Commission (2006)
20. Patel, H., Pettitt, M., Wilson, J.R.: Factors of collaborative working: A framework for a collaboration model. Applied Ergonomics 43(1), 1–26 (2012), http://www.ncbi.nlm.nih.gov/pubmed/21616476 (retrieved)
21. Robillard, P.N., Robillard, M.P.: Types of collaborative work in software engineering. Journal of Systems and Software 53(3), 219–224 (2000), http://linkinghub.elsevier.com/retrieve/pii/S0164121200000133 (retrieved)
22. Martínez-Carreras, M.A., Ruiz-Martínez, A., Gomez-Skarmeta, A.F., Prinz, W.: Designing a generic collaborative working environment. In: IEEE International Conference on Web Services, ICWS 2007, pp. 1080–1087. IEEE (2007)
23. Whittaker, S., Geelhoed, E., Robinson, E.: Shared workspaces: how do they work and when are they useful? International Journal of Man-Machine Studies 39(5), 813–842 (1993)
24. Lafifi, Y., Gouasmi, N., Halimi, K., Herkas, W., Salhi, N., Ghodbani, A.: Trace-based collaborative learning system. Journal of Computing and Information Technology 18(3) (2010)
25. Sehaba, K.: Adaptation of Shared Traces in e-learning Environment. In: 2011 11th IEEE International Conference on Advanced Learning Technologies (ICALT), pp. 103–104. IEEE (July 2011)
26. Bergmann, R.: Experience management: foundations, development methodology, and internet-based applications. Springer (2002)
27. Tautz, C.: Costumizing Software Engineering Experience Management Systems to Organizational Needs. Fraunhofer IRB-Verlag (2001)

28. Steiner, I.D.: Group process and productivity, p. 204. Academic Press, New York (1972), http://books.google.com/books?id=20S3AAAAIAAJ&pgis=1 (retrieved)
29. Ellis, C.A., Gibbs, S.J., Rein, G.: Groupware: some issues and experiences. Commun. ACM 34(1), 39–58 (1991), http://doi.acm.org/10.1145/99977.99987
30. Leblanc, A., Abel, M.-H.: E-MEMORAe2.0: an e-learning environment as learners communities support. International Journal of Computer Science and Applications, Special Issue on New Trends on AI Techniques for Educational Technologies 5(1), 108–123 (2008)
31. Abel, M.-H., Leblanc, A.: An Operationnalization of the Connections between e-Learning and Knowledge Management: the MEMORAe Approach. In: Proceedings of the 6th IEEE International Conferences on Human System Learning, Toulouse, France, pp. 93–99 (May 2008)
32. Abel, M.-H., Leblanc, A.: Knowledge Sharing via the E-MEMORAe2.0 Platform. In: Proceedings of 6th International Conference on Intellectual Capital, Knowledge Management & Organisational Learning, Montreal Canada, pp. 10–19. ACI (2009)
33. Penciuc, D., Abel, M.-H.: Requirement and Modelling of a Workspace for Tacit Knowledge Management in Railway Product Development. In: Proceedings of the 2nd International Joint Conference on Knowledge Discovery Knowledge Engineering and Knowledge Management, pp. 61–70 (2010)
34. Deparis, É., Abel, M.-H., Lortal, G., et al.: Knowledge Capitalization in an Organization Social Network. In: Proceedings of the International Conference on Knowledge Management and Information Sharing, pp. 217–222 (2011)
35. Lehnert, W.G., Lehnert, W.G.: The process of question answering: A computer simulation of cognition. L. Erlbaum Associates (1978)
36. Mollá, D., Vicedo, J.L.: Question answering in restricted domains: An overview. Computational Linguistics 33(1), 41–61 (2007)

Towards Value-Oriented Use of Social Media for Knowledge Management in SME

Ulrike Borchardt

University of Rostock, 18057 Rostock, Germany
ulrike.borchardt@uni-rostock.de

Abstract. Social media is one of the current technology trends in use for market-ing and communication, extending the possibilities for exchange among people. Yet, as every other technology using social media means effort for the organiza-tion using it. However, how can this effort being measured in the workload of a SME. To cover this we visited two SME introducing KM via social media and developed a concept to start KM. The obstacles and measures taken towards a value oriented KM with social media are documented in this paper, showing that a structured process helps and that there are still points missing to estimate the knowledge need to generate a full value oriented approach.

1 Motivation

Continuous knowledge losses between projects and tasks, in customer and partner com-munication etc. strongly influence the long-term business success of each organization. Rather than losing some of the knowledge that has been developed in a project or task, organizations should use relevant knowledge consistently for future projects and tasks in order to solve future problems and tasks faster, in a more efficient and cost effective way. Moreover, they should be able to expand their knowledge instead of having to develop the same knowledge over and over again.

One possible solution to address this problem is the use of social media in the field of Knowledge Management (KM). However, using social media efficiently is associated with a certain effort. For an organization this means employees must have time for activities in the media. As Small and Medium Enterprises (SME) are bound to have the problem as described above as well as limited resources the question arises whether the use of social media is an appropriate solution for them. Already in our earlier work we found by literature research, that only few examples on SME practice in the field of KM are available for research purposes.[1],[2] To close this gap and to gain insight into how the problem is addressed using social media in SME we conducted a case study holding two knowledge intensive SME from Germany to answer the following research questions:

1. How is social media used in the organization?
2. How is social media integrated in the organizations business processes?
3. Which benefits are realized by the use of social media for KM in the organization?
4. Which potentials are not realized within the organizations?

A. Fred et al. (Eds.): IC3K 2012, CCIS 415, pp. 323–336, 2013.

Especially with question 3 and 4 we address the problem of naming concrete benefits of the use of certain technologies for KM, which still remains a problem due to the character of knowledge which hardly can be measured in monetary units, whereas SME with the specific short-termed action horizon look for fast pay-off opportunities. In the case study we have a look at the employees' opportunities to share their opinions, experiences and insights as well as the social media instruments and tools that are already used inside the respective organizations.

The paper is structured as follows. Section 2 provides the fundamentals involved in our work and section 3 introduces the methodology in use to gain the results. Section 4 presents the organizations as well as the individual results. Finally, section 5 provides a conclusion.

2 Fundamentals

This section presents the basic terms in use for our case study as they were also explained to the participating organizations in the case study.

2.1 Social Media

Even though all media can be considered social, social media is nowadays mostly associated with internet-based media services, which are built upon the principles of social interaction and the technological principles provided by web 2.0. The main idea behind this is the transformation of the user from a content consumer only to a content publisher resulting in user generated content. Principally, social media depends upon fundamental interaction between people as the discourse and integration of contents to establish shared-meaning, for using information technology as an instrument. Social media contents examples of collaboration technology, in their conventional form known from the field of CSCW e-mail and Intranet and more modern social networks as well as wikis. In general social media can serve especially as a media for knowledge transfer[3].

Social networks are one facet of social media, representing a kind of online communities, which are based on the technology of web applications and portals[4]. They are also named social networking services (SNS) and include the typical examples of the field as there are Facebook or Xing.

2.2 SME

Though having no standard definition on the term SME, EU guidelines [5] distinguish SME with the help of two characteristics. These two attributes in use are: number of employees and annual turnover. With regard to the annual turnover it is stated that the value of the balance sheet can be used instead, and consequently one or the other has to be fulfilled. However, one of the two and the amount of employees must be met, one attribute does not suffice. Accordingly, an enterprise with less than 10 employees is a micro enterprise. Moreover, the enterprise's turnover is not larger than 2 million Euros per year. This indicates that an enterprise with 10 to 49 employees and a turnover of 10 million Euros (same value holds for the balance sheet) is considered a small enterprise. Consequently, a SME with 50 to 249 employees and an annual turnover of 50 million Euros (balance sheet: 43 million Euros) is considered a medium enterprise. In Germany additionally the term "Mittelstand" exists, which also counts enterprises with less than 500 employees into the group of medium enterprises, as long as their annual turnover does not exceed 50 million Euros[18].

Furthermore, it can be stated with regard to KM that SME apparently suffer stronger from difficulties caused by information overload than larger enterprises do. One explanation for this fact might be the fewer resources, e.g. employees responsible for certain tasks. An employee in a SME usually is not as specialized in a certain field as may be in larger enterprises. Moreover, he or she usually has no backup. Consequently, KM problems arise when the employee is no longer available due to e.g. retirement or illness. One way to address this problem may be to generate a backup for these situations without stressing the resources as far as hiring a new employee, by the electronic capture of knowledge as supported by KMS and KM applications.

2.3 Knowledge Management Systems and Technology Support

Knowledge can be considered the combination of all individual skills and abilities in use to solve a problem[6]. However, as for SME no official definition on the term of KM can be given, since the individual definitions provided in literature vary widely. Assuming that knowledge enables certain actions and problem solving, KM within in the organization is desirable, also with regard to organizational learning. Yet, it should be noted here that two main approaches on KM exist, one human-oriented focusing on the transfer of knowledge between the employees of an organization, whereas the other technical-oriented approach, aiming at supporting the process of KM by the use of information and communication technology (ICT), results in a Knowledge Management System (KMS). Though both approaches have their limitations this paper will focus on the latter, since the first can go without any technical support at all and is not of interest here. As this paper focuses on the use of social media and falls into the line with the technical orientation, following this idea a KMS can be defined according to Maier([7], p.86):

"A knowledge management system is an ICT system in the sense of an application system or an ICT platform that combines and integrates functions for the contextualized handling of both explicit and tacit knowledge, throughout the organization or that part of the organization, that is targeted by a KM initiative. A KMS offers integrated services to deploy KM instruments for networks of participants, i.e. active knowledge workers, in knowledge-intensive business processes along the entire knowledge life cycle. Ultimate aim of KMS is to support the dynamics of organizational learning and organizational effectiveness".

As already described in our former work we assume the model of Maier [9] valid who categorizes a KMS as a holistic approach which is about to fulfill certain services, as depicted in figure 1. Assuming social media and SNS part of a possible KMS we consider them especially interesting for the core KM services "Collaboration" and "Publication" in layer III, supporting the provision and exchange of knowledge between the employees and also with customers.

Fig. 1. Knowledge Management System architecture according to [7]

Though mainly focusing on the technical functionalities with our work, aspects like organization culture cannot be neglected as they for instance strongly influence the attitude of the employees towards the technology[4].

3 Methodology

To visualize the needs for and the benefits social media can have for SME, two SME in Germany were integrated into the case study presented in this paper. We therewith follow the action research approach demanded for the field of KM by Andriessen. [8] Within the study several steps were taken to show the organizations what KM can offer, as beforehand they have not fully realized and implemented the field. Furthermore, a partial requirements analysis took place to find out what was needed in their individual fields of business. Before starting all concerned employees were confronted with the term of KM, as several participants were uncertain whether what they did could be considered KM.

To gather the results for the case study we performed the following steps:

1. Identifying the recent state of KM and the integration of Social Media in the organization. This was done by confronting the employees with some statements about the effects of KM in their organization they should complete. First, they were to name the positive aspects, and afterwards the negative aspects by completing the statements given below in table 1. The statements were provided on paper with white space to be completed. This step also served as a rather informal starting

Table 1. Initial statements on KM in the organization

positive	negative
Ideas for the organization	Problems occurring at work
Information is shared by...	Why don't I say what I know?
I show my knowledge by ..	Why is there a loss of Know-How?
I find information by..	Why is an idea/ message transported?
The distribution of work is..	Why can't I find the knowledge needed?

point, which allowed the individual participant a slow start in the topic without the pressure of having to come up with ideas instantaneously in front of a group.

The statements were completed by each participant individually and collected afterwards. In addition the individuals were observed in their working activities to see which tool they used and which tasks were part of the regular working day.

2. Conducting interviews with the CEO and employees of the company gathering their expectations and demands towards possible solutions. The core questions used in the interviews are provided in the table below.

Table 2. Questions from the interview

Question on SNS
Which tools or SNS are used in the organization?
How long have you applied the tools?
How many employees use the tools actively?
How do you support SNS in the organization?
Why did you decide to introduce SNS?
Which goals are to be fulfilled by using specific SNS?
How much effort is necessary to keep the SNS working?
Was there a training accompanying the introduction?
Which contents are to be found in the SNS?
How is quality and an up to date status guaranteed?
What benefits could be gained from the use?

Before a third step could follow the participants were encouraged to get acquainted with the possibilities of social media in more detail with the help of provided literature, video tutorials as well as exchange with other employees.

3. Brainstorming session to sum up the results and generate the final ideas for further developing and integrating social media into the organizations business processes. This session took place in two phases, first there was an individual meeting with each employee and afterwards one session for all participants in the organization. In this part the moderating part was taken by a member of our research group which was not involved in the organization and had a rather objective point of view on the uttered points. Finally the modus of operation of social media was discussed, to support the successful usage and integration in the business processes since the pure technology is proven to be just a minor part in the KM.

By following this approach we were able to gain the results providing an overview of the situation in the organizations as presented in the following chapter.

4   Case Study

Following we present two knowledge intensive organizations which can be classified as SME and their approaches to use social media for KM.

4.1 The Participating Organizations

This paragraph shortly presents the two participating organizations, which were willing to use social media to address the KM problems in their business processes. For both companies the CEO and two employees were involved in the phases of the case study. The case study took place in February/ March 2012.

Organization 1. This organization was founded in 1992 in Wismar, Germany. The value creating processes of the company are mainly situated in the field of consulting and engineering. Their main tasks lie in consulting services for other SME for e.g. certification/ auditing processes, as well as image processing and technical image construction. The organization employs 10 people (7 full time, 3 students), where the average age of the full time workers lies between 50 and 60 years. The organization covers 50 to 70 orders a year resulting in a net sales volume of approximately 700,000 Euros in 2011. The orders usually are processed by phone and afterwards face-to-face consulting. The organization considers itself in the b2b sector. According to the definition provided in section 2.3 this company is a standard small company.

Organization 2. The second organization in the case study was founded in 2007 and is located in Berlin. The values of the organization are created in the field of Amazon retailing, e-commerce and re-selling. In 2001 the business activities resulted in approximately 1,000,000 Euros net sales volume. To achieve this volume the organization has to cover around 100,000 orders a year, mostly with private customers and consequently it can be classified to be in the b2c sector. The organization has 16 full-time and 10 part-time employees, all under the age of 30. Moreover, a high fluctuation of employees can be noted. They accomplish order processing mainly via e-mail, sometimes by phone calls. Considering the definition this company is volume wise as well as by the employee number a small company.

4.2 Results

Afterwards the results of the case study will be described. Therefore for each of the organizations two partial results will be provided: a) demand identified in KM and b) identified Social Media solutions.

Situation in Organization 1. In organization 1 following demands could be identified: A wish for more information on the decisions made concerning the organization was uttered, as there are CEO opinions as well as possible new positions, orders and tasks. Moreover, a better, improved collaboration with emphasis primarily on employee-to-employee communication should be established. This also included the desire for information sharing by the use of shared documents. In addition, the employees asked for better and especially more opportunities to develop ideas. So an idea and innovation management is needed, which was not yet addressed directly. There is a central weekly meeting but hardly time to talk about new ideas and since an extra permission to use working hours for special activities is necessary, employees do not dare to ask for extra

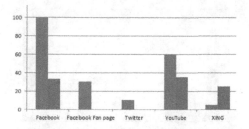

Fig. 2. Percentage of employees using SNS (red: organization 1/ blue: organization 2)

time. Further when new ideas arise and exchanged they are not stored for further actions and therewith forgotten.

After emphasizing the strong social component in KM, which is manifested in the organization's culture it was hinted, that if KM is a topic time has to be spared for arising tasks, though they might not lead directly to an invoice. In the organization there is also rarely any money left to address KM problems. Finally, the need to use knowledge for further acquisition possibilities was uttered.

When looking at the recent state of the art in the organization the situation was described as follows: Though being acquainted with the Internet and the ideas of SNS the organization was not using social media for a long time. They have tried using an instant messenger, which turned out to be unreliable, not freeware, and the complaint was that it did not allow for exchange between more than two people at once. Almost half the employees mentioned that they are not interested in using an instant messenger at all, since they do not see any benefits from its use and feel it would disturb their work. Consequently, emails were mostly used to transfer ideas and information. Individual employees used solutions like Skype, which was in between forbidden by the CEO and afterwards not re-established. The actual use of individual SNS is depicted in figure 2. The time used for activities within the whole company within a week within a certain SNS is given in figure 3.

Concerning the search for information: there is an Intranet solution in use, which was also available from outside of the organization by a static IP address. However, only two employees are actively working with it. Reasons for this behavior for this were identified: it is too complex in access rights, actually finding a piece of information is difficult since the structure of the solution and the names of the documents are not self-explaining. Hence mobile solutions like smartphones were not common among the staff, only two employees own one. And after repeating the question for SNS illustrating it by examples it turned out that there is one employee who started using Xing for acquisition purposes. In the beginning this was not supported by the CEO, yet after it turned out that these activities actually lead to new customers it is agreed on. Still this leads to another problem: there is a need for further information on the abilities of the other employees and their skills. The activities on this field last now for about a year. Moreover, two employees are privately acquainted with Facebook.

Finally, the need for tacit knowledge sharing was uttered in combination with the establishment of electronic workspaces, so one can meet despite being on different locations.

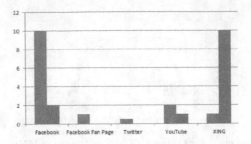

Fig. 3. Hours per week spent on SNS (red: organization 1/ blue: organization 2)

Summing the results gathered here up we gained the following impression: the organization was willing to engage in KM, however was not yet able to come up with concrete ideas on how to do this and if so only considered the technical point of view. But more important there is the actual integration in the business processes and consequently the insight that time has to be spared for KM tasks. With the example of the Xing usage, direct monetary benefits were seen and accordingly that solution was allowed, yet a strong wish for a direct monetary benefit to be named for further activities was uttered. Moreover, we gained the impression that due to the age of employees not much about the possibilities of social media was known, the personal engaged driving employees establishing these media in the company were missing as the CEO was not engaged in this topic himself. Moreover, he did not support possible initiatives due to the sparse resources. This attitude became also obvious when the interviews showed that the employees were interested in the possibilities a wiki can offer but were blocked by the CEO's doubts regarding time lacks and communication overload.

Situation in Organization 2. During the interviews in organization 2 these demands arose: One of the first points mentioned was a necessary idea development and management solution, which should as a second thought also include customers. By now an exchange only took place informally and ideas were easily forgotten or no longer worked on. The participants actually wanted a kind of idea base, offering the possibility to collect, search and comment on ideas including a weekly review. In addition the strong fluctuation among the employees was identified as a problem which needed to be addressed. These resulted in the fact that abilities and skills of colleagues were unknown, which was accompanied by the wish for a stronger communication on work problems between the employees. Additionally the insight was gained that a working calender system was needed as well for coordination purposes. The need for exchange also resulted in the idea of wiki also for customer purposes to allow for a constant provision of information and knowledge combined with the possibility to comment and discuss the available items. It would be best if this were connected to the Facebook fan page and enclosed an internal twitter account.

Moreover, a demand for web 2.0 technologies was uttered to stay up-to-date as a sales company. This should result in a new website including social media like a Xing business page and a twitter account that automatically tweets on product updates and innovations of the organization. The organization could also imagine to use a virtual world to improve video and call conferencing, sales presentation and other remote meetings, they emphasized the advantage of an avatar moving instead of the real person which would save time and money resources.

In contrast to organization 1 this one is already active in social media. Most of the rather young employees privately own a Facebook profile and these are up to 6 years old. The organization also maintains a Facebook fan page for about 3 years now which at the moment has about 200 followers. Furthermore, they run a twitter account which is filled automatically and is now running for 3 years. The percentage of employees having access to different SNS is depicted in figure 2. The organization is also working with Xing, mostly to find new employees or experts they need within a certain field. At the moment they are actually looking for an external web designer also familiar with social media to build a new homepage as desired.

Every employee owns and uses a smartphone, even if it is only for private issues, so they are acquainted with the offered possibilities. Regarding extra tasks the employees mentioned that they need to ask for permission to spare time on things KM since there is a strong need to produce action that can be invoiced. This is especially disturbing for developing new ideas. Moreover, in the field of KM they indicated that there is a need the calenders to be updated to be able to gather information on meetings and staff this way. A calender should be agreed on and the employees were to use it, yet it has to provide suiting access rights to look into colleagues calenders. For pure document exchange they mainly use email. Almost every employee has an instant messenger installed, however mostly for private use, even Skype is in use for private as well as business use.

The overall impression of this organization was that they were interested in almost everything available in the social media sector, which was due to the rather young age of the employees and their personal interest in the available solutions. Yet the problem is to integrate the social media efficiently into the business processes. A strategic plan of how to apply the available solutions was missing as was someone definitely responsible for the tasks. The organization has already realized that they have a certain need for KM and that it is constantly growing, yet by now they have not adapted their organization culture to this need. They could not name the specific benefit of their Facebook activities besides staying in contact with their customers, hence have the idea of using the media for their idea management.

Solutions to Implement. In general these were the main fields of interest for the organizations in the case study: communication among employees as well as with customers, knowledge transfer and innovation management. Table 3 provides an overview on the solutions and activities from the field of Social Media and KM which were considered to be implemented additionally to the available activities. The likelihood of implementation and usage was provided by the CEOs using a Likert scale varying between "very high" to "not interesting".

It can be seen that organization 1 tends to implement a wiki, mainly for idea collection, so ideas can be stored and discussed and are not forgotten. As a means for general knowledge sharing this is only interesting on a long term. However, the organization sees that they are in urgent need for an improved communication, and chooses to improve its instant messaging system. Asked for the reason for their choice they named the extra status information available (e.g. availability, recent activities). The organization

Table 3. Evaluation of possible solutions and actions from the field of Social Media

	Org 1	Org 2
Wiki	very high	middle
Idea collection	very high	very high
Personnel index	middle	high
Calendar use	middle	very high
Virtual World	not interesting	high
Knowledge sharing	middle	middle
Improved instant messenger	very high	very high
Information accuracy activities	low	not interesting
Discuss. forum, blog	not interesting	middle

does not see further use for discussion forums or a blog, as they on the one hand argue that a discussion can be included in a wiki and on the other hand refer to their weekly meeting.

Organization 2 is also interested in using a wiki but on the long term, for them it is more important to improve their Facebook activities and be better informed about their actual available skills, which vary strongly due to the employee fluctuation. They also want to improve the instant messenger use, especially with regard to further information and the separation from private use which lead to a high variety of products in use. They regard the idea collection as important as well but for now are not sure which solution to use.

It was surprising that none of the organizations was willing to work on information accuracy, which from our point of view is due to the fact that both are just beginning to coordinate their activities and are not aware of the problems arising in that field. Even the question of who would be reliable for maintaining the technical solution and giving support in the case of questions did not arise In addition we found that some demands can easily be addressed by changing the organizational culture in relation to social media, as for example the very critical attitude of the CEO towards innovations from that field. Also the wish for quick wins was confirmed by the behaviour of the organizations constantly emphasizing an easy implementation and a fast pay-off.

By extending their IT solutions both organizations can overcome the insecurity of mail exchange they mentioned. Moreover, overhead by mass mails and resent mails can be reduced. But it is not the technology only, there is a need for a change in working mentality from push to pull, yet this needs user-friendly publish and search facilities. In addition both organizations have to be aware of privacy issues, which may arise due to using private profiles or seeing the activities on a wiki. And finally the integration into the business processes must be clarified to be able to determine who owns important knowledge and is bound to enter it into the systems.

4.3   Summary

Returning to the research questions on which our research concentrated we can sum up the results for these two organizations to: For the first question: Social media is mostly used beginning from the private sector, and the more people are in touch with a certain solution the more promoters a solution will have within an organization. Having more customer contact with minor volumes per order a SNS like Facebook seems to be more suitable. This also holds regarding the customer type, SNS seem to be more valuable for the private customer.

On question 2 we found that both organization had problems integrating the solutions available into their business processes, especially in organization 1 with elder employees acting in the b2b sector we found more barriers established than integration activities. An automation for some processes like twittering news is possible as was shown by company 2. Yet, both companies did not show a permanent SNS integration in working processes. On the contrary both organizations emphasized they fear the loss of time through distraction.

Regarding benefits: In organization 1 we found an efficient use of Xing as contact platform for customer acquisition. In organization 2 it is used for the search for external experts or possible staff members. However, this is minor KM related. Although both organizations were looking for possibilities to share knowledge they had not implemented social media for this task. Consequently we can state they have only just begun to see possible potentials like expert networks, knowledge bases and discussion forums. In addition at least organization 2 also saw, yet did not implement, social media as a means for idea management. Furthermore, we gained the impression that SNS are very popular with younger employees, however mostly for private use. The real benefits of a business use are hardly known, though SME see the campaigns of big companies in the media. Hence, it remains unclear for them how to efficiently use it, since there may exist literature and examples on the topic but due to limited time resources a thorough involvement with the topic is not possible. And here the core problem is once again the fact that KM does not deliver invoices short termed and therefore is often not considered interesting with a high priority for SME, even though they might be knowledge intensive like the two in our case study.

5   Conclusions

Resuming the case study committed and presented in this paper and relating it to our former work [1], [2] on KMS in SME several points were confirmed. Once more we could see that a proven systematic approach to introduce KM in SME is still missing, whereas the question for the monetary benefits to be gained from such a solution remains the strongest point while introducing innovations as social media. Moreover, we could see that the attitude towards an active KM in SME still has to be improved, especially with regards to the organization culture allowing time to be used for KM purposes. These relation between technology, organization and human are well document in KM literature [9], yet their practical implementation as well as awareness for them is still insufficient. To allow a better decision making on the introduction of KMS/KM tools in SME reference values and a reliable framework could help to build up a good point of orientation and illustrate the success of such activities, especially if these were free from advertisement and a summarizing subjective illustration.

With regard to social media we found that the solutions from SNS are known, while others, older solutions from the field of CSCW like e.g. instant messenger and discussion forums are almost forgotten. The benefits of those can hardly be estimated, especially not in monetary units. The younger the employees the more involved they are in social media and able as well as willing to use it. However, the mixture with private use also holds the risk of privacy issues and distraction. In addition the private use is not always tolerated, resulting in an antipathetic attitude towards the tools being introduced from private initiatives. Yet, the possibilities for pure content provision within the company were not realized as social media is strongly.only associated with SNS and its communication characteristics. But there is a strong interest in content provision by more than a single employee as well as a constant exchange between all employees and the management. This seems from employees side the possibility to influence the organizations fortune.

However this needs altering the ways of acting and working in the companies, seeing the KM tasks not as an extra but as an essential part of the work. It is (by far) not sufficient to chose the social media instrument - social media usage requires re-thinking the question of corporate knowledge possession, as every new KMS needs motivating the employees and providing them with as detailed information as it is necessary to get acquainted with the technology. A change in operation includes changing the corporate culture, use activation and usage instruction (training as well as the establishment of solutions as a standard). In picking out Social Media instruments, an organization must recall that it can use a bunch of ready and well working Internet applications like Facebook, Twitter or XING, that are somehow restricted to the use of the organization or it can adopt applications that are just designed for the use within its business processes.

5.1 Future Work

Further work on this case study includes a revisit of the organizations to collect feedback on how the solutions have been adopted on the long term. The CEO and employees should be able to verify the solutions to allow further improvements and in the spirit of KM discover and document failures and shortcomings. Within a year a follow up questionnaire is to be conducted to get feedback on the solutions running and the results, as well as the quality they delivered. Certainly this will come down to the question on monetary benefits, however other approaches like the OMIS Success model as introduced by Jennex/Olfman [10] using the perceived benefit idea might deliver a wider and more profound response on the usage of SNS in SME (see figure 4).

By using the dimensions based on the DeLone/ McLean IS Success model the quality and benefits of the social media can be distinguished and therewith provide an idea to allocate possible shortcomings or special success. However, the OMIS Success model is one of the models which is not operationalized (though adapted to SNS in [11]) and therewith leaves the concrete application to the individual researcher. This allows for individual adoptions of it, yet prevents the chance for comparison of the individually gained results. To avoid this and allow for a deeper impact of the model in the practical KM of SME we suggest the establishment of a framework as shown in [1].

Before being able to implement this we had to cope, like described in the introduced case study here, with the topic of knowledge needs analysis. This topic includes various social empirical methods (interviews, questionnaires, observations) but lacks a systematical approach of what to use when since the methods demand high expertise to gain a result to be used for KM implementation in SME. Anyway from our point of view

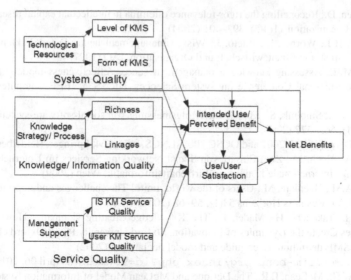

Fig. 4. OMIS Success according to [10]

this analysis is necessary to provide a problem oriented solution for KM, tackling the individual organizations needs and allowing them to chose their focus from the services as suggested in the generic architecture for KMS by Maier (see figure 1). By integrating the services with the dimensions of the OMIS Success model a profound benefit oriented base for recommending a problem based solution should arise. The base for decision making within this framework however is delivered by the needs of the organization, and should therefore be captured methodically to allow a secured entry into decision making within that very framework.

References

1. Borchardt, U.: Towards Value-Driven Alignment of KMS for SME. In: Abramowicz, W., Maciaszek, L., Węcel, K. (eds.) BIS Workshops 2011 and BIS 2011. LNBIP, vol. 97, pp. 220–231. Springer, Heidelberg (2011)
2. Borchardt, U.: Towards A Value-Oriented KMS Recommendation For SME. In: Liu, K., Filipe, J. (eds.) KMIS 2011 - Proceedings of the 3rd International Conference on Knowledge Management and Information Sharing, October 26-29. SciTePress, Paris (2011)
3. Davenport, T.H., Prusak, L.: Working Knowledge: How Organizations Manage What They Know, Knowledge Management. Harvard Business School Press (2000)
4. Danah, M.B., Nicole, B.E.: Social network sites: Definition, history, and scholarship. Journal of Computer-Mediated Communication 13(1), 210–230 (2007)
5. The new SME definition: User guide and model declaration, (2003), http://ec.europa.eu/enterprise/policies/sme/facts-figures-analysis/sme-definition (accessed April 06, 2011)
6. Probst, G., Raub, S., Romhardt, K.: Wissen managen: Wie Unternehmen ihre wertvollste Ressource optimal nutzen, 6th edn. Gabler Verlag, GWV Fachverlag GmbH (2010)
7. Maier, R.: Knowledge Management Systems: Information and Communication Technologies for Knowledge Management, 3rd edn. Springer, Heidelberg (2007)

8. Andriessen, D.: Reconciling the rigor-relevance dilemma in intellectual capital research. The Learning Organization 11(4/5), 393–401 (2004)
9. Bullinger, H.J., Wörner, K., Prieto, J.: Wissensmanagement heute: Daten, Fakten, Trends, Fraunhofer-Inst. für Arbeitswirtschaft und Organisation, IAO (1997)
10. Jennex, M.E.: Assessing knowledge management success/effectiveness models. In: 37 th Hawaii International Conference on System Sciences, HICSS35. IEEE Computer Society (2004)
11. Reisberger, T., Smolnik, S.: Modell zur erfolgsmessung von social-software-systemen. System 103(1), 565–577 (2008)
12. Schuler, D. (ed.): HCII 2007 and OCSC 2007. LNCS, vol. 4564. Springer, Heidelberg (2007)
13. Richter, A., Koch, M.: Zum Einsatz von Social Networking Services im Unternehmen. In: Proceedings Internationale Konferenz Wirtschaftsinformatik, Wien (2009)
14. Kaplan, A.M., Haenlein, M.: Users of the world, unite! The challenges and opportunities of Social Media. Business Horizons 53(1), 59–68 (2010)
15. Nonaka, I., Takeuchi, H., Mader, F.: The Knowledge-Creating Company: How Japanese Companies Create the Dynamics of Innovation. Oxford University Press, Oxford (1995)
16. The new SME definition: User guide and model declaration (2002), http://www.ifm-bonn.org/index.php?id=90 (accessed April 06, 2011)
17. Delone, W.H., McLean, E.R.: The DeLone and McLean Model of Information Systems Success: A Ten-Year Update. J. Manage. Inf. Syst. 19, 9–30 (2003)
18. Statistics agency Germany, Enterpriseregistry (2007), http://www.regionalstatistik.de (accessed April 27, 2011)

Semantically Enriched Obligation Management: An Approach for Improving the Handling of Obligations Represented in Contracts

Barbara Thönssen[1,2] and Jonas Lutz[1]

[1] Institute for Information Systems, University of Applied Sciences and Arts Northwestern Switzerland FHNW, Riggenbachstr. 16, Olten, Switzerland
[2] Dipartmento di Matematica e Informatica, University of Camerino, via Madonna delle Carceri 9, Camerino (MC), Italy
{jonas.lutz,barbara.thoenssen}@fhnw.ch,
barbara.thoenssen@unicam.it

Abstract. Contract Management becomes increasingly important for companies and public administrations alike. Obligations and liabilities are described in contract clauses that are often buried in documents of a hundred pages and more. Although commercial Contract Management Systems (CMS) are available, with a few exceptions relevant information has to be extracted manually which is time consuming and error prone. But even if information extraction is automated and contracts are managed using a CMS, dealing with obligations is still a challenge. Whereas the CMSs deal well with time triggered obligations like periodical payments by setting up corresponding workflows, they fail to trigger obligations based on events, as this knowledge is out of the systems' scope. We introduce an approach to fill the gap as we relate information about the obligations managed in a CMS with background knowledge modelled in an ontology. The ontology is a formal representation of an enterprise architecture extended by top-level concepts. Motivating scenario for the approach is the contract management of a large company. For proof of concept a prototype has been developed.

Keywords: Enterprise Architecture, Enterprise Ontology, Risk Management, Contract Management, Information Extraction, Obligation Management.

1 Introduction

Contracting is a critical issue for enterprises and public administrations alike, as "All organizations depend upon their ability to make, manage, monitor and perform against their business commitments, or to structure and oversee those they receive from their trading partners" [13]. To manage not only the sheer volume of contracts but their growing complexity, The International Association for Contract and Commercial Management (IACCM) believes that automation is fundamental and of critical importance. In a study that IAACM provided [13], benchmarks have shown that the benefits of Contract Management Software are significant for most organizations, despite reservations over the maturity and functionality of the available

A. Fred et al. (Eds.): IC3K 2012, CCIS 415, pp. 337–349, 2013.

options. One software flaw is the weak support of obligation management. Even if a Contract Management System (CMS) is in place, clauses, terms, conditions, commitments and milestones are buried in unstructured text. Thus, a first step can be document segmentation and information extraction. In the Swiss national funded project DokLife contract documents are analysed and information like contract partners, contract date, jurisdiction, etc. is automatically extracted. Moreover, the documents are divided into disjoint sections for further analysis, e.g. classification and section specific information extraction. The original contract document, its segments and the extracted metadata are imported into a commercial CMS.

However, although with such an approach contract *document* management is improved, *obligation* management remains a challenge. Whereas time triggered obligations (e.g. sending a report at a certain time) can be supported by the workflow functionality of a Contract Management System, event triggered obligations cannot. The problem is that these events occur outside the CMS and cannot be foreseen. If for example a company goes bankrupt it is of vital interest to know whether a relationship exits with this company, what kind of relationship that is, whether the relationship is contracted, and whether obligations are to be met.

Our approach aims at providing the context for contract management by linking the extracted information, i.e. contract metadata to business objects described in an enterprise ontology. In the ontology (background) knowledge about business objects and business events are made explicit. Hence, this knowledge can be used to automatically assess business events, for example reported in external sources of commercial information or newspaper reports, and to identify affected obligations for which the respective contract(s) can be retrieved.

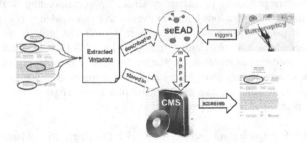

Fig. 1. Obligation Management Overview

Figure 1 gives an overview of the approach: information is automatically extracted from contract documents, the metadata is stored in the relational database of an CMS and mapped to the respective concepts in an ontology, which is a semantically enriched Enterprise Architecture Description. A business event (e.g. bankruptcy) triggers the search for affected contracts in the ontology and if found, the related documents can be accessed via the CMS.

The remaining paper is structured as follows: Section two describes the motivating scenario for semantically enriched obligation management. In section three background and related work is presented. Section four introduces the three pillars of our approach and in section five we give proof of concept by describing an implemented prototype. We conclude in section six and give an outlook on future work.

2   Motivating Scenario

"Contracts exist because separate legal entities that decide to form a relationship with each other need some formal record of understanding regarding the nature of the relationship and their respective rights and obligations" [13]. Motivating scenario for our approach is a large machine manufacturer with several regional subsidiaries all over the world. The enterprise as a whole as well as each subsidiary has hundreds of contracts with suppliers, customers, sub-contractors, etc. Contract Management is considered one of the top ten risks an enterprise must face [11]. Therefore, all contracts are managed with a CMS. The company's own contracts are created with the CMS, but the biggest number by far is issued externally and thus paper copies are scanned and then stored in the CMS. Many of the contracts are voluminous and identification of relevant information, such as obligations and conditions is cumbersome and time consuming. However, if this metadata is entered the CMS provides simple workflow functionality to trigger time related obligations like periodic payments (to make or to receive), submissions of reports, etc.

Besides managing contracts with a CMS, the machine manufacturer mandated an information provider for monitoring news about business partners, competitors, suppliers, etc. A risk manager is in charge of evaluating the news with respect to their impact on the company. This task is also time-consuming as well as time-critical and error-prone (with respect to completeness of evaluation).

To improve contract capturing, metadata generation and obligation management, the tasks should be automated.

3   Background and Related Work

Although the importance of contract management is well-known in the economy, little scientific research has been done on that topic. Contract management has been addressed in the European integrated project TrustCoM [22], that aims to develop a framework for trust, security and contract management in dynamic Virtual Organisations [6]. "Trust between VO [Virtual Organisations] members can be supported by each being transparently aware of the obligations and performance of others, so that business risks are both mitigated, and monitorable" [22]. There are two main differences to our approach: Firstly, TrustCoM obligation (policies) are modelled in the form of Event-Condition-Action rules which define how the VO should adapt to failures, changes in requirements, security events, etc. Secondly, TrustCoM starts from contract creation based on (structured) templates. We consider that ideal but oversimplified. Many companies already have templates, approved by their legal department, which they are obliged to use. Even so, many commitments remain imprecisely described in contracts, for example what efforts the mitigation duty requires or what point in time the obligation arises [10]. In our approach we address these issues as we start from unstructured contract documents and describe obligations semantically instead of transforming them.

In a well-known approach for describing entities of an enterprise semantically in terms of an ontology Bertolazzi et al. [4] analysed and compared existing ontologies, namely the Toronto Virtual Enterprise (TOVE) and The (Edinburgh) Enterprise Ontology (EO) with their own proposal for a Core Enterprise Ontology (CEO). Leppänen [14] introduced a context-based enterprise ontology and referred in his contribution in addition to TOVE and EO to the REA Enterprise Information System. Most recently, [18] introduced a ContextOntology for dealing better with change in enterprises.

Whereas the term 'ontology' emerged in the context of Artificial Intelligence and later in the World Wide Web, particularly of the Semantic Web [8], roughly at the same time enterprise architecture became generally known as a management topic in the end of the 1980ies [21]. Contrary to enterprise ontologies, which are concrete representations of (generalized) enterprise architectures developed to be re-used in enterprises [27], adopted and enhanced to the enterprise's specific needs, Enterprise Architecture Frameworks (EAF) "provide the guidance and rules for developing, representing, and understanding architectures" ([7], p 4).

Kang et al. [18], Hinkelmann et al. [16] and Thönssen [26]) suggested relating an enterprise ontology to an Enterprise Architecture Framework in order to increase the quality of the enterprise ontologies, for example with respect to completeness. Our approach is based on ArchiMate, a standard that not only provides guidance but comprises a lightweight and scalable modelling language for architecture descriptions [17]. Since ArchiMate's representation language is based on the UML 2.0 notation for class diagrams the 'meaning' of the entities is not described. "A meaning [...] is a specialized description that aims to clarify or stipulate a meaning. [...] Typical examples of meaning descriptions are definitions, ontologies, paraphrases, subject descriptions, and tables of content" [17].

Despite the consent about using an ontology for describing meaning no agreement has been achieved yet on the appropriate level of formalization and the degree of formality [8]. Fox & Grüninger [8] regard an enterprise model as the *computational* representation of the structure, activities, processes, information, resources, people, behaviour, goals, and constraints of an enterprise. This means that the enterprise ontology should be represented in such a way that a machine can *process* it. Because ArchiMate's original representation language is not expressive enough to achieve that, a more formal representation is needed for our purpose, like for example RDF(S) or OWL [23].

Having enterprise objects represented in an ontology, for example the *business object* 'contract' and related context information like *business actor* 'contract partner' and *product* 'Service Level Agreement', allows the addition of background information to a contract. In other words, a contract document and its metadata can be stored in a CMS but the *business object* 'contract' and its context are stored in the enterprise ontology. That leads to the issue of relating databases to ontologies, a reqirement that has been investigated from the very beginning of the semantic web [16]. Approaches to combine relational databases (RDB) and ontologies have become known as 'database to ontology mapping problem', or more generally characterized as 'object-relational impedance mismatch problem' [16]. The problem that has to be solved lies in the structural difference of the relational and object-oriented models. It has been studied from different points of view for various kinds of reason [3], like semantic annotation of dynamic web pages, heterogeneous database integration, mass generation of Semantic Web data or ontology learning [16]. Sahoo et al. [15] distinguish between 'Automatic Mapping Generation' and 'Domain Semantics-driven Mapping Generation'. Whereas the first method directly maps RDB and RDF schemas, the latter considers "domain semantics that are often implicit or not captured at all in the RDB schema" ([15], p 5). As the focus of our research here is not on the mapping, we refer to some excellent papers on the subject (e.g. [4], [19], [11]). We also refer to the W3C RDB2RDF Working Group who provides many publications on automatic mapping, for example a strategy for directly mapping relational data to RDF [2], and a language specification (R2RML) to express customized mappings from relational databases to RDF datasets [7].

For our approach, Direct Mapping (DM) is sufficient as it is performant, simple and easy to implement. Direct Mapping defines an RDF graph representation of the data in any relational database with a set of common datatypes [2]. The DM technique entails the transformation of relational database data and schema into an RDF graph which is called the direct graph. The relational tables are mapped to classes in an RDF vocabulary, and the attributes of the tables to properties in the vocabulary [12]. This technique has been introduced by the RDB2RDF working group [19].

Our approach carries research further with respect to modelling the context of a contract in terms of an ontology that represents enterprise objects in a way that is machine executable. Taking a standard of an enterprise architecture description as a basis, i.e. formalizing and enhancing ArchiMate, is a new approach to represent the background information needed for event based obligation management.

4   Enhancing Contract Management

Our approach comprises three parts: contract segmentation and classification, metadata creation, and obligation monitoring. Because of the limited space we focus on obligation management and only briefly introduce the other two parts; they are simply pre-requisites but our approach would work as well with manually entered metadata.

4.1   Automatic Contract Segmentation and Classification

In order to facilitate the problem of metadata creation especially for large contract documents a contract is automatically split into disjoint segments. Paragraph numbers and titles are recognized and the separated parts are stored for further processing.

Figure 2 shows the different parts of a contract document.

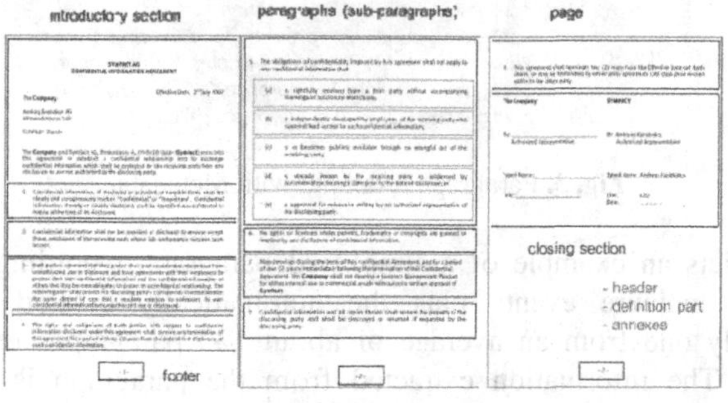

Fig. 2. Contract Segments

Contract segmentation is implemented in a commercial third party product for layout analysis to identify paragraph numbering based on regular expressions. After that, each of the paragraphs is classified with respect to the obligation type, e.g. *finance, report, notification.* Classification is done by a Support-Vector-Machine (libSVM, cf. http://www.csie.ntu.edu.tw/~cjlin/libsvm/). Paragraphs that are not numbered, like introduction and closing section or header and footer, are omitted as it is unlikely that they describe obligations.

4.2 Semi-automatic Metadata Creation

For the contract as a whole as well as for its paragraphs metadata are automatically extracted, e.g. contract partner, contract beginning, contract end, applicable law. For text analysis GATE (an open source solution for text processing; cf. http://gate.ac.uk/overview.html) is used and some JAVA web-services.

In addition to metadata for the whole contract additional metadata is created for single paragraphs, e.g. obligation type, trigger, dates and conditions. For this we also use regular expressions to extract due dates, conditions and triggers (time, period or event).

The analysed contract document, its paragraphs plus the created metadata are stored in an XML-file for further editing. In the DokLife research project for example, which is funded by the Federal Office for Professional Education and Technology OPET (Project KTI Nr. 10902.2 PFES-ES des Bundesamtes für Berufsbildung und Technologie) the file is imported into a commercial CMS (cf. to the project's Web-Site URL: http://www.doklife.ch for more information). Within the CMS for each obligation the text-segment and its metadata are displayed and the user can accept or correct it.

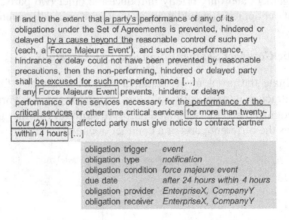

Fig. 3. Paragraph Example with Metadata

Figure 3 depicts an example of a simplified paragraph recording the obligation regarding force majeure events plus the automatically created metadata. The paragraph is only one from an average of about 120 paragraphs extracted from a single contract. The information extracted from the paragraph is indicated by a rectangle. As shown, some of the obligation's metadata can be extracted (obligation trigger, obligation condition, due date), obligation type is provided by the SVM, and obligation provider and receiver are created by resolving the information 'a party' (which can be the contractor, here Enterprise X, or the contractee, here Company Y).

Contract segmentation and classification, and information extraction build the basis for improving obligation management.

4.3   A Broader View on Contract Management

To improve obligation management a contract document is considered a representation that realizes a business object (The Open Group 2009) and "A business object is defined as a unit of information that has relevance from a business perspective". The quotation is taken from the ArchiMate framework, standardized by The Open Group (2009). We adopted that notion and consider a contract document a representation that realizes a contract business object and one or more obligation business objects. In ArchiMate, business objects are related to business behaviour elements, such as business events, to business roles and business actors. With that all context information necessary for dealing with event-based obligations, recorded in contract are met.

For our purpose we refined the existing ArchiMate concepts with respect to creating sub-concepts and additional relations. Figure 4 depicts a simplified excerpt of concepts and relations, depicted in the ArchiMate notation, relevant for our approach. We added two `BusinessObjects` `Contract` and `Obligation` that are realized in the `ContractDocument` (which is a `Representation`). A `BusinessObject` is associated to a `BusinessEvent`; `Bankruptcy`, `MoratoriumPetition`, `Injunction` and `Merger&Acquisition` are (instances of) business events. A `Contract` is concluded with a `BusinessActor` with the assigned `BusinessRole` of a `ContractPartner`.

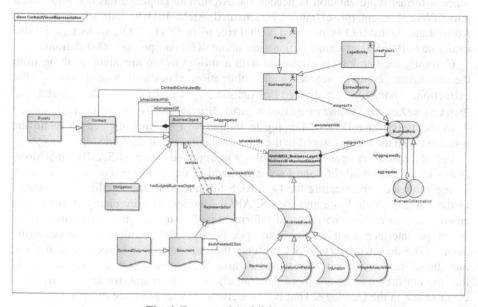

Fig. 4. Excerpt of ArchiMate concepts

Conceptually we add three top-level concepts (with its sub-concepts) time, location and event. With this we are able to model events, externally to an enterprise, for example ActOfGod (Hurricane, Tsunami, etc.) or ManMade (war, riots, etc.), and to relate it to BusinessObjects in order to infer obligations triggered by force majeure events.

Further development of ArchiMEO followed Fox & Grüninger (1997) methodology on elaborating competency questions.

Example of an informal Competency Question:

```
Given an environmental disaster (earthquake, hurricane,
tsunami, etc.) and some constraints (location or about a
business relationship), which business partners are
concerned?
```

From the Competency Questions content of ArchiMEO has been validated and further elaborated as shown in Figure 4. The Competency Questions are then rewritten in a formal way using SPARQL (see chapter 5).

5   SERO

As a proof of concept we developed a prototype for Semantically EnRiched Obligation management (SERO). Pre-requisite for SERO is the enterprise ontology and the results of contract segmentation and metadata creation for ontology population. This can be done automatically or manually.

The enterprise ontology used in SERO fully comprises the ArchiMate entities but is enhanced where necessary, as suggested in the standard [17], and detailed above. Since a formal representation is needed the ArchiMate language has been translated into OWL. The enterprise ontology is named ArchiMEO to indicate its roots in ArchiMate. ArchiMEO is modeled with Protégé in OWL 2 QL dialect (it is also available in RDFS) and actually comprises about 400 concepts and 600 relations.

Currently the ontology is populated with a sub-set of the metadata resulting from the automatic contract segmentation, obligation classification and information extraction, namely ContractDocument, Contract, Obligation, BusinessActor, BusinessRole and BusinessRelationship.

In the next phase Direct Mapping technique will be implemented to relate the instances with the entities stored in the relational database of the CMS.

For obligation management the ontology is imported into a SESAME triple store. Querying of the ArchiMEO ontology is realized in the SERO prototype by a JAVA Swing Interface implementing the OpenRDF Sesame API. The Application connects to the Sesame Triple Store and uses SPARQL queries, as exemplarily illustrated in chapter 5.1, to receive the required information. To simulate real world events the prototype interface contains fields to specify type and location of the occurring events. The depicted query in 5.1 presents the data related to a specific Use Case, introduced below, namely 'Earthquake' and 'Fukushima'. Of course, in SERO the implemented queries are more generic, allowing the search for any location and event stored in the Ontology. That is, via a graphical user interface one can ask for an event to search for affected business partners and if any contracted obligations.

The resulting information is added to JAVA objects and lists of objects such as `ContractDocument`, `Contract`, `Obligation` and `BusinessActor`, are displayed as tangent contracts by the event in the GUI.

To illustrate our approach for improving the handling of obligations represented in contracts, in the following two use cases examples are provided, based on the motivating scenario given above.

5.1   Use Case 1: Environmental Disaster

Initial Situation: In a newspaper an environmental disaster is reported and the machine manufacturer – as global player – wants to know if business partners are affected and if so, if obligations are due.

Analysis procedure: News, e.g. published on a newspaper web-site, are analysed using the same text analysis methods used for analysing the contract documents. For simplification in the prototype, such information is entered via a Graphical User Interface (GUI); in use case 1 this could be 'earthquake' and 'Fukushima'. Since in ArchiMEO events, e.g. earthquake, tsunami, flood are modelled and also information about locations, e.g. part of countries as Fukushima or Tōhoku are stored, knowledge about the type of the event (here: act of god) and the location (here: Japan) can be inferred. Based on the address of a business partner's production plants, headquarter, etc. it can easily be found out if he or she is affected or not. The query can be refined with respect to the closeness of a partner's location to the site of the disaster by exploiting geographical data for distance determination.

If a business partner is affected the kind of existing business relations can be inferred, respectively, what role that partner plays, e.g. supplier or consultant. Again, this information can be used to refine the query, as for example it would be more important for the manufacturer if the production plant of a supplier were in the epicentre of the earthquake than the headquarters of a consultant.

To find out whether obligations are due, the business object `obligation` is queried for obligation receiver and obligation provider. As depicted in Figure 3 each party has the duty to notify the contract partner if a force majeure event affects its performance. With our approach the business partner affected by the earthquake in Fukushima is automatically identified, his or her obligations can be checked and, as an act of god like an earthquake is a force majeure event, this obligation can be identified also automatically. As result a warning can be issued and the contract document in which the obligation is represented can be retrieved from the CMS.

Example of the query in SPARQL:

```
PREFIX eo:<http://ch.fhnw.eo#>
PREFIX rdf:<http://www.w3.org/1999/ 02/22-rdf-syntax-ns#>
PREFIX top:<http://ch.fhnw.top#>
PREFIX rdfs:<http://www.w3.org/2000/ 01/rdf-schema#>

SELECT ?businessActor ?eventType ?eventParent ?obligation
?role ?location ?partCountry ?locationCountry ?contract
WHERE {
Eo:Earthquake rdf:type ?eventType.
?eventType rdfs:subClassOf ?eventParent.
```

```
{ ?businessActor eo:businessActorIsSituatedInLocation
?locationCountry.
?locationCountry rdf:type top:Country.
} UNION { ?businessActor eo:businessActorIsSituatedInLocation
?location
{ { ?location top:cityIsLocatedInCountry ?locationCountry }
UNION {
?location top:cityIsLocatedInPartOfCountry ?partCountry.
?partCountry top:partOfCountryBelongsToCountry
?locationCountry. } }
}
?businessActor eo:businessActorHasAssignedBusinessRole ?role.
?businessActor eo:ibusinessActorPerformsObligation
?obligation.
?obligation eo:obligationIsAgreedInContract ?contract.

FILTER(?location = eo:Fukushima|| ?partCountry = eo:Fukushima
|| ?locationCountry = eo:Fukushima)
}
```

5.2   Use Case 2: Bankruptcy and Dislocation

Initial situation: An information service provider gives notice that a company filed bankruptcy in one country but at the same time opens a new production plant in another country. The machine manufacturer wants to know what consequences this has for him.

Analysis procedure: The provided information is analysed and ArchiMEO is queried for obligations due in case of bankruptcy for that very business partner. Whereas this part of the query is easy to execute, finding out obligations related to the production site is not. In SERO it can be done by inferring information from the location. Assume the production site has been in Singapore and is now moved to Vietnam; the manufacturer is located in Switzerland. Since Switzerland has a free trade agreement with Singapore but not with Vietnam, the reason for the supply contract is not valid any more. In ArchiMEO legal bases for contracts can be described and thus, in use case 2 not only the obligation to notify the machine manufacturer about the business event could be identified automatically but also contracts affected by a change of legal pre-requisites.

As a side effect, records management for contract documents can be improved as well. If a contract partner goes bankrupt all contracts become invalid. The documents' retention period starts by the end of the current year which can be triggered automatically based on ArchiMEO.

With our approach we can improve the handling of obligations represented in contract documents by providing the missing link to external business events (e.g. bankruptcy) or force majeure events (e.g. earthquake). We can automatically identify affected business partners, business relations, business roles inferring the enterprise ontology and trigger event-based obligations represented in contract documents which can be stored in a Contract Management System.

Our approach has been presented and reviewed by two focus groups, namely the user group of a commercial CMS and the consortium of the APPRIS project. APPRIS (Advanced Procurement Performance and Risk Indicator System) is a Swiss national funded research project (KTI-Nr. 12102.1 PFES-ES) with the goal of integrating risk, procurement and knowledge management into one early warning system.

6   Conclusions and Further Work

"According to 42% of enterprises in a new study the top driver for improvements in the management of contracts is the pressure to better assess and mitigate risks" [1]. With our approach, a first step in this direction is taken, as the risk of overseeing obligations and contracts affected by events (business, acts of god or man-made) can be minimized.

With the help of the SERO prototype the feasibility can be (and has been) demonstrated that obligations and contracts, triggered by business events, can be identified automatically and thus, contract management can be improved significantly.

Our approach will be used and further developed in the APPRIS project. The functionality of SERO will be used to detect warning signals based on risk indicators related to events in order to address the contract management risk.

Technologically, SERO will be improved with respect to instance management. Whereas at present the ontology is not yet physically linked to the contract documents stored in a Contract Management System, in the next version of SERO instances of ArchiMEO will be related to entities of a CMS's database via Direct Mapping.

Currently, all background information is fully stored in ArchiMEO. In the future, Linked Open Data sources like GeoNames will be integrated. The GeoNames geographical database for example covers all countries, and contains over eight million place names (Wick n.d.). Also promising is the Open Government Data initiative in Switzerland (opendata.ch) with respect to integrating business administration issues like laws and regulations. Research will focus on how these sources can be truly integrated into ArchiMEO instead of simply mashing them up and displaying them alongside each other [5].

References

1. Anon: Contract Management: Optimizing Revenues and Capturing Savings. Contract Lifecycle Management and the CFO (2007),
   http://findarticles.com/p/articles/mi_pwwi/is_200705/ai_n19058810/ (accessed April 19, 2012)
2. Arenas, M., et al.: 2011. A Direct Mapping of Relational Data to RDF. W3C Working Draft 20 September 2011, pp. 1–11 (September 2011)
3. Auer, S., et al.: Use Cases and Requirements for Mapping Relational Databases to RDF. W3C Working Draft 2 June 2010, pp. 1–18 (June 2010),
   http://www.w3.org/2001/sw/rdb2rdf/use-cases/ (accessed January 26, 2012)
4. Barrasa, J., Corcho, Ó., Gómez-Pérez, A.: R2O, an Extensible and Semantically Based Database- to-ontology Mapping Language. In: Proceedings of the Second Workshop on Semantic Web and Databases (SWDB 2004), Toronto, Canada (2004)

5. Bertolazzi, P., et al.: An Approach to the Definition of a Core Enterprise Ontology: CEO. In: International Workshop on Open Enterprise Solutions: Systems, Experiences, and Organizations - OES-SEO 2001, Rome, pp. 104–115 (2001)
6. Bizer, C., Heath, T., Berners-Lee, T.: Linked Data - The Story So Far. International Journal on Semantic Web and Information Systems (IJSWIS) 5(3), 1–22 (2009)
7. Das, S., Sundara, S., Cyganiak, R. (eds.): R2RM: RDB to RDF Mapping Language. W3C Working Draft 20 September 2011, pp. 1–32 (November 2011), http://www.w3.org/TR/r2rml/ (accessed January 26, 2012)
8. Dietz, J.L.G.: Enterprise Ontology. Theory and Methodology. Springer, Heidelberg (2006)
9. Dimitrakos, T., Golby, D., Kearney, P.: Towards a Trust and Contract Management Framework for dynamic Virtual Organisations. In: Proceedings of the eChallenges Conference (eChallenges 2004), Vienna, Austria (2004)
10. DoDAF. DoD Architecture Framework Volume I: Definitions and Guidelines. Architecture, I, pp. 1–46 (April 2007)
11. Fox, M.S., Grüninger, M.: Enterprise Modeling. AI Magazine 19(3), 109–121 (1998)
12. Fox, M.S., Grüninger, M.: On Ontologies and Enterprise Modelling. In: International Conference on Enterprise Integration Modelling Technology, vol. 97. Springer (1997), http://www.eil.utoronto.ca/enterprise-modelling/papers/fox-eimt97.pdf
13. Goetz, C.J., Scott, R.E.: Mitigation Principle: Toward a General Theory of Contractual Obligation. Virginia Law Review 69(6) (1983), https://litigation-essentials.lexisnexis.com/webcd/app?action=DocumentDisplay&crawlid=1&srctype=smi&srcid=3B15&doctype=cite&docid=69+Va.+L.+Rev.+967&key=2c499ddc896515ef43456ba87a993188
14. Grosse-Ruyken, P.T., Wagner, S.M.: APPRIS Project Report, Zürich (2011)
15. Hert, M., Reif, G., Gall, H.C.: A Comparison of RDB-to-RDF Mapping Languages. In: Proceedings of the 7th International Conference on Semantic Systems, Graz, Austria, pp. 25–32 (2001), http://dl.acm.org/ft_gateway.cfm?id=2063522&ftid=1054499&dwn=1&CFID=81605256&CFTOKEN=14608054
16. Hinkelmann, K., Merelli, E., Thönssen, B.: The Role of Content and Context in Enterprise Repositories. In: Proceedings of the 2nd International Workshop on Advanced Enterprise Architecture and Repositories - AER (2010)
17. IACCM, Contract Management Software. Market Sizing and Status Report (2007)
18. Kang, D., et al.: An ontology-based Enterprise Architecture. Expert Systems With Applications 37(2), 1456–1464 (2010), http://dx.doi.org/10.1016/j.eswa.2009.06.073
19. Kontchakov, R., et al.: The Combined Approach to Query Answering in DL-Lite. In: Proceedings of the Twelfth International Conference on the Principles of Knowledge Representation and Reasoning (KR 2010), Toronto, Canada, pp. 247–257 (2010)
20. Leppänen, M.: A Context-Based Enterprise Ontology. In: Guizzardi, G., Wagner, G. (eds.) Proceedings of the EDOC International Workshop on Vocabularies, Ontologies and Rules for the Enterprise (VORTE 2005), pp. 17–24. Springer, Enschede (2005)
21. Matthes, D.: Enterprise Architecture Frameworks Kompendium. Springer, Heidelberg (2011)
22. Sahoo, S.S., et al.: A Survey of Current Approaches for Mapping of Relational Databases to RDF, pp. 1–15 (2009)
23. Spanos, D.-E., Stavrou, P., Mitrou, N.: Bringing relational databases into the Semantic Web: A survey. Semantic Web – Interoperability, Usability, Applicability, 1–41 (2011)

24. The Open Group, ArchiMate 1.0 Specification (2009),
    http://www.ncbi.nlm.nih.gov/pubmed/20087110
25. Thönssen, B.: An Enterprise Ontology Building the Bases for Automatic Metadata
    Generation. In: Sánchez-Alonso, S., Athanasiadis, I.N. (eds.) MTSR 2010. CCIS, vol. 108,
    pp. 195–210. Springer, Heidelberg (2010)
26. Thönssen, B., Wolff, D.: A broader view on Context Models to support Business Process
    Agility. In: Smolnik, S., Teuteberg, F., Thomas, O. (eds.) Semantic Technologies for
    Business and Information Systems Engineering: Concepts and Applications (2010)
27. Uschold, M., Williamson, K., Clark, P.: Ontology Reuse and Application. In: Guarino, N.
    (ed.) Proceedings of the First International Conference FOI 1998, Trento, Italy,
    pp. 179–192 (1998)
28. W3C Working Group (2011), RDB2RDF Implementations,
    http://www.w3.org/2001/sw/rdb2rdf/wiki/Implementations (accessed
    January 25, 2012)
29. Wick, M.: GeoNames, http://www.geonames.org/ (accessed April 22, 2012)
30. Wilson, E.M.: TrustCoM Project Final Report V1 (2007),
    ftp://ftp.cordis.europa.eu/pub/ist/docs/st/
    trustcom-finalreport_en.pdf
31. Wilson, M., Arenas, A., Schubert, L.: The TrustCoM Framework for trust, security and
    contract management of web services and the Grid - V2, (2006),
    http://www.clrc.ac.uk/Activity/
    ACTIVITY=Publications;SECTION=225
32. Yu, L.: Developer's Guide to the Semantic Web online. A. Springer (2011),
    http://lib.myilibrary.com?ID=308236

Knowledge Management for Innovation and Product Development in Supply Chains

Lixin Wang and Athanassios Kourouklis

Business School, University of the West of Scotland, Paisley Campus, Paisley PA1 2BE, U.K.
{Lixin.wang,thanos.kourouklis}@uws.ac.uk

Abstract. New Product Development (NPD) has been viewed as critical to a firm's competitiveness and development. This paper presents a theoretical framework that utilizes existing approaches to facilitate effective use of knowledge management techniques in the NPD process. The development of this framework is based on the integration of relevant theoretical fields such as supply chain management, and open innovation. More specifically, it has accurately reflected on the nature of existing knowledge management systems and captured the core issues of NPD in a three stage approach of knowledge audit, knowledge calibration and knowledge absorption.

Keywords: Customer Knowledge Management, Supplier Involvement, Open Innovation, New Product Development, Product Lifecycle Management.

1 Introduction

1.1 Background

In today's business environment, firms are continually challenged by shorter product lifecycles, faster technological changes, demanding and sophisticated customers and the abiding trends of internationalisation, globalisation and convergence of industries. In response to these changes, increasingly, innovation and new product development (NPD) has been viewed as critical to a company's success [1-2]. However, the customer's rapidly changing preferences, the heterogeneity of their demands and resulting micro segmentation of product categories [3] have severely challenged a firm's capability to introduce new products. Most importantly, due to the paucity or deficiency of internal knowledge assets, firms have to rely on external knowledge to foster innovation and to enhance their performance [4]. This is also echoed by a shift in research on open innovation, where the purposive inflows and outflows of knowledge, as the impetus of accelerating innovation process, can be effectively managed [5]. Furthermore, speeding up creative operations will rely on a firm's ability to formulate a competitive strategy, co-ordinate with the supply chain, and compensate for intrinsic deficiencies by optimally leveraging external knowledge resources [4].

Knowledge has become the primary resource for the new economy and has been advocated by some researchers that knowledge will become the only source of competitive advantage [6-8]. It is therefore becoming of strategic importance that firms

A. Fred et al. (Eds.): IC3K 2012, CCIS 415, pp. 350–376, 2013.

constantly improve their ability to effectively manage knowledge flows, ensure success of NPD and increase competitiveness [9-12]. However, although knowledge is a decisive element for NPD, much of that required knowledge and needed information resides outside the firm's boundaries, specifically with customers, suppliers, business partners, and even competitors [13-14]. It therefore becomes imperative for firms to identify valuable knowledge sources within both internal and external environment and foster innovation quicker than the competitors [15], provided that they can accurately identify, grasp and embody the relevant technical and market knowledge within the NPD process [16]. Specifically, the onus to collect information on customers' needs through various means, as the principal component for NPD, and improve "the ability to import knowledge from the market" lies with manufacturers [17]. Particularly, effective integration of internal and external knowledge is a way of decreasing the possibility that "knowledge could be underutilized" [18-20]. In this setup, internal and external collaboration can play a decisive role in transferring tacit knowledge and building collective know-how [5], [20], [21].

Proponents of Supply Chain Management (SCM) have argued that the change in the nature of competition is becoming the momentum which shifts the competition from individual firms to supply chains [22]. Moreover, recent research has indicated that the supply chain is becoming the major source of external knowledge, skills, ideas and added value through collaborative efforts across the chain members [23-24], simultaneously influencing the present and future [25]. Collaborative innovation, in the supply chain, is about value co-creation through effective integration of all interested parties and an established way of doing business [26-28]. This shift has urged the managerial practice to re-audit and re-build the value-adding system and forced firms to re-evaluate and re-structure their value chains networks by adopting a holistic view. It has been widely discussed and accepted that valuable knowledge could be obtained and exploited through collaboration and cooperation across SC networks and optimally add value for the end customers. In this setup, customers as the "prosumer" [29] together with suppliers have been increasingly considered as the key drivers, co-innovators, co-developers and primary resources to NPD and as external actors, are increasingly influencing the process of innovation [30]. Value co-creation, recently, as customer engagement behaviour [31], is becoming the key to NPD [18] through generation of ideas and active contributions for fulfilling customer and market needs [32-33]. Rapidly changing technologies have provided customers with more opportunities to play a greater role by exchanging information with firms in the process of value co-creation [33-34]. Additionally firms have to simultaneously acquire solution knowledge scattered within both in the internal and external environment.

Increasingly, the end-user, and particularly lead users, has also been identified as key participants in product innovation [30], and in the formation of innovative networks [35]. Lead users can provide latent knowledge on product development, as well as improvement suggestions and solutions which can be embedded into the innovative process and then embodied into the new products [36-37]. Current research efforts in utilizing KM techniques for innovation have to overcome several problems such as reconcilement of perspectives between knowledge management and innovation, heterogeneity and distribution of knowledge within and across firms, and effective balance between knowledge exploration and exploitation for innovation[38] [39]. Essentially, how to effectively integrate KM into the innovative process and significantly increase the return on investment has increasingly been regarded as a challenge. Unfortunately, these issues have never been simultaneously discussed, particularly for NPD, or in a much broader context, and unsurprisingly there are no comprehensive and reliable theoretical reference frameworks. A number of gaps that hinder the endeavours of integrating these relevant theoretical perspectives into a creditable and reliable framework have been identified and are discussed below.

Inter-firm KM effort is increasingly becoming an emerging research field wherein the potential value of KM might be optimally explored and exploited. However, tacit knowledge extraction coupled to the intricate technical requirements of KM, even within an intra-firm context, has become a barrier in pursuing value gain from inter-firm KM implementation. In an inter-firm context KM requires a sophisticated understanding of issues in the value chain as well as of employable techniques. Integrating KM into the NPD process requires an understanding of the relationship between KM and NPD and of the means by which the firms can deal with sequential aspects of this relationship, these being:

- how to identify solution knowledge based on accurate product need information;
- how to successfully acquire that solution knowledge and embed into the existing knowledge pool;
- how to effectively exploit the embedded knowledge to develop new products and increase the ROI.

So gaps exist on how to compensate for the intrinsic deficiencies of knowledge management systems and on how to identify and employ most suitable KM techniques to facilitate innovative efforts in an inter-firm and open innovation context. The shift from closed innovation to open innovation requires firms to do more than just "open the book" and have to consider various factors emerging from the competitive business environment and complexity of synthesizing external and internal knowledge assets. Particularly, the challenges are not only to elaborate and employ reliable techniques for managing external knowledge, but also how to seek need information and solution knowledge from external actors who possess valuable knowledge, open cultures and are willing to engage in learning and knowledge creating activities with a certain degree of risk [40]. Such a process should be founded upon a clear understanding of potential innovation paths through collaboration and competition in the SC network and most importantly on the effective utilisation of KM techniques to facilitate integration. Exploring and exploiting potential innovation paths relies on effective supply chain network relationships with partners that possess sharable knowledge. Utilising this knowledge, in a relative time frame through PLM, will create knowledge loops in a SC network context. The perspectives of lifecycle management and SC networks have provided theoretical foundations as to how to integrate knowledge creation into different phases of the innovation process [41]. However the difficulty is how to evaluate the value of knowledge and the feasibility of embedding knowledge specially if there is a shift in value distribution across organisational boundaries in the SC network.

1.2   Aims of Research

Following the discussion above, the design of a comprehensive and creditable approach to bridge these gaps therefore becomes a prerequisite to the harmonious combination of these relevant theoretical fields. Although the main focus of the research is to develop and present a preliminary, theory-based framework which can facilitate the process of managing knowledge flows for NPD and outline a novel approach for innovation in a SC network context, the adopted approach to do so becomes part of the contribution of this work as well. The developed framework employs knowledge audit, knowledge calibration and knowledge absorption techniques to control knowledge flow across three collaborative innovation phases, pre-acquisition, in-acquisition and post-acquisition.

Following the introductory section, this paper is presented in six sections. Section two gives a brief overview of the adopted research approach. The results of an extended literature review, presented in section three, are followed by an evaluation of extant models and frameworks related to the integration of KM and NPD in a supply chain network context. The foundations of the proposed theoretical framework are presented and discussed in section five, explained in more detail in section six and concluded upon in the final section.

2   Methodology

The methodological approach, presented in (figure 1) is the result of the research aims. In order to design a theory-based framework, gaps specification has been positioned as the prerequisite of formulating the basic research aims. It is followed by a relevant literature review that highlights not only the main building blocks of extant frameworks but also uncovers potential disadvantages which need to be solved within this research. It is believed that these aspects, if explored and developed effectively, provide one approach to bridging the theoretical gaps.

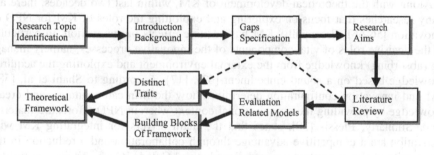

Fig. 1. Research Methodology

3   Literature Review

3.1   Knowledge Management

Increasingly, innovating firms have to improve their abilities to meet the never-ending requirements, from demanding customers and fierce competition, by effectively employing KM techniques to manage internal and/or external valuable knowledge resources. Therefore, in order to innovate, firms need to create an inventory of knowledge assets and make it more visible, accessible, sharable and measurable [42-43]. The firms can benefit from successful KM implementation by enhancing their competitive advantage, customer focus, employee relations and development, innovation, and lower costs [43-44].

KM is a kind of strategy which involves delivering the right knowledge to the right persons at the right time (APQC). As a strategy, effective KM implementation for innovation can be solely realized by aligning with the overall business strategy not only within internal environment, but also across the external supply chain [45-47]. Accordingly, KM efforts has been understood, no longer merely as an option but rather as a core issue that has to be thoroughly dealt with for firms to surpass the global competition [48].

As a process [49], KM can improve the learning abilities and increase effectiveness and efficiency of organizational performance by systematically coordinating processes of knowledge internalization, externalization, socialization and combination [50-51]. This viewpoint about KM has been advocated by Davenport et al., [52] who place the attention on the process of "knowledge import" and "knowledge export". The underlying principle is to "export" imported knowledge to the rest of firm for the purpose of solving problems or encouraging innovation.

KM can be seen as an effective coordinating mechanism which ultimately enables the resource to be converted into capabilities [53-54]. Similarly, Earl [55] suggests that KM can be regarded as central to product and process innovation or improvement, executive decision-making and organizational adaptation and renewal. In terms of innovation, the KM perspective of NPD is about how to seek optimal ways of controlling the valuable knowledge assets. In essence, this process is an effort of utilizing these mechanisms to coordinate the conversion process, namely from the embedded knowledge to embodied knowledge [56].

Along with the theoretical development of KM, within last two decades, there are many researches that focus on exploring and exploiting the roles of KM for NPD or innovation [56-66]. Essentially, KM has positive effects on innovation, in particular on the decisive roles of vital components of the innovative processes; namely managing (absorbing) knowledge from the external environment and exploiting the acquired knowledge based on a sound embedment [67], [17]. According to Shani et al, [58], KM and innovation configuration determines how the firm can capitalize and create knowledge by providing the organisational context wherein NPD efforts are materialised. Similarly, Plessis [20], states that the main drivers for integrating KM with innovation are a competitive advantage through collaboration and a reduction in the complexity of the innovation process. Basadur and Gelade [63] combined knowledge capture and utilization with the four phases of the KM cycle, namely generation, conceptualisation, optimisation and implementation. They place more attention on the relationship between KM and organization learning and they address ways by which the firms can become a "thinking organization." Abou-zeid and Cheng [65] presented a modified SECI model for knowledge creation and utilization that differentiates the K-creation process, which involves tacit knowledge transfer, from the K-utilization process affected by the explicit and simple internalization and combination processes. Similarly an ontology-based methodology, to solve knowledge sharing problems in the NPD process was proposed by Bradfield and Gao [64].

Essentially, tacit nature of knowledge [69], stickiness of knowledge [70] and "knowledge that is located, embedded and invested in practice" [71-72] have been viewed as the main barriers which hinder processes of knowledge creation. It is worth noting that knowledge, as the key component of all forms of innovation [39], is rooted in organizational activities and practices and is embedded in multiple elements and sub-networks [73]. Therefore it can only be transferred when there are no barriers between senders and recipients [74]. Moreover, due to their knowledge-intensive and multi-disciplinary nature, NPD processes require KM techniques to transfer, integrate and regenerate knowledge from both the internal and external environments. More specifically, an effective combination of two theoretical fields, KM and NPD, should facilitate knowledge creation as well as identification of knowledge deficiencies, knowledge detection in the value chain and subsequent collaboration.

3.2   Innovation

Today's business environment is fiercely competitive. Globalization, ever fast changing technologies and increasingly demanding customers are constantly pushing the performance bar upward. Becoming an innovator is the only way to be a winner. Unsurprisingly, successful firms have to innovate at the global frontier better than their rivals [75]. Particularly, success of NPD will help the firms to siege new opportunities and actually propel into new business fields and gain first-mover advantages or surpass the competitors in term of responsiveness or innovativeness. If companies fail to continually innovate, they die [5]. A survey conducted by the Product Development and Management Association (PDMA) has shown that successful new products contribute 50% to 60% of sales in most companies [76]. Similarly, Simth [77] states that 75% of the revenues in successful firms are generated from new products or services that never existed five years ago.

Innovation is, basically, aimed at producing new knowledge that can be developed into doable solutions for society through distinctive and idiosyncratic market acceptable products and services. [78-79]. According to CBI [23], innovation is considered as being 'the successful exploitation of new ideas' across industrial networks that collaborate in a SC context to stimulate the creation of these ideas. Consequently, this process will rely on the decision to exploit and develop the power of effective KM implementation to support innovation and creativity [25]. By doing so, firms become much more prepared to innovate and perform successfully to meet the requirements from the customer and market faster and better than the competitors. DTI [80] also concluded that there is a need to take a broader view of the innovatory process and to tap into a network environment, because that individual actor is seldom capable to innovate independently. Networks through establishment of "weak and strong ties" [81] and bridging of "structural holes" [82] can greatly enhance the processes of knowledge creation. This viewpoint has been advocated by Antonio who states that the knowledge needed for innovation is often a product of the configuration and combination of different fields of knowledge from heterogeneous resources [83]. Therefore, it has become a strategic attractive option to acquire knowledge from external sources to compensate for scarcity of internal resources [87]. Accordingly, open innovation [5], as an emerging innovation strategy, has been regarded as the essential element to revitalize in-house innovation or closed innovation and to accelerate the

innovation process [5], [84], [85], [86]. Gassmann et al., [88] presented three archetypes of the open innovation process:

- Outside-in: integrating external knowledge, customers and suppliers;
- Inside- out: bring ideas to market;
- Coupled processes: couple outside-in and inside-out processes and work in al-
liances in a complementary manner.

It is undeniable that innovation depends on knowledge [20] and the foremost purpose
of KM is to facilitate innovation [89], change and product development [90-93].
Firms can engage in collaborative relationships to identify knowledge capability,
knowledge reliability and richness, and develop receptors to absorb external knowledge to improve organizational competitive advantage [94], [20].

3.3   Supply Chain Management

SC as a value chain can offer the opportunities to simultaneously improve the individual
firm's performance and increase the possibilities to achieve common goals of "growing
the pie" [95]. Meanwhile, it provides firms with a way to optimally leverage core competences and unique skills and strategically outsource non-core activities to external
networks [96-98]. Basically, SCM covers all business processes between vertically
linked entities within three dimensions, action, relationships and processes [99-103].
The Council of Supply Chain Management Professionals (2006) concluded that SCM
"encompasses the planning and management of all activities involved in sourcing
and procurement, conversion and all logistics management activities. In essence,
supply chain management integrates supply and demand management within and across
companies."

The relation-oriented definition [104] of SCM suggests that relationships, cooperation and mutuality are vital in improving effectiveness, efficiency and overall performance. Meanwhile, the resource-based view of firm perspective explains why
relationships between buyers and sellers are the most important intangible resources
[105], [54]. Consequently, appropriate relationships with channel members are not
only the antecedent of successful "outsourced activities" but also the consequence of
fruitful collaboration. Therefore, it is core to improve the abilities and create a mechanism, by which the intricate relationship can be enforced [99], [106].

SCM presents the effective integration of key business processes that add value to
end-customer, from upstream suppliers, manufacturers, distributors and retailers
[107]. So it can be seen as means to coordinate functions and processes and respond
to the requirements of customers through effective management of information and
knowledge across the network [108]. Cooper et al. [100] state that SCM encompasses
three closely inter-related elements: the SC network structure, the SC business
processes, and the SCM components as depicted in figure2. Such a configuration
provides the basis to identify and consider all exogenous and endogenous variables
related to NPD and create strategically [110-112], operationally and technologically
long-term stable relationships [104].

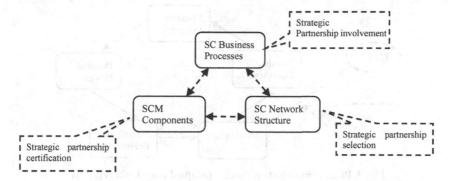

Fig. 2. SCM Framework (adapted from [185], [100])

Based on the extant literature, it is widely accepted that the suppliers' innovative capabilities are the major determinants for collaboration. More specifically, Burton [113] argues that suppliers accounted for approximately 30% of quality problems and 80% of product lead-time problems. Increasingly, the suppliers' role, particularly for NPD, has shifted from simple component providers to systematic development partners that possess and can provide manufacturers with valuable knowledge related to the final product. Recently, most research in this field focuses on the timing of supplier's involvement. Petersen et al., [114] state that early supplier integration is an important coordinating mechanism for decision making that links product design, process design and supply chain design together. Moreover, early supplier involvement (ESI) in product development has been increasingly regarded as the vital means by which the manufacturers will be able to leverage the maximum value of the suppliers' knowledge assets, cut down development risks and ensure the success of innovative activities [115].

Supplier-involved Collaborative Product Development (CPD) is viewed as an effective way to create wider collaborative networks and absorb external knowledge. Essentially, supplier integration into NPD process is a social [116-117] and systematic process [114]. Traditionally, research on supplier-involved CPD addresses the key factors, such as, collaboration method, platform, tools and standards [119-122]. However three elements are critical in supplier integration: (1) the extent to which the supplier influences decision-making; (2) the amount of control the buyer retains over the design; and (3) the frequency of design-related communications [114], [123]. Meanwhile, the compatibility, reputation and fit between parties are also critical elements [124-127]. Supplier integration offers a number of advantages. It provides "sticky" information regarding new ideas and feasibility, and supplementary product and process solutions [128-130], reduces the complexities and risk of the innovation process [129-132], and it facilitates communications, information exchange through networks and quality improvements [133-129]. Notably, a few researchers believe that there is no link between supplier involvement and key innovative performance [134] and argue that it is difficult to accomplish the "seemingly positive outcomes" of supplier involvement in NPD [135-137]. However, negative arguments and lack of studies on this process do not diminish the significant roles of suppliers in the innovative process.

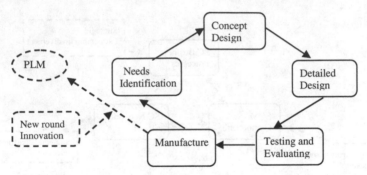

Fig. 3. Product introduction process (adapted from Jenkins [200])

3.4 Customer Knowledge Management

Traditionally, a NPD project (see Figure 3) needs to effectively coordinate R&D and marketing, to identify potential product opportunities, market requirements and configure an operational capability to produce. This combination can be regarded as the process of leveraging complementary knowledge resources. Success of NPD projects requires companies to develop competences by creating an external knowledge sharing ecosystem, which can not only ensure the success of NPD, but also it is hard to simulate [138]. In practice the rate of failure of NPD is high, estimated to be between 40-75% [139], with lack of a fit between new products attributes and customer requirements being a major cause [140-141]. Integrating the identification of customer needs in the product development efforts is a prerequisite to successful NPD [142], [3]. Accordingly, capturing customers and market requirements along with knowledge for solving problems is becoming an imperative for innovating firms. Absorbing customer knowledge through customer integration into the NPD process strengthens a company's core competences [143], as customers will cooperate and creatively contribute to the different phases of the innovation process [144]. Particularly in a B2B business environment, customers are becoming co-producers [145] and "customer relationship management" is becoming an attractive and academically interesting mechanism, for knowledge creation and innovation [146], [147], [30]. Customers, in the 21st century, are not anymore passive recipients of NPD but they are demanding to play a more active role [159].

Customer knowledge, as an important knowledge asset for an organization, can be broadly defined as the combination of external consumer knowledge and supply chain knowledge [148-149]. According to Wallace [150], to manage customer knowledge is to capture knowledge in need and solution information [151-152], as such knowledge will contribute to the development of the "right" products [152]. However, "sticky information" [91], could jeopardise such a process, as customers are often not able to express their requirements [153-154]. Customer involvement affects all phases of NPD and influences the effectiveness and efficiency of the NPD process [146], [155]. This process is referred to as "interactive value creation" [152] or "value co-creation" [144] in an open innovation context [5]. According to Prahalad et al., [147] effective interaction with customers is a prerequisite to value co-creation and the means of encouraging customer involvement in innovation [156-157]. Meanwhile, leveraging external knowledge effectively and proactively, through KM based inter-firm collaboration, is becoming a sine qua non for a firm's development [13].

Recently, customer involvement has been widely discussed as customer knowledge management (CKM), a compelling approach to harness valuable customers' knowledge and capability [159-161]. Generally, CKM is described as ongoing process of generating, disseminating and re-using customer knowledge to satisfy consume requirements [162]. Similarly, CKM refers to processes of involving customer into innovative performance improvement activities by sharing valuable knowledge within the network environment. This process actually increases the firms' competitive advantages by encouraging a two-way exchange of knowledge that benefits both parties [163]. However this process is considered to be relatively passive and even tacit by most of researchers and practitioners and the challenge is to create mechanism for managing the relationships [159], [161], [164]. According to Gibbert et al, [161], there are five styles of CKM:

- Prosumerism that stems from the expression "prosumer" [29] and indicates that customers can play the key roles as co-innovators, as in Bosch and Mercedes-Benz, Quicken, IKEA practice [161].
- Team-based co-learning that focuses on embedding customer knowledge into a platform which can facilitate the process of embodying the shared knowledge into new product or service, Amazon and Toyota as the most typical examples.
- Mutual innovation, initially identified by Von Hippel [128] that describes the end-user's decisive role in innovation. Rider Logistics is a quoted example; a trucking company that developed to a logistics solution provider through mutual innovation with its customers [161].
- Communities of creation that differ from traditional communities of practice and where interaction of customer groups achieves the common goal of knowledge creation [28]. Examples are Beta created by Microsoft and Netscape and Antenna shops from Sony and Panasonic.
- Joint Intellectual Property is the most intense form of cooperation between companies and their customers that share ownership of NPD [161]. For example, Skandia Insurance and Kooperativa Forbundet.

3.5   Users as Innovators

Since at least Adam Smith's example (1776) of "a boy's innovation for saving his own labour", we have witnessed a significant transition on the role of users in innovation documented by expressions such as "users play an important but peripheral role" [165-166], "users are the sources of innovation, not just a helper" [128] and more recently, "users as innovators" [158], [167]. The concept "users as innovators" has been investigated through cases in open-source software development [168-169] where they are "user-entrepreneurs" [170-171]. Other studies include the "Collective Customer Commitment" method [3] successfully employed by companies such as Threadless, and Enos's [181] research in oil refining; Freeman's [179] study in chemical industry; Hollander 's [180] report on DuPont's rayon plants and von Hippel's [151] research in scientific instrument innovation process. These studies have shown that knowledge from users or customers [172], [174] and customer knowledge management are indispensable for developing innovative products [173]. More recently Cooper and Dreher [199] conducted a survey of 150 firms and concluded that the Voice of Customer (VoC) has been adopted as the most popular and the most effective source of new product ideas.

3.6   Product Lifecycle Management

Improvements relating to the effectiveness and efficiency of NPD across the product lifecycle are becoming key business factors [175-176]. Moreover, CKM has been positioned as an effective approach which is closely linked to and supported by Product Lifecycle Management (PLM) [177]. According to Stark [178], PLM is "the activity of managing a company's products all the way across their lifecycle (from cradle to grave) in the most effective way". In essence, PLM is the starting point of the innovative process which consists of continual knowledge identification and knowledge acquisition from customers or market. Therefore, the effectiveness of PLM will dramatically influence further processes of knowledge creation and ultimate success of NPD projects. Gathering information and knowledge from the customer and market therefore will be the key start of efforts in synthesizing knowledge loops for successful NPD. Particularly, in terms of managing knowledge across the product lifecycle, different approaches may be required in the different stages of the cycle [177] There are a number of issues that arise and need to be addressed and these are discussed later in the presentation of the framework.

4   Evaluation of Extant Frameworks

According to the extant literature, there are numerous studies, (a comparison is presented in Table 1) which have focused on developing the methods of integrating customers or suppliers into the NPD process by utilizing KM techniques [30], [130], [114], [134], [94], [182], [13], [164], [119], [183], [184], [58], [115], [187], [188]; [189]). Although, these studies have shown idiosyncratic understanding and provided various means by which the firms might link KM with NPD process, the common characteristic is that the firms need to effectively leverage knowledge from external actors and then optimally internalize and exploit that knowledge within the innovative processes. What is additionally evident is that most of these frameworks do not address a number of issues:

1. Lack of Studies about KM Techniques for Inter-firm Knowledge Flow's Control in NPD Activities

Most commonly discussed and employed KM techniques are not problem-free and impose challenges in an intra-firm KM context and unsurprisingly will directly affect inter-firm KM activities. What kinds of techniques can be suitable for managing knowledge across organizational boundary? Can the SECI (socialization, externalization, combination, and internalization) processes occur easily within inter-firm KM activities? Therefore, a requirement will be to explicitly analyze the relationship between NPD and KM and elaborate on feasible and reliable techniques which can embody the essential and vital linkages between these two fields.

2. Lack of a Holistic View of Knowledge Flows for NPD

In essence, KM based NPD is about control of knowledge flows. Accordingly, the key as to formulating a comprehensive framework is to identify knowledge sources, create knowledge flows and effectively manage them. The starting point is the identification of need information. This is followed by retrieval of solution knowledge that needs to be effectively embodied into new products or services. However, most frameworks offer one-sided solutions by focusing on either the customer side [30], [182], [184] or the suppliers side [114], [115], [130].

3. Lack of Discussions on How to Formulate Knowledge Flows in a Supply Chain Context

Although, there are a few studies which have been focused on expanding knowledge flow formulation into a supply chain context, such as [13], [164], [115], [119], they merely provide a basic outline which addresses the importance of integration. More specifically there is a need to effectively understand and conduct NPD in a more systematic and comprehensive manner by considering the intrinsic elements of relevant theoretical fields. Therefore the framework has to focus on how to effectively leverage the potential value of external actors by creating a closed knowledge loop for NPD to bridge need information and solution knowledge.

4. Lack of Studies of Multidimensional Factors for Control of Knowledge Flows

The main components of the framework should focus on not only indentifying and grasping the need information and along with solution knowledge, but also paying attention to the techniques as to how to effectively and efficiently internalize and exploit the solution knowledge to improve the quality of products and services. Accordingly, there are various factors, which will actually affect this process and need to be deeply investigated, in creating a reliable framework. However, most research has overlooked or failed to closely explore what will influence the process of embedding solution knowledge into an existing knowledge pool without any chaotic consequences and little attention has been focused on the process of embodiment of embedded knowledge.

5   Foundations of a Theoretical Framework

Just as discussed above, this paper aims to expand NPD into a SC context by utilizing KM techniques. Accordingly, it is vital to structure a workable framework based on effective combination of the NPD, SCM and KM concepts. Inspired by the strategic supply chain model [185], [100] this theoretical framework is constructed into three phases, pre-acquisition, in-acquisition and post-acquisition. Moreover, in order to embrace open innovation strategy [5], [143], the framework especially focuses on two aspects: firstly analyzing internal and external knowledge assets regarding to internal availability and external complementary credibility (collaborative sharing). This process is extraordinarily linked with Knowledge (Management) Audit Approach [186] and has to be conducted within the pre-acquisition phase. Secondly, as a process of transferring embedded knowledge to embodied knowledge [56], innovation is

Table 1. Relevant models and frameworks

Models	Authors	Main characteristics	Description of Models
A new approach to developing customer products	Thomke and von Hippel (2002)	♦customer-as-Innovator; ♦forward the interface location between supplier and customer;	It forwards the location of interface between supplier (manufacturer) and customer and the trail-and-error iterations by employing Customer-as-Innovators Approach.
Supplier integration based model	Ragatz et al., (2002)	♦exogenous variables: needs and alignment; technology uncertainty; ♦endogenous variables: integrative strategies and team processes;	It focuses on synthesizing two exogenous and two endogenous variables that affect successful supplies integration into the NPD process.
Customer as Innovators Approach	Thomke and von Hippel (2002)	♦form a user-friendly toolkit; ♦increase the flexibility of production process; ♦customer selection; ♦evolvement of toolkits;	They develop a Customer as Innovators Approach and present five steps for turning customers into innovators.
Conceptual model for NPD organization knowledge system	Mohrman et al., (2003)	♦contextual organizational elements; ♦knowledge worker behaviours; ♦knowledge outcomes;♦knowledge effectiveness;	Organization knowledge system consists of four constructs, namely contextual organizational elements, knowledge worker behaviours, knowledge outcomes and effectiveness.
A new business model for collaborative product services	Ming et al., (2003)	♦knowledge resources come from stakeholder, customer, manufacturer and suppliers;	The model focuses on synthesizing the knowledge resources from stakeholder, customer, manufacturer and supplier into a Collaborative Product Services pool which is controlled by developer.
Three steps for systematic process of involving supplier into NPD	Petersen et al., (2003)	♦deep and accurate understanding about suppliers knowledge; ♦effective communication with suppliers; ♦fully exploit the value of supplier's knowledge and capability;	There are three steps to fully exploit the value of suppliers' knowledge and capability. Furthermore, they develop a simplified structural equation model that integrates the major activities required for integrating suppliers into the NPD process.
Contingency factors for supplier involvement	Wagner et al., (2006)	♦contingency factors on the organizational level;♦contingency factors at project level;	It is vital to match the product architecture and the type of design based on effective interaction with "right" suppliers, from "none" to "black box" supplier integration. Meanwhile, criteria for selecting suppliers, enhancement of buyer-supplier relationship, timing of involvement etc.
Knowledge-based Product Model for NPD	Dutt (2006)	♦reacting with response from customer and market; ♦knowledge based product;	The model focus on creating organizational memory, which possesses embedded knowledge and embodies learning, to satisfy dynamic needs and requirements from customers through reaction with response from customer.
Strategic contingency model	Ju, Li and Lee (2006)	♦knowledge management;♦knowledge characteristics;♦organizational learning and innovation;	They present a strategic contingency model which mainly focuses on the effective integration of knowledge management, knowledge characteristics, organizational learning and innovation.
Marketing and KM based framework	Kohlbacher (2007) (2008)	♦relationship with supply chain; ♦customer relationship management ♦product development management and marketing research; ♦knowledge co-creation with customer;	The framework depicts the relationship between supply chain management, market research, customer relationship management and product development management. It addresses the vital roles of relationship between marketing process and customer, supplier and partners etc, particularly knowledge co-creation with customer.

Table 2. Relevant models and frameworks

Models	Authors	Main characteristics	Description of Models
Conceptual model for adopting internal and external resources	Tessarolo (2007)	◆internal cross-functional organizational structure; ◆externally involving customers and suppliers;	It addresses that both internally adopting cross-functional organizational structures for development and externally involving customers and suppliers into the process can be powerful drivers.
An alliance oriented framework	Jiang and Li (2009)	◆KM for innovation within alliance context; ◆the broader scope of alliance and the greater opportunity for sharing knowledge;	It portrays the relationships between KM and alliance characteristic in term of innovative performance. They investigate and conclude that the broader scope of alliance and the greater opportunity for interaction of sharing knowledge.
Conceptual framework of Consumer Co-creation	Hoyer et al., (2010)	◆importance of theoretical synthesis; ◆introduce four phases new product introduction; ◆three vital factors for consumer co-creation;	Theoretical integration along with degree of co-creation is the central of four-phase of NPD, namely ideation, product development, commercialization and post-launch. Meanwhile, it highlights three vital factors for this process, namely consumer motivators, firm stimulators and firm impediments.
Five phases based continuous innovation process	Xu et al., (2010)	◆idea generation ; ◆research and development; ◆implementation; ◆commercialization ◆internalization; ◆ notably it address the phases about pre-creation and post usage;	They develop five-stage NPD process and propose a macro model of KM for continuous innovation and depict knowledge lifecycle phases: pre-creation, creation, intermediate, usage and post usage. And it can fit into knowledge assets from physical view, human view and technological view for the purpose of NPD.
A socio-technical systems based framework	Shani and Sena (2010)	◆business environment cluster; ◆social subsystem; ◆technological subsystem; ◆KM system; ◆NPD processes and outcomes	It portrays sustainability of NPD process through effective combination of five aspects, namely business environment cluster, social subsystem and technological subsystem, knowledge management system etc.
Supplier-involved collaborative product development framework	Wang et al., (2010)	◆collaborative business model; ◆collaborative process model; ◆collaborative operation model;	Essentially, three models respectively focus on the strategic analysis regarding with collaborative strategy, market information analysis and target product status analysis; on coordination and integration of customer and supplier in CPD; on the detailed activities at operational level.
Conceptual framework	Akram et al., (2011)	◆integrate KM with technology, KM activities and human capital into NPD; ◆collaboration between internal and external environment, learning and culture etc;	It focuses on integrating KM and other factors, such as technology, KM activities, knowledge assets, human capital etc. It positions SECI model as the central of the innovation and is supported by collaboration between internal and external environment, learning and culture etc.
Knowledge flows oriented model for NPD process	Assima-kopoulos and Chapelet (2012)	◆key factors are people, technology, systems and products; ◆knowledge flows control is to id-entify knowledge source, obtain knowledge and exploit knowledge;	The model puts technology, systems and products at the central and depicts NPD cycle: design phase, analysis phase, development phase and full launch. Notably, it is based on the knowledge flows control, namely identify knowledge source, obtain knowledge and exploit knowledge.

subject to coordination of the two actions. In its essence, Knowledge embedding can be matched with Knowledge Calibration [190] and ultimate knowledge embodying will be closely linked with Knowledge Absorption and absorptive capability [191]. Subsequently, as key drivers of innovation, these two KM techniques can be utilized to facilitate the process of knowledge embedding and knowledge embodying across two phases: in-acquisition and post-acquisition. In fact, this framework might compensate for the deficiencies of existing Open Innovation studies and expand the related research into a broader context

6   Preliminary Theoretical Framework

Based on the literature review, a preliminary theory-based ACA (Audit, Calibration and Absorption) framework (See Figure 4) is formalized as following:

6.1   Pre-acquisition and Knowledge Audit

Audit Approach is a critical part of a KM framework and an effective first step of internal KM efforts [186]. Effectiveness of Knowledge Audit is a determinant which directly affects the decision "can I do it" (Knowledge Management Audit) and further activities related to managing knowledge resources. Essentially, it can provide an outline by systematically investigating and evaluating the "health" of a firm's knowledge and ability and readiness of further KM implementation. Most importantly, it encourages two fundamental and philosophical conversions; from we do not know what we know to know what we don't know and from know what we don't know to know how to know. The main objectives of Knowledge Audit or knowledge management audit in the pre-acquisition phase are:

- What we know and what we don't know (knowledge and abilities gaps identification)
- Who knows and can we cooperate (partnership selection)
- How to make it happen (knowledge loop creation)
- Are we ready to embrace activities of KM? (ability audit)
- To formulate an innovative strategy (outside-in, inside-out or coupled model);
- To analyze data or information from customers and market (need information)
- To locate and evaluate the valuable external knowledge assets and select strategic partnerships;
- To create and disseminate strategic goals through mutual collaboration and cooperation;

6.2   In-acquisition and Knowledge Calibration

According to Pillai & Goldsmith [190], knowledge calibration is a measure of the degree of agreement between knowledge accuracy and confidence within the knowledge acquiring and embedding processes. Meanwhile, capability to calibrate knowledge acts as a facilitator or a valuable catalyst which can support firms to make judgement about strategic partners in term of abilities, characteristics, potential development and criteria of meaningful interaction. The main objectives in this phase constitute the requirement for the following actions:

- Building up trustworthy relationships and enhancing mutuality;
- Addressing shared goals and consistently improving routine activities;
- Harmoniously integrate and optimally utilize IT-based hard infrastructure and people-based soft mechanisms;
- Improving leaning awareness and abilities;
- Cooperative Chain Culture Creation (C4);

6.3   Post-acquisition and Knowledge Absorption

Employing KM techniques will aim to facilitate not only sharing of knowledge between providers and receivers but also embodiment or absorption of the acquired knowledge into the new products or services. Accordingly, there are numerous factors that will affect the success of knowledge absorption, but amongst them, absorptive capability and the learning processes are the decisive determinants [191-192]. Absorptive capability is the ability to use prior knowledge to recognize the value of new knowledge and to assimilate and apply it to create new knowledge and capabilities [191]. Cohen & Levinthal [191] described "absorptive capacity" as a key factor in the innovative process. Generally, it refers to the ability not only to acquire and assimilate information, but also to exploit it [191], [193]. Improvement of "absorptive capacity" is closely linked to interaction with the external environment [191] and therefore interaction and openness have become the determinants which affect the effectiveness and efficiency of the knowledge creation [194]. Interaction, at individual level in particular, at inter-firm level is critical to articulate and amplify knowledge [7] and to establish channels for embedding and embodying knowledge flows for innovation. Openness may also speed up the pace of innovation by effectively leveraging outside sources of cognition and competence in the development of new products or services.

The learning processes are the mechanisms and key impetus that effectively compensate for the firms' ability deficiencies [192]. The capability of a firm to absorb external knowledge and information has been viewed as one of the pillars related to transformation of knowledge and its conversion into new products [195]. How to benefit from the dynamic interaction with external actors, in increasingly competitive market, has been regarded as the bottleneck that influences new products and services but also the development of the firm itself. According to Souitaris [196], there are two aspects which affect the interaction process: scanning for external information and cooperating with external firms. In a dynamic environment, focusing on continuous leaning to enhance the organizational knowledge capability and strengthen the knowledge transformation process will improve quantity and quality of knowledge accumulation [197], enhance the organizational knowledge base and the ability of knowledge creation [198]. The key issues of this phase should be addressed as following:

- Continuously improve the strategic partnership;
- Evolve from knowing firm to learning firm(encouraging individual creative activities; indirectly or directly customer involvement etc);

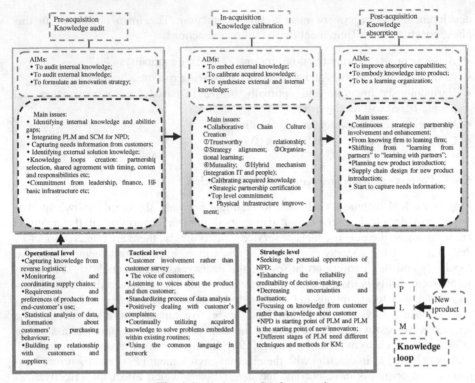

Fig. 4. Preliminary ACA framework

- Migrate from learning from partners to learning with partners (acting with suppliers as a whole by joint activities and optimal use of knowledge to reduce uncertainties);
- Embody technology knowledge and market knowledge into NPD and plan product introduction (integrating related factors, such as distributors, retailers, inventory and marketing etc);

7   Concluding Remarks

As part of a wider research project, this paper mainly focuses on presenting a theoretical framework which can be considered as a reference model for further research. The proposed framework focuses on systematically synthesizing relevant theoretical fields and utilizes existing approaches to facilitate effective use of knowledge management techniques in the NPD process in a supply chain by adopting an OI strategy. More specifically, it has accurately reflected on the nature of existing knowledge management systems and captured the core issues of NPD in a three-phase approach. It logically integrates three KM techniques into the process of innovation from the wider perspective of SCM. In essence, by referencing this framework, researchers and practitioners can easily manage NPD projects in terms of optimally leveraging knowledge resources and accurately self-positioning, detecting and employing solutions in the realisation of new product or services. As a result, a firm can achieve shorter "time to the market", gaining from first mover advantages and satisfying the end customer requirements. At this stage, the proposed framework needs to be evaluated from a practical point of view. The process of validation will be undertaken to verify and improve the creditability and feasibility of this framework.

References

1. D'Aveni, R.: Hyper competition: Managing the Dynamics of Strategic Manoeuvring. The Free Press, New York (1994)
2. Veliyath, R., Fitzgerald, E.: Firm capabilities, business strategies, customer preferences, and hypercompetitive arenas: the sustainability of competitive advantages with implications for firm competitiveness. Competitiveness Review 10(1), 56–82 (2000)
3. Ogawa, S., Piller, F.T.: Reducing the Risks of New Product Development. Sloan Management Review 47, 65–72 (2006)
4. Ireland, R.D., Hitt, M.A., Vaidyanath, D.: Managing strategic alliances to achieve a competitive advantage. Journal of Management 28(3), 413–446 (2002)
5. Chesbrough, H.: Open Innovation. The New Imperative for Creating and Profiting from Technology. Harvard Business School Press, Boston (2003)
6. Druker, P.F.: Managing for the Future. Butterworth-Heinemann, Oxford (1992)
7. Nonaka, I., Takeuchi, H.: The Knowledge Creating Company, How Japanese Companies Create the Dynamics of Innovation. Oxford University Press, Oxford (1995)
8. Nonaka, I.: The dynamic theory of organizational knowledge creation. Organization Science 5(1), 14–37 (1994)
9. Tidd, J., Hull, F.M.: Services Innovation: Organizational responses to technological opportunities and market imperatives. Imperial College Press, London (2003)
10. Karmarkar, U.: Will you survive the services revolution? Harvard Business Review 82(6), 100–107 (2004)
11. Schulttze, U., Stabell, C.: Knowing what you don't know? Discourses and contradictions in knowledge management research. Journal of Management Studies 41(4), 549–573 (2004)
12. Chesbrough, H.W., West, J., Vanhaverbeke, W.: Open Innovation: Researching a New Paradigm. Oxford University Press, Oxford (2006)
13. Kohlbacher, F.: Knowledge-based New Product Development Fostering Innovation Through Knowledge Co-Creation. Int. J. Technology Intelligence and Planning 4(3), 326–346 (2008)
14. Thomke, S.H.: Experimentation Matters: Unlocking the Potential of New Technologies for Innovation. Harvard Business School Press, Boston (2003)
15. Darroch, J.: Knowledge management, innovation and firm performance. Journal of Knowledge Management 93, 101–115 (2005)
16. Aoshima, Y.: Transfer of system knowledge across generations in new product development: empirical observations from Japanese automobile development. Industrial Relations 41(4), 605–628 (2002)
17. Leonard, D.: Wellsprings of Knowledge: Building and Sustaining the Sources of Innovation. Harvard Business School Press, Boston (1998)
18. Baddi, A., Sharif, A.: Information management and knowledge integration for enterprise innovation. Logistics Information Management 16(2), 145–155 (2003)
19. Chen, J., Zhaohui, Z., Xie, H.Y.: Measuring intellectual capital. Journal of Intellectual Capital 5(1), 195–212 (2004)

20. Du Plessis, M.: The role of knowledge management in innovation. Journal of Knowledge Management 11(4), 20–29 (2007)
21. Cavusgil, S.T., Calantone, R.J., Zhoa, Y.: Tacit Knowledge Transfer and Firm Innovation Capability. Journal of Business and Industrial Marketing 18(1), 6–21 (2003)
22. Christopher, M.: Logistics & Supply Chain Management. Pitmans, London (1992)
23. Confederation of British Industry (CBI): Innovation Survey, CBI (November 2005)
24. Sainsbury, L.: Race to the top: A review of the Government's science and innovation policies, Independent HM-Treasury Report, HMSO (October 2007)
25. IfM and IBM: Succeeding through Services Innovation: a discussion paper, Cambridge, University of Cambridge Institute for manufacturing, United Kingdom (2007)
26. Doz, Y., Santos, J., Williamson, P.: From Global to Metanational: How Companies Win in the Knowledge Economy. Harvard Business School Press, Boston (2001)
27. Lawer, C.: On Customer Knowledge Co-creation and Dynamic Capabilities. Working Paper, Cranfield School of Management (2005)
28. Sawhney, M., Prandelli, E.: Beyond customer knowledge management: customers as knowledge co-creators. In: Malhotra, Y. (ed.) Knowledge Management and Virtual Organizations, pp. 258–282. Idea Group Publishing, Hershey (2000)
29. Toffler, A.: The Third Wave. Morrow, New York (1980)
30. Thomke, S., von Hippel, E.: Customers as Innovators. Harvard Business Review, 5–11 (April 2002)
31. van Doorn, J., Katherine, N., Lemon, V., Mittal, S., Doreén Pick, P.P., Peter, V.: Customer Engagement Behavior: Theoretical Foundations and Research Directions. Journal of Service Research 13(3), 253–266 (2010)
32. O'Hern, M.S., Aric, R.: Customer Co-Creation: A Typology and Research Agenda. In: Malholtra, N.K. (ed.) Review of Marketing Research, vol. 6, pp. 84–106. M.E. Sharpe, Armonk (2009)
33. Ernst, H., Hoyer, W.D., Krafft, M., Jan-Henrik, S.: Consumer Idea Generation. working paper, WHU, Vallendar (2010)
34. Bolton, R.N., Shruti, S.-I.: Interactive Services: A Framework, Synthesis and Research Di-rections. Journal of Interactive Marketing 23(1), 91–104 (2009)
35. Furukawa, I.: Deai no 'ba' no kosoryoku – Maketingu to shouhi no 'chi' no shinka (The imagination of the meeting 'ba' – The Evolution of Marketing and Consumption 'Knowledge'), Yuhikaku, Tokyo (1999)
36. Eisenberg, I.: Lead-User Research for Breakthrough Innovation. Research-Technology Management 54(1), 50–58 (2011)
37. Reichwald, R., Piller, F.: InteraktiveWertschöpfung: Open Innovation, Individualisierung und neue Formen der Arbeitsteilung, 2nd edn., Wiesbaden (2009)
38. Goh, A.: Harnessing knowledge for innovation: an integrated management framework. Journal of Knowledge Management 9(4), 6–18 (2005)
39. Chapman, R., Magnusson, M.: Continuous innovation, performance and knowledge management: an introduction. Knowledge and Process Management 13(3), 129–131 (2006)
40. Desouza, K.C., Awazu, Y.: Gaining a competitive edge from your customers: exploring three dimensions of customer knowledge. KM Review 7(3), 12–15 (2004)
41. Cantisani, A.: Technological innovation processes revisited. Technovation 26(11), 1294–1301 (2006)
42. Jarrar, Y.F.: Knowledge management: learning for organizational experience. Managerial Auditing Journal 17(6), 322–328 (2002)
43. Skyme, D., Amidon, D.: The knowledge agenda. Journal of Knowledge Management 1(1), 27–37 (1997)

44. Dykeman, J.B.: Knowledge management moves from theory toward practice. Managing Office Technology 43(4), 12–13 (1998)
45. Mudge, A.: Knowledge management: do we know what we know? Communication World 16(5), 24–29 (1999)
46. Okunoye, A., Karsten, H.: Where the global needs the local: variation in enablers in the knowledge management process. Journal of Global Information Technology Management 5(3), 12–31 (2002)
47. Dan, H.: (June 2011), http://www.informationweek.com/762/know.htm
48. Singh, M., Kant, R.: Knowledge management as competitive edge for Indian engineering industries. In: Proc. of the International Conference on Quality and Reliability, Chiang Mai, Thailand, November 5-7, pp. 398–403 (2007)
49. OECD: Conclusions from the Results of the Survey of Knowledge Management Practices for Ministries/Departments/Agencies of Central Government in OECD Member Countries. February 3-4, GOV/PUMA/HRM (2003)
50. Mockler, R.J., Dologite, D.G.: Expert systems to support strategic management decision-making. In: Hussey, D.E. (ed.) International Review of Strategic Management, vol. 3, pp. 133–148 (1992)
51. Nonaka, I.: A Dynamic Theory of Organizational Knowledge Creation. Organization Science 55(1), 14–37 (1994)
52. Davenport, T.H., Prusak, L., Wilson, J.H.: Who's bringing you hot ideas and how are you responding? Harvard Business Review 81(2), 59–64 (2003)
53. Nelson, R.R., Winter, S.G.: An Evolutionary Theory of Economic Change. Harvard University Press (1982)
54. Darroch, J., McNaughton, R.: Examining the links between knowledge management practices and types of innovation. Journal of Intellectual Capital 3, 210–222 (2002)
55. Earl, M.: Knowledge management strategies: toward a taxonomy. Journal of Management Information Systems 18(1), 215–233 (2001)
56. Madhavan, R., Grover, R.: From Embedded Knowledge to Embodied Knowledge: New Product Development as Knowledge Management. Journal of Marketing 62(4), 1–12 (1998)
57. Pitt, M., MacVaugh, J.: Knowledge management for new product development. Journal of Knowledge Management 12(4), 101–116 (2008)
58. Shani, A.B., Sena, J.A., Olin, T.: Knowledge management and new product development a study of two companies. European Journal of Innovation Management 6(3), 137–149 (2003)
59. Popadiuk, S., Choo, C.W.: Innovation and knowledge creation: How are these concepts related? International Journal of Information Management 26, 302–312 (2006)
60. Soderquist, K.E.: Organizing Knowledge Management and Dissemination in New Product Development Lessons from 12 Global Corporations. Long Range Planning 39, 497–523 (2006)
61. Caloghirou, Y., Kastelli, I., Tsakanikas, A.: Internal Capabilities and External Knowledge Sources: Complements or Substitutes for Innovative Performance? Technovation 24(1), 29–39 (2004)
62. Ramesh, B., Tiwana, A.: Supporting Collaborative Process Knowledge Management in New Product Development Teams. Decision Support Systems 27, 213–235 (1999)
63. Basadur, M., Gelade, G.A.: The role of knowledge management in the innovation process. Creativity and Innovation Management 15(1), 45–62 (2007)
64. Bradfield, D.J., Gao, J.X.: A methodology to facilitate knowledge sharing in the new product development process. International Journal of Product Research 45(7), 1489–1504 (2007)

65. Abou-zeid, E., Cheng, Q.Z.: The effectiveness of innovation: a knowledge management approach. International Journal of Innovation Management 8(3), 261–274 (2004)
66. Ju, T.L., Li, C.Y., Lee, T.S.: A Contingency Model for Knowledge Management Capability and Innovation. Industrial Management and Data System 106(6), 855–877 (2006)
67. Nesta, L., Saviotti, P.P.: Coherence of the Knowledge Base and the Firm's Innovative Performance: Evidence from the U.S. Pharmaceutical Industry. The Journal of Industrial Economics 53(1), 123–142 (2005)
68. Basadur, M., Gelade, G.A.: The role of knowledge management in the innovation process. Journal Compilation, Blackwell Publishing 15 (2006)
69. Polanyi, M.: The Tacit Dimension. Anchor Day Books, New York (1966)
70. Von Hippel, Tyre: The mechanics of learning by doing: Problem discovery during process machine use. Tech. and Culture 37, 312–329 (1996)
71. Bourdieu, P.: Outline of a Theory of Practice. Cambridge University Press, Cambridge (1977)
72. Lave, J.: Cognition in Practice: Mind, Mathematics, and Culture in Everyday Life. Cambridge University Press, Cambridge (1988)
73. Cumming, J.L., Teng, B.S.: Transferring R&D Knowledge: They Key factors affecting knowledge transfer success. Journal of Engineering and Technology Management 20, 39–68 (2003)
74. Moreland, R.L., Argote, L., Krishnan, R.: Socially shared cognition at work: transitive memory and group performance. In: Nye, J.L., Brower, A.M. (eds.) What's So Social about Social Cognition? Social Cognition Research in Small Groups, pp. 57–84. Sage, Thousand, Oaks (1996)
75. Porter, M.E., Stern, S.: Innovation: location matters. MIT Sloan Management Review 42(4), 28–36 (2001)
76. Hustad, T.P.: Reviewing current practices in innovation management and a summary of selected best practices. In: The PDMA. Handbook of New Product Development, pp. 489–51. Wiley and Sons, New York (1996)
77. Smith, D.: Designing an innovative Britain. ESRC: The Edge (22), 2 (2006)
78. Harkama, S.: A complex perspective on Learning within Innovation Projects. The Learning Organization 10(6), 340–346 (2003)
79. Gloet, M., Terziovski, M.: Exploring the relationship between knowledge management practices and innovation performance. Journal of Manufacturing Technology Management 15(5), 402–409 (2004)
80. Department of Trade and Industry: Innovation in Services, DTI Occasional Paper No. 9, HMSO (June 2007)
81. Granovetter, M.: The strength of weak ties. American Journal of Sociology 78(6), 1360–1380 (1973)
82. Burt, R.S.: Structural holes: The social structure of competition. Harvard University Press, Cambridge (1992)
83. Antonio, M.P.: External knowledge sources and proximity. Journal of Knowledge Management 13(5), 301–318 (2009)
84. Tether, B.: Who co-operates for innovation, and why. An empirical analysis. Research Policy 31, 947–967 (2002)
85. Coombs, R., Harvey, M., Tether, B.: Analyzing distributed processes of provision and innovation. Industrial and Corporate Change 12, 1125–1155 (2003)
86. Howells, J., James, A., Malik, K.: The sourcing of technological knowledge: distributed innovation processes and dynamic change. R&D Management 33, 395–409 (2003)

87. Freeman, L., Romney, K., Freeman, S.: Cognitive structure and informant accuracy. American Anthropologist 89, 310–325 (1987)

88. Gassmann, O., Enkel, E.: Towards a Theory of Open Innovation: Three Core Process Archetypes. In: R&D Management Conference (RADMA), Lisabon, Portugal (2004)

89. Parlby, D., Taylor, R.: The Power of Knowledge: A Business Guide to Knowledge Management (2000), http://www.kpmgconsulting.com/index.html

90. Nonaka, I., von Krogh, G., Voelpel, S.: Organizational Knowledge Creation Theory: evolutionary paths and future advances. Organization Studies 27(8), 1179–1208 (2006)

91. Von Hippel, E.: "Sticky Information" and the Locus of Problem Solving: Implications for Innovation. Management Science 40(4), 429–439 (1994)

92. Aoshima, Y.: Transfer of system knowledge across generations in new product development: empirical observations from Japanese automobile development. Industrial Relation 41, 605–628 (2002)

93. Cusumano, M., Nobeoka, K.: Thinking beyond Lean. Free Press, New York (1998)

94. Ju, T., Li, C.Y., Lee, T.S.: A contingency model for knowledge management capability and innovation. Industrial Management & Data Systems 106(6), 855–877 (2006)

95. Harwick, T.: Optimal Decision-Making for the Supply Chain. APICS – The Performance Advantage 7(1), 42–44 (1997)

96. Cox, A.: Power, value and supply chain management. Supply Chain Management: an International Journal 4(4), 167–175 (1999)

97. Laseter, T.: Balanced Sourcing: Cooperation and Competition in Supplier Relationships. Jossey-Bass Publishers, San Francisco (1998)

98. Quinn, F.J.: The clock speed chronicles. Supply Chain Management Review 3(4), 60–64 (2000)

99. Bowersox, D.J., Closs, D.J., Stank, T.P.: 21st Century Logistics: Making Supply Chain Integration a Reality. Council of Logistics Management, Oakwood (1999)

100. Cooper, M.C., Lambert, D.M., Pagh, J.D.: Supply Chain Management: More Than a New Name for Logistics. The International Journal of Logistics Management 8(1), 1–13 (1997)

101. Lambert, D.J., Cooper, M.C., Pagh, J.D.: Supply chain management, implementation issues and research opportunities. International Journal of Logistics Management 9(2), 1–19 (1998)

102. Bask, A.H., Juga, J.: Semi-integrated supply chain: towards the new era of supply chain management. International Journal of Logistics: Research and Applications 4(2), 137–152 (2001)

103. Persson, G.: Supply chain management, a multidisciplinary study of integrated supply chains. In: Persson, G., Grønland, S.E. (eds.) Research Report No. 9. Norwegian School of Management BI, Oslo (2002)

104. Aitken, M.J.: An Introduction to Optical Dating. Oxford University Press, Oxford (1998)

105. Penrose, E.T.: The theory of the growth of the firm. Basil Blackwell, Oxford (1959)

106. Drucker, P.F.: Management's New Paradigms. Forbes Magazine, 152–177 (October 5, 1998)

107. Richard, J., Arend, W., Joel, D.: Small business and supply chain management: is there a fit? Journal of Business Venturing 20, 403–436 (2005)

108. Shaw, N.C., Meixell, M.J., Tuggle, F.D.: A case study of integrating knowledge management into supply chain management process. In: Proceedings of the 36th Hawaii International Conference on System Sciences (2003)

109. Hult, G., Thomas, M., Ketchen, D.J., Slater, S.F.: Information processing, knowledge development and strategic supply chain performance. Academy of Management Journal 47(2), 241–254 (2004)

110. Burt, D.N., Soukup, W.R.: Purchasing role in new product development. Harvard Business Review 63, 90–97 (1985)

111. Swink, M.: Threats to new product manufacturability and the effects of development team integration processes. Journal of Operations Management 17, 691–709 (1991); Brainpower. Fortune, 44–56, (1999)

112. Shin, H., Collier, D.A., Wilson, D.D.: Supply management orientation and supplier/buyer performance. J. Oper. Manage. 18, 317–333 (2000)

113. Burton, T.T.: JIT/repetitive sourcing strategies: tying the knot with your suppliers. Production and Inventory Management Journal 29(4), 38–41 (1988)

114. Petersen, K.J., Handfield, R.B., Ragatz, G.L.: Supplier integration into new product devel-opment: coordinating product, process and supply chain design. Journal of Operations Management 23(3/4), 371–388 (2005)

115. Wang, X.H., Fu, L.W., Ming, X.G., Kong, F.B., Li, D.: Supplier-involved collaborative product development in PLM. International Journal of Computer Applications in Technology 37 (2010)

116. Griffin, A., Hauser, J.: Patterns of Communication among Marketing, Engineering, and Manufacturing—A Comparison between Two Product Teams. Management Science 38(3), 360–373 (1992)

117. Bensaou, M.: Interorganizational Cooperation: The Role of IT. Information Systems Research 8(2), 107–124 (1997)

118. Adam, L., Finn, W.: Value in business markets: What do we know? Where are we going? Industrial Marketing Management 34, 732– 748 (2005)

119. Tessarolo, P.: Is the integration enough for fast product development? An empirical investigation of the contextual effects of product vision. Journal of Product Innovation Management 24(1), 69–82 (2007)

120. Grebici, K., Blanco, E., Rieu, D.: Framework for managing preliminary information in collaborative design processes. In: Proceedings of the International Conference on Product Lifecycle Management (PLM 2005), Lyon,France (2005)

121. Lam, P.K., Chin, K.S.: Identifying critical success factors for conflict management in collaborative new product development. Industrial Marketing Management 34, 761–772 (2005)

122. Chen, Z., Siddique, Z.: A model of collaborative design decision making using timed petrinet. In: ASME 2006, International Design Engineering Technical Conference and Computers and Information Engineering Conference, Philadelphia, Pennsylvania, USA (2006)

123. Wasti, S., Liker, J.: Collaborating with Suppliers in Product Development: A U.S. and Japan Comparative Study. IEEE Transactions on Engineering Management 46(2), 245–257 (1999)

124. Singh, K.: The Impact of Technological Complexity and Inter-firm Cooperation on Business Survival. Academy of Management Journal 40(2), 339–367 (1997)

125. Hartley, J., Zirger, B.J., Kamath, R.: Managing the Buyer–Supplier Interface for On-Time Performance in Product Development. Journal of Operations Management 15(1), 57–70 (1997)

126. Jemison, D.B., Sitkin, S.B.: Corporate Acquisitions: A Process Perspective. Academy of Management Review 11(1), 145–163 (1986)

127. Nooteboom, B., Berger, H., Noorderhaven, N.G.: Effects of Trust and Governance on Rela-tional Risk. Academy of Management Journal 40(2), 308–338 (1997)
128. von Hippel, E.: The sources of innovation. Oxford University Press, New York (1988)
129. Handfield, R., Ragatz, G., Monczka, R., Petersen, K.: Involving Suppliers in New Product Development. California Management Review 42(1), 59–82 (1999)
130. Ragatz, G., Handfield, R., Petersen, K.: Benefits associated with supplier integration into new product Development under conditions of technology uncertainty. Journal of Business Research 55, 389–400 (2002)
131. Smith, P., Reinertsen, G.: Developing Products in Half the Time, pp. 1–320. Van No strand, New York (1991)
132. Trygg, L.: Concurrent Engineering Practices in Selected Swedish Companies: A Movement or an Activity of the Few? Journal of Product Innovation Management 10(5), 403–416 (1993)
133. Dyer, J., Ouchi, W.: Japanese-Style Partnerships: Giving Companies a Competitive Edge. Sloan Management Review 35(1), 51–63 (1993)
134. Wagner, S.M., Hoegl, M.: Involving suppliers in product development: Insights from R&D directors and project managers. Industrial Marketing Management 35, 936–943 (2006)
135. Wagner, S.M.: Intensity and managerial scope of supplier integration. The Journal of Supply Chain Management 39(4), 4–15 (2003)
136. Monczka, R.M., Handfield, R.B., Scannell, T.V., Ragatz, G.L., Frayer, D.J.: New product development: Strategies for supplier integration. ASQ Quality Press, Milwaukee (2000)
137. Primo, M.A.M., Amundson, S.D.: An exploratory study of the effects of supplier relationships on new product development outcomes. Journal of Operations Management 20(1), 33–52 (2002)
138. Carlile, P., Rebentisch, E.S.: Into the black box: The knowledge transformation cycle. Management Science 49(9), 1180–1195 (2003)
139. Stevens, G., Burley, A., Piloting, J.: Piloting the rocket of radical innovation. Research Technology Management 46(2), 16–25 (2003)
140. Mansfield, E., Schwartz, M., Wagner, S.: Imitation Costs and Patents: An Empirical Study. Economic Journal 91(364), 907–918 (1981)
141. Zirgir, B.J., Maidique, M.A.: A Model of New Product Development: An Empirical Test. Management Science 36(7), 867–883 (1990)
142. Hauser, J., Gerard, J.T., Griffin, A.: Research on Innovation: A Review and Agenda for Marketing Science. Marketing Science 25, 686–717 (2006)
143. Gassmann, O., Kausch, C., Enkel, E.: Integrating Customer Knowledge in the Early Innovation Phase. In: 6th Europe. Conf. on Organizational Knowledge Learning and Capabilities, pp. 1–22 (2005)
144. Zwass, V.: Co-Creation: Toward Taxonomy and an Integrated Research Perspective. International Journal of Electronic Commerce 15(1), 11–48 (2010)
145. Gummesson, E.: Relationship marketing in the new economy. Journal of Relationship Marketing 1(1), 37–57 (2002)
146. Griffin, A., John, R.: Integrating R&D and Marketing: A Review and Analysis of the Literature. Journal of Product Innovation Management 13(3), 191–215 (1996)
147. Prahalad, C.K., Ramaswamy, V.: The Future of Competition: Co-Creating Unique Value with Customers, Harvard Business School Press, Boston (2004)

148. Paquette, S.: Customer knowledge management available at, `ftp://ftp.eng.umd.edu/:/home/glue/s/p/spaquett/pub/docs/Paquette%20-%20Customer%20Knowledge%20Management.pdf`

149. Bennet, R., Gabriel, H.: Organizational factors and knowledge management within larger marketing departments: an empirical study. Journal of Knowledge Management 3, 212–225 (1999)

150. Wallace, D.P.: Knowledge management: historical and cross-disciplinary themes, London (2007)

151. von Hippel, E.: The dominant role of users in the scientific instrument innovation process. Research Policy 5, 212–239 (1976)

152. Reichward, P., Piller, F.: Interakitivewertschopfung: Open Innovation, Individuallisierung, und neueFormen der Arbeitsteilung, 2nd edn. Wiesbaden (2009)

153. Füller, J., Matzler, K.: Virtual product experience and customer participation—A chance for customer-centered, really new products. Technovation 27(6-7), 378–387 (2007)

154. Jeppesen, L.B.: User Toolkits for Innovation: Consumers Support Each Other. Journal of Product Innovation Management 22(4), 347–362 (2005)

155. Lettl, C., Cornelius, H., Hans, G.G.: Learning from Users for Radical Innovation. International Journal of Technology Management 33(1), 25–45 (2006)

156. Gassman, O., Sandmeier, P., Wecht, C.H.: Extreme Customer Innovation in the Front End Learning from a New Software Paradigm. International Journal of Technology Management 33(1), 46–66 (2006)

157. Baker, W.E., Sinkila, J.M.: Market Orientation and the New Product Paradox. Journal of Product Innovation Management 22(6), 483–502 (2005)

158. Prahalad, C.K., Ramaswamy, V.: The future of competition: Co-creating unique value with customers. Harvard Business School Press, Boston (2003)

159. Davenport, T.H., Harris, J.G., Kohli, A.K.: How do they know their customers so well? MIT Sloan Management Review, 63–73 (winter, 2001)

160. Desouza, K.C., Awazu, Y., Yamakawa, S., Umezawa, M.: Facilitating knowledge management through market mechanism. Knowledge and Process Management 12(2), 99–107 (2005)

161. Gibbert, M., Leibold, M., Probst, G.: Five styles of customer knowledge management, and how smart companies put them into action, `http://archive-ouverteunige.ch/downloader/vital/pdf/tmp/h84ltht483penmb83ol4turhq1/out.pdf`

162. Sofianti, et al.: Customer knowledge management in new product development (2009), `http://scholar.google.co.uk/scholar?q=Sofianti+et+al.%2C+2009&hl=en&as_sdt=0%2C5` (February 8, 2012)

163. Stewart, T.A.: Intellectual Capital: The New Wealth of Organizations (1997)

164. Kohlbacher, F.: International Marketing in the Network Economy: A Knowledge-based Approach. Palgrave Macmillan, Basingstoke (2007)

165. Burns, T., Stalker, G.M.: The management of innovation. Tavistock, London (1961)

166. Rothwell, R.: The characteristics of successful innovators and technically progressive firms. R&D Management 3, 191–206 (1977)

167. Baldwin, C., Hienerth, C., von Hippel, E.: How user innovations become commercial products: A theoretical investigation and case study. Research Policy 35, 1291–1313 (2006)

168. von Krogh, G., von Hippel, E.: Special issue on open source software development. Research Policy 32, 1149–1157 (2003)

169. von Krogh, G., von Hippel, E.: The promise of research on open source software. Management Science 52, 975–983 (2006)

170. Harhoff, D., Henkel, J., von Hippel, E.: Profiting from voluntary information spillovers: How users benefit by freely revealing their innovations. Research Policy 32, 1753–1769 (2003)
171. Shah, S.K., Tripsas, M.: The accidental entrepreneur: The emergent and collective process of user entrepreneurship. Strategic Entrepreneurship Journal, 1, 123–140 (2007)
172. Bergman, J., Jantunen, A., Saksa, J.-M.: Enabling Open Innovation Process through Interactive Methods: Scenarios and Group Decision Support Systems. IJIM 13(1), 139–156 (2009)
173. Su, C., Chen, Y., Sha, D.: Linking innovative product development with customer knowledge: A data-mining approach. Technovation 26(7), 784–795 (2006)
174. Leonard, D., Sensiper, S.: The role of tacit knowledge in group innovation. California Management Review 40(3), 112–132 (1998)
175. Svensson, D., Malmqvist, J.: Strategies for Product Structure Management of Manufacturing Firms. Transactions of the ASME, Journal of Computing and Information Science in Engineering 2, 50–58 (2002)
176. Ciocoiu, M., Nau, D.S., Gruninger, M.: Ontologies for Integrating Engineering Applications. Transactions of the ASME, Journal of Computing and Information Science in Engineering 1, 12–22 (2001)
177. Ameri, F., Duta, D.: Product lifecycle management: closing the knowledge loops. Computer-Aided Design & Applications 2(5), 577–590 (2005)
178. Stark, J.: Product lifecycle management. In: 21st Century Paradigm for Product Realization. Decision Engineering Series, p. 441. Springer, London (2005)
179. Freeman, C.: Chemical process plant: innovation and the world market. National Institute Economic Review 45, 29–57 (1968)
180. Hollander, S.: 77 M Source of Increased Efficiency: A study of Dupont Rayon Plants. MIT Press, Cambridge (1965)
181. Enos, J.L.: Petroleum Progress and Profits. MIT Press (1962)
182. Dutt, H.: Role of Knowledge Manager in Knowledge Product Development. Delhi Business Review 7(1) (2006)
183. Jiang, X., Li, Y.: An empirical investigation of knowledge management and innovative per-formance: The case of alliances. Research Policy 38, 358–368 (2009)
184. Hoyer, W.D., Chandy, R., Dorotic, M., Krafft, M., Singh, S.S.: Customer Creation in New Product Development. Journal of Service Research, 13(3), 283–296 (2010)
185. Lambert, D.M., Cooper, M.C.: Issues in Supply Chain Management. Industrial Marketing Management 29, 65–83 (2000)
186. Liebowitz, J. (ed.): The knowledge Management Handbook. CRC Press, Boca Raton (1999)
187. Akram, K., Siddiqui, S.H., Nawaz, M.A., Ghauri, T.A., Cheema, A.K., Amjad, K.H.: Role of Knowledge Management to Bring Innovation: An Integrated Approach. International Bulletin of Business Administration (11) (2011) ISSN: 1451-243X
188. Xu, J., Houssin, R., Caillaud, E., Cradoni, M.: Macro process of knowledge management for continuous innovation. Journal of Knowledge Management 14(4), 573–592 (2010)
189. Assimakopoulos, D.G., Chapelet, B.: Knowledge flows in an NPD Team from the Semi-conductor Industry (February 2013), http://download.springer.com/ static/pdf/257/chp%253A10.1007%252F978-1-4614-0248- 0_3.pdf?auth66=1362930317_0941a919688334fa4eb283a6d2a3669d& ext=.pdf
190. Pillai, K.G., Goldsmith, R.E.: Calibrating managerial knowledge of customer feedback measures: a conceptual model. Marketing Theory 6(2), 223–243 (2006)

191. Cohen, W., Levinthal, D.: Absorptive capacity: A new perspective on learning and innovation (1990)
192. Helfat, C., Finkelstein, S., Mitchell, W., Peteraf, M.A., Singh, H., Teece, D.J.: Winter SG (eds), pp. 19–29. Blackwell, Oxford (2007)
193. Kim, L.: Building technological capability for industrialization: analytical frameworks and Korea's experience. Industrial and Corporate Change 8(1), 111–136 (1999)
194. Foray, D.: Generation and distribution of technological knowledge: incentives, norms and institutions. In: Edquist, C. (ed.) Systems of Innovation. Technologies, Institutions and Organisations, pp. 64–85. Pinter (1997)
195. Caloghirou, Y., Kastelli, I., Tsakanikas, A.: Internal capabilities and external knowledge sources: complements or substitutes for innovation performance. Technovation 24(1), 29–39 (2004)
196. Souitaris, V.: Strategic influences of technological innovation in Greece. British Journal of Management 12(2), 131–147 (2001)
197. Drucker, P.F.: Post-capitalist society. Harper, New York (1993)
198. Wijnhoven, F.: Acquiring organizational learning norms: a contingency approach for understanding deuteron learning. Management Learning 32(2), 181–200 (2001)
199. Cooper, R.G., Dreher, A.: Voice-of-Customer Methods: what is the best source of new product ideas? Marketing Management Magazine, 38–48 (winter, 2010)
200. Jenkins, S., Forbes, S., Durrani, T.S.: Managing the product development process (Part 1: an assessment). Int. J. Technology Management 13(4), 359–377 (1997)

Soft Systems Methodology: A Conceptual Model of Knowledge Management System Initiatives

Nor Hasliza Md Saad[1], Hasmiah Kasimin[2], Rose Alinda Alias[3], and Azizah Abdul Rahman[3]

[1] School of Management, Universiti Sains Malaysia, 11800, Penang, Malaysia
[2] Faculty of Economics and Management, Universiti Kebangsaan Malaysia Selangor, Malaysia
[3] Faculty of Computer Science and Information System,
Universiti Teknologi Malaysia Johor, Malaysia
norhasliza@usm.my, miah@ukm.my,
{alinda,azizahar}@utm.my

Abstract. This paper demonstrates how Soft Systems Methodology (SSM) can be used to consider a broad range of issues relevant to implement KMS initiatives including the technical, social and organizational context. The focus of applying the SSM is to develop a conceptual model that provides the basis of identifying the process involved and exploring the factor influencing the KMS initiatives. A multiple case study approach is used to get an extensive picture of and analyze the experience of four Malaysian Public Universities (MPUs). The findings provide comprehensive understanding on how KMS initiatives are being implemented in MPUs through the development of a conceptual model. The conceptual model provides the guidelines which can be used as a theoretical framework and practical tool to determine the important activities integrated within the process and the influencing factors in the KMS initiatives implementation.

Keywords: Knowledge Management Systems, Soft Systems Methodology, Conceptual Model, Malaysian Public Universities, Multiple Case Studies.

1   Introduction

KMS initiative is considered as a multidisciplinary field that involves the technical disciplines such as computer science and information technology and the social science disciplines of management and sociology [2]. The implementation of KMS is related to the applying appropriate information system to manage organizational knowledge for the improvement and benefit of the organization. This study argues that KMS initiatives are considered as socio-technical system comprises of technology, processes, and human involvement to function in the whole system in meeting the desired objective.

Over the last two decades the implementation of KMS initiatives has been received the widespread attention from various type of organization both business sector and public sector. In the beginning, the trend of practice and research area of KMS initiatives has been prevalent in the business sector .In recent years, higher education

A. Fred et al. (Eds.): IC3K 2012, CCIS 415, pp. 377–392, 2013.

sector around the world is started implementing KMS initiatives while others considering or working towards implementing them. According to many researchers, the
KMS initiatives need to be implemented in the higher education sector to change its
classical paradigm to confront changes in the external environment change and provide effective services to meet market demand and enhance the organisation [8, 27].
A variety of different approach of KMS initiatives can be introduced not only in the
areas of teaching and learning, but also in the area of administration to support a wide
range of business processes. Many potential KMS initiatives have been identified to
support administrators, faculty, researchers, students and alumni. Most research in
KMS initiatives in the higher education sector have been applied in the developing
countries such as US and UK. However, there is little knowledge on how MPUs implement KMS initiatives and what challenges they confront.

The objective of the study is to demonstrate how the SSM approach can be applied
to investigate the implementation of KMS initiatives. The intention of using the SSM
is to develop a holistic framework to describe the fundamental process of the implementation KMS initiatives through the development of conceptual model. Furthermore, the analysis identifies influencing factors in the surrounding environment which
are facilitating to the implementation of KMS initiatives and which may impede the
effectiveness of KMS initiatives. Thus, this study proposes a holistic conceptual model to provide a better understanding on the nature of KMS initiatives from the
experience of the four MPUs.

2   Application of Soft Systems Methodology

The SSM emerged in response to the limitations of the hard systems approach to ade-
quately address complex real world problems that involve human issues. The emergence of a hard systems approach was influenced by systems engineering and system
analysis, which use a systematic approach to problem-solving in relation to the design, development and operation of a machine to achieve predefined objectives [15].
Thus, to solve a problem they use reductionists concepts that divide the problem into
smaller pieces and manageable fragments, without emphasising human or organisational issues [4]. This approach is highly appropriate in clearly structured and well-
defined problems but is insufficient to address complex unstructured problems
that deal with social, culture and political issues. In the early 1970s, the soft systems
methodology was incorporated into a practical methodology by Professor Peter
Checkland in collaboration with his colleagues at Lancaster University [9]. SSM is a
well-established problem-solving approach to handle complex problem situations,
which concerns the wider contextual aspect of the problem involving human activities
and contextual issues. The SSM is applicable in a wide range of problem situations
which has been used in different research field such as management science,
education research, sociology and health care[17].

The approach used to apply SSM to research can vary, but the basic feature corresponding to the conceptual basis remains constants. In general, SSM typically has
three primary concerns in process of improving problem situations. First, SSM is
concerned with examining complex problems involving the socio-technical system,
which involves human intervention. A strong emphasis is placed on understanding the

different perceptions of multiple stakeholders involved in the problem situations. Second, SSM highlights the importance of creating a purposeful human activity model relevant to the problem situations as a device to identify appropriate changes that could be made to improve the problem situation. Finally, SSM strives to create a learning system to identify methods for improvement by providing with one or more alternative solutions rather than an optimisation approach [23].

The traditional SSM approach based on seven-stage model of analysis consists of following stages: - Stage 1: The problem situation is viewed as unstructured, Stage 2: The problem situation is expressed, Stage 3: The root definition of relevant systems is defined, Stage 4: This stage involves the development of conceptual models as defined by the description in the root definition and being treated as a system, Stage 5: Comparison of stage 2 (real world situation) and stage 4 (conceptual model), Stage 6: Feasible and desirable change are identified and assessed to improve the problem situations, and Stage 7: Action to improve the problem situation.

The first part of this research is to find out the problem situation and development of root definition that generated concern of KMS initiatives in MPUs, which involve the analysis of stage 1, stage 2 and stage 3. In this part, this research adopts a multiple case study approach to investigate four MPUs representing two major characteristics of the higher education environment in Malaysia: the older universities established before 1990, where generally larger in size; the newer universities established after 1990, where commonly representing smaller in size. This variation was aimed to enhance the understanding of each phenomenon under study in multi-faceted manner to represent the uniqueness embedded in the social and organisational context. The four cases were divided into two groups for the purposes of comparison and contrast. One group consists of University A(UA) and University B(UB), which are categorised as older universities. The other group consists of University C(UC) and University D(UD), which represent newer universities.

An interpretive approach is used as a mode of inquiry to allow the researcher to establish meaning from the complex problem of the real-world situation. The analysis of the cases results in the classification of the main activities involved in the process and the identification of the main factors influencing the process. Finally, cross-case analysis is used to explore similarities and differences among cases in terms of themes and issues of concern. Data collection came mainly from interviews and document analysis, and each of these methods offers important insights and understanding into the cases. First, the interviews were held with the KMS champions who were highly involved in major activities in the implementation of KMS initiatives at the university level. They included information technology (IT) department directors and managers, Chief Librarians and other related administrative directors. Second, relevant documents were collected from government publications, annual reports, institutional websites, business manuals and slide presentations. Data were collected on the issues related to (i) the champion of KMS initiatives; (ii) the process involved; and (iii) the influencing factors of the implementation of KMS initiatives

This paper attempts to concentrate on presenting the research findings based on the development of conceptual model of human activity system in the stage 4 of SSM, but does not discuss the whole range of process involve in SSM. At this stage a conceptual model is develop to represent a relevant human activity systems through a set of connected activities implies from the problem situation of KMS initiatives. Every conceptual model must have a management or a monitoring and control activity in order to achieve its objective. The management activity is used as a basis to enquire about each activity involved has a monitoring and control activity. Section 3 below provides findings of the development of a conceptual model.

3   Findings: A Conceptual Model of KMS Initiatives

This section compares the findings of the four cases, highlighting the similarities and differences to find the common patterns of activities in a conceptual model for implementing KMS initiatives. The analysis revealed that this conceptual model has six related activities, and each activity was then analysed with the influencing factors together with key highlighted. The overall of conceptual model is depicted in Figure 1.

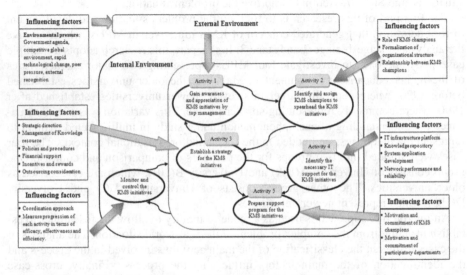

Fig. 1. The Conceptual Model of KMS initiatives in MPUs

This study shows that all four cases are slowly undergoing changes to embrace the challenge of the implementation of the KMS initiative. The cases differ in their priority and scope of bringing the KMS initiative to the university context. However, they have common activities that indicate they are operating in the similar context of MPUs.

3.1   Activity 1: Gain Awareness and Appreciation of KMS Initiatives by Top Management

This activity deals with the awareness of and appreciation for KMS initiatives by top management. In each case, the appreciation of KMS initiatives by top management was found to be crucial for creating more formal and conscious attempts to implement KMS initiatives at the university level. They were important instruments for bringing KMS initiatives to the forefront of the university agenda [29, 30]. The universities' decision to implement KMS initiative is generally triggered by a specific event or circumstance, which they are taken place. In general, the summary of activity 1 is listed in Table 1.

Table 1. The Summary of Activity 1

Influencing Factor	Key Issues Highlighted	Case
External pressure	Government agenda, competitive and global education environment, and rapid technological change	UA, UB, UC, UD
	External recognition	UB,UC
	Peer community pressure	UC, UD

According to the findings of the analysis, the primary factor influences for this activity is environmental Pressure. The external environmental context within the social contexts of the MPU represents the influence of various factors that contributed to the implementation of KMS initiatives. In all cases, the government agenda stood out as one of the most important external factors for bringing KMS initiatives into universities. Many respondents recognised the contribution of this factor in influencing top-level decision makers. Another external factor that emerged from the data analysis concerns advances in technology. All of the case universities struggle to keep pace with rapid rate of technological changes. In several cases, external recognition seems to be an effect of continuously maintaining the effective utilisation of technology for facilitating KMS initiatives. Awards and public recognition have been possible incentives for influencing the implementation of KMS initiatives. An example of external recognition is the government rewards for excellent utilisation of technology in improving business performance. The university's interest in the KMS initiatives also arises because of its desire to follow other MPUs that are already embarking on KMS initiatives.

3.2 Activity 2: Identify and Assign KMS Champions to Spearhead the KMS Initiatives

In activity 2, it is important to consider that the university's top management contributed strongly to the selection and assignment of KMS champions. The university's top management plays a critical role in identifying specific KMS champions with multidisciplinary expertise and representatives from the core business departments that would enhance the implementation. A summary of activity 2 is listed in Table 2.

Table 2. The Summary of Activity 2

Influencing Factors	Key Issues Highlighted	Case
a) Identification of KMS Champion	Diversity of expertise domains (IT department and Library).	UA,UB, UC,UD
b) Formalisation of the Organisational Structure	Clear specialized department (KM center) and create new role(CIO/CKO).	UA, UC
	Lack of clear specialized department and create new role.	UB,UD
c) Relationship between KMS Champions	Close collaboration between KMS champions	UC,UD
	Lack of collaboration between KMS champions	UA,UB

According to the findings of the analysis, the primary influences for this activity can be classified as follows:

a)   Identification of KMS Champion

With regards to the characteristics of the KMS champions, an interesting element has emerged. In all of the cases, the two main domains of expertise have emerged in making the KMS initiatives successful: the technical professionals from the IT department and information management professionals from the library domain. The KMS champions contribute their expertise to support KMS initiatives. The IT department is more strongly driven by their efforts in preparing for the advanced technical requirements of KMS initiatives. The library is concerned with upgrading traditional library services and information resources into a more digital environment. The selection of KMS champions should be based on expertise that can effectively assist in ensuring better implementation. Many studies on KMS initiatives within the HE context have reported that either the library or the IT department indeed play a major role in the projects and activities of KMS [5-7, 11].

b)   Formalisation of the Organisational Structure

The concern for this activity is to specify a clear role and responsibility of the KMS champion for KMS initiatives. The top management is in a position of authority to delegate responsibility and the setting of the organisational structure. The presence of an organisational structure setting by creating specific positions and structure would help to make KMS initiatives highly visible organization [19], as demonstrated in the case of UA and UC. Daft [12] argued that a lesser degree of formalisation of the organisational structure, such as creating a task force function and responsibility, seems to solve the short-term problem and lack of sustainability. However, the issue of a lack of a clear organisational structure emerged in the other two cases. These cases were concerned with establishing a taskforce to address KMS initiatives rather than adjusting their existing organisational structure.

c)   Relationship between KMS Champions

The coordination of the relationship of KMS champions was another common issue that drew attention in the analysis of the four cases. From the findings, collaboration between KMS champions is critical to achieve a comprehensive and unified direction to support university-wide KMS initiatives. This need is often due to the normal practice of these KMS champions, which have different business practices and services. The close relationships among these key players were achieved by the skilful coordination and monitoring of the university's top management. The close relationship of KMS champions in the case of UC was able to contribute to the standardisation of the KMS project and initiatives;hence, they utilised the scarce resources available in the most efficient manner to support the project. This accomplishment is achieved through coordination by top management [5], and without this formality, the management of collaboration is unlikely.

3.3   Activity 3: Establish a Strategy for the KMS Initiatives

This activity is considered to be very crucial that would serve as a platform to effectively guide the overall implementation of KMS initiatives. A strategy needs to be established to decide what important elements should be included in the KMS initiatives, as presented in Table 3.

Table 3. The Summary of Activity 3

Influencing Factors	Key Issues Highlighted	Case
a) Strategic direction	Lack of clear strategy direction	UA,UB,UD
	Clear strategy direction and comprehensive	UC
b) Management of Knowledge resource	Systematic procedure to manage knowledge resource	UC
	Lack of systematic procedure to manage knowledge resource	UA,UB,UD
	Less emphasis on leveraging tacit knowledge	UA,UB,UC,UD
c) Policies and procedures	Lack of clear policies and procedures	UA,UB,UD
	Clear policies and procedures	UC
d) Outsourcing consideration	Speed up project development	UA, UC
	Gain knowledge and skills	UA,UC
	Avoid bias decision	UA
e) Financial support	Insufficient financial support	UA, UB
	Sufficient financial support	UC
	Constraints in financial planning	UA,UB
f) Incentives and rewards	Lack of incentive and reward	UA,UB, UD

According to the findings of the analysis, the primary influences for this activity can be classified as follows:

a)   Strategic Direction

The strategic direction is an essential activity, as it ultimately addresses the overall purpose and direction of focus in KMS initiatives. The identification of a strategy approach plays a significant role in the early stages of implementation. Many of these cases do not have a clear comprehensive strategy to support university-wide KMS initiatives. However, they are aware of the importance of KMS, and the strategy is acknowledged to be embedded in other strategies. This is one of the reasons why all the cases are still in the infancy stage KMS initiatives implementation. Interestingly, it seemed that two broad approaches drove the strategic objective of KMS initiatives. First, the technology-driven strategy focused on the important responsibility of the IT department. From the perspective of IT-driven strategy, the "IT Master Plan" became a generic concept that supported the strategy of the KMS initiatives. Second, the library-driven strategy centred on the library department playing an important role in the development of the digital library. The emphasis of this strategy was perceived to revolutionise traditional library services and information resources into a more digital environment.

b)   Management of Knowledge Resource
The central issue for initiating KMS is to identify the types of potential knowledge resources that can offer strategic value and outcomes to the organisation. The essence of this activity is to make knowledge resources more accessible and available online. In all of the cases, priority was according to leveraging explicit knowledge resources that exist in terms of business documents and reports or reside in various resources, including in the core databases. In addition, this knowledge resource is somehow lacking in a standardisation procedure to manage. To make this matter more problematic, knowledge is scattered across the university and exists in a variety of formats. However, most cases demonstrated less effort to leverage tacit knowledge sources from human experience and business activities. This tacit knowledge is usually considered to be more difficult to leverage than explicit knowledge, and there is poor understanding of the proper way to manage tacit knowledge.

c)   Policies and Procedures
In most cases, the KMS champions are concerned with the lack of policies and procedures for clearly regulating and controlling the related activities. Furthermore, it is clearly stated that very little effort has been made by top management to put appropriate policies and procedures in place to support KMS initiatives. It is particularly challenging for KMS champions to induce participating departments to participate in their KMS initiatives because the introduction of KMS initiatives somehow changes the current practice of business processes to encourage the adoption of IT applications and enhance knowledge-sharing activities. However, only the case of UC provided policies and procedures to guide all departments within the university to ensure the adoption of KMS initiatives. Notwithstanding, a number of studies have discussed the importance of creating well-documented policies and procedures to address core activities in the process of KMS initiatives [22, 25, 31].

d)   Outsourcing Consideration
The role of KMS champions is ultimately to be responsible for managing KMS initiatives according to the plan. Some of the cases naturally underestimated the complexity of preparing and managing KMS initiatives to be completed according to the project schedule and desired outputs. Unfortunately, these cases did not consider effective decisions in gaining the benefits of outsourcing (e.g., expertise, cost, and time). Among the four institutions, two cases stressed the importance of employing outsourced support. They believed that this approach might influence the process of KMS initiatives in a positive way. These two cases highlighted their conscious decisions to hire external consultants during the initial stage of introducing KMS initiatives. This effort was particularly considered to be a method for gaining the advantage of the specialised skills of the consultants and to accelerate project development.

e)   Financial Support
The issue of financial support appeared to have an important influence on the selection and development of new technological and innovative solutions at the institutional level. The first primary concern brought up in all of the cases was the time constraint related to financial IT planning. The three cases agreed that to continue

updating the system to keep abreast of rapid rate of technological change, they needed to upgrade the comprehensive archival systems and introduce new systems to keep them in compliance with their current technological functions. The problem of allocation financial support for KMS initiatives is many facetted, especially is in the long term and involved various interrelated projects. The allocation of financial support tended to be concerned with priorities need to be addressed. According to Wong [34], decision maker(s) should develop a realistic scope of the project, according to available financial support.

f)   Incentive and Reward

Reward and incentive are another important consideration for effective KMS initiatives. The focus of this activity is to encourage participation in KMS initiatives. One case identifies the importance of preparation incentives and rewards to KMS champions and staff participants which found that the incentives help individuals to increase their willingness to participate and feel appreciated for their contribution. Many studies have posited that the essence of incentives and rewards is to support changes to employee attitudes and behaviours such that they will contribute and participate in KMS initiatives [28].

3.4   Activity 4: Identify the Necessary IT Support for the KMS Initiatives

This activity concerns efforts to decide on and prepare the necessary IT solutions to support the desired KMS initiatives' objectives, as listed in Table 4. This activity is essential for effective implementation of KMS initiatives. In this activity, the IT professional provides a crucial role in identifying and guiding on the requirements for IT capabilities and functionality that can support and enhance the process for capturing, storing and disseminating knowledge [32].

Table 4. The Summary of Activity 4

Influencing Factors	Key Issues Highlighted	Case
a) IT infrastructure and platform	Lack of coordination and standardisation	UA,UB,UD
	Effective Coordination and standardisation	UC
	Complexity of maintaining the mixture of legacy and new technology	UA,UB
	Robust IT infrastructure	UC
b) Knowledge repository	Lack of coordination and standardisation	UA,UB,UD
	Complexity of managing large amounts of databases	UA,UB,
	Large collection of resources involved in digitalising	UA,UB,
	Lack of information security control	UA,UB,UD
	Complexity of digital document categorisation	UA,UB,UC, UD
	Less complexity of managing a small amount of resources.	UC
c) System application development	Isolated development and focus on departmental needs	UA,UB,UD
	Integrated and interoperable	UC
d) Network performance and reliability	Extensive network security threats and poor performance.	UA,UB
	Lack of network interruption	UD,UC

There are five major component issues that should be addressed for effective KMS initiatives.

a)   IT Infrastructure and Platform

In each case, IT infrastructure influenced the preparation of technology requirements for KMS initiatives. There are two major barriers experienced by several cases for moving towards the mission of preparing IT platforms for facilitating KMS initiatives. The first barrier is the lack of a standard and common IT infrastructure; current platforms are heterogeneous and controlled by different departments. Within this situation, the equipment and infrastructure are often poorly managed, which leads to inefficient use of resources. Another concern regarding IT infrastructure is the maintenance of insufficient technical requirements that are outdated or lack capable technologies. This issue reflects concerns about the challenge of preparing an appropriate IT infrastructure in which some components of the installation-based infrastructure are subject to upgrades or replacement. The other potential problem raised was incompatibility and complexity with the installation-based infrastructure. The well preparation of IT infrastructure was especially apparent in the case of UC, especially with well-planned state-of-the-art technology in providing a coordinated and standardised approach. This factor seems to facilitate better technology management without much concern for the various conflicts of multiple standards of equipment and outdated technology. The development of a well-planned architecture of an IT infrastructure for the entire university environment is an important consideration to facilitate coordination, management, and connectivity among different departments[35].

b)   Knowledge Repository

In all cases, database resources, with their various challenges and opportunities, were brought into play because they are considered to be at the heart of the knowledge resources that can be better utilised. There are six major barriers that were experienced by these cases in moving towards the mission of preparing technology platforms for facilitating KMS initiatives. First, a lack of coordination and standardisation of database resources has a negative influence on the complex process of data integration, data availability and data accessibility. The cause of the problem was identified as being either the incompatibility of heterogeneous platforms or that the database resources were placed at dispersed locations that often lacked common data definitions and poor data documentation. On the contrary, UC had an encouraging experience with enterprise database solutions and centralised data management for the entire knowledge repository. This university acknowledged that this database approach was designed to enforce consistency and facilitate database management across different resources.

Second, the huge amount of database resources has also increased the complexity of managing a knowledge repository for the cases of UA and UB. These universities have undertaken the task of collecting their archive resources together with current data resources into a more manageable effort. Third, they also have large collections of databases resources and paper-based documents requiring effective electronic management to make them more accessible and available. Fourth, throughout the four cases, the lack of systematic categorisation of digital documents is widely recognised as one of the earliest and most crucial efforts in managing digital documents. Finally, the issue of information security was the most pressing concern in all of the cases. There is a crucial need for better information security in terms of user access control and document confidentiality.

c)   System Application Development

According to the cross-case analysis, there is a significant difference between UC and the other three universities. In many cases, the continuing effort to develop multiple applications for different purposes in an uncoordinated manner worsens the lack of information shared and increases duplication efforts. UC attempted to take advantage of offering integrated and interoperable applications for business usage. The benefit of this approach is that it would tremendously streamline business processes, enhance information flow across departments, and reduce the usage of paper. Cain et al.(2008) suggested that universities should focus on applications for supporting the streamlining of business processes by understanding and developing integration to meet the range of business function needs. Furthermore, many cases focus on user-friendly applications but ignore the importance of customisation and personalisation of the user interface.

d)   Network Performance and Reliability

The issue of network performance concerns the network speed and connectivity of system applications. The major current network issues emphasise the concern regarding threats from hackers, intruders or viruses. This focus is due to a lack of coordination in controlling fragmented server locations across the university and frequent service interruptions. Network performance is another concern that supports the effectiveness of KMS initiatives, particularly in the cases of UA and UB. Specifically, network connection problems, such as network failure or a slow connection, tend to erode the efficiency of information flow and decrease user satisfaction. Centralised policy management and network interface provisioning are powerful strategies to regulate the network and control traffic load for performance, efficiency and security [16].

3.5   Activity 5: Prepare Support Programs for the KMS Initiatives

The preparation of appropriate support programs is another important stage to address in the process of implementing KMS initiatives. Each case study has its own way to make not only KMS champions but also participating departments aware of the current KMS initiatives being conducted and to attempt to clearly spread KMS initiatives. The summary of the activity 5 is listed in Table 5.

a)   Motivation and Commitment of KMS Champions

In the three cases, motivation emerged as an influencing factor that encouraged the key players to effectively implement the KMS initiatives. The motivating factors might be in terms of the sponsorship of required resources and leadership from top management in pursuing the wider scope of KMS initiatives across the university. For these KMS champions, the top management was fundamental in its position of

Table 5. The Summary of Activity 5

Influencing Factors	Key Issues Highlighted	Case
a) Motivation and commitment of KMS champions	Lack of clear understanding of the KMS concept	UA,UB,UD
	Lack of IT skills and knowledge	UA,UB,UC,UD
	Perceived benefit	AU,UB,UC
	Lack of top management support	UD
b) Motivation and commitment of participatory departments	Lack of perceived benefit of the project	UA,UB,UD
	Prefer an individual department approach	UA,UB,UD
	Lack of trust and confidence	UA,UB,UD
	Perceived benefit of the project	UC
	Communication between KMS champions and business departments	UC
	Trust and confidence to share their information	UC

authority to set the direction of KMS initiatives and to delegate resources to drive the KMS initiatives forward. Motivation and commitment of the KMS champions influences the effectiveness of KMS initiatives [14]. This study found that the KMS champions perform an important role in distributing KMS messages and activities consistently across the university. In addition, the three cases also demonstrated that the KMS champions were motivated to spearhead KMS because they perceived that KMS initiatives would provide a new opportunity to enhance business processes and decision making. In the majority of the cases, it was clear that there was a problem with the process of firmly understanding the concept of KMS initiatives. Many studies have identified that understanding the concept of KMS plays a major role in preparing and identifying effective approaches to supporting KMS initiatives [1, 24]. Furthermore, the KMS champions need to be equipped with IT knowledge and skills that will help them make decisions or develop effective IT support. Several studies have revealed that public sector is confronted by a severe technology skills deficit, mostly in the form of a lack of proper training.[20, 21].

b) Motivation and Commitment of Participating Departments

The key objective of this activity is to create awareness about KMS initiatives over the entire university and attract other departments' participation. The finding indicates that the involvement and participation of the business departments is crucial to give appreciation to their ideas and comments, including creating a sense of ownership and perceived benefit of the project. In many cases, they were concerned about the lack of a knowledge-sharing culture due to a lack of communication and connectivity among the various business departments because of differences in their operations and services. Furthermore, the KMS champion faces another problem concerning a lack of trust and confidence in the information security flow in the digital environment. This issue was partly due to the absence of a formal approach and procedural guidelines to deal specifically with this problem. Thus, it became difficult for KMS champions to stimulate and motivate the various departments to have a favourable attitude towards knowledge sharing. This finding is consistent with the literature on IT project development, which found that the key players of a project should not underestimate the stakeholder's capacity to influence either the escalation or failure of the project [18, 33].

3.6   Management Activity: Monitoring and Controlling the Implementation of the KMS Initiatives

The management activity is also considered a very important activity in SSM, as it has become essential to monitor and control with the issues that prevent effective KMS initiatives. A summary of management activity is shown in Table 6. This activity is very important, as once the process of implementing KMS initiatives is underway, it becomes essential to constantly monitor and control the progress and performance of KMS initiatives. It is interesting to note that the finding indicates that most cases do not have institutional monitoring and controlling mechanisms of the implementation of KMS initiatives at the university level. The central coordination of monitoring and controlling KMS would provide a more unified and collaborative approach. There is a need to develop a monitoring and controlling system that would constantly assess the implementation processes [3].

Table 6. The Summary of Management Activity

Influencing Factors	Key Issues Highlighted	Case
a) Coordination approach	Lack of standard coordination approach	UA,UB,UD
	Standard coordination approach	UC
b) Measure progression	Lack of performance measurement	UA,UB,UC,UD

4   Discussion

In general, KMS initiatives are MPUs are in the beginning stage to implement KMS initiatives. It is acknowledged that there are variation of influencing factors and key issues in each case. However, these cases have common activities that indicate there are in the similar organizational setting of MPUs context.. It is clear that each activity has a different influencing factor that can be classified into multiple perspectives analysis. Activities 1, 2 and 3 and management activity were concerned with the influential factors of the organizational perspectives. Consequently, activity 4 was influenced by technical perspectives. Activity 5 was affected by personal perspectives. Age and size of the university can be part of the important factors to determine the complexity of KMS initiatives. The older and the larger universities in the cases had acknowledged the wider scope to cover the overall implementation of KMS initiatives, in terms of resources, number of departments, and staff behavior.

As such, the age and the size of the university indicated that the older MPUs have more challenging issues in dealing with the technical, organisational, and personal perspectives, compared to the newer MPUs. From the technical perspectives, the existence of several established technologies acts as a barrier for the older MPUs because they must be considered if they want to utilise newer technologies. In addition, the size of these old universities makes them more complex and more costly to maintain compared to newer universities. It is important for older universities to consider changing IT in a slow and incremental process rather than choosing a radical change [26]. The analysis of organizational perspectives shows that the older MPUs appear to be more challenging of developing unified strategic direction, managing of large knowledge resource, creating appropriate policies and procedures, and getting adequate financial support. The older the organisation, the more stable the structure and have a greater number of departments and thus generally exhibit greater complexity in managing change[10]. In many cases, the older MPUs were more strongly affected by personal perspective constraints than were newer MPUs. The factors for this trend were knowledge-sharing attitudes among KMS champions and business departments. The older organisations have more established organizational behaviour that has become institutionalised and business activities that have become routinised compared to newer MPUs [13].

5　Conclusions

The main focus of this study was to get an overall understanding of the current practice of KMS initiatives at MPUs. The SSM is presented as a tool for analyzing and understanding unstructured problems that deal with the complexity of the KMS initiatives in MPUs. This approach proved to be suitable for analyzing the KMS initiatives, since their implementation are not only concerned about technical aspect, but also put emphasis on the contextual environment in with they are embedded. In particular, the proposed conceptual model contribute to a better understanding of current KMS initiatives through the identification of necessary activities needed and the relationship between these activities. Consequently, the identification of influencing factors of each activity within the conceptual model is useful in better understanding of problem situations within the activities involved. In addition, the conceptual model also illustrates key issues within the activities systems which provide insight into what should be concerned to improve the problematic situation. Then, through the comprehensive conceptual model the appropriate action can be taken to improve the whole process of KMS initiatives. In this way, the proposed conceptual model of KMS initiatives can be used as an analytical tool to guide the analysis of the process of implementing KMS initiatives in higher education and can also be applied as a guideline to support the introduction of KMS initiatives, especially in the context of MPUs.

Acknowledgements. The first author would like to thanks the Universiti Sains Malaysia for supporting this paper under *"Fund for Overseas Conferences (1001/JPNP/ AUPRM003)"*.

References

1. Ajmal, M., Helo, P., Keka, T.: Critical factors for knowledge management in project business. Journal of Knowledge Management 14, 156–168 (2010)
2. Alias, R.A., Md. Saad, N.H.: A Multiple Perspectives Review of Knowledge Management Literature. Journal of Advancing Information and Management Studies 1, 17–32 (2004)
3. Baudoin, P., Branschofsky, M.: Implementing an Institutional Repository: The DSpace Experience at MIT. Science & Technology Libraries 24, 31–45 (2003)

4. Bennetts, P., Wood-Harper, A., Mills, S.: An Holistic Approach to the Anagement of Information Systems Development: A Review Using Soft Systems Approach and Multiple Viewpoints. Systemic Practice and Action Research 13, 189–205 (2000)
5. Cain, T.J., Branin, J.J., Sherman, W.M.: Knowledge Management and the Academy: Strategies and Solutions at the Ohio State University are changing how expertise and knowledge are documented and shared. EDUCAUSE Quarterly 31, 26–33 (2008)
6. Chan, D.L.H., Kwok, C.S.Y., Yip, S.K.F.: Changing Roles of Reference Librarians: The Case of the HKUST Institutional Repository. Reference Services Review 33, 268–282 (2005)
7. Chang, S.-H.: Institutional repositories: the library's new role OCLC Systems & Services, vol. 19, 77-9(3) (2003)
8. Checkland, P.: New Maps of Knowledge Some Animadversions (Friendly) on: Science (Reductionist), Social Science (Hermeneutic), Research (Unmanageable) and Universities (Unmanaged), System Research and Behavioral Science. System Research 17, S59–S75 (2000)
9. Checkland, P.: Systems Thinking, Systems Practice. John Wiley & Sons, Chichester (1981)
10. Cranfield, D.J., Taylor, J.: Knowledge Management and Higher Education: A UK Case Study. The Electronic Journal of Knowledge Management 6, 88–100 (2008)
11. Cronin, B., Davenport, E.: Knowledge management in higher education. In: Bernbaum, G. (ed.) Knowledge management and the Information Revolution EDUCAUSE Leadership Strategies Series, vol. 3. Josey-Bass Inc., San Franciso (2000)
12. Daft, R.L.: Organization Theory and design. Thomson Higher Education, OHIO (2007)
13. Hannan, M.T., Freeman, J.: Structural inertia and organizational change. American Sociological Review 49, 149–164 (1984)
14. Holsapple, C.W., Joshi, K.D.: An Investigation of Factors that Influence the Management of Knowledge in Organisations. Journal of Strategic Information Systems 9, 235–261 (2000)
15. Ingram, H.: Using Soft Systems Methodology to Manage Hotels: A Case Study. Managing Service Quality 10, 6–10 (2000)
16. Joshi, J., Ghafoor, A., Aref, W., Spafford, E.: Digital government security infrastructure design challenges. IEEE Computer 34, 66–72 (2001)
17. Kotiadis, K.: Using Soft Systems Methodology to Determine the Simulation Study Objectives. Journal of Simulation 1, 215–222 (2007)
18. Markus, M.L.: Power, Politics and MIS Implementation. Communications of the ACM 26, 430–444 (1983)
19. McDermott, R., O'Dell, C.: Overcoming cultural barriers to sharing knowledge. Overcoming Cultural Barriers to Sharing Knowledge 5, 76–85 (2001)
20. Moon, M.J.: The Evolution of E-Government Among Municipalities: Rhetoric or Reality? Public Administration Review 62, 424–433 (2002)
21. Norris, D.F., Fletcher, P.D., Holden, S.: Is your local government plugged in? In: Highlights of the 2000 Electronic Government Survey (2001)
22. Ordóñez de Pablos, P.: Measuring and Reporting Structural Capital: Lessons from European Learning Firms. Journal of Intellectual Capital 5, 629–647 (2004)
23. Petkov, D., Petkova, O., Andrew, T., Nepal, T.: Systems Thinking Techniques for Decision Support in Complex Situations. Decision Support Systems 43, 1615–1629 (2007)
24. Pieris, C., David, L., William, M.: Excellence in knowledge management: an empirical study to identify critical factors and performance measure. Measure Business Excellence 7, 29–45 (2003)

25. Ronald, D.F., Kulkarni, U.: Knowledge Management Capability: Defining Knowledge Assets. Journal of Knowledge Management 11, 94–109 (2007)
26. Ronnback, L., Holmstrom, J.: IT-adaptation challenges in the process industry: an exploratory case study. Industrial Management and Data Systems 107, 1276–1289 (2007)
27. Serban, A.M., Luan, J.: Overview of Knowledge Management: New Directions for Institutional Research. In: Serban, A.M., Luan, J. (eds.) Knowledge Management: Building a Competitive Advantage in Higher Education, vol. 113, pp. 5–16. Jossey-Bass, San Francisco (2002)
28. Singh, M.D., Kant, R.: Knowledge management barriers: An interpretive structural modeling approach. International Journal of Management Science and Engineering Management 3, 141–150 (2008)
29. Singh, M.D., Kant, R.: Knowledge management barriers: An Interpretive Structure Modeling Approach. International Journal of Management Science and Engineering Management 3, 141–150 (2008)
30. Soliman, F., Spooner, K.: Strategies for implementing Knowledge Management: Role of Human Resources Management. Journal of Knowledge Management 4, 337–345 (2000)
31. Syed Omar Sharifuddin, S.I., Rowland, F.: Benchmarking Knowledge Management in a Public Organisation in Malaysia. Benchmarking: An International Journal 11, 238–266 (2004)
32. Tseng, S.: The effects of information technology on knowledge management systems. Expert Systems with Applications 35, 150–160 (2008)
33. Walsham, G.: Interpreting Information Systems in Organizations. Wiley, Chichester (1993)
34. Wong, K.Y.: Critical Success factors for implementing knowledge management in small and medium enterprises. Industrial Management and Data Systems 105, 261–279 (2005)
35. Zakareya Ebrahim, Z.I.: E-government Adoption: Architecture and Barriers. Business Process Management Journal 11, 589–611 (2005)

Towards a Procedure for Assessing Supply Chain Risks Using Semantic Technologies

Sandro Emmenegger[1], Knut Hinkelmann[1], Emanuele Laurenzi[2],
and Barbara Thönssen[1]

[1] Institute for Information Systems,
University of Applied Sciences and Arts Northwestern Switzerland FHNW, Olten, Switzerland
[2] Institute of Information and Process Management,
University of Applied Science St. Gallen FHS, St. Gallen, Switzerland
{sandro.emmenegger,knut.hinkelmann,barbara.thoenssen}@fhnw.ch,
emanuele.laurenzi@fhsg.ch

Abstract. In the APPRIS project an Early-Warning-System (EWS) is developed applying semantic technologies, namely an enterprise ontology and an inference engine, for the assessment of procurement risks. Our approach allows for analyzing internal resources (e.g. ERP and CRM data) and external sources (e.g. entries in the Commercial Register and newspaper reports) to assess known risks, but also for identifying 'black swans', which hit enterprises with no warning but potentially large impact. For proof of concept we developed a prototype that allows for integrating data from various information sources, of various information types (structured and unstructured), and information quality (assured facts, news); automatic identification, validation and quantification of risks and aggregation of assessment results on several granularity levels. The motivating scenario is derived from three business project partners' real requirements for an EWS.

Keywords: Early-Warning-System, Supply-Chain-Management, Risk-Management, Enterprise Ontology, Semantic Technology.

1   Introduction

According to a recent survey of PricewaterhouseCoopers [18] more than 70% of the investigated companies have an Enterprise Risk Management (ERM) in operation. These ERM systems focus on management of known risks (e.g. alerting a delay in delivery based on the analysis of ERP data), but fall short "in dealing with unknown external threats, such as those arising from geopolitical events, economic developments, new technologies, talent, global trade, commodity costs, and terrorism. According to our [PwC's] survey, over 50% of companies feel they are not doing a good job of managing these risks" ([18], p5).

Globalization of the economy, on-going change of the market situation and ever-increasing cost pressure cause new business models to take up the challenges. In the manufacturing industry networked, virtual or extended enterprises have emerged [17] allowing for global sourcing without the necessity of owning all the players of the

A. Fred et al. (Eds.): IC3K 2012, CCIS 415, pp. 393–409, 2013.

supply chain [4]. Whereas that strategy brings down the costs it increases the effort on managing business relations, particularly with respects to the supply chain.

In parallel dynamism of the economic environment increases and therefore, the risk factors that affect the performance of the supply chain, too. Studies of Volatier et al. [29] show that the risk portfolio can change significantly within a period of three months (factor 8 more critical suppliers), and thus greatly increase the vulnerability of the own enterprise. The Global Risks Barometer presents an overview of 37 risks analysed in 18 workshops by more than 500 leading experts and decision-makers [6]. The survey not alone identified risks and assessed the likelihood to occur in the next 10 years but also show how risks are interconnected. To look not only at direct suppliers but on the whole supply chain is a trend identified in the latest annual survey by PRTM Management Consultants about Global Supply Chain Trends 2010–2012 with 350 participating manufacturing and service companies [10].

As shown in the survey of PricewaterhouseCoopers [18], extreme and unexpected risk events grew over the past few years, e.g. the eruption of Iceland's Eyjafjallajökull volcano in April 2010, the Japanese tsunami in 2011 and the Arab Spring. "These threats— often referred to as black swans (a term coined by Nassim Nicholas Taleb)—are nearly impossible to predict and can have catastrophic consequences" ([18], p 9). According to PricewaterhouseCoopers ([18], p 9) most ERM systems do not have the ability to identify external events and certainly cannot weight their impacts on the organization.

Although data about such external events, as well as data on exchange rates, commodity prices, company data, etc. are published on the Internet, the availability of information does not solve the problem. Without a systematic methodology and efforts to remain informed one drowns in the flood of information which is also the problem of lack of selectivity [19]. Albeit, today risk management in procurement is barely supported with tools and appropriate methods are missing. According to a study of [30] more than 70% of the surveyed companies command "unstructured" (18%) or "re-active" (55%) risk management in procurement.

The APPRIS approach seeks to remedy this. It aims at integrating risk, procurement and knowledge management into one early warning system. To do so, an enterprise ontology is used for knowledge representation stored in a triple store, risk assessment is implemented in Java and a graphical user interface is realized to simulate the necessary enhancements of a project partner's commercial Supply-Management-System. The developed prototype gives proof of concept and is used for evaluation.

The paper is structured as follows: In chapter two the APPRIS-approach is introduced. The approach illustrates the project principles based on risks and indicators, introduces an enterprise ontology for the risk domain and provides an insight into implementation details and technologies used. In chapter three we highlight related research. Chapter four describes the evaluation process. We close in chapter five with a conclusion and an outlook.

Supply Disruption Risks				
	Stakeholder Crisis	Strategic Crisis	Operational Crisis	Financial Crisis
Organizational signals	- Ownership changes in the supplier firm and the successor has no interest in running the business per se; - High employee turnover in the supplier firm - Supplier's order books;	- Lack of information regarding supplier's inventory; - Higher prices are voluntarily paid to keep the supplier alive; - Supply is unsecure; - No disaster planning at key suppliers;	- Frequent obsolescence of parts; - Plant fire/plant shut down or any similar unprecedented situation; - Loss of volume/product/mix flexibility; - Capacity issues; - No production stock available for ramp-up phase;	- Lack of information regarding supplier's insurance (ex ante contract signing); - Supplier requests for reduced payment terms, factoring etc; - Profit slump (although it is not always transparent); - Invitation to talks with banks (it is already too late by then);
Environmental signals	- Changes in political landscape; - Changes in regulations; - Changing import/export custom rules; - Natural hazards; - Labor/transportation strikes;	- Disruptions via technology levers;	- External disturbances like political movements (for example Egypt) causing disruption in the supply chain;	- Factors like economic crisis that impact each and every member of the supply chain causing perturbations in the flow of materials, information or/and money;
Network-related signals	- Supplier is (over)dominated by competitors;	- Single/sole sourcing market conditions; - Supplier is highly dependent on a handful of suppliers and/or customers;	- Lack of flexibility in the suppliers' supply-network; - Frequent stock-outs within the network;	- A major player within the supply chain is facing financial distress;

Fig. 1. Supply Disruption Risks' Matrix [11]

2   The APPRIS Approach

2.1   Principles

The APPRIS approach is based on a study by Grosse-Ruyken & Wagner [11] who identified ten top procurement risks. Grosse-Ruyken & Wagner [11] developed a matrix for each of the ten risks characterizing the sources of a risk (organizational risk sources, environmental risk sources and network-related risk sources) and four crises (stakeholder crisis, strategy crisis, operational crisis and financial crisis). Fig. 1 depicts the matrix for the Supply Disruption Risk. For each of the top ten risks warning signals have been identified and classified into the matrix. We took these matrixes as starting point and determined risk indicators for warning signals, which have been considered most important by the project's business partners. For 10 out of a total of approximately 180 warning signals, risk indicators have been derived.

Risk indicators can be very different since one can be a number (e.g. of force majeure events per year), another one can be mode (e.g. the transportation mode of a deliverer) and third one can be a specific business event (e.g. the production manager leaves the supplier). All indicators need different scales of measure.

In order to have the best possible basis different kinds of information sources and types are considered: data extracted from a company's ERP system, data delivered by a service provider like Dun & Bradstreet (who is a project partner), information allocated by a news provider like LexisNexis (who is also a project partner), information extracted from web sites (e.g. company sites or commercial registers) and user-generated input, as some information isn't available publicly.

Results of risk identification and assessment must be displayed in an easy-to-understand way. Therefore monitor suspension system is developed simulating the enhancements of the graphical user interface of a project partner's commercial Supply-Management-System.

2.2 Knowledge Representation

Using an ontology for enterprise modelling is a well-known and accepted approach and several models have been developed, for example the Toronto Virtual Enterprise (TOVE) by Fox et al. [8], the Enterprise Ontology (EO) by Uschold et al. [28], the Core Enterprise Ontology (CEO) by Bertolazzi et al. [1], the Enterprise Ontology by Dietz [5] and more recently the ContextOntology by Thönssen & Wolff [27]. Despite the consent about using an ontology for describing enterprise entities no standard or even an agreement has been achieved yet on the appropriate representation language for an enterprise ontology.

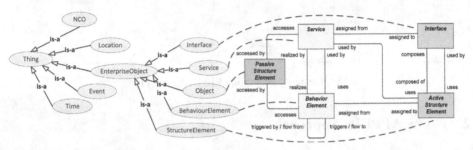

Fig. 2. ArchiMEO concepts derived from the ArchiMate Standard [22]

For the APPRIS approach we derived the following requirements: The enterprise ontology must

- be formally represented in a language which is understood by humans and machines alike,
- be linked to external data sources to integrate information of already existing applications (e.g. ERP Systems, Supply-Chain-Management Systems),
- be based on standards to ensure exchangeability and re-use,
- be easy to use to allow enhancements and adaptations by business users.

As none of the existing ontologies mentioned above meets these requirements, an ontology has been developed based on the ArchiMate standard and represented in RDFS[1]. ArchiMate is an enterprise architecture framework, providing a modelling notation which intentionally resembles the UML notation [22]. Since ArchiMate is not formalized enough to be machine understandable, we transformed its concepts and relations into an ontological representation [26]. The ontology is called ArchiMEO to indicate these roots and its purpose as Meta Enterprise Ontology.

Fig. 2 depicts ArchiMEO's top-level concepts and its sub-concepts. As shown, the ArchiMate concepts are all considered sub-concepts of the top-level concept *EnterpriseObject*. As ArchiMate focuses on the inter-domain relationships but risks evolve from external events, additional top-level concepts have been introduced, namely *time*, *event*, *location* and *NCO*. *NCO* is top-level concept introduced for

---

[1] http://www.w3.org/TR/rdf-schema/

'non-categorized objects', i.e. concepts of general interest. Moreover, concepts and relations of the ArchiMate business layer have been detailed as the granularity level of the standard was not sufficient enough for risk modelling. Table 1 shows the enhancements which have been made to ArchiMate for that reason.

Table 1. ArchiMate concepts and its ArchiMEO sub-concepts

ArchiMate	ArchiMEO
BusinessObject	*Top10ProcurementRisk*
	CrisisPhase
	WarningSignal
	RiskIndicator
BusinessEvent	*RiskEvent*
BusinessActor	*Person*
	LegalEntitiy
BusinessCollaboration	*BusinessRelationship*
BusinessRole	*Supplier*
	Customer

2.3   Implementation

The APPRIS approach is implemented as a prototype of an Early Warning System (EWS). The EWS prototype is built as loosely coupled extension to an existing Supply-Chain-Management-System of a project partner. Hence the EWS can draw upon complex visualisation components and focus on functionality. Enterprise internal data, like extracts of ERP (Enterprise-Resource-Planning) systems, are stored in a relational database. The prototype provides three functional modules (Fig. 3) with semantically enriched risk management capabilities:

- Source processing engine,
- Risk assessment, and
- Risk Monitor.

Source Processing Engine
The source processing engine of the EWS monitors internal and external information providers and creates and assembles risk events, which are further processed during the risk assessment.

Sources are integrated through web-services (e.g. provided by LexisNexis, Dun&Bradstreet, Twitter), via a batch import from an ERP (i.e. SAP) to the relational database or via direct access of internet resources based on HTTP.

The sources are either actively monitored or if a notification service is available, the source processing engine is triggered. In both cases queries resp. filters are applied to retrieve only the information of interest. Terms used in the filters and queries are for example "Earthquake", "Bankruptcy", "Location changes", etc.

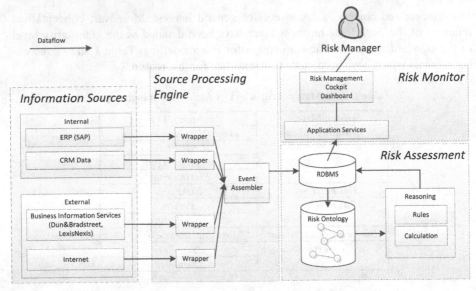

Fig. 3. EWS system context diagram with functional modules

If notable information has been identified, relevant terms for the risk detection are extracted, e.g. the name of a supplier or the location of an event. Based on this information a risk event for the internal processing is created and stored in the knowledge database. For example: A key supplier is located in Japan and we receive the news about an earthquake in Japan from Twitter. The source processing engine extracts relevant information about this disaster: location, magnitude, time, etc. and creates a specific risk event (*NaturalDisasterEvent*), which is further processed in the risk assessment module.

Risk Assessment

The risk assessment module can be seen as the core part of the early warning system. This module is based on the semantic model, the risk indication and the risk evaluation components.

The semantic risk model is an extension of ArchiMEO as described in chapter 2.2. The integrated development environment used for modelling the risk ontology is TopBraid[2]. The core risk model is based on the concepts *RiskEvent*, *RiskIndicator*, *CrisisPhase*, *WarningSignal* and *Top10ProcurementRisk*. For simplification the system is explained based on these concepts and relationships shown in Fig. 4. Starting point is the risk event, depicted at the very right hand side of Fig. 4.

RiskEvent

A risk event in our context is either a business – or a force majeure event with a potential impact on the company's supply-chain risks. An example for a business event

---

[2] http://www.topquadrant.com/products/TB_Composer.html

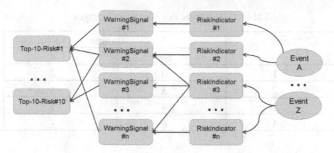

Fig. 4. Core risk concepts

is the information that a supplier has financial problems and is close to go bankrupt. A force majeure event might be a flood disaster. In our context this can have an impact on suppliers located in the area of this natural disaster.

A risk event has properties like temporal information (creation time, effective date), the source (information provider) and a reliability value. The reliability value depends on the reliability of the different sources (ERP, newspapers, blog etc.) and the time. For instance, master data provided by the internal ERP-System has a higher reliability than a newspaper message or a post on a social media platform.

Semantic Risk Model
The aspect of time is considered to differ between news and facts. News are statements made about the future, like the news that a company plans to buy a competitor. Facts are statements provided by official sources (e.g. company registries) or master data systems like the internal ERP system. Since we express this all in one reliability value, the handling and the risk event evaluation in a risk indicator becomes quite generic. The reliability calculation is done with the following formula:

$$\text{Reliability}_{\text{Facts}} = \text{Reliability}_{\text{Source}} * 1.0 \tag{1}$$

$$\text{Reliability}_{\text{News}} = \text{Reliability}_{\text{Source}} * 0.7 \tag{2}$$

Table 2. Examples for source reliabilities

Source	Reliability
ERP (ex. SAP)	1.0
Serious Newspaper X	0.7
Social Media (ex. Twitter)	0.4
Government service	1.0

For example: Consider an event in the future where the information source is the newspaper X from Table 2. Applying the formula (2) we can expect the following result:

$$\text{Reliability}_{\text{News}} = 0.7 * 0.7 = 0.49 \tag{3}$$

The reliability values for the source can be defined by the risk manager when setting up the early warning system. For each detected risk event the assigned risk indicators are checked.

Table 3. RiskIndicator scores and ranges

Score	Ranges	
	>=	<
1	0	2
2	3	3
3	4	5
4	6	-

RiskIndicator
According to The Institute of Operational Risk [21] risk indicators are metrics used to monitor identified risk exposures over time and these indicators must be capable of being quantified as an amount, percentage, ratio, number or count. In the EWS we either count the number of events (e.g. number of earthquakes in the last year in a certain area) or we consider the latest event and its value (e.g. the latest company rating delivered from Dun & Bradstreet). In both cases, the result value is rated based on pre-defined ranges. Table 3 gives an example of a metric for a risk indicator, that is used to assess the number of *NaturalDisasterEvent*.

Assume, in the last six month LexisNexis reported four times about earthquakes in a certain area. According to Table 3, the number of earthquakes would be rated with score 3. The score is a value of 1-4 (1=Low risk, 2=Medium risk, 3=High risk, 4=Extreme risk). Whereas the metric is the same for all risk indicators boundaries differ depending on the type of event. Scores and boundaries of the *RiskIndicator*'s can be defined by the risk manager, too.

Taking into account the reliability of the risk event source, we applied the following formula:

$$weightedScore = score * reliability \qquad (4)$$

After the weighted scores are associated with the risk indicators, a warning signal is substantiated if a certain threshold is exceeded.

WarningSignals
Warning signals are pointers to risks. Depending on the crisis phase they belong to, they lead to different risk importance.

To take into account the different importance of the crisis phases, from being only stakeholder-related to being critical to the very survival of the firm [11], phases are weighted differently. Table 4 shows the four different values the warning signals can get (value 0.2 – 1).

Table 4. Crisis phase and their values

Crisis Phase	Value
Stakeholder Crisis Phase	0.2
Strategic Crisis Phase	0.5
Operational Crisis Phase	0.8
Financial Crisis Phase	1

Each warning signal is assigned to one or more of the top 10 risks according the matrix of [11].

Top 10 Risks

[11] have determined 10 procurement risks to be the most relevant for businesses today. These top 10 risks have been implemented in the semantic model as instances of *Top10Risk*:

- Supplier default risk
- Supply quality risk
- Contract management risk
- Pricing risk
- Logistics/transportation risk
- Supply disruption risk
- Supplier capacity risk
- Sourcing management risk
- Socio-political risk
- E-procurement technology, process, and infrastructure risk

As more than one warning signal may trigger the same risk, we would need a formula to somehow aggregate the two warning signals' values to get the overall top ten risk value. If we aggregated both values by means taking an average, the final risk's outcome would drastically decrease its importance. For instance let's take "1" (warning signal belonging to the financial crisis) and "0.2" (warning signal belonging to the stakeholder crisis):

$$(1 + 0.2) / 2 = 0.6 \tag{5}$$

In order to avoid this problem, a formula has been proposed and validated by the APPRIS team as well as their business partners:

$$1 - \prod_{i=1}^{n}(1 - P_i) \tag{6}$$

Where "*P*" is the value of an early warning signal, and "n" is the number of early warning signals in a top ten procurement risk.

This formula is based on the independent events in the theory of probability. It regards a rather general and established concept which can be found in many textbooks and paper such as the fourth chapter of [2]. The formula is appropriate for this case because it satisfies the following conditions:

- The warning signals are independent, i.e. one signal does not affect another one
- The formula assigns an increasing value to each potential warning signal based to the crisis phase it belongs to, i.e. a warning signal belonging to the Stakeholder Crisis phase would get a value less than one belonging to the Financial Crisis.

    Also the warning signals' values that are not triggered (with a value of "0") can be considered in the formula because they do not decrease the final importance of the top ten risks.

In the evaluation step, the formula is applied and the respective result is then shown on the monitor suspension system by means of a colored flag.

Technical Implementation

To work smoothly with the risk ontology in Java, an ontology to object mapping framework has been evaluated. Here we had to choose between two approaches the currently available frameworks are based on. Either we generate the objects out of the semantic model or we use Java Annotations. We decided to go with the annotation approach, since this one integrates smooth in the Java environment and provides more flexibility. With Empire[3] we have even found a JPA (Java Persistence API) implementation which fits our requirements best. JPA is well known by experienced Java programmers and it makes it easy to work with the Ontology.

The risk calculation and more knowledge resp. risks are inferred through SPIN[4] rules. SPIN is based on SPARQL, the RDF[5] query language. SPIN fits well with the integrated development environment TopBraid[6]. In the operational environment we use the Jena TDP[7] triple store.

The calculated risk values are aggregated in Java on different levels (supplier, sub-supplier, product etc.) and are made available via a web service interface. The customized supply-chain management system frontend reads the risk values and makes them available in different views to users like the management or the procurement manager.

Use Case Example

To illustrate the overall approach of the risk assessment in the following an example is given. The general risk assessment procedure, depicted in Fig. 5, underpins the use case example along its description.

Let's assume company "Becker AG" is an automaker that uses vacuum pumps for generating negative pressure to the brake booster of passenger cars and light trucks. The company has two suppliers of the pumps (let's call them Supplier A and Supplier B). Supplier A is located in Singapore. We assume that Supplier A will move from Singapore to Vietnam in the near future. The news of changing location is delivered by an information provider electronically. At this point in the first functional module "Source Processing Engine" (Fig. 3), values of relevant terms effected by this news are instantiated. Relevant terms might be the following:

- Name of the company
- Old/New Location of the company
- Presence/Absence of free-trade agreement in the company's country

---

[3] https://github.com/mhgrove/Empire
[4] http://spinrdf.org/
[5] http://www.w3.org/TR/rdf-primer/
[6] http://www.topquadrant.com/products/TB_Composer.html
[7] http://jena.apache.org/documentation/tdb

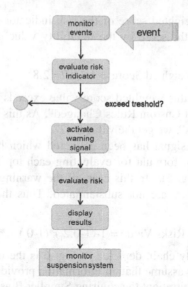

Fig. 5. The Risk Assessment Procedure

The value of such terms will affect the (business) rules concerning the risk event "Location Changes".

```
SELECT (COUNT(?freeTradeAgr))AS ?value) WHERE {
   ?locChangeEvent a risk:LocationChangeEvent .
   ?locChangeEvent risk:hasNewCountry ?newCountry .
   ?newCountry risk:hasFreeTradeAgr ?freeTradeAgr.}}.
BIND (IF(((?value >= ?low4) && (?value < ?up4)), 4.0
   ,IF(((?value >= ?low1) …) AS ?score) .
BIND ((?score * ?reliability) AS ?weightedScore).}.
```

The outcome of this rule infers the new fact that there will not be a free-trade agreement anymore for the Supplier A (as it will be in Vietnam). Thus, the risk event "*Location Changes*" feeds the value of the risk indicator "*Free Trade Agreement into Force*". As it is shown in Table 5 applying the normalization to such risk indicator enables to obtain the score "4".

Table 5. Normalization

Score	Free Trade Agreement Into Force
1	Yes
2	-
3	-
4	No

Since the risk event we are taking into consideration is a *News*, its reliability is calculated through the following formula (see chapter *Semantic Risk Model.*):

$$\text{Reliability}_{News} = \text{Reliability}_{Source} * 0.7 = 1.0 * 0.7 = 0.7 \qquad (7)$$

In order to have the weighted score of the risk indicator "*Free Trade Agreement into Force*", we multiply the score by the reliability value of its risk event (second box of Fig. 5).

$$\text{Weighted Score} = 4 * 0.7 = 2.8 \tag{8}$$

Assuming the fact that this weighted score value exceeds the threshold, it triggers the warning signal "Import Custom Rules Changed". As this warning signal is classified as "Stakeholder Crises", we get the value "0.2".

So far only a warning signal has been activated which belongs to the "Supplier Disruption Risk". Now the formula for evaluating each top ten risk value is applied (box "evaluate risk" of Fig. 5). In this case all the warning signals, except the one mentioned, have '0' as they are not substantiated. Thus the results of the formula appear as follows:

$$\text{Supply Disruption Risks Value} = 1-(1-0.2_1)*(1-0_2)\ldots*(1-0_{30})=0.2 \tag{9}$$

Now, assume the supply chain depicted in Fig. 6 is the one the company Becker AG is involved in. Further assume that an information provider delivers to our system the news that Supplier A.1 is about to acquiring Supplier B as shown in Fig. 7.

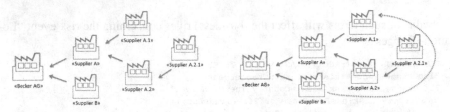

Fig. 6. Supply Chain   **Fig. 7.** Merge and Acquisition

In this situation comes into consideration the advantage to have an ontology concerning the recognition of those risks that affect suppliers belonging to other tiers of the supply chain. This in turns might be very damaging for our company's health.

The rule below shows a part of the whole rule inferring that the company "Becker AG" might soon face a single supplier situation.

```
SELECT DISTINCT ?businessRelationship WHERE {{
  SELECT ((?product) AS ?singleSourcedProduct) WHERE {
  ?product risk:prodIsInvolvedInBR ?businessRelationship}
  GROUP BY ?product
  HAVING (COUNT(?businessRelationship) = 1)}.
  ?singleSourcedProduct risk: prodIsInvolvedInBR ?businessRelationship.}}.
```

This time, the warning signal that is triggered along the risk assessment procedure is "Single/sole sourcing market conditions". This warning signal is classified as "Strategic Crisis", thus it gets the value "0.4". As in the previous scenario, also this warning signal belongs to the Supply Disruption Risk. Therefore the overall final value of this risk will change including the last warning signal as well. The formula appears as follows:

$$\text{Supply Disruption Risks Value}=1-(1-0.2_1)*(1-0.4_2)\ldots*(1-0_{30})=0.52 \tag{10}$$

In the last step of the risk assessment procedure (Fig. 5), the risk values are passed to the risk monitor and the display now shows a yellow flag for the Supply Disruption Risks.

Risk Monitor
Detecting and assessing risks are one part of an early warning system, the presentation of the results is another. The calculated risk values shall be shown to the users on an aggregation level appropriate to their role. The management board is interested in overall figures on the company level, whereas the individual procurement manager is mostly interested to see the risks of suppliers, resp. products he is responsible for. Another view can be based on the location of risk events and suppliers as shown in Fig. 9.

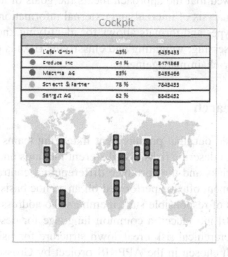

Fig. 8. Risk cockpit and dashboard

In the APPRIS project we provide a graphical user interface to simulate the enhancements one of the project partner will develop for his existing Supply-Chain-Management System.

This system is integrated through a relational database and web services. It allows to view indicators on different aggregations levels. The location of supplier and its risk value can be shown on a map. The system provides also notification service and allows sending emails triggered by risk value changes. This alerting service is a first step towards an active monitoring system, instead of a simple risk reporting dashboard. But monitoring means more. Events and risks should be integrated in the enterprise risk management processes. An advanced system might run automated workflow processes and support the management in the active risk mitigation and handling.

2.4   Evaluation

The approach is currently evaluated by the APPRIS business partners. Evaluation is done on the basis of the prototype. That is, the prototype is assessed with respect to the provided functionality of risk identification, focusing on 'black swans'. Non-functional aspects like performance or user-friendliness are not considered in this evaluation. They will be covered in the enhancements of the commercial supply-chain-management system developed by the software vendor in the APPRIS consortium.

Goal of the evaluation is to determine the appropriateness, capability and applicability of our approach for risk management. The prototype is used to illustrate the approach and thus, to make it easier for the evaluators to assess the underlying scientific concepts.

The evaluation showed that the approach meets the goals of an early warning system. Combining the analysis of internal and external information sources, supporting different information types and formats allows for identifying known risks and 'black swans'. The way of assessing, weighting and classifying risks is considered appropriate and viable in supply-chain-management.

3   Related Research

Tah & Carr [20] figured out, that procurement managing teams of an enterprise use different terminology to describe risks, use different methods and techniques for analyzing and managing risks and thus produce different and contradicting results. Furthermore, risk management often is performed on an ad hoc basis and is depending on individual assessments of responsible staff members. To address the afore mentioned issues, Tah & Carr [20] introduced a common language for describing risks. Therefore they provide a hierarchical risk breakdown structure for risk classification quite similar to the approach chosen in the APPRIS project by Grosse-Ruyken & Wagner [11]. The class diagram for project risk management suggested by Tah & Carr [20] provided valuable input for modelling ArchiMEO, too. However, ArchiMEO goes beyond their approach by formalizing the knowledge in a machine understandable and executable way.

Xiwei et al. [31] suggest the use of linguistic techniques for risk evaluation. To cope with the problem of fuzzy information about risks the authors presented a method, based on linguistic decision analysis to assess an overall risk value and suggest ways of mitigating risks. Whether the approach of Xiwei et al. [31] could be re-used or adapted for APPRIS is subject to further research.

The use of ontologies for modelling supply chains (interoperability) has been investigated by Grubic & Fan [12]. Based on a literature review six supply chain ontology models were identified. Although the authors explain method and search criteria, the selection seems somehow arbitrary. Ontologies were evaluated that have not been specifically designed for supply chain issues, for example the Enterprise Ontology [28] and TOVE [7], but are general approaches for representing enterprise architecture (description). Other ontologies developed for a similar purpose but less well-known, like REA [9], CEO [1], the Context-based Ontology [15] or the Context Ontology [27] were not considered. However, Grubic & Fan [12] developed a comparison framework and identified nine gaps in existing supply chain ontology models. Five of them are addressed by the ontology used for APPRIS [25].

However, for operational use – as APPRIS strives for – knowledge representation is not enough but an enterprise ontology must be enhanced to an enterprise repository as suggested by Hinkelmann et al. [13], Thönssen [24, 23]. In our approach we go in this direction by mapping entities of the Supply Management System's database to ontological concepts.

Chi [3] developed a rule-based ontological knowledge base for monitoring partners across a supply network. Although Chi [3] provides a sound methodology for modelling the domain of the supply network in an ontology, its content remains application specific since no standard is considered. But most important is that the approach is not integrated in the daily operations but an isolated task.

4   Conclusions / Further Work

Detecting risks as early as possible is of vital interest for all enterprises. At present risk management is performed – if ever – on the basis of in-house information, e.g. extracted from ERP systems like delays in delivery. More and more information would be available, either offered by information providers like Dun & Bradstreet or LexisNexis or publicly available on the web. Yet, risks are often detected too late due to late publication, not recognized importance or hidden impacts.

Our approach of an early warning system addresses this problem by combining the analysis of different information sources, types and formats in order to early identify and assess risks in the supply-chain. We showed how an enterprise ontology is used to represent domain knowledge and how ontology to object mapping based on Java annotations works. We also showed that identifying risks is just a first step. After this, risks are to be classified (e.g. regarding reliability), weighted (with respect to crisis phases) and correlated to other risks. We provide a prototype with graphical user interface that simulates the visualisation of risk evaluation results.

Our approach contributes significantly to improving risk management in the supply chain and thus is of considerable economic importance.

The EWS is formally evaluated by the APPPRIS project's business partners and the technical partner will implement the prototype's functionality in his Supply-Management-System.

However, there are still several aspects not considered yet, for example how the EWS could be improved to identify not only risks but also opportunities, and how decision making can be supported after risks were identified. Therefore the various possibilities for dealing with risks should be provided and consequences made explicit. Furthermore a learning component could be applied to improve suggested decisions automatically during time.

References

1. Bertolazzi, P., Krusich, C., Missikoff, M., Manzoni, V.: An Approach to the Definition of a Core Enterprise Ontology: CEO. In: International Workshop on Open Enterprise Solutions: Systems, Experiences, and Organizations - OES-SEO 2001, pp. 104–115, Rome(2001)
2. Billinton, R., Allan, R.N.: Reliability Evaluation of Engineering Systems. Springer (1992)
3. Chi, Y.-L.: Rule-based ontological knowledge base for monitoring partners across supply networks. Expert Systems with Applications 37(2), 1400–1407 (2010), doi:10.1016/j.eswa.2009.06.097
4. Chung, W.W.C., Yam, A.Y.K., Chan, M.F.S.: Networked enterprise: A new business model for global sourcing. International Journal of Production Economics 87(3), 267–280 (2004), doi:10.1016/S0925-5273(03)00222-6
5. Dietz, J.L.G.: Enterprise Ontology. Springer, Heidelberg (2006)
6. Emmerson, C.: Global Risks, Sixth Edition, pp. 1–55 (2011), http://reports.weforum.org/wp-content/blogs.dir/1/mp/uploads/pages/files/global-risks-2011.pdf
7. Fox, M.S., Grüninger, M.: Enterprise Modeling. AI Magazine 19(3), 109–121 (1998), doi:10.1147/sj.372.0170
8. Fox, M.S., Barbuceanu, M., Grüninger, M., Lin, J.: An Organization Ontology for Enterprise Modelling. Simulating Organizations: Computational Models of Institutions and Group, pp. 131–152. AAAI/MIT Press (1996)
9. Geerts, G.L., McCarthy, W.E.: The Ontological Foundation of REA Enterprise Information Systems. Business, 1–34 (2000)
10. Geissbauer, R., D'heur, M.: 2010-2012. Global Supply Chain Trends, pp. 1-28 (2011)
11. Grosse-Ruyken, P.T., Wagner, S.M.: APPRIS Project Report. Integration The Vlsi Journal, 1–124 (2011)
12. Grubic, T., Fan, I.-S.: Computers in Industry Supply chain ontology: Review, analysis and synthesis. Computers in Industry 61(8), 776–786 (2010), doi:10.1016/j.compind.2010.05.006
13. Hinkelmann, K., Merelli, E., Thönssen, B.: The Role of Content and Context in Enterprise Repositories. In: Proceedings of the 2nd International Workshop on Advanced Enterprise Architecture and Repositories - AER 2010 (2010)
14. Horrocks, I., Patel-schneider, P.F., Boley, H., Tabet, S., Grosof, B., Dean, M.: SWRL: A Semantic Web Rule Language Combining OWL and RuleML. Syntax (2004)
15. Leppänen, M.: A Context-Based Enterprise Ontology. In: Guizzardi, G., Wagner, G. (eds.) Proceedings of the EDOC International Workshop on Vocabularies, Ontologies and Rules for the Enterprise (VORTE 2005), Enschede, Netherlands, pp. 17–24. Springer, Berlin (2005)
16. O'Connor, M., Das, A.: SQWRL: a Query Language for OWL (2011)
17. Park, K.H., Favrel, J.: Virtual Enterprise - Information System and Networking Solution. Computers & Industrial Engineering 37, 441–444 (1999)
18. PricewaterhouseCoopers, Risk in review. Coping with the unknown: risk management strategies for an uncertain world (2012), http://www.pwc.com/us/en/risk-assurance-services/publications/assets/pwc-risk-in-review-aug-2012.pdf
19. Priddat, B.P.: E-Government als Virtualisierungsstrategie des Staates. Demokratisierung der Wissensgesellschaft und professioneller Staat. TECHNIKFOLGENABSCHÄTZUNG Theorie und Praxis 11(3/4) (2002)

20. Tah, J.H.M., Carr, V.: Towards a framework for project risk knowledge management in the construction supply chain. Advances in Engineering Software 32(10-11), 835–846 (2001), doi:10.1016/S0965-9978(01)00035-7
21. The Institute of Operational Risk, Key Risk Indicators (2010)
22. The Open Group (2012), ArchiMate 2.0 Specification. The Open Group, http://pubs.opengroup.org/architecture/archimate2-doc/
23. Thönssen, B.: An Enterprise Ontology Building the Bases for Automatic Metadata Generation. In: Proceedings of the 4th International Conference on Metadata and Semantics, MTSR1200, pp. 195–210 (2010)
24. Thönssen, B.: Formalizing low - level governance instruments for a more holistic approach to automatic metadata generation. In: Proceedings of the 5th International Conference on Methodologies, Technologies and Tools enabling e-Government, Camerino, Italy, pp. 1–12 (2011)
25. Thönssen, B.: Turning Risks Into Opportunities. Electronic Government, tbp. (2012), http://www.inderscience.com/browse/index.php?journalID=72
26. Thönssen, B.: Automatic, Format-independent Generation of Metadata for Documents Based on Semantically Enriched Context Information. Knowledge Creation Diffusion Utilization. University of Camerino. (2013), http://ecum.unicam.it/429/
27. Thönssen, B., Wolff, D.: A broader view on Context Models to support Business Process Agility. In: Smolnik, S., Teuteberg, F., Thomas, O. (eds.) Semantic Technologies for Business and Information Systems Engineering: Concepts and Applications (2010)
28. Uschold, M., King, M., Moralee, S., Zorgios, Y.: The Enterprise Ontology. The Knowledge Engineering Review 13 (1997)
29. Volatier, L., Cordon, C., Gallery, C.: Suppliers and vendors first. International Association For Contract & Commerical Management (IACCM) (2009), https://www.iaccm.com
30. Wyman, O.: Studie: Risikomanagement im Einkauf 2010. Einkauf und Management. Das Online-Magazin des Forum Einkauf (2010), http://www.einkauf-und-management.at/index.php/einkauf/more/studie_risikomanagement_im_einkauf_2010/ (accessed October 20, 2013)
31. Xiwei, W., Stößlein, M., Kan, W.: Designing knowledge chain networks in China — A proposal for a risk management system using linguistic decision making. Technological Forecasting and Social Change 77(6), 902–915 (2010), doi:10.1016/j.techfore.2010.01.002

Pervasive Ensemble Data Mining Models to Predict Organ Failure and Patient Outcome in Intensive Medicine

Filipe Portela[1], Manuel Filipe Santos[1], Álvaro Silva[2],
António Abelha[3], and José Machado[3]

[1] Centro Algoritmi, University of Minho, Portugal
[2] Serviço de Cuidados Intensivos, Centro Hospitalar do Porto, Portugal
[3] CCTC, University of Minho, Portugal
{cfp,mfs}@dsi.uminho.pt,
moreirasilva@gmail.com,
{jmac,abelha}@di.uminho.pt

Abstract. The number of patients admitted to Intensive Care Units with organ failure is significant. This type of situation is very common in Intensive Medicine. Intensive medicine is a specific area of medicine whose purpose is to avoid organ failure and recover patients in weak conditions. This type of problems can culminate in the death of patient. In order to help the intensive medicine professionals at the exact moment of decision making, a Pervasive Intelligent Decision Support System called INTCare was developed. INTCare uses ensemble data mining to predict the probability of occurring an organ failure or patient death for the next hour. To assure the better results, a measure was implemented to assess the models quality. The transforming process and model induction are both performed automatically and in real-time. The ensemble uses online-learning to improve the models. This paper explores the ensemble approach to improve the decision process in intensive Medicine.

Keywords: Data Mining, Intensive Care, Organ Failure, Patient Outcome, INTCare, Ensemble, Real-time, Pervasive Health Care.

1 Introduction

This work is the culmination of the progress achieved in the INTCare project - an Intelligent Decision Support System (IDSS) for Intensive Medicine (IM). The first approach used offline-learning [1] and some patient data to predict organ failure and patient outcome with a good *accuracy* [1, 2]. To shift this concept to a real environment using real data and in real-time was a huge challenge.

Further work used some data acquired automatically, however the data was processed and transformed manually [3]. The results achieved were interesting; even though the agility of the process wasn't very good due to the high number of tasks which required human efforts. At this level, the solution required a changing on the

A. Fred et al. (Eds.): IC3K 2012, CCIS 415, pp. 410–425, 2013.
© Springer-Verlag Berlin Heidelberg 2013

environment [4, 5] and on the way of the data is collected, processed and transformed. Using a set of intelligent agents [6] the entire Knowledge Discovery in Database (KDD) process [7] was automated. Nowadays the system prepares the data in real-time for the data mining tasks.

More recently, real-time data acquisition has been combined with online-learning [8]. The results were assessed in terms of *sensitivity, specificity* and *accuracy*. In the opinion of the physicians using only one measure wasn't the best option. Also, the way of how the models were compared is not the best. In order to overcome this problem, ensemble Data Mining (EDM) techniques were adopted and a data quality measure combining *sensitivity, accuracy* and *total error* was introduced. This paper presents the latest results. After a first experience where only 129 patients were evaluated a second one was made using more patient data, in this case the data comprises 335 patients. The main objective of this work is to understand if increasing the number of data available also increases the models' quality, in parallel, the benefits of using ensembles are studied.

This paper is divided into seven sections. After an introduction of the subject the principal concepts and the related work are described. The sections three and four present the initial phases of Knowledge Discovery in Database. Next, the fifth section presents the pervasive ensemble data mining and the sixth chapter makes an evaluation of the results achieved. Finally, some conclusions are outlined.

2   Background

2.1   Offline Vs. Online

The previous results obtained using EURICUS database [1] was the main motivation to develop this work. In EURICUS based work the variables used were collected manually in an offline mode: "The data was monitored, collected and registered manually, every hour, all Intensive Care Units (ICU) patient biometrics were recorded in a standardized sheet form by the nursing staff. The adverse events were also assigned in a specific sheet at an hourly basis."[1]. The variables used were: Age, Critical Events, Admission Variables, Outcome, and Sepsis-related Organ Failure Assessment (SOFA) [9, 10].

Now, the objective is to obtain all of those variables automatically and at same time to induce data mining models using an online approach in order to predict the organ failure and patient outcome in real-time. The greater challenge is the development of some procedures using all the values obtained by the data acquisition system instead of using hourly values. This change allows for a continuous data monitoring.

2.2   INTCare System

INTCare system is composed by four subsystems [11]: Data acquisition, knowledge management, Inference and Interface. Each subsystem is autonomous and uses intelligent agents [12] to perform automatically some tasks. For data acquisition were used the following agents: Gateway, Vital Signs Acquisition agent, Electronic Nursing Record (ENR) agent, Laboratory Results (LR) agent and Electronic Health Record (EHR / AIDA). Furthermore, it was used a pre-processing agent for the data validation and data transformation. Finally, the induction of data mining models is ensured by the Data Mining (DM) agent. The ensemble is induced automatically and in real-time by DM agent, whenever some request is done.

2.3   Knowledge Discovery Process

Knowledge Discovery from Databases (KDD) process is recognized as a process which can obtain new knowledge using some data. This process is composed by five stages: Selection, Pre-Processing, Transformation, Data Mining and Interpretation [7]. Figure 1 shows the KDD process for the ICU data. The database is populated with data from seven major sources. The data are selected from the data warehouse to be processed or transformed, depending on the goal of each one of the variables. After this task, the data are available to be presented by the Electronic Nursing Record (ENR) and prepared for creating Data Mining Models. Finally, all models are evaluated and the obtained knowledge is presented in the INTCare system.

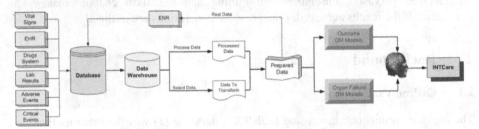

Fig. 1. ICU Knowledge Discovery in Database Process

2.4   Ensemble Data Mining

The use of Data Mining (DM) techniques in the medical area has been gaining an increasing interest by researchers [13]. Being this work a typical DM classification problem [14] and bearing in mind the idea of having a pervasive and real-time IDSS, a set of DM approaches were explored.

In the last experience ensemble data mining has been adopted. The reason of using ensembles has to do with the established principle that: the *sensitivity* can usually be improved by using ensembles of predictive models instead of a single model [15]. The ensemble-learning methodology consists in two sequential phases: the training and the testing phase [16]. In the training phase several different predictive models are generated from the training set. In the test phase the ensemble is executed and aggregates the outputs for each predictive model [16]. In this project was followed the Stacked Generalization methodologies [17, 18] and the learning procedure was divided into four steps [16]. To split the dataset the stratification technique is used. For each target a different dataset was considered with the same distribution on the target classes (0, 1).

The use of Oracle technology, especially Oracle Data Mining (ODM) facilitated the ensemble induction [19, 20]. Classification techniques used are some of the most used in DM [21] such as Support Vector Machine (SVM), Decision Trees (DT) and Naïve Byes (NB).

2.5   Pervasive Health Care

During the development of INTCare, some features were added according to the pervasive health care concept [22, 23], having as main purpose making the system available anywhere and anytime [5]. According to Varshney [24] pervasive health care can be defined as "conceptual system of providing healthcare to anyone, at any time, and anywhere by removing restraints of time and location while increasing both the coverage and the quality of healthcare". This approach is based on information that is stored and available online [25]. In order to turn the system into pervasive system pervasive computing features should be considered. Satyanarayanan [26] characterizes the pervasive computing as an evolutionary step resulting from previous two steps: first distributed computing and then mobile computing. The main characteristics are smart spaces, invisibility, localized scalability and uneven conditioning. INTCare uses pervasive computing features and it is fed by the probabilities provided by the ensemble data mining process. The results attained are displayed in situated devices.

2.6   Related Work - Forth Approach (Ensemble Data Mining)

The models were induced in real-time using online-learning by the DM agent [27]. To evaluate the ensemble three measures were considered: *Sensitivity, Accuracy,* and *Total Error (Terror)*. The average and the standard deviation of each one of the measures were estimated considering 10 runs. The use of ensemble helps to select the best model in the cases where more than one model present good results (e.g. outcome, hepatic and the respiratory systems). For each target / fold a separate dataset has been considered. The data corresponds to:

- Period in analysis: 105 days;
- Number of patients: 129.

Table 1. Ensemble Results

Target	Accepted by quality measures	Sensitivity	Accuracy	Specificity	Terror
Cardiovascular	YES	**97,95** ± 0,31	**76,81** ± 2,35	**41,81** ± 5,75	**23,19** ± 2,35
Coagulation	YES	**91,20** ± 3,57	**65,69** ± 3,83	**49,61** ± 6,15	**34,31** ± 3,84
Hepatic	NO	69,24 ± 9,41	**82,89** ± 2,57	87,34 ± 3,22	**17,10** ± 2,57
Outcome	YES	**99,77** ± 0,33	63,58 ± 3,11	49,58 ± 4,90	**36,42** ± 3,11
Renal	NO	77,17 ± 12,41	43,08 ± 4,66	43,08 ± 4,66	49,09 ± 5,39
Respiratory	NO	67,11 ± 5,67	**63,86** ± 4,27	**60,39** ± 6,75	**36,14** ± 4,27

Table 1 presents the performance achieved by the ensemble for each target. The values correspond to the average of the measures obtained during ten runs of the ensemble. Each average has associated the standard deviation. Respiratory, hepatic and renal systems don't meet the measures established and aren't considered by the pervasive system.

3   Data Selection and Pre-processing

The two initial phases of Knowledge Discovery in Database (KDD) process [4] use the data acquisition system presented in the background section (Figure 1) to obtain the data. The first phase is concerned to the data selection from database and it is in agreement with data necessary to feed the DM Models:

ICU_HL7 $\subseteq$ {Vital Signs}
ICU_HL7_T $\subseteq$ {Vital Signs auto validated (real values)}
ICU_PARAM $\subseteq$ {ICU Limits (max, min) values}
ICU_LR $\subseteq$ {All Lab Results}
ICU_DRUGS $\subseteq$ {All Patient Drugs administrated}
ICU_ENR $\subseteq$ {Data validated and provided from ENR}
ICU_CEVENTS $\subseteq$ {ICU Critical Patient Events}
EHR_ADMIN $\subseteq$ {ICU Patient Admission}
EHR_OUT $\subseteq$ {ICU Patient Outcome}

The second phase is responsible for the automatic data validation and patient identification. In this phase it is ensured that all data collected are valid and are correctly identified, i.e. all values collected are within the normal ranges of ICU values, and they have a valid patient identification (PID) [6].

At pre-processing phase, other procedures are executed to prepare the Data Mining input table. For instance, only the values collected during the first five days are used. When the patient is admitted into the ICU, an agent prepares automatically the table adding 120 rows for that patient. When the patient goes out, if he/she leaves before 120 hours, the rows in excess are deleted. This table is used as a temporary table for DM input. This table stores: the case mix values for all 120 lines (hours), the number of Critical Events and SOFA (0, 1) values for each hour. DM agent gets all variables present in temporary tables and, in addition, calculates the values in fault.

4   Transformation

The third phase of the KDD process is autonomous requiring no manual actions. All the tasks are performed automatically and in real-time by the INTCare intelligent agents. The variables in use are:

SOFA Cardio, Respiratory, Renal, Liver, Coagulation, neurologic = $\{0, 1\}$

Case Mix = {Age (1-4), Admission type (U or P), Admission from (1-6)}

Critical Events Accumulated (ACE) = {ACE of Blood Pressure (BP) (IN), ACE of Oxygen Saturation (SpO2) (IN), ACE of Heart Rate (HR) (IN), ACE of Urine Output (Ur) (IN)}

Ratios1 (R1) = {ACE of BP/elapsed time of stay (Q+), ACE of SO2/elapsed time of stay (Q+), ACE of HR/elapsed time of stay (Q+), ACE of Ur/elapsed time of stay (Q+), Total of ACE / elapsed time of stay (Q+)}}

Ratios2 (R2) = {ACE of BP / max number of ACE of BP (Q+), ACE of SO2/ max number of ACE of SO2 (Q+), ACE of HR / max number of ACE of HR (Q+) , ACE of Ur / max number of ACE of Ur (Q+), Total of ACE (Q+), Total of ACE / Total ACE max (Q+)}

Ratios (R) = R1 U R2

ACE of HR. Sum of values in a hourly base for each event type, i.e. if in the first hour has 1 event and in the second hour 2, the ACE for the second hour is 3.

Total of ACE is the sum of all ACE for the hour.

Max Number of ACE is the maximum number of each variable present in Table 5.

Elapsed Time of Stay. Total number of hours elapsed since the patient admission in the moment when ratio is calculated.

Total ACE max is the maximum value for ACE obtained by a patient in a specific hour (Table 5).

Outcome = {0, 1}

Table 2 presents the values considered by Data Mining models.

As referred in the previous work [8], the first transformation process is a simple task for analysing the values collected and for transforming them according to some rules (if then else). This process is applied to the variables presented in Table 2. When there is a case mix, all variables are inserted in the database. When a patient comes into the ICU, a procedure is executed. Regarding to the age parameter, the procedure verifies the patient age. For the admission type and origin, the admission form is consulted in the EHR. In all the cases the values are processed and the value is inserted into DM_INPUT table. In the case of the SOFA, the approach is a little bit different, the values are collected in real-time and in a continuous way.

The data mining models only use one value per hour. All collected values are considered and then the final value is assigned. If it is verified more than one result by hour, only the worst value of the hour it is considered. For example, in the case of cardiovascular system, there are five different possibilities to be categorized as 1 (BP, Dopamine, Dobutamine, Epi and Norepi). The SOFA values are then transformed into binary variables, where 0 describes normality and 1 describes dysfunction/failure and comprises the original SOFA value. By default the SOFA value variable is 0 and, if some condition is verified (e.g. for coagulation, platelets <=150) the values are

Table 2. Variables transforming (example)

ID		Variable	Min	Max	Value
Age		-	18	46	1
		-	47	65	2
		-	66	75	3
		-	76	130	4
Admission Type		Urgent	-	-	u
		Programed	-	-	p
Admission From		Chirurgic	-	-	1
		Observation	-	-	2
		Emergency	-	-	3
		Other ICU	-	-	4
		Other Hospital	-	-	5
		Other Situation	-	-	6
SOFA	Cardio	BP (mean)	0	70	1
		Dopamine	0,01	-	1
		Dobutamine	0,01	-	1
		Epi / Norepi	0,01	-	1
	Renal	Creatinine	1.2	-	1
	Resp	Po2/Fio2	0	400	1
	Hepatic	Bilirubin	1.2	-	1
	Coagul	Platelets	0	150	1
	Neuro	Glasgow	3	14	1

updated to 1. This update has effect in the starting date when the value was measured. The outcome value (live or died) is updated according to the patient discharge condition, when the patient die all of the values in the input table are updated to 1.

The second transformation phase uses critical events. Firstly, a set of procedures are executed in order to understand if a value is critical and if the event is adverse. Table 3 presents the variables in study and the *min* and *max* values for each case.

Table 3. Data Ranges

EvId	Descr	Min EC	Max EC	Min Val	Max Val	Min Any	Max Any
1011	BP	90	180	0	300	60	
3000	O2	90	100	0	100	80	
2009	HR	60	120	0	300	30	180
DIU	UR	30	1000	0	1000	10	

The values are validated and then the system determines if they are critical and how critical are they. According to Table 4, a value can be considered *normal* (0), *critical* (1) or *too critical* (2). If a value is critical (1), the event will be considered critical only if the values collected are maintained during a period of time. If the value is spontaneous and is too bad (2), the event will be always considered critical independently the extent of the event. Then, each collected value will be inserted in the events table according to the event type and if the predecessor event is or not the same. To know if this event type is of the same type a flag is used. A procedure is used to understand if the critical values collected may or may not represent a critical event. To this end the Table 4 is used. To consider an event as a critical event two

Table 4. The protocol for the out of range physiologic measurements (adapted by Álvaro [1])

	BP (mmHg)	SpO2 (%)	HR (bpm)	UR (ml/h)
Normal range	90 - 180	>= 90	60 - 120	>= 30
Critical event $_a$	>= 1h	>= 1h	>= 1h	>= 2h
Critical event $_b$	< 60	<80	<30 V> 180	<= 10

a Defined when continuously out of range.

b Defined anytime.

main constraints should be satisfied. For example, in the case of SpO2 the values should be between 80 and 90 for more than one hour, or less than 80 during some period.

The next procedure reviews the values according to the event time and event type: if it is verified a critical event, the value will be inserted into the critical events table. Then a procedure is executed hourly. This procedure calculates the Accumulated Critical Events (ACE) – to reflect the patients' clinical evolution/severity of illness by hour.

The next step consists in obtaining the ACE ratios. To implement this process a set of calculations are executed in the exact moment when the value was collected. This process requires more memory and processing time and can delay the other procedures. For the ratios is used the maximum number of occurrences verified in the past to a specific hour (Table 5). The maximum values are updated according to the maximum number of events verified for a patient for each variable in a specific hour.

The maximum number of *ACE* is automatically determined by hour. A procedure is executed to verify if the number of *ACE* is higher than the values present in the database. Table 5 presents an example of the maximum *ACE* values for each variable / hour verified in a patient. For example, in the case of blood pressure the maximum number of *ACE* observed in a patient at the 20th hour is six. However, at the 30th hour it is five. This result signifies that the patient who had a bigger number of BP CE at the 20th hour was discharged before the 30th hour of hospitalization. The total events are the maximum number of events occurred for all the categories and verified only in a patient during a specific hour.

Table 5. Critical Events daily number (example)

Variables	Max number of accumulated critical events										
Hour	1	2	3	4	5	10	15	20	30	40	50
blood pressure events	0	1	1	1	1	1	3	6	5	3	3
heart rate events	0	1	1	1	2	2	4	3	3	7	8
oxygen events	0	2	2	2	2	4	6	7	9	10	11
urine events	0	0	0	0	0	0	0	0	0	0	0
Total events	0	2	2	2	3	6	8	10	16	15	16

The next procedure calculates the ACE and all ratios for DM model. During all the processes described above, a procedure is responsible to get all data generated and store them into a specific table for the DM task. Finally, and after having all values correctly inserted into DM input table, another procedure cleans the inconsistent values. This procedure is responsible to delete all rows having null / incorrect values.

Table 6 presents the discretization rules defined to code continuous values (values ∈ {|R0+}). The ranges were created using a 7-point-scale adapted from Clinical Global Impression - Severity scale (CGI-S) [28]. The goal of *CGI-S* is to allow the clinician to rate the severity of illness [29]. The boundaries of each set were defined by the ICU doctors considering the significance of each set of values. More severe cases are assigned to levels 6 and 7. At the top of the table is the identification of the set. The left column identifies the variable. In the middle of the table are defined the ranges for each set.

The *R1min* and *R1max* are used by *R1* (max number of ACE). According to the percentage of the value it is categorized. For example, for the *R1* attribute '*ACE of BP / max number of ACE of BP*', if a patient has 7 *ACE* at the sixth hour and the maximum verified in the past for this time is 10 *ACE*, the respective set, according to table IV, will be 4 (7/10 = 0,7).

For all attributes of R1, the ranges of the set are equal. For the *R2* (ratios that use the elapse time) attributes, it is used the rows (*R2 BP min to R2 TOT max*) to determine the set. In this case, each attribute has a different range. For the attribute '*ACE of O2/elapsed time of stay*' if, for example, a patient has 10 ACE of O2 at the 50th hour, the ratio value will be 0,2 and the DM set will be 5 (0,1<0.2<=0,3).

Finally, all ACE values are grouped in accordance with their importance and number. For example, a patient that has 8 *ACE*, using table 6 and ace row, corresponds to the set 3 (5 < 8 <=8) in the DM input table.

Table 6. Discretization set of Data Mining Inputs

SET		0	1	2	3	4	5	6	7
R1	Min	-0,1	0	0,2	0,4	0,6	0,8	1	-
	Max	0	0,2	0,4	0,6	0,8	1	+00	-
R2	Min	-0,1	0,000	0,020	0,040	0,075	0,100	0,300	0,500
BP	Max	0,000	0,020	0,040	0,075	0,100	0,300	0,500	1
R2	Min	-0,1	0,000	0,020	0,040	0,075	0,100	0,300	0,500
O2	Max	0,000	0,020	0,040	0,075	0,100	0,300	0,500	1
R2	Min	-0,1	0,000	0,001	0,003	0,006	0,010	0,030	0,100
HR	Max	0,000	0,001	0,003	0,006	0,010	0,030	0,100	1
R2	Min	-0,1	0,000	0,020	0,050	0,080	0,100	0,300	0,500
UR	Max	0,000	0,020	0,050	0,080	0,100	0,300	0,500	1
R2	Min	-0,1	0,000	0,020	0,050	0,080	0,100	0,300	0,300
TOT	Max	0,000	0,020	0,050	0,080	0,100	0,300	0,500	1
ACE	Min	-0,1	0	3	5	8	10	12	15
	Max	0	3	5	8	10	12	15	50

5   Pervasive Ensemble Data Mining

Fig. 2 provides an overview of Data Mining modulation. In this figure it can be observed that the data preparation module is executed using the data stored in the database. After the transformation phase, data are stored into DM_INPUT_DB table. This table is then used to predict the value of the six target variables. Afterwards, the obtained results are applied into the prediction table of the patients admitted into ICU (UCI_PATIENT_5DAYS_AG). Then, six new columns are added containing the prediction of value 1 (occur an organ failure or patient die) for each target. In the data mining engine, each target constitutes an individual process.

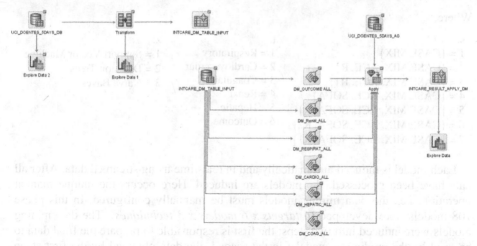

Fig. 2. DM Model

The fourth approach encloses the ensemble Data Mining techniques; in this case this model is executed 10 times for each target.

In order to automate this process, some researches has been done to find how to induce DM models automatically. As result it was possible to develop a procedure to execute the DM engine in real-time. The DM agent is responsible to run the engine whenever a request is made. The process of inducing Data Mining models is divided into two steps:

- Predictive Models – 126 models are induced combining seven scenarios (S1 to S7), six targets and three different techniques (SVM, DT and NB);
- Ensemble – the models are assessed in terms of the *sensitivity, accuracy, total error* and *specificity*. The best model for each target (t) is then selected.

The main purpose of the ensemble is to select the most suited model from a set of candidates. In order to evaluate the models, a quality measure was defined. This measure is based in the results obtained by the models in terms of *sensitivity, accuracy* and *total error*. The selected models are used by the pervasive system only if they satisfy the following conditions:

- *Total Error* <= 40%
- *Sensitivity* >= 85%
- *Accuracy* >= 60%

These thresholds were defined in order to assure a minimum level of quality in models. The measure was defined in accordance with ICU doctors. The values can be adjustable anytime and are commonly accepted in the medical community.

The ensemble can be defined as a three-dimensional matrix M composed by $s=7$ scenarios ($s1$ to $s7$) x $t=6$ targets ($t1$ to $t7$) x $z=3$ techniques ($z1$ to $z3$). Each element of M corresponds to a particular model and can be defined as:

$$M_{s,t,z} = \begin{cases} s = 1 \dots 7 \\ t = 1 \dots 6 \\ z = 1 \dots 3 \end{cases}$$

Where,

s:	t:	z:
1 = {CASE MIX}	1 = Respiratory	1 = Support Vector Machine
2 = {CASE MIX, ACE, R}	2 = Cardiovascular	2 = Decision Trees
3 = {CASE MIX, ACE, R1}	3 = Coagulation	3 = Naïve Bayes
4 = {CASE MIX, ACE, SOFA}	4 = Renal	
5 = {CASE MIX, ACE, SOFA, R}	5 = Hepatic	
6 = {CASE MIX, ACE, SOFA, R2}	6 = Outcome	
7 = {CASE MIX, ACE, SOFA, R1}		

Each model is induced automatically and in real-time using streamed data. After all data have been processed, the models are induced. Here occurs the unique manual operation, i.e., the data mining models must be manually configured. In this phase 108 models were developed (*6 targets x 6 models x 3 techniques*). The data mining models were induced into two steps: the first is responsible to prepare the final data to be used by the prediction models; in the second, the data obtained by the first stage are used by DM techniques to predict the probability of failure of each organ and patient outcome.

In the first step, the data stored in the DM input table is loaded. The numbers (ACE and ratios) are distributed using the presented discretization techniques. The other values are maintained as they are, and a final table is generated. During the DM modulation the neurologic system was not considered due to the high number of data in fault.

6   Evaluation

In order to evaluate the models, a test phase using online-learning was performed. The Data Mining techniques were applied on the following dataset:

Data Description:

Collection Time:	367 days
Patients Number:	335
Data Considered:	Values of five first days
Exclusion criterion I:	Patient with data collected intermittently, i.e., the collection system failed at least more than one hour in a continuous way;
Exclusion criterion II:	Existence of null values;

6.1   Data Input Distribution of the Inputs and Targets

Tables 7 and 8 present the distribution of the attributes considered. For the test phase the original dataset (DMIT) was divided into two different datasets using the holdout sampling method: 70% of the data were considered for training and 30% for testing (stratified by the target). Each target has a different dataset.

Table 7. Input Case Mix Variables distribution

Attribute	1	2	3	4	5	6	U	P
AGE	12,40	41,10	18,85	27,65	-	-	-	-
ADMIN_FROM	50,69	0,22	16,81	11,86	2,21	18,21	-	-
ADMIN_TYPE	-	-	-	-	-	-	72,58	27,42

Table 8. Input ACE and Ratios Variables distribution (%)

Attribute	0	1	2	3	4	5	6	7
ACE_BP	82,06	14,56	1,75	0,91	0,31	0,06	0,24	0,11
ACE_BP_MAX	73,57	1,61	7,05	5,78	3,51	1,21	7,26	-
ACE_BP_TIME	82,06	1,33	4,74	5,79	2,49	1,30	1,24	1,05
ACE_HR	80,62	13,76	3,35	1,25	0,38	0,37	0,23	0,04
ACE_HR_MAX	80,62	1,78	5,98	6,75	2,12	0,74	2,01	-
ACE_HR_TIME	80,62	2,64	2,86	2,77	1,79	1,66	2,85	4,81
ACE_O2	73,44	16,51	3,63	2,77	1,13	1,01	0,38	1,12
ACE_O2_MAX	73,40	6,17	6,21	7,37	2,79	1,13	2,94	-
ACE_O2_TIME	73,39	3,76	5,42	6,00	3,92	2,51	2,53	2,46
ACE_UR	82,06	14,56	1,75	0,91	0,31	0,06	0,24	0,11
ACE_UR_MAX	82,06	1,33	4,74	5,79	2,49	1,30	1,24	1,05
ACE_UR_TIME	73,57	1,61	7,05	5,78	3,51	1,21	7,26	-
ACE_TOT	61,50	21,06	5,07	4,18	1,94	1,55	1,37	3,32
ACE_TOT_MAX	61,42	11,91	11,78	8,00	3,61	1,10	2,18	-
ACE_TOT_TIME	61,42	5,22	7,50	9,11	5,69	2,95	3,93	4,19

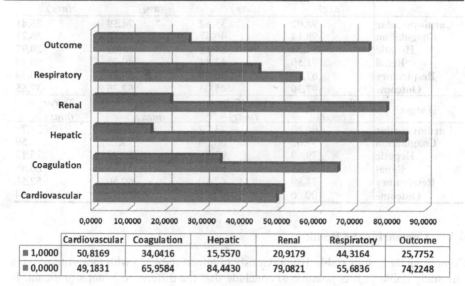

	Cardiovascular	Coagulation	Hepatic	Renal	Respiratory	Outcome
■ 1,0000	50,8169	34,0416	15,5570	20,9179	44,3164	25,7752
■ 0,0000	49,1831	65,9584	84,4430	79,0821	55,6836	74,2248

Fig. 3. Distribution of the classes

Fig. 3 presents the distribution of the classes (in percentage) for each one of the targets. For example, in the input dataset 15,56 % of the rows has the value of renal variable equal to 1. This value does not means that 25,78% of the patients has an renal failure, because the values are related to the rows and not with the patients.

This means that 25,78% of the records correspond to patients with renal failure. In this case, only the cardiovascular system present a high level of cases with organ failure (cardiovascular result = 1): 50,82%. Respiratory system presents a difference of 10% between the two values, in this case 44,32% of the records present a value equal to 1. In all of the other targets the numbers of positive cases (final result = 1) are substantially lower.

6.2   Ensemble Results

After the DM engine has been run, the best results obtained for each target are compared. Table 9 presents the performance achieved by the ensemble for each target. The values correspond to the average of the measures obtained during ten runs of the ensemble. The hepatic, respiratory and renal system didn't meet the measures established. In some executions of the renal system the results, reached the measure established, however having into account the average of the 10 runs, the results weren't satisfactory. Table 9 shows these results, as can be seen, the maximum *sensitivity* of renal system was 91,89.

Table 9. Ensemble Results by organ systems and outcome (%)

Target	Sensitivity (avg)	Specificity (avg)	Accuracy (avg)	Terror (avg)
Cardiovascular	**98,02**	35,52	**74,59**	**25,41**
Coagulation	**90,14**	46,47	**63,79**	**36,21**
Hepatic	57,34	84,18	**79,03**	**20,97**
Renal	71,50	42,44	49,21	50,79
Respiratory	67,03	42,09	59,26	40,74
Outcome	**97,30**	45,03	**62,35**	**37,65**
Target	Sensitivity (max)	Specificity (max)	Accuracy (max)	Terror (max)
Cardiovascular	99,60	43,72	77,80	27,79
Coagulation	94,62	60,56	73,22	41,59
Hepatic	79,69	93,38	83,77	25,12
Renal	91,89	54,70	57,18	60,48
Respiratory	73,89	67,50	69,50	52,56
Outcome	100,00	57,61	70,52	43,64

6.3   Comparing the Results

Comparing the results (table 10) previously obtained (table 1) with those observed in this study (table 9) it is possible to conclude that the difference is not significant. In general the results are widely satisfactory, for example the cardiovascular system presents a average *sensitivity* 0,07 upper, however the *accuracy* and *total error* are worst. The models are chosen according to the results obtained in terms of the measure, at the moment when data mining engine requires a model to answer to some target. Table 10 presents the variation (Δ) between the previous results and the newer.

Table 10. Variation of the obtained results

Target	Sensitivity OLD	Accuracy OLD	TERROR OLD	Sensitivity Δ	Accuracy Δ	TERROR Δ
Cardiovascular	97,95	76,81	23,19	0,07	-2,22	2,22
Coagulation	91,20	65,69	34,31	-1,06	-1,9	1,9
Hepatic	69,24	82,89	17,10	-11,9	-3,86	3,87
Renal	77,17	43,08	36,42	-5,67	6,13	14,37
Respiratory	67,11	63,86	49,09	-0,08	-4,6	-8,35
Outcome	99,77	63,58	36,14	-2,47	-1,23	1,51

For each one of the six targets was used three different DM techniques (SVM, DT and NB). Table 11 describes the experimental settings considered for each technique. In addition, this table indicates whether the used value corresponds to a default value or to a user-defined value.

Table 11. Techniques Configurations

DM Technique	Parameter Name	Parameter Value	Parameter Type
	Minrec Node	10	Input
	Max Depth	7	Input
	Minpct Split	.1	Input
DT	Impurity Metric	Gini	Input
	Minrec Split	20	Input
	Minpct Node	.05	Input
	Prep Auto	On	Input
NB	Pairwise Threshold	0	Input
	Singleton Threshold	0	Input
	Conv tolerance	.001	Input
	Active learning	Al Enable	Input
SVM	Kernel function	Linear	Default
	Complexity factor	0.142831	Default
	Prep auto	On	Input

7   Conclusions

The development of autonomous and real-time IDSS is a big challenge. The most difficult is the data transformation task (the process of obtaining critical events and their ratios). Ensemble data mining using online-learning and real-time data is another obstacle. This work goes in this direction, proving that it is possible to implement an IDSS in critical health environments minimizing the human intervention.

After automate the entire KDD process, in particular, the data transformation and the data mining processes, a new approach was experienced using ensembles.

INTCare is now an autonomous system and it is able to automatically and in real-time to predict the organ failure and patient outcome for the next 24 hours. The DM engine operates autonomously running the models and broadcasting the results through the INTCare system.

Experimental work was conducted in order to determine whether the use of online-learning and ensembles are or not a good option. The results are in favour of the last approach. The tasks can now be performed automatically and in real-time with a reduced human efforts. Additionally, the better models are used in the precise moment the ICU professionals need to take a decision. Furthermore, the performance of the models doesn't decrease over the time. The results observed in both situations were satisfactory corroborating the hypothesis of adopting ensemble approach.

Future work will include more data to optimize the DM models, in particularly the urine output and Glasgow Score.

Acknowledgements. This work is supported by FEDER through Operational Program for Competitiveness Factors – COMPETE and by national funds though FCT – Fundação para a Ciência e Tecnologia in the scope of the project: FCOMP-01-0124-FEDER-022674. The authors would like to thank FCT (Foundation of Science and Technology, Portugal) for the financial support through the contract PTDC/EIA/72819/ 2006 (INTCare) and PTDC/EEI-SII/1302/2012 (INTCare II). The work of Filipe Portela was supported by the grant SFRH/BD/70156/2010 from FCT.

References

1. Silva, Á., Cortez, P., Santos, M.F., Gomes, L., Neves, J.: Rating organ failure via adverse events using data mining in the intensive care unit. Artificial Intelligence in Medicine 43, 179–193 (2008)
2. Silva, A., Cortez, P., Santos, M.F., Gornesc, L., Neves, J.: Mortality assessment in intensive care units via adverse events using artificial neural networks. Artificial Intelligence in Medicine 36, 223–234 (2006)
3. Boas, M.V., Santos, M.F., Portela, F., Silva, A., Rua, F.: Hourly prediction of organ failure and outcome in real time in Intensive Care Medicine (2010)
4. Portela, F., Santos, M.F., Vilas-Boas, M.: A Pervasive Approach to a Real-Time Intelligent Decision Support System in Intensive Medicine. In: Fred, A., Dietz, J.L.G., Liu, K., Filipe, J. (eds.) IC3K 2010. CCIS, vol. 272, pp. 368–381. Springer, Heidelberg (2013)
5. Portela, F., Santos, M.F., Silva, Á., Machado, J., Abelha, A.: Enabling a Pervasive approach for Intelligent Decision Support in Critical Health Care. In: HCist 2011 – International Workshop on Health and Social Care Information Systems and Technologies, Algarve, Portugal, p. 10. S.-A.T. Publication (2011)
6. Portela, F., Santos, M.F., Gago, P., Silva, Á., Rua, F., Abelha, A., Machado, J., Neves, J.: Enabling real-time intelligent decision support in intensive care. In: 25th European Simulation and Modelling Conference- ESM 2011, 446 pages (2011)
7. Fayyad, U.M., Piatetsky-Shapiro, G., Smyth, P.: From data mining to knowledge discovery: an overview (1996)
8. Portela, F., Pinto, F.: Data Mining Predictive Models For Pervasive Intelligent Decision Support In Intensive Care Medicine. In: KMIS 2012 - nternational Conference on Knowledge Management and Information Sharing, Barcelona, INSTICC (2012)
9. Vincent, J., Mendonça, A., Cantraine, F., Moreno, R., Takala, J., Suter, P., Sprung, C., Colardyn, F., Blecher, S.: Use of the SOFA score to assess the incidence of organ dysfunction/failure in intensive care units: Results of a multicenter, prospective study. Critical Care Medicine 26, 1793–1800 (1998)

10. Vincent, J.L., Moreno, R., Takala, J., Willatts, S., De Mendonca, A., Bruining, H., Reinhart, C.K., Suter, P.M., Thijs, L.G.: The SOFA (Sepsis-related Organ Failure Assessment) score to describe organ dysfunction/failure. Intensive Care Medicine 22, 707–710 (1996)

11. Portela, F., Santos, M., Vilas-Boas, M., Rua, F., Silva, Á., Neves, J.: Real-time Intelligent decision support in intensive medicine. In: KMIS 2010- International Conference on Knowledge Management and Information Sharing, Valência, Espanha, p. 7 (2010)

12. Santos, M.F., Portela, F., Vilas-Boas, M., Machado, J., Abelha, A., Neves, J.: INTCARE - Multi-agent approach for real-time Intelligent Decision Support in Intensive Medicine. In: 3rd International Conference on Agents and Artificial Intelligence (ICAART). Springer (2011)

13. Bellazzi, R., Zupan, B.: Predictive data mining in clinical medicine: current issues and guidelines. International Journal of Medical Informatics 77, 81–97 (2008)

14. Han, J., Kamber, M.: Data mining: concepts and techniques. Morgan Kaufmann (2006)

15. Dietterich, T.: Ensemble methods in machine learning. Multiple Classifier Systems, 1–15 (2000)

16. Kantardzic, M.: Data mining: concepts, models, methods, and algorithms. Wiley-IEEE Press (2011)

17. Ting, K.M., Witten, I.H.: Issues in stacked generalization. Arxiv preprint arXiv:1105.5466 (2011)

18. Wolpert, D.H.: Stacked generalization*. Neural networks 5, 241–259 (1992)

19. Concepts, O.D.M.: 11g Release 1 (11.1). Oracle Corp. 2007 (2005)

20. Tamayo, P., Berger, C., Campos, M., Yarmus, J., Milenova, B., Mozes, A., Taft, M., Hornick, M., Krishnan, R., Thomas, S.: Oracle Data Mining. In: Data Mining and Knowledge Discovery Handbook, pp. 1315–1329 (2005)

21. Wu, X., Kumar, V., Ross Quinlan, J., Ghosh, J., Yang, Q., Motoda, H., McLachlan, G.J., Ng, A., Liu, B., Yu, P.S.: Top 10 algorithms in data mining. Knowledge and Information Systems 14, 1–37 (2008)

22. Varshney, U.: Pervasive healthcare and wireless health monitoring. Mobile Networks and Applications 12, 113–127 (2007)

23. Orwat, C., Graefe, A., Faulwasser, T.: Towards pervasive computing in health care - A literature review. BMC Medical Informatics and Decision Making 8, 26 (2008)

24. Varshney, U.: Pervasive Healthcare Computing: EMR/EHR, Wireless and Health Monitoring. Springer-Verlag New York Inc. (2009)

25. Mikkonen, M., Väyrynen, S., Ikonen, V., Heikkila, M.O.: User and concept studies as tools in developing mobile communication services for the elderly. Personal and Ubiquitous Computing 6, 113–124 (2002)

26. Satyanarayanan, M.: Pervasive computing: vision and challenges. IEEE Personal Communications 8, 10–17 (2002)

27. Portela, F., Santos, M.F., Silva, A.: Towards Pervasive and Intelligent Decision Support in Intensive Medicine - A Data Stream Mining Approach. In: AIM - Artificial Intelligence in Medicine (2013)

28. Guy, W.: ECDEU assessment manual for psychopharmacology. Rockville, Md (1976)

29. Guy, W., Rush, J., et al.: Clinical Global Impressions (CGI) Scale. Psychiatric Measures, APA (2000)

Watch Out and Improve IT: Adapting COBIT 5.0 Framework Based on External Context Discovery

Eduardo Costa Ramos, Flávia Maria Santoro, and Fernanda Baião

NP2Tec, Department of Applied Informatics,
Federal University of the State of Rio de Janeiro (UNIRIO), Rio de Janeiro, Brazil
{eduardo.ramos,flavia.santoro,fernanda.baiao}@uniriotec.br

Abstract. The governance and management of enterprises need to be designed, in line with issues arising from the internal and external environment. Especially in the Information Technology (IT) area, it is important to control and work up existing enterprise governance enablers. COBIT 5 is a framework for governing and managing enterprise Information Technology (IT) that supports enterprise executives and management staff in defining and achieving business goals and related IT goals. The specification of COBIT 5 highlights events in the enterprise's internal and external environment that can typically signal or trigger a focus on the processes related to IT governance and management. However, it lacks to follow a knowledge perspective. Therefore, we propose an adaptation to the COBIT 5 framework to use competitive intelligence on its processes. We specifically propose to apply the BPECREL (Business Process External Context Relevance) method, which identifies and prioritizes external variables that impact on the execution of a process and of its specific activities. We evaluated our proposal through an example that illustrates how external variables may impact the process activities.

Keywords: COBIT 5.0, External Context, Knowledge Management, Competitive Intelligence, KDD, Business Process, Data Mining.

1 Introduction

Enterprises must design their governance and management strategies based on internal and external environment concerns. For example, issues such as ethics and culture; applicable laws; regulations and policies; mission; vision and values; governance policies and practices; business plan and strategic intentions; operating model and level of maturity; management style; risk appetite; capabilities and available resources; and industry practices should be considered [5]. Particularly, in the Information Technology (IT) area, it is important to control current enterprise governance enablers.

COBIT 5 is a framework that aims at governing and managing enterprise Information Technology (IT) and supports enterprise executives and management staff in defining and achieving business goals and related IT goals [5]. However, approaches for effectively adopting and adapting COBIT 5 should consider the specific context

A. Fred et al. (Eds.): IC3K 2012, CCIS 415, pp. 426–439, 2013.

requirements for each enterprise. Besides, COBIT is often sustained by other frameworks, good practices and standards, and these will also need to be adjusted [5]. In this sense, the Chapter 3 of the COBIT 5 [6] specification presents some examples in the enterprise's internal and external environment that might trigger a focus on the processes related to IT governance and management: merge, acquisition or divestiture; a shift in the market, economy or competitive position; a change in the business operating model or sourcing arrangements; a new regulatory or compliance requirements; a significant technology change or paradigm shift; an enterprise-wide governance focus or project; an external audit or consultant assessments.

Despite the fact that COBIT 5 calls attention to processes that consider the external environment, it does not provide an understanding on how integrate them; moreover, it does not consider a knowledge perspective. These perspectives are essential for an effective governance development. Therefore, we propose an adaptation to the COBIT 5 framework to apply Competitive Intelligence concepts on its processes.

The goal of our approach is to allow organizations to be proactive (rather than reactive) against external changes. In this scenario, we specifically propose to apply the BPECREL (Business Process External Context Relevance) method [13][11], which identifies and prioritizes external variables that impact on the execution of a process and of its specific activities. The BPECREL method is based on Competitive Intelligence concepts and Data Mining techniques. We evaluated our proposal through an example, which provided evidences that the external variables pointed out by the method application influenced the process execution and its specific activities.

This paper is structured as follows: Section 2 presents the COBIT 5 framework. Section 3 presents work related to context-aware business processes and defines context concepts. Section 4 summarizes the BPECREL Method. Section 5 shows an example scenario. Section 6 analyses and discusses this work and section 7 concludes it with final considerations and points to promising evolutions.

2   COBIT5 - A Business Framework for the Governance and Management of Enterprise IT

According to [5], in COBIT 5 specification the "COBIT" (formerly known as Control Objectives for Information and related Technology) is used only as its acronym. COBIT 5 is a complete, internationally accepted framework for governing and managing enterprise information technology (IT) that supports enterprise executives and management in their definition and achievement of business goals and related IT goals [5][8]. COBIT describes five principles and seven enablers that support enterprises in the development, implementation, and continuous improvement and monitoring of good IT-related governance and management practices [5][8]. COBIT 5 is generic and useful for enterprises of all sizes, whether commercial, not-for-profit or in the public sector [5]. It helps enterprises create optimal value from IT by maintaining a balance between achieving benefits and optimizing risk levels and resource use.

COBIT 5 is based on five key principles for governance and management of enterprise IT   [5]: (1) Meeting Stakeholder Needs; (2) Covering the End-to-end

Enterprise; (3) Applying a Single, Integrated Framework; (4) Enabling a Holistic Approach; (5) Separating Governance From Management.

The COBIT 5 process reference model succeeds the COBIT 4.1 process model and aggregates the Risk IT and Val IT process models to it [7]. In this model, the IT-related practices and activities of the enterprise are subdivided into two main areas: governance and management.

Governance ensures that enterprise objectives are achieved by evaluating stakeholder needs, conditions and options; setting direction through prioritization and decision making; and monitoring performance, compliance and progress against direction and objectives (EDM) [5]. Management plans, builds, runs and monitors activities in alignment with the direction set by the governance to achieve the enterprise objectives (PBRM) [5].

Management is divided into domains of processes. The four Management domains are in line with the PBRM responsibility areas (plan, build, run and monitor) [7]. These domains are: APO-Align, Plan and Organize; BAI-Build, Acquire and Implement; DSS-Deliver, Service and Support; and MEA-Monitor, Evaluate and Assess [5]. Some Enterprise IT Management processes are [7]:

- APO03 Manage enterprise architecture;
- APO04 Manage innovation;
- APO05 Manage portfolio;
- APO06 Manage budget and costs;
- APO08 Manage relationships;
- APO13 Manage security;
- BAI05 Manage organizational change enablement;
- BAI08 Manage knowledge;
- BAI09 Manage assets;
- DSS05 Manage security service;
- DSS06 Manage business process controls.

COBIT 5 processes cover end-to-end business and IT activities, i.e., a full enterprise-level view [7]. COBIT 5 covers all functions and processes required to govern and manage enterprise information and related technologies wherever that information may be processed [5]. **COBIT 5 addresses all the relevant internal and external IT services, as well as internal and external business processes [5]. The COBIT 5 Framework relates COBIT 5 processes to IT goals, and these ones to enterprise's goals.** For example, the COBIT 5 Enterprise Goal "8-Agile responses to a changing business environment" primarily depends on the achievement of the following IT-related goals [6]:

- 01-Alignment of IT and business strategy;
- 07- Delivery of IT services in line with business requirements;
- 09- IT agility;
- 17- Knowledge, expertise and initiatives for business innovation.

These IT-related goals primarily depend on the achievement of other COBIT 5 Processes. As an example, the IT-related goal "09- IT agility" primarily depends on the achievement of the following COBIT 5 processes [6]:

- EDM04-Ensure Resource Optimization;
- APO01-Manage the IT Management Framework;
- APO03-Manage Enterprise Architecture;
- APO04-Manage Innovation;
- APO10-Manage Suppliers;
- BAI08- Manage Knowledge.

Some COBIT 5 processes that need to identify external variables are [5]:

- EDM01- Ensure Governance Framework Setting and Maintenance: this process has an activity to analyze and identify the internal and external environmental factors (legal, regulatory and contractual obligations) and trends in the business environment that may influence governance design.
- APO04- Manage Innovation: this process comprises two Key Management Practices (KMPs), one to maintain an understanding of the enterprise environment (APO04.02) and the other to monitor and scan the technology environment (APO04.03).
- MEA03- Monitor, Evaluate and Assess Compliance with External Requirements: this process is more detailed in section 4.

However, none of these COBIT 5 processes define how to identify the external variables. The next section discusses the concept of context as external variable relevant to business processes.

3   Context-Aware Business Processes

Context is defined as any information that can be used to characterize the situation of an entity [1]. In a business process scenario, context is the minimum set of variables containing all relevant information impacting the design and implementation of a business process [14]. The concept of context has recently revealed its relevance in business process management area. Identifying, documenting and analyzing contextual issues might help to make clear how changes in the environmental setting of an organization should lead to adaptations in processes.

According to Rosemann et al. [15], the external context is related to elements that are part of an even broader system whose behavior is beyond the sphere of control of an organization. Those elements exist within the business network where the organization operates. Besides, the environmental context lies ahead of the business network and is related to categories such as society, nature, technology and economy. Even if this context is not in close proximity to the day-to-day business of the organization, it still represents a high impact in the way the organization defines and executes business processes.

There are some proposals that deal with context associated to business process [9] [15], [16], defining the relevance of external information (or external context) for the execution of a process in an organization is still an open research issue. The manipulation of organizational knowledge, as well as environmental and external information, requires the application of knowledge discovery techniques so as to automatically handle and extract patterns from it.

A taxonomy for context, described by [16], which is composed of the most usual contextual information (location, time, resource and organization) aims at supporting context elicitation. Nunes *et al.* [9] also presented a model for context to support knowledge management within the scenario of a business process. The model developed by these authors is an ontology that establishes a representation for context elements associated with process activities. Based on this model, process instances and their context are stored and further could be re-used. The types of context elements presented are: (i) information that exist during the execution of an activity (time, artifacts), (ii) information about individuals or groups that perform an activity, (iii) information to spell out the interaction between individuals within the activity performed. Both proposals do not provide explicit methods for context elicitation and neither consider external environment context.

Another approach for bringing out context is stated by [17] with the goal of learning and gradually improving business processes considering three elements: process paths, context and goals. They argue that the success of a process instance can be affected not only by the actual path performed, but also by environmental conditions, not controlled by the process. Their work is based on an experience base, including data of past process instances: actual path, achieved outcome, and context information.

Our paper is based in the BPECREL method that identifies and prioritizes external variables that impact the execution of specific activities of a process [13]. This method was evaluated in a case study, which showed how the discovered variables influenced specific activities of the process. Next section details the BPECREL method.

4   The BPECREL Method

This section presents BPCREL, a method that identifies and prioritizes external variables that impact the execution of specific activities and outcomes of a process, applying Competitive Intelligence (CI) concepts and Data Mining (DM) techniques. BPCREL was previously proposed and evaluated by some study cases in [11], [12], [13] and is illustrated in Figure 1.

Steps 1 to 7 of BPCREL are responsible for applying two CI methods – Key Intelligence Topics (KIT) [3] and Critical Success Factors (CSF) – to systematically define information at the strategic level. Key Intelligence Topics (KITs) support the specification, definition and prioritization of information needs at the strategic level of the organization. They represent items that must be constantly monitored to guarantee business success.

The specification of the contents of each KIT is more detailed by several KIQs (Key Information Questions). For example, the KIT "Strategic Investment Decisions" may consist of the following KIQs: "What is the involvement of other investors in competitors?" and "What are the critical investments from competitors?" [19]. According to BPCREL, KITs are identified through interviews with managers, asking open questions. Each KIT may fall into one category, and grouped into surveillance areas. The three KIT categories are: (i) strategic decisions and actions; (ii) topics for early warning, considering threats and issues that decision makers want to know

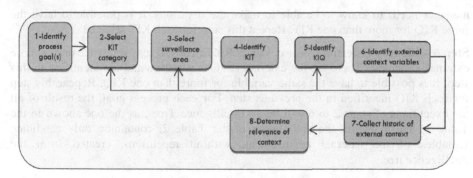

Fig. 1. Method for external context variables identification

previously, and (iii) major players in the market, such as customers, competitors, suppliers and partners [3].

Finally, Step 8 of BPCREL follows a KDD process to predict the process goal (defined in step 1) and to determine the relevance of the external context variables (identified by the KIT/KIQ approach) to the process outcomes and to the process activities outcomes.

The 8 BPCREL steps are detailed as follows.

Step 1 – Identify Process Goal(s). Identify all the goals related to a given process and their appropriate measures [15].

Step 2 – Select KIT Category. Herring [3] has divided KITs into three categories: 1) Strategic Decisions and Issues, 2) Early-warning KITs, considering threats and issues on which decision makers do not want to be surprised and 3) Key player KITs (such as customers, competitors, suppliers and partners). Each of the three categories is selected iteratively, to conduct the subsequent steps.

Step 3 – Select Surveillance Area. Steps 3 to 6 are part of a top-down approach, in which top level areas must be considered to give support to the next step. General surveillance areas are: social, technology, economic, ecology, political, legal and competitors, due to all industries are influenced by them. However, they can be selected from any framework, such as Five Forces model [10], SLEPT or STEEP Analysis [18], or from a combination of them. BPCREL focuses on events that occur externally to the process (or ultimately to the organization where it runs) and somehow interfere within this process. Rosemann *et al.* [15] propose that the external layer of their model is composed of the following types of context: suppliers, capital providers, workforce, partners, customers, lobbies, states, competitors. This step is repeated for each of the three KIT categories.

Step 4 – Identify KIT. Key Intelligence Topics (KITs) are identified by interviewing the main decision-makers and asking them open-ended, non-directive questions [4]. An interview protocol is very useful to ensure the consistency of results [3]. Repeat this step for each of the surveillance area selected.

Step 5 – Identify KIQ. Key Intelligence Questions (KIQs) should be identified for each KIT. KIQs represent the information needs listed in the KIT, i.e. what the manager needs to know to be able to make the decisions. It is possible to have the same KIQ for more than one KIT. Repeat this step for each KIT selected.

Step 6 – Identify External Context Variables. Each KIQ may reference one or more external variables. These are the external context variables and are identified in this step. It is possible to have the same variable for more than one KIQ. Repeat this step for each KIQ identified in the previous step. For each process goal, the result of all the executions of steps 2 to 6 will be an Intelligence Tree, as the one shown in the Table 1, and a list, as the one shown in the Table 2, containing only candidate variables of the external environment, without repetitions, created from the intelligence tree.

Table 1. Example of a part of an Intelligence Tree after all the executions of steps 2 to 6 of BPECREL method for a process goal to maximize approvals of clients that will pay all the debt for the Credit Analysis Process of an European Bank

KIT category	Surveillance Area	KIT	KIQ	External Context Variable
Early-warning	Economic	Possible causes that prevent a client to pay his/her debt	Did the unemployment rate at the moment he/she did not honor his debt interfere in the client not paying or paying?	**Unemployment Rate** at the moment the client signed the contract ; **unemployment rate** at the moment the client did not honor his debt;
			Inflation rate at the moment the client signed the contract and at the moment he/she did not honor his debt	**Inflation Rate** at the moment the client signed the contract; **Inflation Rate** at the moment the client did not honor his debt;

Table 2. Example of candidates variables of the external context

Candidates Variables of the External Context
Unemployment Rate
Inflation Rate

Step 7 – Collect Past Information of the External Context. In this step, historical data of each identified external context variable is collected and stored in the organizational memory. This data should refer to the same period of which data about the process (and its activities) outcome is also available.

Step 8 – Determine Relevance of the External Context to the Process Outcomes and to the Process Activities Outcomes. This step applies data mining techniques to prioritize which context variable to capture and store in the Organizational Memory, according to its relevance. A KDD process [2] is followed that automatically finds the most relevant subset of external context variables that best predict the process (and its activities) outcome. Before applying any KDD process, it is necessary to understand the application domain and identify what is expected to be discovered from the customer's viewpoint [2]. Each KDD step is detailed as follows:

Step 8.1 (Selection) - this step consists on creating a target data set (or focusing on a subset of variables or data samples), on which discovery is to be performed. In this step, the historical data of the external context is associated to the process activities outcomes and to the process execution results, for the same period.

Step 8.2 (Pre-processing) - this step consists on cleaning and pre-processing the target data set in order to obtain consistent data;

Step 8.3 (Transformation) - this step consists in finding useful features to represent data so that the target data set is reduced. By applying dimensionality reduction or transformation methods, the effective number of variables under consideration can be reduced or invariant representations for the data can be found [2].

Step 8.4 (Data Mining - DM) – generically, data mining consists of searching for patterns of interest in a particular representational form. Many models can be created to allow comparing which one has the best accuracy for predicting a target attribute. BPCREL applies feature selection and decision tree data mining techniques to discover which subset of external context variables more precisely lead to a specific process outcome.

Step 8.5 (Interpretation/Evaluation) - this step consists on the interpretation and evaluation of the mined patterns. The decision tree constructed in the previous step provides an adequate representation for knowledge managers to understand and interpret the circumstances in which a change in an external variable impacted the results of process instances in the past and, further, which activities of the processes were impacted. We argue that this interpretation will enable the decision maker to prepare process adaptations required to handle future modifications in the external variables, or at least to quickly react to those changes in the environment, when they occur.

5   Applying BPECREL Method to Support a COBIT 5 Process

This section illustrates an example of how we propose to apply competitive intelligence and data mining for supporting COBIT 5 processes that need to monitor external variables. With this objective in mind, we propose to apply the BPECREL method (described in Section 3) to every main business process of the organization.

This will allow the organization to know the relevance of each identified external variable to each business process (and to their specific activities).

This Section describes an example scenario in which we have applied the BPECREL method on a fictitious organization with 3 main business processes. Our focus is to support the COBIT 5 process "Monitor, Evaluate and Assess Compliance with External Requirements process (MEA03)".

5.1   The COBIT 5 MEA03 Process

The COBIT 5 MEA03 process ("Monitor, Evaluate and Assess Compliance with External Requirements") – which will be referred to as the MEA03 process from now on – evaluates that "IT processes and IT-supported business processes are compliant with laws, regulations and contractual requirements". MEA03 also obtains "assurance that the requirements have been identified and complied with, and integrates IT compliance with overall enterprise compliance". It belongs to the Management Area and to the Monitor, Evaluate and Assess Domain.

According to the COBIT 5 specification [5], the MEA03 process is primary related to two COBIT 5 IT goals: "02-IT compliance and support for business compliance with external laws and regulations"; and "04-Managed IT-related business risk". These two COBIT 5 IT goals are primary related to the following COBIT 5 Enterprise goals: "04-Compliance with external laws and regulations"; "15- Compliance with internal policies"; "03- Managed business risk (safeguarding of assets)"; "07- Business service continuity and availability"; and "10- Optimization of service delivery costs".

The MEA03 process is formed by four Key Management Practices (KMPs):

1.   Identify external compliance requirements;
2.   Optimize response to external requirements;
3.   Confirm External Compliance;
4.   Obtain assurance of external compliance.

The first KMP (MEA03.01) identifies and monitors changes in local and international laws, regulations and other external requirements that must be complied with from an IT perspective, on a continuous basis. This KMP has six activities and the first one is to assign responsibility for identifying and monitoring any changes of legal, regulatory and other external contractual requirements relevant to the use of IT resources and to the processing of information within the enterprise business and IT operations.

Since COBIT 5 does not specify how to identify external requirements, we propose to apply the BPECREL method for this purpose on every main business process.

5.2   Business Processes

Our example scenario assumes an Organization with 3 main business processes (BP): BP1, BP2 and BP3. Each BP has its goals. For every BP, the organization is interested in achieving its BP goals. A BP could be a software development process in a software house organization, for example. In this case, the organization must make decisions, such as whether or not to authorize the beginning of a software development project; what to do to maximize the chances of an ongoing project to be concluded; and whether to deactivate a project or to continue with it [13], [11].

5.3   Application of the Method

In this example scenario, we executed the 8 steps of the BPECREL method to define relevant external variables that influenced the main business processes (BP) of our

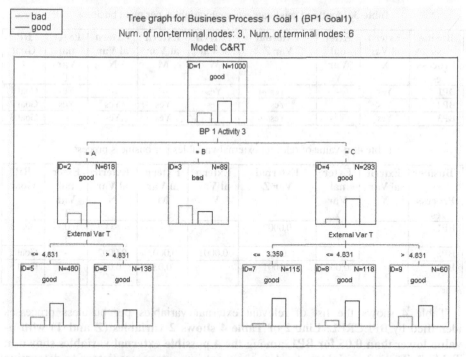

Fig. 2. Part of the decision tree C&RT for the Business Process 1 considering the best predictor variables to the dependent variable Goal 1 of BP1

hypothetic organization, as shown in section 4.2. The result after applying steps 1 to 7 of BPECREL is a list of possible relevant external variables, illustrated in Table 3. Step 8 outputs a list of the relevant external variables and a decision tree showing the relation among them and with the BP activities and goals (Figure 2).

Step 1 to 7. For the defined process goal of every BP (BP1, BP2 and BP3), the result of all the executions of steps 1 to 7 was a list of possible relevant external variables that can impact the goals of BP1, BP2 and BP3, illustrated in Table 3. The first column discriminates the 3 BPs. The last column lists the goals for each BP. Intermediate columns show fictitious candidate external variables identified through the CI questionnaires. A cell with "yes" states that this external variable was identified for this BP. The list of candidate relevant external variables per BP is composed by all the variables referring to columns with a "yes" cell in the BP table line. For example, the list of candidate relevant external variables identified for BP1 and its Goal1 is {X, Z, W, T}. According to BPCREL specification, the historical data of each identified external context variable is collected and stored in the organizational memory.

Step 8. In this step we followed the KDD process for BP1, BP2 and BP3. We applied the Feature Selection technique to filter variables according to their relevance. The C&RT (Standard Classification Trees with Deployment) technique is used to build a decision tree that explicitly shows the relation among the relevant external variables, the process outcomes and the process activities outcomes for predicting the BP goal.

Table 3. Identified candidate external variables per main business

Business Processes	External Var X	External Var Y	External Var Z	External Var W	External Var M	External Var N	External Var T	BP Goal
BP1	Yes	-	Yes	Yes	-	-	Yes	Goal 1
BP2	Yes	-	Yes	Yes	Yes	Yes	Yes	Goal 2
BP3	Yes	Yes	Yes	-	Yes	Yes	-	Goal 3

Table 4. p-value of relevant external variables per business process

Business Processes	External Var X	External Var Y	External Var Z	External Var W	External Var M	External Var N	External Var T	BP Goal
BP1	-	-	0,0002	-	-	-	0,0001	Goal 1
BP2	-	-	0,04	0,0001	0,0003	0,0002	-	Goal 2
BP3	-	0,0002	0,0001	-	0,03	0,04	-	Goal 3

Table 4 shows the list of relevant external variables per business processes identified by BPECREL. **Line 2 of Table 4 shows 2 variables (Z and T) with p-value lower than 0.05 for BP1, among the 4 possible external variables shown in Table 3. These 2 variables stand as the most important predictors to determine the BP1 goal**. The p-value of a variable is inversely proportional to its relevance, that is, the lower a p-value of a variable is for a specific process BPi, the more relevant is this variable for this process. For example, in Table 4, the most relevant external variable for BP1 is T, the 2nd most relevant is Z; in the case of BP2, the most relevant external variable is W, followed by N, M and Z. For BP3, the most relevant variable is Z, followed by Y, M and finally N.

Step 8.4 (Data Mining). Decision trees are powerful tools for classification and prediction that can be generated by data mining techniques. In this step, we show an example of a decision tree C&RT (Figure 2).

Step 8.5 (Interpretation/Evaluation). The decision tree C&RT of Figure 2 shows the relation between the relevant external context variables, the relevant process activities outcomes and the relevant process outcomes of BP1 and its Goal 1. The Goal 1 of BP1 is represented by a binary outcome (dependent) variable "Goal 1" that can have 1 of 2 values: "bad" or "good". Each box in the tree of Figure 2 shows the number of instances classified at that node and the distribution of the dependent variable values (Goal1). The root node (ID=1) contains 1,000 instances (i.e. all the instances in the dataset), and splits data to 3 descendants based on the outcome of "BP 1 Activity 3" (A, B or C). The central root descendant (ID=3) contains 89 instances, where most of them lead to Goal 1 as "bad". This node is not split further because most instances have the same value of Goal 1 ("bad"). The other 2 root descendants split further based on the "External Var T" predictor variable. These split resulted in 5 leaf nodes. The second leaf node (ID=6) clearly shows the relevance of the external variable to the process activity 3, by evidencing that when the "External Var T" raises above 4.831 then both Goal 1 outcomes from BP1 occur with almost the same probability. This may trigger a change during the process execution, with the process manager taking actions for maximizing the chances of the goal of BP1 being "good" (or, otherwise, minimizing the chances of being "bad").

6   Analysis and Discussion

The COBIT Framework relates COBIT 5 processes to IT goals, and these to enterprise goals. The use of BPECREL method in COBIT allowed not only to discover which external variables should be monitored, but also to know which business processes where impacted by these variables. The BPECREL method can go even deeper until discovering the activities of business processes that were impacted by these external variables.

The models generated by the BPECREL method allow the decision makers to take agile and proactive actions, such as quickly updating a business activity or a whole business process when there is a specific change in an external variable, since the models show the effects that this change has produced in the past.

It is important to note that a possible external variable identified on BPECREL may not have its relevance identified by the method. This may happen when the variable was not enough relevant to the BP or when the dataset did not have enough information characterizing its relevance. The decision makers may choose to monitor or not the external variables that were not relevant after applying BPECREL. In the case that a variable is related to many BPs (as with External VarX in Table 3 and Table 4), it may be worthy monitoring it even if its relevance was not detected. Moreover, the number of impacted BP by an external variable may increase its relevance. For example, even though VarZ is not considered the most relevant, it impacts the greatest number of BPs.

We can also observe that BPECREL is associated with some other activities of MEA03. It is possible to assess the impacts of variations on external variables such as legal aspects that biased the process even positively or negatively and thus helping to make previsions while tendencies to new deviations occur.

The BPECREL method was evaluated in study cases [11], [12], [13] that resulted in some remarks, as for example, that the discovered knowledge depends on the amount of detailed information available in the processes logs and in the external environment. Thus, it is essential to collect the appropriate information about the execution of activities within the enterprise and about the external context. Another observation alerted that transforming KIQs to external variables may not always be a trivial task and it can take more time to decide how to make this transformation. Lastly, the decision makers should take into account experts' judgment to choose the variables to be monitored.

7   Conclusions

The application of BPECREL processes, such as illustrated in this paper, highlights opportunities for integration with other COBIT 5 processes. For example, knowledge management initiatives may store all knowledge gained in applying BPECREL to facilitate their subsequent reuse in decision making; moreover, innovation management initiatives may adopt BPECREL to come up with opportunities for process improvements.

Supporting the identification and prioritization of variables to be considered in the context of the external environment allows that changes in those variables might trigger decision making, by adapting the process (new requirements) or implementing new business rules. Those are typical issues that IT processes should also understand and provide fast answers to. Therefore, the knowledge gained on business are reflected and aligned with actions in IT processes.

As future work we suggest conducting case studies to validate the proposal of improving COBIT by applying the BPECREL method in the MEA03 process and in other COBIT 5 processes, such as innovation management and risk management; improving COBIT by creating a new COBIT 5 process or adapting a exiting one to centralize all competitive intelligence activities used by COBIT 5, such as defining external variables in the innovation management and in MEA03 process. This could be done by adapting the knowledge management process. We also suggest applying the BPECREL method in other COBIT 5 processes using all surveillance areas proposed in the BPECREL method and not the surveillance areas indicated in some COBIT processes.

References

1. Dey, A.K.: Understanding and using context. Personal and Ubiquitous Computing 5(1), 4–7 (2001)
2. Fayyad, U.M., Piatetsky-Shapiro, G., Smith, P.: Advances in Knowledge Discovery and Data Mining. AAAI/MIT Press (1996)
3. Herring, J.P.: Key Intelligence Topics: A Process to Identify and Define Intelligence Needs. Competitive Intelligence Review 10(2) (1999)
4. Herring, J.P., Francis, D.B.: Key Intelligence Topics: A Window on the Corporate Competitive Psyche. Competitive Intelligence Review 10(4) (1999)
5. ISACA, COBIT5: A Business Framework for the Governance and Management of Enterprise IT. United States of America (April 2012), http://www.isaca.org/COBIT ISBN 978-1-60420-237-3
6. ISACA, COBIT5: Enabling Processes. United States of America (April 2012), http://www.isaca.org/COBIT ISBN 978-1-60420-241-0
7. ISACA, Comparing COBIT 4.1 and COBIT 5. Documents/COBIT5-Compare-With-4.1.ppt (April 2012), https://www.isaca.org/COBIT/
8. ISACA, Governance, risk management and compliance (GRC) (April 2012), http://www.isaca.org/COBIT/Documents/COBIT5-and-GRC.ppt
9. Nunes, V.T., Santoro, F.M., Borges, R.B.: A Context-based Model for Knowledge Management embodied in Work Processes. Information Sciences 179, 2538–2554 (2009)
10. Porter, M.E.: How competitive forces shape strategy. Harvard Business Review (March/April 1979)
11. Ramos, E.C.: BPECREL: A Method for Discovering Relevant External Context Variables Associated with Business Processes. MSc thesis. NP2Tec, Department of Applied Informatics, Federal University of the State of Rio de Janeiro (UNIRIO), Rio de Janeiro, Brazil, In Portuguese (2011)

12. Ramos, E.C., Santoro, F.M., Baião, F.: Process improvement based on External Knowledge Context. In: ACIS 2010, Brisbane, Australia (2010)
13. Ramos, E.C., Santoro, F.M., Baião, F.A.: A Method for Discovering the Relevance of External Context Variables to Business Processes. In: International Conference on Knowledge Management and Information Sharing (KMIS), Paris, France (2011)
14. Rosemann, M., Recker, J.C.: Context-aware Process Design Exploring the Extrinsic Drivers for Process Flexibility. In: Latour, T., Petit, M. (eds.) The 18th International Conference on Advanced Information Systems Engineering, Luxembourg, Grand-Duchy of Luxembourg. Proceedings of Workshops and Doctoral Consortium (2006)
15. Rosemann, M., Recker, J., Flender, C.: Contextualization of Business Processes. International Journal of Business Process Integration and Management 3, 47–60 (2008)
16. Saidani, O., Nurcan, S.: Towards Context Aware Business Process Modelling, Workshop on Business Process Modelling, Development, and Support (BP MDS), Trondheim, Norway (2007)
17. Soffer, P., Ghattas, J., Peleg, M.: A Goal-Based Approach for Learning in Business Processes, In: Nurcan, et al. (eds.), Intentional Perspectives on Information Systems Engineering. Springer (2010)
18. The Times, SLEPT analysis. 100 edn. (2010), http://www.thetimes100.co.uk (last accessed April 2000)
19. Vuori, V., Pirttimäki, V.: Identifying of Information Needs in Seasonal Management. Frontiers of E-business Research, 588–602 (2005)

Integration of Event Data from Heterogeneous Systems to Support Business Process Analysis

Alejandro Vera Baquero and Owen Molloy

National University of Ireland, Galway, Ireland

Abstract. Business Intelligence (BI) systems have traditionally been warehouse based, and have not been sufficiently process-aware to support the needs of process improvement type activities. It has been a challenge to leverage BI (and increasingly Analytics) functionality within the context of an overall process model. The ability to drill down into process data, track specific chains of process events, perform what-if type analysis, as well as monitoring overall aggregate performance is where process-aware Business Activity Monitoring (BAM) systems can play a significant role in improving performance. This paper presents a system prototype with the capabilities of integrating event data flowing through different heterogeneous systems such as business process execution language (BPEL) engines, enterprises resource planning (ERP) systems, workflows, legacy systems, etc., as well as storing this data into a global process execution repository. A new language for querying the stored event information is presented.

Keywords: Business Intelligence, Business Activity Monitoring, Business Performance Management, Business Process Analytics, Event Modelling, Business Process Execution Language.

1 Introduction

In many (if not most) process improvement initiatives we are faced with the problem of distributed heterogeneous systems and standards, often linked by bespoke software. Furthermore, process improvement initiatives need to be agile, rapidly moving from modelling to measurement and analysis without investing in massive data transfer, cleaning and modification.

Unfortunately BPEL (Business Process Execution Language) systems implement at best just some of the distributed business process activities. Business Intelligence (BI) systems come at a high cost in terms of time and resources. Data warehouses are not typically process-aware and must be re-engineered in response to changes in the process design. Process-aware systems allow querying directly on the process data itself, while maintaining knowledge about the process design or model. In this paper we present a flexible, lightweight, BPEL-agnostic solution for business activity (process) monitoring.

A. Fred et al. (Eds.): IC3K 2012, CCIS 415, pp. 440–454, 2013.

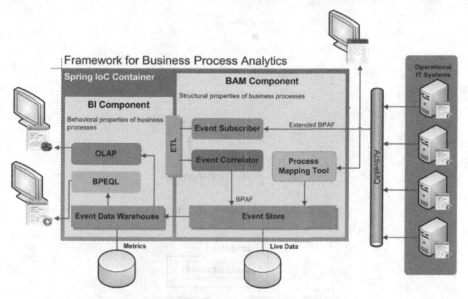

Fig. 1. Architectural approach of the framework

2   The Framework

This research provides two main contributions: firstly a generic event model construct that can represent the execution data of any business process regardless of the environment in which it is executed. Secondly, an IT infrastructure with the ability to monitor business processes from operational systems and analyse their execution outcomes.

Fig. 1 shows the architectural approach of the framework which is broken down into two main components. A BAM component that is responsible for providing event stream processing capabilities and a BI component which is the functional unit that produces the analytical information on business process performance.

2.1   An Event Model for Business Activity Monitoring and Business Process Analysis

The event model presented in this paper is built upon the BPAF (Business Process Analytics Format) standard, specified in [10], combined with some important features of the iWISE model discussed in [4]. The BPAF standard has been extended in order to accommodate the event correlation features defined by the iWISE software.

The iWISE system is fully described in [4] and defines an event-based model to represent the results of business process executions as business events supplied from heterogeneous environments where processes cross both organizational and

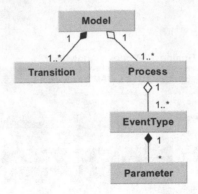

Fig. 2. High-level event-based model [4]

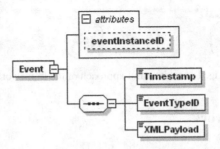

Fig. 3. The iWISE Event element [4]

software boundaries. The main modelling constructs are depicted using a simplified UML (Unified Modelling Language) class diagram depicted in Fig. 2.

A significant contribution to the event model presented in this work from iWISE is the structure used to represent a business event (Fig. 3). This element establishes the relationship between the events and their respective process instances.

EventTypeID specifies the event type information. *Timestamp*: the element that contains the time at which the event occurred. The XMLPathExpression element, located in the *EventType* data, is used to identify an element or attribute within the XML document contained in the *XMLPayload* of the event.

The BPAF Model

BPAF is a standard format published by the Workflow Management Coalition to support the analysis of audit data across heterogeneous business process management systems [10]. It enables the delivery of basic frequency and timing information to decision makers, such as the cycle times of processes, wait time, number of process instances completed against the failed ones, etc. [11]. BPAF is designed as an XML schema and consists of a generic design for a process analytics system which provides an event format independent of the underlying process model.

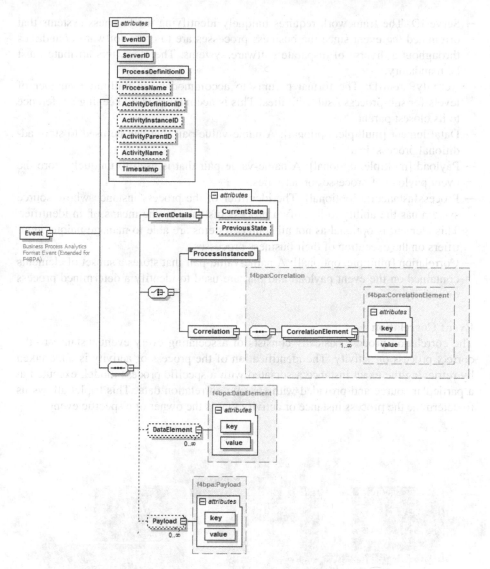

Fig. 4. Extended BPAF Model

The Extended BPAF Model

The BPAF has been modified with the purpose of achieving two main aims. The first aim is to provide the framework with the data required to correlate the events produced by the execution of cross-organizational business processes. And the second aim is to accommodate the structural properties of process or activity instances that are of relevance to business analysts. The extended and modified elements are specified below.

- ServerID: The framework requires uniquely identifying the business systems that originated the event since the business processes are to cross software boundaries throughout a diverse of disparate software systems. Therefore, this attribute must be mandatory.
- ActivityParentID: The format permits to accommodate an unlimited number of levels for sub-process / sub-activities. This is accomplished by keeping a reference to its closest parent.
- DataElement [multiple, optional]: A name-value-pair that can be used to store additional process data.
- Payload [multiple, optional]: A name-value-pair that is used to uniquely store the event payload of processes or activities.
- ProcessInstanceID [optional]: The identifier of the process instance whose source system has the ability to identify a process instance by the means of an identifier. This element is optional as not all source systems are able to manage unique identifiers on the execution of their business processes.
- Correlation [multiple, optional]: A name-value-pair that stores a subset of elements contained on the event payload and that are used to identify a determined process or activity.

Event Correlation
The correlation process basically consists of associating every event instance to the correct process or activity. The identification of the process or activity is undertaken by retrieving the exact instance associated with a specific process model, executed at a particular source and provided with specific correlation data. This triplet allows us to determine the process instance or activity that is the owner of a specific event.

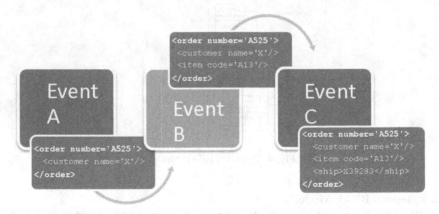

Fig. 5. Event Correlation

In source systems that have the ability to generate and manage identifiers on their source instances, such as BPEL engines, it is not necessary to provide any correlation information on the event message. In such cases, the instance identifier is provided instead, and in turn, this is used to correlate the subsequent events.

2.2   Analytics Requirements

The majority of process metrics are obtained by analysing the timestamp of a set of correlated events which are associated to a determined process instance or activity [11]. The use of such metrics provides business analysts with an understanding of the behavioural aspects of business processes.

The framework captures and records the timestamp of events containing the time at which they occurred on the source system, not when they are packaged or delivered. This property is essential in order to identify and analyse the correct sequence of process instances, as well as ensuring that the generation of metrics produces precise information on its outcomes.

Based on the event timing information and the BPAF state model presented in previous sections, it is possible for an analyst to determine the measurement of different behavioural aspects of business processes with the aim at "evaluating what happened in the past, to understand what is happening at present and to develop an understanding of what might happen in the future" [11].

2.3   Architecture of the Framework (F4BPA)

The architecture of the framework for business process analytics (F4BPA) is illustrated in Fig. 6 and consists of two main subsystems, the Business Activity Monitoring subsystem and the Business Intelligence subsystem. They are both built upon the Spring Framework version 3, and hence, they are Java-based enterprise applications.

The BAM subsystem is composed of a set of listener software modules (Event Publisher) that collects the events from business systems and publishes them through a message broker platform, an Event Subscriber module that listens and processes the

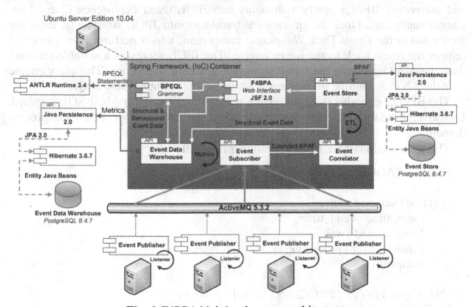

Fig. 6. F4BPA high-level system architecture

incoming events, an Event Correlator module that identifies and correlates consecutive events, and an Event Store module which persists the event data.

The BI subsystem is composed of an Event Data Warehouse and a Business Process Execution Query Language (BPEQL) module. The Event Data Warehouse is responsible for the generation and persistence of metrics, as well as serving as a data interface for querying the data warehouse containing metrics. The BPEQL module basically parses, executes and returns the results of query statements.

Business Process Execution Query Language

Many query languages for business processes have been proposed, using a variety of different approaches such as SQL-like languages, languages based on graphs and ontologies.

The BP-Ex query language proposed in [2] is a user-friendly interface based on a graph representation for querying business process execution traces.

The FPSPARQL is a query language for analyzing event logs of process-oriented systems based on the concepts of folders and paths. FPSPARQL extends the SPARQL graph query language by implementing progressive techniques in a graph processing engine [3].

The EP-SPARQL (Event Processing SPARQL) is an extension of the SPARQL querying language for event processing and stream reasoning that enables stream-based querying [1].

The query language proposed in this work resembles the SARI-SQL language discussed in [8]. SARI-SQL defines a language comparable to ANSI-SQL from a declarative perspective. The advantage of the languages based on an SQL-like syntax is that SQL is an industry standard that is widely used in business environments.

The BPEQL component provides a query engine that processes query statements formulated in our query language. The query engine works as a translator by parsing and converting BPEQL query statements into JPQL (Java Persistence Query Language) statements. Once the queries are translated into JPQL statements, these are forwarded to the Event Data Warehouse component, which performs the query and returns the result back to the query engine. The JPQL serves as a suitable intermediary layer for accessing the metrics stored at the data warehouse. The query engine uses the ANTLR runtime for parsing and translating the queries into JPQL.

The BPEQL grammar is based upon a reduced version of the ANSI-SQL standard. Likewise, it incorporates new features to adapt the language for a business process domain.

The specification of the BPEQL grammar is as follows:

```
SELECT [AGGREGATE]
   (
  (*) | (id | name  | source
        | start_time  | end_time
        | turn_around | wait
        | change_over | processing
        | suspend)
   )
FROM (ACTIVITY | PROCESS | MODEL | MAP)
[WHERE condition]
```

Grammar Specification

The grammar definition is broken down into two main components, a lexer and a parser. The lexer is specified by a set of rules that defines the lexical analysis. The parser is a grammar specification that determines if an input is syntactically correct with respect to the formally defined grammar. This is also known as syntactical analysis. Furthermore, the grammar features some semantic rules which make the recognizer also work as a translator. Fig. 7 illustrates how the BPEQL translator works internally by generating the parse-tree and evaluating the semantic rules at the parse tree nodes against the input statement. The translation is carried out on the following input query:

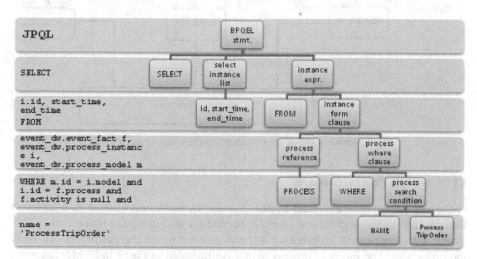

Fig. 7. BPEQL syntactic tree

3   Evaluation

The initial evaluation strategy to date has been based on a qualitative analysis in terms of effectiveness, completeness and usability. An implementation of the framework presented in this paper has been created along with a set of tests to assess the framework. The results of the test executions were used to evaluate the framework against the above quality attributes. For capturing the events, three different instances of BPEL engines have been deployed on a local network. These engines recreate the business process scenario specified in the next section, and which is used for testing purposes. Obviously any system capable of outputting the event format information could be used instead of these test BPEL instances.

The BPEL vendor of choice is the Apache ODE (Orchestration Director Engine) 1.3.5 of the Apache Software Foundation. Every Apache ODE instance corresponds to a determined organizational unit, and under every unit is executed a particular BPEL process. A specific plug-in (F4BPA-ODE) captures the business events produced by the Apache ODE servers. This plug-in is attached to every BPEL engine and uses the own Apache ODE API to access to the persisted data. Once the data is re-trieved, the events are sent to the network in the BPAF extended format (F4BPA-BPAF), after being converted by the means of ETL processes.

The architecture of the prototype previously described is graphically depicted in Fig. 8.

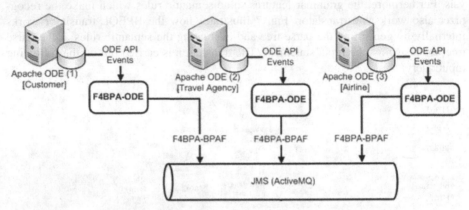

Fig. 8. Architecture of the prototype on the event capturing side

3.1   Sample Process Scenario

The event information managed by the framework must be enclosed in a business domain, thus a sample process scenario is needed for evaluating the framework. This sample business process is based on a travel planner where the customers can book and order trips. The business process model is illustrated in Fig. 9 in BPMN notation. The business process is launched upon a plan trip requested action. The root process interacts with other sub-processes which are part of third party systems that represents the organizational boundaries of the business process. Three different pools have been established, a Customer, a Travel Agency and an Airline, where each defines a different organization, and whose processes are part of the trip planning process.

In order to simulate a distributed environment on a real test case, each sub-process has been implemented as BPEL processes which are executed in a separated BPEL instance. Likewise, there is a single BPEL instance per pool representing the system boundaries, while the BPEL engines are fully accessible throughout the network.

3.2   Tests

Several tests have been carried out over the framework aiming to produce a volume of event data large enough as to obtain fair results. The storage of a large amount of event data, produced by the continuous execution of the business process, generates plenty of valuable information that enables analysts to gain insight into business

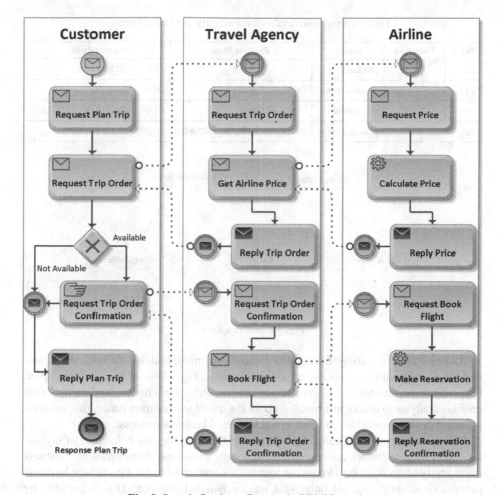

Fig. 9. Sample Business Process in BPMN notation

performance. The event model allows drilling down the business process execution outcomes into multiple levels of detail, whereby it is possible to either know the waiting time of a determined activity or the overall execution time of a cross-organizational process. Table 1 outlines a sample of the execution result of the activity getAirlinePrice associated to the ProcessTripOrder process.

Whilst the live data outlined above give an insight into the business process execution flow, they do not provide measurable information about business performance. Therefore, it is desirable to provide a fact table per process instance or activity. Consequently, a dimensional model has been settled for this purpose. Fig. 10 shows the UML star diagram used for storing and accessing the behavioural information of process and activity instances.

Table 1. Execution results of the activity 'getAirlinePrice'

Evt.Id	Timestamp	Activity	Previous State	Current State
22	2012-03-07 21:51:20.86	getAirlinePrice		OPEN_NOT_RUNNING
23	2012-03-07 21:51:20.86	getAirlinePrice	OPEN_NOT_RUNNING	OPEN_RUNNING
26	2012-03-07 21:51:20.86	getAirlinePrice	OPEN_RUNNING	OPEN_RUNNING_IN_PROGRESS
42	2012-03-07 21:51:22.07	getAirlinePrice	OPEN_RUNNING_IN_PROGRESS	CLOSED_COMPLETED

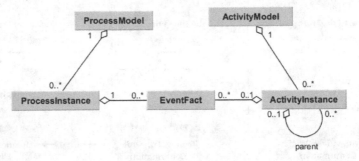

Fig. 10. Star diagram

As depicted at the above figure, this dimensional model allows the retrieval of any fact associated with any process or activity instance regardless of their nesting level. Likewise, it is possible to retrieve accumulative metrics for a determined model. This enables analysts to obtain information about the average execution times, rate failures, standard deviation, etc. for any part of a distributed business process.

The fact table provides valuable information for analysing the behaviour of organizational processes which are represented in single instances. However, what business users really often need is to know the average execution time for a particular business process or activity, and not only for a unique isolated instance. This is possible by grouping rows in the fact table and applying an aggregate function over the metrics. A filter around a set of processes or activities, which correspond to a specific model, will achieve the desired result.

3.3   Evaluation

The execution of test cases has shown that the prototype meets the research challenges. The prototype collected the data from the Apache ODE server, correlated and stored the business events in a central repository. Additionally, the metrics were generated successfully as per previous points.   The framework has provided a knowledge base that enables analysts to track business processes using the BPEQL language. For instance, the language can construct queries such as "What is the average execution time taken for the process 'X'?" or "What is the suspending time for the activity 'Y'?". However it cannot currently answer questions such as "At what time the

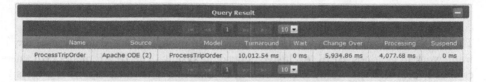

Fig. 11. Screenshot of the BPEQL query result

activity 'Y' failed for the last time?" or "What is the rate failure for the process 'Z'?"
These last questions require the ability to drill down to the event level to identify a
failed state.

The query language can retrieve over most common process metrics and also aggregate metrics over an evaluation function. This enables the framework to retrieve
the average execution time of a determined process, activity or map. Fig. 11 illustrates
how a BPEQL statement retrieves the average times for a particular process as per the
query below.

SELECT AGGREGATE * FROM PROCESS WHERE NAME = 'ProcessTripOrder'

The usability of the language is high since it can query any nested level of a business process by just specifying the desired level on the FROM clause. Furthermore, it
can also refer to a specific process or activity instance by filtering by an instance ID,
or even grouping similar instances of a determined model in a simple manner by specifying the process name. In this regard, the payload data play an important role in
fetching a determined process. This would enable end-users to determine a specific
instance by providing such data instead of dealing directly with instance identifiers
which are complex to deal with.   For example, from a business analyst perspective, it
is more adequate to formulate questions such as "What is the execution time of instance which order number is equal to 'A525'?" rather than "What is the execution
time of the instance whose identifier is 2834768?".   In terms of usability, these features will be added in future versions.

4   Conclusions and Future Work

The BPEQL language has the capability of retrieving behavioural and structural information from any level of a business process, but we need to next add the ability to
drill down to the event level. This will be quite important in certain business
processes, for example tracking individual patients through a care pathway. In spite of
not providing this functionality in the language, it can be easily extended to support
this feature since this information is stored and managed by the framework.

A framework for monitoring and analysing business process performance has been
presented in this paper. An event-based model was devised for supporting the data
required for analysing business processes. The framework has adopted a centralized
approach for monitoring the operational activities, collecting the business events and
inferring knowledge from the gathered information. The system provides significant
capabilities for analysing business process performance through the use of a query
language developed for this purpose.

The framework, which was prototyped using an event-driven architecture, uses a combination of event models that takes advantage of two complementary approaches. The iWISE [4] event model features cross-functional event sequences and permits the framework to be a non-BPEL exclusive dependent system. The BPAF model [10], in contrast, provides powerful capabilities for enabling the analysis of business processes behaviour.

In the absence of standards for querying business processes, a query language has been developed. The successful implementation and evaluation of the prototype has demonstrated that it is possible to monitor and query the structural and behavioural properties of business processes through the construct of a general purpose event model. Moreover, the business data can be unified and centralized seamlessly regardless of the underlying source systems.

In future works, the BPEQL grammar will be extended to improve its expressive power. Additionally, its usability will also be improved by incorporating references to business data without using identifiers, so that query construction will be significantly eased. The framework is sufficiently flexible to easily incorporate the extensions mentioned above. There are plenty of possibilities for incorporating metrics and key performance indicators (KPI) without affecting the normal functionality of the existing system. Consequently, the BPEQL grammar can also be improved by incorporating these new elements gradually, thus improving the power and expressiveness of the language.

Other potential further research using the framework includes support for predictive analysis and integration with simulation and optimisation techniques and systems. This would pave the way for enabling the user to augment existing data with hypothetical information in order to perform what-if analysis over simulated scenarios.

Behavioural patterns recognition is another technique that could be leveraged by the system in order to detect undesirable business process behaviours that are experienced frequently or on a continuous basis.

Finally, event data centralization is not the only option to store and analyse distributed business data. This approach presents some drawbacks in terms of distributed analytics that need to be addressed in order to enable business analysts to access analytical information in an adequate response time basis. Firstly, the continuously execution of distributed business processes produces a very large volume of event data that cannot be efficiently managed by a centralized solution. Secondly, centralization entails a significant latency from the time the event occurs on source to the time the event is recorded in the central repository. And thirdly and last, relational databases are not adequate to manage a number of event data in the order of hundred millions records. These shortcomings prevent the framework from providing business analytics at or near real time on highly distributed environments.

Current efforts are being addressed to provide an extension of the framework. Handling collaborative analytics on a fully distributed BI environment is a challenging task. Nonetheless, a cloud-based infrastructure built around an extension of the framework can be complemented with a federative approach, presented in [7], in terms of data warehousing and distributed query processing. The BI subsystem component can be attached to every operational business system along with their own local event repository. The event-based model presented herein represents the

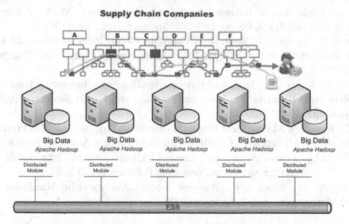

Fig. 12. Cloud-based infrastructure to support distributed business process analytics

global schema proposed by Rizzi's approach. Thus, business process analytics is carried out collaboratively in each organization independently by performing distributed queries along the collaborative network (See Fig. 12).

To conclude, the previous figure illustrates how the use of Big Data technology on source systems can enhance the framework to deal with very large volumes of data. Likewise, the use of Big Data Analytics on source systems can significantly reduce the response time of analytic processing on environments that must deal with a very high load of data such as the present one. And finally, the combination of a distributed architecture with a cloud-based computing environment will ease the provision of a real collaborative.

References

1. Anicic, D., Fodor, P., Stojanovic, N., Rudolph, S.: EP-SPARQL: A Unified Language for Event Processing and Stream Reasoning. In: WWW 2011 Proceedings of the 20th International Conference on World Wide Web (2011)
2. Balan, E., Milo, T., Sterenzy, T.: BP-Ex: A uniform query engine for Business Process. In: EDBT 2010 Proceedings of the 13th International Conference on Extending Database Technology (2010)
3. Behesti, S., Benatallah, S., Motahari-Nezhad, H., Shakr, S.: FPSPARQL: A Language for Querying Semi-Structured Business Process Execution Data. UNSW-CSE-TR-1103, School of Computer Science and Engineering. University of New South Wales, Australia (2011)
4. Costello, C.: Incorporating Performance into Process Models to Support Business Activity Monitoring. National Universisty of Ireland, Galway (2008)
5. Kang, J., Han, K.: A Business Activity Monitoring System Supporting Real-Time Business Performance Management. In: Convergence and Hybrid Information Technology, ICCIT 2008, pp. 473–478 (2008)
6. Parr, T.: (n.d.). ANTLR Parse Generator, http://www.antlr.org (retrieved June 11, 2012)

7. Rizzi, S.: Collaborative Business Intelligence. In: Aufaure, M.-A., Zimányi, E. (eds.) eBISS 2011. LNBIP, vol. 96, pp. 186–205. Springer, Heidelberg (2012)
8. Rozsnyai, S., Schiefer, J., Roth, H.: SARI-SQL: Event Query Language for Event Analysis. In: Proceedings of the 2009 IEEE Conference on Commerce and Enterprise Computing (2009)
9. Seufert, A., Schiefer, J.: Enhanced Business Intelligence - Supporting Business Processes with Real-Time Business Analytics. In: Database and Expert Systems Applications, Copenhagen, pp. 919–925 (2005)
10. WfMC, Workflow Management Coalition - Business Process Analytics Format Specification, Workflow Management Coalition - Business Process Analytics Format Specification (2009), http://www.wfmc.org/Download-document/Business-Process-Analytics-Format-R1.html (retrieved February 8, 2012)
11. ZurMuehlen, M., Shapiro, R.: Business Process Analytics. In: Handbook on Business Process Management, vol. 2, Springer (2009)

Author Index

Abdelazziz, Lamiaa 149
Abdullin, Artur 37
Abel, Marie-Hélène 308
Abelha, António 410
Akiyoshi, Masanori 297
Alias, Rose Alinda 377
Alirezaie, Marjan 179
Almeida, Mário 228
Avros, Renata 19

Baião, Fernanda 426
Baquero, Alejandro Vera 440
Barthès, Jean-Paul A. 308
Borchardt, Ulrike 323
Bresso, Emmanuel 84
Buraga, Sabin 165

Cheng, Chao 130
Conrad, Stefan 101

da Silva, Miguel Mira 228, 243
Devignes, Marie-Dominique 84
Dias, David Galego 243
Dinsoreanu, Mihaela 68

Emmenegger, Sandro 393
Exman, Iaakov 271

Fernández, Susel 194
Fraga, Anabel 284

Greis, Noel P. 211
Grisoni, Renaud 84

Habib, Mena B. 113
Hamada, Keiichi 297
He, Liang 130
Heckmann, Paul 260
Hinkelmann, Knut 393

Kasari, Melissa 3, 130
Kasimin, Hasmiah 377
Kourouklis, Athanassios 350

Laurenzi, Emanuele 393
Li, Qiang 308
Llorens, Juan 284

Loutfi, Amy 179
Lutz, Jonas 337

Machado, José 410
Marsa-Maestre, Ivan 194
Md Saad, Nor Hasliza 377
Mendes, Carlos 228, 243
Molloy, Owen 440

Nagi, Khaled 149
Napoli, Amedeo 84
Nasraoui, Olfa 37
Negru, Stefan 165
Neumann, Günter 53
Nguyen, Thi Thuy Anh 101
Nogueira, Monica L. 211

Portela, Filipe 410
Potolea, Rodica 68

Rahman, Azizah Abdul 377
Ramos, Eduardo Costa 426
Robles, Karina 284
Ruiz, Alejandro 284

Salvador, Nuno 228
Samejima, Masaki 297
Santoro, Flávia Maria 426
Santos, Manuel Filipe 410
Schmeier, Sven 53
Silva, Álvaro 410
Smail-Tabbone, Malika 84
Soffer, Avi 19
Speicher, Daniel 260

Teng, Yue 130
Thönssen, Barbara 337, 393
Timonen, Mika 3, 130
Toivanen, Timo 130

van Keulen, Maurice 113
Velasco, Juan R. 194
Volkovich, Zeev 19

Wang, Lixin 350

Yagel, Reuven 271
Yahalom, Orly 19